Virtual Nonlinear Multibody Systems

NATO Science Series

A Series presenting the results of scientific meetings supported under the NATO Science Programme.

The Series is published by IOS Press, Amsterdam, and Kluwer Academic Publishers in conjunction with the NATO Scientific Affairs Division

Sub-Series

I. Life and Behavioural Sciences	IOS Press
II. Mathematics, Physics and Chemistry	Kluwer Academic Publishers
III. Computer and Systems Science	IOS Press
IV. Earth and Environmental Sciences	Kluwer Academic Publishers
V. Science and Technology Policy	IOS Press

The NATO Science Series continues the series of books published formerly as the NATO ASI Series.

The NATO Science Programme offers support for collaboration in civil science between scientists of countries of the Euro-Atlantic Partnership Council. The types of scientific meeting generally supported are "Advanced Study Institutes" and "Advanced Research Workshops", although other types of meeting are supported from time to time. The NATO Science Series collects together the results of these meetings. The meetings are co-organized bij scientists from NATO countries and scientists from NATO's Partner countries – countries of the CIS and Central and Eastern Europe.

Advanced Study Institutes are high-level tutorial courses offering in-depth study of latest advances in a field.
Advanced Research Workshops are expert meetings aimed at critical assessment of a field, and identification of directions for future action.

As a consequence of the restructuring of the NATO Science Programme in 1999, the NATO Science Series has been re-organised and there are currently Five Sub-series as noted above. Please consult the following web sites for information on previous volumes published in the Series, as well as details of earlier Sub-series.

http://www.nato.int/science
http://www.wkap.nl
http://www.iospress.nl
http://www.wtv-books.de/nato-pco.htm

Series II: Mathematics, Physics and Chemistry – Vol. 103

Virtual Nonlinear Multibody Systems

edited by

Werner Schiehlen

Institute B of Mechanics,
University of Stuttgart, Germany

and

Michael Valášek

Department of Mechanics,
Czech Technical University in Prague, Czech Republic

Kluwer Academic Publishers

Dordrecht / Boston / London

Published in cooperation with NATO Scientific Affairs Division

Proceedings of the NATO Advanced Study Institute on
Virtual Nonlinear Multibody Systems
Prague, Czech Republic
23 June–3 July 2002

A C.I.P. Catalogue record for this book is available from the Library of Congress.

ISBN 1-4020-1339-6 (HB)
ISBN 1-4020-1340-X (PB)

Published by Kluwer Academic Publishers,
P.O. Box 17, 3300 AA Dordrecht, The Netherlands.

Sold and distributed in North, Central and South America
by Kluwer Academic Publishers,
101 Philip Drive, Norwell, MA 02061, U.S.A.

In all other countries, sold and distributed
by Kluwer Academic Publishers,
P.O. Box 322, 3300 AH Dordrecht, The Netherlands.

Printed on acid-free paper

NATO SCIENCE PROGRAMME
ADVANCE STUDY INSTITUTE

VIRTUAL NONLINEAR MULTIBODY SYSTEMS

PRAGUE, CZECH REBUPLIC
23 JUNE – 3 JULY 2002

MAIN SPONSORS

North Atlantic Treaty Organization, Brussels, Belgium

European Research Office of the U.S. Army, London, UK

OTHER SPONSORS

Department of Mechanics, Faculty of Mechanical Engineering, Czech Technical University in Prague, Prague, Czech Republic

Institute B of Mechanics, University of Stuttgart, Stuttgart, Germany

DIRECTORS

Werner Schiehlen, University of Stuttgart, Germany
Michael Valášek, Czech Technical University in Prague, Czech Republic

SCIENTIFIC COMMITTEE

Jorge Ambrósio, Instituto Superior Técnico, Lisboa, Protugal
Edwin Kreuzer, Technical University Hamburg-Harburg, Germany
John McPhee, University of Waterloo, Ontario, Canada
Ahmed A. Shabana, University of Illinois at Chicago, Illinois, USA

CONTENTS

LIST OF PARTICIPANTS

Lecturers

Ambrósio, Jorge
IDMEC/IST, Instituto Superior Técnico,
Av. Rovisco Pais, 1049-001 Lisboa, Portugal

Berbyuk, Victor
Depart. of Machine and Vehicle Systems, Chalmers University of Technology
412 96 Gothenburg, Sweden

Bestle, Dieter
Lehrstuhl für Maschinendynamik, BTU Cottbus,
Universitätsplatz 3/4, 03044 Cottbus, Germany

Blajer, Wojciech
Institute of Applied Mechanics, Technical University of Radom
ul. Krasickiego 54, 26-600 Radom, Poland

Chernousko, Felix
Institute for Problems in Mechanics, Russian Academy of Science
Pr. Vernadskogo 101, 117526 Moscow, Russia

Eberhard, Peter
Institute of Applied Mechanics, University of Erlangen-Nuremberg
Egerlandstrasse 5D, 91058 Erlangen, Germany

Funk, Kilian
Lehrstuhl für Angewandte Mechanik, Technische Universität München
85747 Garching, Germany

Haug, Edward
NADS and Simulation Center,
2401 Blvd. Coralville, Oakdale, IA 52241, USA

Kreuzer, Edwin
Technische Universität Hamburg-Harburg,
Eissendorfer Strasse 42, 21073 Hamburg, Germany

Lankarani, Hamid
Mechanical Engineering Dept., Wichita State University
Wichita, KS 67260-0133, USA

McPhee, John
Systems Design Engineering, University of Waterloo
200 University Avenue West, Waterloo, Ontario N2L 3G1,Canada

Nikravesh, Parviz
Dept. of Aerospace & Mechanical Engineering, University of Arizona
Tucson, AZ 85721, USA

Pascal, Madeleine
LMM, Université Pierre et Marie Curie
Tour 66, Case 162, 4 Place Jussieu, 75252 Paris Cedex 05, France

Pereira, Manuel
Mechanical Engineering Department, IDMEC, Instituto Superior Técnico,
Av. Rovisco Pais, 1049-001 Lisboa, Portugal

Schiehlen, Werner
Institute B of Mechanics, University of Stuttgart
Pfaffenwaldring 9, 70550 Stuttgart, Germany

Shabana, Ahmed
Dept. of Mechanical Engng., University of Illinois at Chicago
842 West Taylor Street, Chicago, IL 60607-7022, USA

Valášek, Michael
Dept. of Mechanics, Faculty of Mechanical Eng., CTU in Prague
Karlovo nam 13, 121 35 Praha 2, Czech Republic

Zakhariev, Evtim
Institute of Mechanics, Bulgarian Academy of Sciences
Acad. G. Bonchev Street, bl. 4, 1113 Sofia, Bulgaria

Participants

Agrawal, Sunil
Department of Mechanical Engineering, University of Delaware
222 Spencer Laboratory, Newark, DE 19716, USA

Anitescu, Mihai
Department of Mathematics, University of Pittsburgh,
Thackeray 301, Pittsburgh, PA 15213, USA

Belda, Květoslav
Institute of Information Theory and Automation AS
Pod vodárenskou věží 4, 182 08 Prague 8 - Libeň, Czech Republic

Beneš, Karel
MECAS
Uslavska 10, 32600 Plzeň, Czech Republic

Boikov, Ilia
Penza State University
Krasnay 40, 440026 Penza, Russia

Boikova, Alla
Penza State University
Krasnay str 40, 440026 Penza, Russia

Brüls, Olivier
LTAS - Aérospatiale, Mécanique et Matériaux, Institut de Mécanique et de Génie Civil
Bât. B52, Chemin des Chevreuils 1, 4000 Liege, Belgium

Camlibel, M. Kanat
Dept. of Mathematics, University of Groningen
P.O. Box 800, 9700 AV Groningen, The Netherlands

Carrarini, Antonio
DLR - Vehicle System Dynamics Group
P.O. Box 1116, 82230 Wessling, Germany

Czaplicki, Adam
Academy of Physical Education in Warsaw
Ul. Akademicka 2, 21-500 Biala Podlaska, Poland

Dequidt, Antoine
ENSIMEV, Universite acute de Valenciennes,
Le Mont Houy, 59313 Valenciennes 9, France

Dias, João
IDMEC/ISTI, Instituto Superior Técnico
Av. Rovisco Pais, 1049-001 Lisboa, Portugal

Dmitrotchenko, Oleg
Dept. of Applied Mechanics, Bryansk State Technical University
bulv. im. 50-letiya Oktyabrya 7, 241035 Bryansk, Russia

Eiber, Albrecht
Institute B of Mechanics, University of Stuttgart
Pfaffenwaldring 9, 70550 Stuttgart, Germany

Ellermann, Katrin
Mechanik und Meerestechnik, TU Hamburg-Harburg,
Eissendorfer Strasse 42, 21071 Hamburg, Germany

Escalona, José L.
Department of Mechanical Engineering, University of Seville
Camino de los Descubrimientos , 41092 Seville, Spain

Figurina, Tatiana
Institute for Problems in Mechanics, RAS
101 prospect Vernadskogo, 117526 Moscow, Russia

Flores Fernandes, João Paulo
Departamento de Engenharia Mecanica, Universidade do Minho
Campus de Azurem, 4800-058 Guimaraes, Portugal

Fraczek, Janusz
Institute of Aeronautics and Applied Mechanics
Nowowiejska 24, 00-665 Warsaw, Poland

Fuchs, Andreas
Vehicle System Dynamics Group, Institute of Aeroelasticity, DLR German Aerospace Center,
P.O.Box 1116, 82230 Wessling, Germany

Georgiev, Krassimir
Mechatronic Systems Department, Institute of Mechanics, BAS
acad.G.Bonchev street, blok 4, 1113 Sofia, Bulgaria

Hyncik, Ludek
New Technologies Research Centre in the West-Bohemia
Univerzitní 8, 306 14 Plzeň, Czech Republic

Jatsun, Sergey
Kursk State Technical University
50 Let Oktyabrya 94, Kursk 305040, Russia

Kalkan, Nevin
Faculty of Science, Physics Department, University of Istanbul
Vezneciler - Ýstanbul, Turkey

Kolesnikov, Maxim
Moscow State Institute of Steel and Alloys
Mkr.Vostochny 17-31, Stary Oskol, 309502 Belgorodska Obl., Russia

Kolodziejczyk, Krzysztof
Institute of Applied Mechanics, Technical University of Radom
ul. Krasickiego 54, 26-600 Radom, Poland

Kostadinov, Kostadin
Bulgarian Academy of Sciences
Acad. G. Bonchev St., Block 4, 1113 Sofia, Bulgaria

Kovalev, Roman
Dept. of Applied Mechanics, Bryansk State Technical University
bulv. 50-let Oktyabrya 7, Bryansk 241035, Russia

Kövecses, József
Space Technologies, Canadian Space Agency
6767 route de l'Aéroport Saint-Hubert, Québec J3Y 8Y9, Canada

Kübler, Lars
Institute of Applied Mechanics, University of Erlangen-Nuremberg
Egerlandstrasse 5, 91058 Erlangen, Germany

Lache, Simona
Transilvania University of Brasov
29 B-Dul. Eroilor, 2200 Brasov, Romania

Lampariello, Roberto
Institute of Robotics, DLR German Aerospace Center
P.O.Box 1116, 82230 Wessling, Germany

Lefeber, Dirk
Faculty of Applied Science, University of Brussels
Pleinlaan 2, 1050 Brussels, Belgium

Lidberg, Mathias
Department of Applied Mechanics, Chalmers University of Technology
412 96 Gothenburg, Sweden

Micu, Ioan
University Transilvania of Brasov,
B-dul Eroilor 29, 2200 Brasov, Romania

Mikkola, Aki
Department of Mechanical Engineering, Lappeenranta University of Technology
PL 20, 53851 Lappeenranta, Finland

Müller, Andreas
Institute of Mechatronic at TU Chemnitz
Reichenhainer 5tr. 88, 09126 Chemnitz, Germany

Muth, Beate
Institute B of Mechanics, University of Stuttgart
Pfaffenwaldring 9, 70550 Stuttgart, Germany

Naudet, Joris
Department of Mechanical Engineering, Faculty of Applied Sciences, Free University of Brussels
Pleinlaan 2, 1050 Brussels, Belgium

Negrut, Dan
Mechanical Dynamics, Inc.
2300 Traverwood Drive, Ann Arbor, MI 48105, USA

Neto, Maria Augusta
Dep. de Eng. Mecanica, Universidade de Coimbra
Pinhal de Marrocos (polo II), 3030 Coimbra, Portugal

Pathak, Kaustubh
Mechanical Systems Lab., Dept. of Mechanical Engineering, University of Delaware
134 Spencer Lab, Newark, DE 19716, USA

Pedersen, Sine Leergaard
Technical University of Denmark,
Bygn 404, 2800 Lyngby, Denmark

Pogorelov, Dmitri
Dept. of Applied Mechanics, Bryansk State Technical University
b. 50 let Oktyabrya 7, 241035 Bryansk, Russia

Polakova, Zuzana
KAME, Sj.F. KOSICE
Gocovo 179, 049 24 Roznava, Slovakia

Pombo, João
IDMEC, Instituto Superior Técnico
Av. Rovisco Pais, 1049-001 Lisboa, Portugal

Sayapin, Sergey
Mechanical Engineering Research Institute
Russian Flat 188,75 Marx St., 249020 Obninsk, Kaluga Region, Russia

Schmitke, Chad
System Design Engineering, University of Waterloo
200 University Ave. W., Waterloo, Ontario N2L 3G1, Canada

Schupp, Gunter
Fahrzeug-Systemdynamik, Institut für Aeroelastik, DLR - Deutsches Zentrum für Luft- und Raumfahrt
Postfach 1116, 82230 Wessling, Germany

Schwab, Arend L.
Delft University of Technology,
Mekelweg 2, 2628 CD Delft, The Netherlands

Sciacovelli, Donato
ESA-ESTEC
Keplerlaan 1, 2200 AG Nordwijk, The Netherlands

Šika, Zbynek
Dept. of Mechanics, Faculty of Mechanical Eng., CTU in Prague
Karlovo nam 13, 12135 Praha 2, Czech Republic

Steinbauer, Pavel
Dept. of Mechanics, Faculty of Mechanical Eng., CTU in Prague
Karlovo nam 13, 121 35 Praha 2, Czech Republic

Stroe, Ion
Politehnica University of Bucharest
Splaiul Independentei 313, 77206 Bucharest, Romania

Talaba, Doru
Product Design and Robotics Department , University Transilvania of Brasov
Bd. Eroilor 29, 2200 Brasov, Romania

Taralova, Ina
Institut de Recherche en Communications et Cyberne, Ecole Centrale de Nantes
1 Rue de la Noe, 44321 Nantes, France

Tavares da Silva, Miguel
Departamento de Engenharia Mecânica, IDMEC, Instituto Superior Técnico
Av. Rovisco Pais, 1049-001 Lisboa, Portugal

Terze, Zdravko
Faculty of Mech. Eng. and Na., University of Zagreb
Ivana Lurica 5, 10000 Zagreb, Croatia

Tobolar, Jakub
Institut für Aeroelastik, DLR
P.O.Box 1116, 82230 Weßling, Germany

Valentini, Pier Paolo
Dipartimento di Ingegneria Meccanica, Università di Roma Tor Vergata,
Via del Politecnico 1, 00133 Roma, Italy

Valverde, Juan
Department of Mechanical Engineering, University of Seville
Camino de los Descubrimientos, 41092 Seville, Spain

Vassileva, Daniela
Institute of Mechanics,
Acad. G. Bonchev St., Block 4, 1113 Sofia, Bulgaria

Vita, Leonardo
Dipartimento di Ingegneria Meccanica, Università degli Studi di Roma Tor Vergata
Via del Politecnico 1, 00133 Roma ,Italy

Vlase, Sorin,
University Transilvania of Brasov
b-dul Eroilor 29, 2200 Brasov, Romania

von Dombrowski, Stefan
Institute of Robotics and Mechatronics, DLR - German Aerospace Center Research Center Oberpfaffenhofen
82234 Weßling, Germany

Wojtyra, Marek
Institute of Aeronautics and Applied Mechanics, Warsaw University of Technology
ul. Nowowiejska 24, 00-665 Warszawa, Poland

Wray ,Will
Mathengine Plc.
60 St. Aldates, Oxford OX11ST, United Kingdom

Yan, Jin
Mechanical Systems Lab, Department of Mechanical Engineering, University of Delaware
134 Spencer, Newark, DE 19716, USA

Yazykov, Vladislav
Dept. of Applied Mechanics, Bryansk State Technical University
bulv. im. 50-letiya Oktyabrya, 7, 241035 Bryansk, Russia

Zanevskyy, Ihor
Lviv State Institute of Physical Culture
Kostyushko str. 11, 79000 Lviv, Ukraine

PREFACE

This book contains an edited version of lectures presented at the NATO ADVANCED STUDY INSTITUTE on VIRTUAL NONLINEAR MULTIBODY SYSTEMS which was held in Prague, Czech Republic, from 23 June to 3 July 2002. It was organized by the Department of Mechanics, Faculty of Mechanical Engineering, Czech Technical University in Prague, in cooperation with the Institute B of Mechanics, University of Stuttgart, Germany. The ADVANCED STUDY INSTITUTE addressed the state of the art in multibody dynamics placing special emphasis on nonlinear systems, virtual reality, and control design as required in mechatronics and its corresponding applications. Eighty-six participants from twenty-two countries representing academia, industry, government and research institutions attended the meeting. The high qualification of the participants contributed greatly to the success of the ADVANCED STUDY INSTITUTE in that it promoted the exchange of experience between leading scientists and young scholars, and encouraged discussions to generate new ideas and to define directions of research and future developments.

The full program of the ADVANCED STUDY INSTITUTE included also contributed presentations made by participants where different topics were explored, among them: Such topics include: nonholonomic systems; flexible multibody systems; contact, impact and collision; numerical methods of differential-algebraical equations; simulation approaches; virtual modelling; mechatronic design; control; biomechanics; space structures and vehicle dynamics. These presentations have been reviewed and a selection will be published in this volume, and in special issues of the journals Multibody System Dynamics and Mechanics of Structures and Machines.

This book brings together, in a tutorial and review manner, a comprehensive summary of current work in the field. It is therefore suitable for a wide range of interests, ranging from the advanced student to the researcher and implementator concerned with advanced issues in multibody dynamics. The applicational aspects will help the reader to appraise the different approaches available today and their usefulness as efficient tools. The book is organized into four parts, each one of them addressing techniques and methods in one of the principal areas of the ADVANCED STUDY INSTITUTE:

- Nonlinear Multibody Dynamics,
- Virtual Multibody Systems,
- Control Design and Mechatronics,
- Challenging Applications.

Altogether 18 invited lectures and 9 selected contributed papers are published in this volume resulting in a text-book on Virtual Nonlinear Multibody Systems following the nature of an ADVANCED STUDY INSTITUTE.

Without the sponsorship and financial support of the NATO Scientific and Environmental Affairs Division and the European Research Office of the US Army, this ADVANCED STUDY INSTITUTE and this book would not have been possible. The support of the University of Stuttgart and the Czech Technical University in Prague is also gratefully acknowledged.

The ADVANCED STUDY INSTITUTE Directors wish to express their gratitude to the members of the Organizing Committee, Prof. J. Ambrósio, Prof. E. Kreuzer, Prof. J. McPhee and Prof. A. A. Shabana for their help and support in organizing the ADVANCED STUDY INSTITUTE.

We would like to thank all lecturers and participants in the ADVANCED STUDY INSTITUTE for their presentations and for the lively and stimulating discussions. The great success of the ADVANCED STUDY INSTITUTE is also a result of the competent work of the staff of the Directors, in particular the most professional work of Dr.-Ing. Albrecht Eiber from the Institute B of Mechanics, University of Stuttgart.

Stuttgart and Prague, January 2003

Werner Schiehlen and Michael Valášek

ADDRESS AT THE OPENING SESSION

NATO ASI on Virtual Nonlinear Multibody Systems
Prague, Czech Republic, 23 June 2002

WERNER SCHIEHLEN
University of Stuttgart, Germany

What is virtual in multibody dynamics?

Well, the term "virtual reality" has taken on the meaning of the "computer generated perception of reality controlled by an involved human". The term "virtual reality" motivated the emerging use of the term "virtual prototype", suggesting both computer and human involvement in the process of design and testing of a prototype that does not exist as real object. Complex multibody systems composed of rigid and flexible bodies performing spatial motion and various complex tasks are up-to date objects of the virtual prototyping. The simulation of the multibody system action and reliability demands for adequate dynamic models taking into account various phenomena and requirements. Standard dynamics does not regard all nonlinear effects that appear as a result of the action of multibody systems as well as their mutual interaction. The virtual prototyping and dynamic modeling of such systems are, from economical point of view, perspective fields of scientific investigations having in mind the huge expenses for the design and manufacturing of prototypes.

Some historical remarks show that multibody system dynamics is based on classical mechanics and its engineering applications ranging from mechanisms, gyroscopes, satellites and robots to biomechanics. Multibody system dynamics is characterized by algorithms or formalisms, respectively, ready for computer implementation. As a result simulation and animation are most convenient. Future research fields in multibody dynamics are identified as standardization of data, coupling with CAD systems, parameter identification, real-time animation, contact and impact problems, extension to electronic and mechatronic systems, optimal system design, strength analysis and interaction with fluids. Further, there is a strong interest on multibody systems in analytical and numerical mathematics resulting in reduction methods for rigorous treatment of simple models and special integration codes for Ordinary Differential Equations and Differential Algebraic Equations representations supporting the numerical efficiency. New software engineering tools with modular approaches promise better efficiency still required for the more demanding needs in biomechanics, robotics and vehicle dynamics. Scientific research in multibody system dynamics has been devoted to improvements in modeling considering nonholonomic constraints, flexibility, friction, contact, impact, and control. New methods evolved with respect to simulation by recursive formalism, to closed kinematic loops, reaction forces and torques, and to pre- and post-processing by data models, CAD coupling, signal analysis, animation and strength evaluation. Multibody system dynamics is applied to a broad variety of engineering problems from aerospace to civil engineering, from vehicle design to micromechanical analysis, from robotics to biomechanics. The fields of application are steadily increasing, in particular as multibody dynamics is considered as the basis of mechatronics, e.g. controlled mechanical systems. These challenging applications require more fundamental research on a number of topics which are presented in the following.

1. Datamodels from CAD

Within the multibody system community many computer codes have been developed, however, they differ widely in terms of model description, choice of basic principles of mechanics and topological structure so that a uniform description of models does not exist. A most desirable data exchange would permit the alternate use of validated multibody system models with different simulation codes.

2. Parameter identification

The parameter identification is an essential part of multibody dynamics. The equations of motion of mechanical systems undergoing large displacements are highly nonlinear, however, they remain linear with respect to the system parameters.

3. Optimal Design

Due to development of faster computing facilities the multibody system approach is changing from a purely analyzing method to a more synthesizing tool. Optimization methods are applied to optimize multibody systems with respect to their dynamic behaviour.

4. Dynamic strength analysis

The results obtained in research on strength analysis of material bodies can be applied and combined with the multibody system approach.

5. Contact and impact problems

Rigid and/or flexible bodies moving in space are subject to collisions what mechanically means impact and contact. Contact problems usually include friction phenomena which may be modelled by Coulomb's law.

6. Extension to control and mechatronics

The applied forces and torques acting on multibody systems may be subject to control. Then, the multibody system is considered as the plant for which a controller has to be designed. Today, mechatronics is understood as an interdisciplinary approach to controlled mechanical systems usually modelled as multibody systems.

7. Nonholonomic systems

The nonholonomic systems are of engineering interest in vehicle dynamics and mobile robots. The contact problem may also be a problem for the nonholonomic systems.

8. Integration codes

The dynamic equations of motion are presented as ODE or DAE. Efficient algorithms for numerical integration of these equations are of major importance.

9. Real time simulation and animation

Efficient and fast simulation is always desirable in computational dynamics but it is really necessary for hardware-in-the loop and operator-in-the-loop applications. There are two approaches to achieve real time simulation: high speed hardware and efficient software. Multibody system dynamics is called to contribute to the efficiency of the software by recursive and/or symbolic formalism and fast integration codes.

10. Challenging applications

Multibody system dynamics has a broad variety of applications, some of which will be mentioned. In biomechanics the walking motion is an important topic for some while. However, there are much more problems in biomechanics which can be modeled and solved by multibody dynamics ranging from vehicle occupants to sport sciences. Multibody dynamics is also a solid basis for nonlinear dynamics. In particular, impact and friction induced vibrations show chaotic behaviour. The noise generation in railway wheels due to rail-wheel contact forces can be also considered as a highly nonlinear phenomenon. An up-to-date development include also the control of chaos. The control aspects in multibody dynamics are getting more and more important. Vehicle, aircraft and spaceship dynamics and reliability have always been challenging applications. In relation to transportation systems a contemporary application of multibody dynamics is the structural and occupant crashworthiness.

Having in mind the state-of-the-art and the contemporary needs of science and engineering practice, all this is described by the title of the NATO ASI beginning today: VIRTUAL NONLINEAR MULTIBODY SYSTEMS.

NON-LINEAR DYNAMICS OF MULTIBODY SYSTEMS WITH GENERALIZED AND NON-GENERALIZED COORDINATES

AHMED A. SHABANA
Department of Mechanical Engineering
University of Illinois at Chicago
842 West Taylor Street
Chicago, Illinois 60607-7022, U.S.A.

ABSTRACT For efficient dynamic simulation of mechanical and aerospace systems, the use of different sets of coordinate types may be necessary. The components of multibody systems can be rigid, flexible or very flexible and can be subject to contact forces. Examples of challenging problems encountered when multibody systems are considered are crashworthiness, problems of cables used in rescue operations and heavy load handling, belt drives, leaf spring system design, tire deformations, large deformation of high-speed rotors, stability problems, and contact problems. Most large displacement problems in structural mechanics are being solved using incremental solution procedures. On the other hand, general purpose flexible multibody computer tools and methodologies in existence today are not, in general, capable of systematically and efficiently solving applications that include, in addition to rigid bodies and bodies that undergo small deformations, bodies that experience very large deformations. The objective of this paper is to discuss the development of new computational algorithms, based on non-incremental solution procedures, for the computer simulation of multibody systems with flexible components. These new algorithms that do not require special measures to satisfy the principles of work and energy and lead to optimum sparse matrix structure can be used as the basis for developing a new generation of flexible multibody computer programs. The proposed non-incremental algorithms integrate three different formulations and three different sets of generalized coordinates for modeling rigid bodies, flexible bodies with small deformations, and very flexible bodies that undergo large deformations. The implementation of a general contact model in multibody algorithms is also presented as an example of mechanical systems with non-generalized coordinates. The kinematic equations that describe the contact between two surfaces of two bodies in the multibody system are formulated in terms of the system-generalized coordinates and the *non-generalized surface parameters*. Each contact surface is defined using two independent parameters that completely determine the tangent and normal vectors at an arbitrary point on the body surface. In the contact model presented in this study, the points of contact are determined on line during the dynamic simulation by solving the non-linear differential and algebraic equations of the constrained multibody system. The augmented form of the equations of motion expressed in terms of the generalized coordinates and non-generalized surface parameters is presented in this paper.

W. Schiehlen and M. Valášek (eds.), Virtual Nonlinear Multibody Systems, 1–16.

1. Introduction

A new generation of flexible multibody simulation tools is required for the analysis of modern, high-speed, light-weight multibody system applications. Existing multibody simulation tools are not suited for modeling the dynamic phenomena that are encountered in the analysis of modern systems. The existing tools are designed to solve problems of mutlibody systems that consist of rigid components or flexible components that experience small deformations. Furthermore, existing algorithms fail to produce efficient and accurate solutions for some basic vehicle elements, such as *leaf springs*, and flexible cables used in rescue operations and heavy load handling. Most existing flexible multibody computational algorithms, with few exceptions, employ two different dynamic formulations. The first formulation is based on the classical *Newton-Euler* or *Lagrangian equations* for rigid bodies, while the second formulation is based on the *floating frame of reference approach.* In the rigid body formulation [1-5], six independent coordinates are initially used to describe the body configuration. These six coordinates include three translation coordinates that define the location of the origin of a centroidal body reference, and three orientation coordinates that define the orientation of the body reference. The generalization of this formulation to study the dynamics of flexible bodies that experience small deformations required introducing an additional set of generalized coordinates in order to be able to define the body deformation with respect to its reference [6-10]. In the early eighties, this generalization led to the development of a new generation of general-purpose flexible multibody computer codes that are currently widely used in industry and research laboratories. Such codes, that have been used as the basis for developing new computational design and analysis approaches, have been used in the design of many mechanical and aerospace system applications, and have resulted in tremendous saving of resources that would have been otherwise wasted. Preliminary designs are made in a computer environment before building the actual prototypes, while existing designs are continuously improved by utilizing the first generation of flexible multibody computer codes.

The general treatment of the rigid body contact problem requires the use of a set of parameters that describe the geometry of the contact surfaces. Two parameters are introduced for each surface. Using the four surface parameters associated with each contact, five algebraic contact constraint equations are formulated. The kinematic contact constraint equations are expressed in terms of the generalized coordinates of the two bodies as well as the non-generalized surface parameters. In order to be able to satisfy the nonlinear contact constraint conditions at the position, velocity, and acceleration levels, third partial derivatives of these constraint equations with respect to the surface parameters must be evaluated.

2. Multibody System Solution Procedures

Multibody system applications are inherently nonlinear due to the large reference displacements and the constraints that restrict the motion of the system components. For this reason, multibody system applications are different from the structural dynamics applications and require certain solution procedures in order to obtain accurate solutions. Linearization of the dynamic equations can be the source of serious problems when multibody system applications are considered. As a result, multibody formulations, in the most part, are highly nonlinear and require computer implementation based on a non-incremental procedure. In order to further elaborate on some of the difficulties that can be encountered when the equations of motion of a simple system are linearized, we consider the simple pendulum example shown in Fig. 1. The nonlinear equation of motion of this single degree of freedom rigid body system is given by

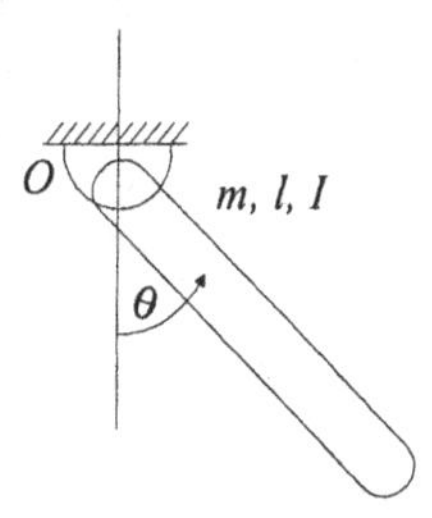

Figure 1 Simple Pendulum

$$I_O\ddot{\theta} + mg\frac{l}{2}\sin\theta = 0 \tag{1}$$

In this equation, *2* is the angle of rotation of the pendulum, I_O is the mass moment of inertia of the pendulum about the fixed point, m and l are, respectively, the pendulum mass and length, and g is the gravity constant. Recall that $\ddot{\theta} = \dot{\theta}\frac{d\dot{\theta}}{d\theta}$ and substituting into the preceding equation, one obtains the following constant *energy integral of motion* for this conservative system:

$$\frac{1}{2}I_O\dot{\theta}^2 - mg\frac{l}{2}\cos\theta = c_1 \tag{2}$$

where c_1 is a constant. The preceding equation, that states that the change in the system kinetic energy is equal to the change in the system potential energy, shows that the total energy has an upper limit. Therefore, an accurate numerical solution of the nonlinear relationship of Eq. 1 must satisfy the principle of work and energy as defined for the pendulum system in Eq. 2. Most multibody algorithms do satisfy the principle of work and energy since the equations of motion are not linearized.

A linearization of Eq. 1 produces the following equation:

$$I_O\ddot{\theta} + mg\frac{l}{2}\theta = 0 \tag{3}$$

Using again the fact that $\ddot{\theta} = \dot{\theta}\frac{d\dot{\theta}}{d\theta}$ into the preceding equation and integrating leads to the following constant of motion:

$$\frac{1}{2}I_O\dot{\theta}^2 + mg\frac{l}{4}\theta^2 = c_2 \tag{4}$$

where c_2 is a constant. The solution of Eq. 3 satisfies Eq. 4 and does not satisfy the energy constant of Eq. 2. Equation 4 is an approximation for the principle of work and energy only when the reference rotation as defined in the global system is small. If the reference rotation is finite, the integral of motion of Eq. 4 is no longer an accurate representation of the principle of work and energy. As a result, any solution, obtained incrementally or non-incrementally, that satisfies Eq. 3 must satisfy Eq. 4 and can lead to *energy drift* as the angle *2* increases since such a solution is not required to satisfy Eq. 2.

The simple pendulum example demonstrates that set of dynamic equations defines an integral or constant of motion. Correctly formulated, an integral of these equations defines the principle of work and energy. Accurately solved, these equations should not lead to energy drift. The simple pendulum example demonstrates that violation of the principle of work and energy can be an indication that the differential equations of motion are not correctly formulated due to inaccurate description of the finite rotation. If the dynamic equations are correctly formulated, the solution of these equations must satisfy the principle of work and energy. The simple pendulum example also explains the reason for introducing the concept of the floating frame of reference formulation that leads to accurate modeling of the reference rotations.

3. Past Challenges in Flexible Multibody System Dynamics

Development of the first generation of computational flexible multibody algorithms was successfully accomplished as result of extensive research efforts focused on solving several computational and fundamental problems. Examples of the challenging problems that were encountered are the following:

1. ***Finite Rotation Problem*** As demonstrated by the simple pendulum example presented in the preceding section, in multibody system applications accurate representation of the large rotation in particular and the rigid body dynamics in general is necessary for the development of robust and accurate simulation tools. In the finite element literature [11-15], two types of conventional finite elements can be recognized; isoparametric and non-isoparametric finite elements. Isoparametric finite elements such as brick, rectangular, and triangular elements are capable of describing arbitrary rigid body displacements. On the other hand, non-isoparametric elements such as conventional beam and plate elements are not capable of accurately describing arbitrary rigid body displacements and produce non-zero strains when the finite element rotates as a rigid body. However, it is to be noted that these

conventional beam, plate and shell elements are widely used in modeling mechanical, aerospace and structural systems. Therefore, one of the first challenges encountered in the early eighties in developing general purpose flexible multibody computer programs is to develop a formulation that allows using the widely used conventional beam and plate elements and at the same time obtain an exact representation of the rigid body dynamics. This problem was solved using the concept of the *intermediate element coordinate system* [16-17]. The use of the intermediate element coordinate system in the framework of the floating frame of reference formulation leads to a formulation that produces zero strain under an arbitrary rigid body motion when conventional beam and plate elements are used. The use of the non-linear floating frame of reference formulation does not lead to *energy drift* since the reference rotation is accurately described using non-linear dynamics relationships.

2. ***Joint and Constraints*** The displacements of multibody system components are restricted due to the existence of kinematic constraints that represent mechanical joints and specified motion trajectories. In flexible multibody dynamics, these algebraic constraint equations depend on the reference motion and the elastic deformation and influence the type of the elastic coordinates selected for formulating the equations of motion. Efficient solution of the small deformation problems in multibody system applications requires the use of modal reduction techniques to represent the deformation of the flexible bodies with respect to their coordinate system using a minimum set of elastic coordinates. Selection of the deformable body coordinate systems and its relationship with the selected set of mode shapes is a fundamental problem that has been recognized and addressed when developing computational algorithms for flexible multibody systems.

3. ***Coupling Between Reference and Elastic Displacements*** Due to the large rotation of the finite element in the multibody floating frame of reference formulation, the resulting expression for the element inertia forces is highly nonlinear and leads to a strong coupling between the reference motion and the elastic deformation. This dynamic inertia coupling was neglected in many of the investigations on flexible mechanical systems in the seventies and eighties [18-22]. It was demonstrated that ignoring this inertia coupling in high-speed mechanical and aerospace systems leads to inaccurate solutions. One of the challenging problems encountered in the early eighties in developing general methodologies for flexible multibody dynamics is to obtain a formulation that accounts for this dynamic coupling and all inertia nonlinearities.

4. ***Coupling with Commercial Finite Element Codes*** Another important issue that was the subject of investigations is establishing an interface between existing commercial finite element computer programs and the newly developed general purpose flexible multibody codes. In the finite element floating frame of reference formulation introduced in the early eighties, it was demonstrated that constant integrals, called *inertia shape integrals*, in addition

to the element stiffness matrix must be evaluated in order to construct the nonlinear dynamic equations of motion of the flexible components. These inertia shape integrals can be evaluated in a preprocessor finite-element computer program, thereby reversing the sequence of computations adopted in the seventies and eighties. By establishing such an interface between finite element and flexible multibody computer programs, the rich library of existing finite element codes can be systematically exploited, allowing modeling a wide range of flexible multibody applications.

5. ***Numerical Solution Procedures*** Numerical algorithms developed for flexible multibody systems have been successfully used for solving the resulting system of differential and algebraic equations that are functions of the coupled reference and elastic degrees of freedom. Special numerical procedures that require the integration of two sets of variables that have different frequencies and different scales were proposed and used in the analysis of many multibody applications.

Extensive research allowed solving many of these fundamental problems. Such research efforts made flexible multibody computer programs efficient and user friendly to the point that, in the most part, every major industry or research laboratory conducting research in aerospace or mechanics is currently using one of the general purpose multibody computer codes.

4. New Challenges

While the first generation of flexible multibody computer codes has been widely used and the capabilities and special features of these codes are being exploited, existing flexible multibody codes are not suited for solving many important problems that are encountered in the design of mechanical, aerospace and structural systems. Examples of these problems are crashworthiness, cables used in heavy load handling and rescue mission, belt drives, very flexible rotor systems, tire problems, flexible suspension leaf spring systems, stability problems, etc. It is recognized that computational multibody tools developed decades ago are no longer effective in solving many of the new challenging flexible multibody dynamics problems. Therefore, more research is needed in order to develop new computational algorithms that will lead to the development of a new generation of flexible multibody computer programs. In order to achieve this objective, the implementation of a new large deformation formulation in flexible multibody computer programs is necessary. For consistency with the existing multibody solution procedures, the new large deformation finite-element formulation will be based on a non-incremental solution strategy that does not utilize typical finite element co-rotational formulations [23-25]. The new finite element formulation and solution procedures have the following features [26-30]:

1. Unlike large rotation vector formulations, the proposed large deformation formulation does not require interpolation of rotations or slopes.

2. No infinitesimal or finite rotations are used as nodal coordinates. The element nodal coordinates consist of global displacement and slope coordinates.

3. The solution procedure used is non-incremental and does not require the use of typical co-rotational finite element procedures. Preliminary results indicated that highly nonlinear large deformation problems can be efficiently solved using explicit numerical integration methods [31].

4. The method does not require the use of specific measures to satisfy the principles of work and energy. For conservative systems, energy is conserved using available first order differential equation solvers [31].

5. The proposed formulation guarantees the continuity of displacement gradients, thereby leading to more accurate calculations of the strains and stresses. It also captures the effect of the geometric centrifugal stiffening.

6. The proposed method leads to a constant mass matrix for two- and three-dimensional applications, and at the same time it relaxes many of the assumptions currently employed in many large deformation models. As a result, the Coriolis and centrifugal inertia forces are identically equal to zero.

As compared to the significant developments made in the early eighties, it is expected that the implementation of a large deformation formulation in flexible multibody algorithms will have a greater impact and will require solving new fundamental and computational problems. In this investigation, it is proposed to use a new procedure based on the *absolute nodal coordinate formulation* to develop new non-incremental computational algorithms that can serve as the basis for developing a new generation of flexible multibody computer programs. Preliminary results indicate that this formulation can be effective, efficient and accurate in modeling large rotation and large deformation problems. This formulation, as previously pointed out, leads to constant inertia matrix in two- and three-dimensional applications, and as a consequence, the centrifugal and Coriolis inertia forces are identically equal to zero. The absolute nodal coordinate formulation does not require the use of special measures to satisfy the principle of work and energy and does not employ infinitesimal or finite rotations as nodal coordinates, yet all the information about the nodal rotations can be obtained. Accurate results for highly nonlinear large deformation problems can be efficiently obtained using explicit integration methods as demonstrated by the preliminary results obtained.

Some important issues that need to be addressed in developing large deformation and contact solution algorithms for multibody system applications are summarized as follows:

1. ***Evaluation of Existing Large Deformation Formulations*** Different large deformation formulations used in the dynamic modeling of flexible multibody vehicle systems need to be evaluated in order to identify the advantages and

drawbacks of each method. This evaluation will also define the limitations and domain of applicability of each method in solving particular applications.

2. ***Nonlinear Constrained Kinematic Equations*** As previously demonstrated using the simple pendulum example, linearization of the kinematic relationships can lead to wrong solutions for systems that experience large displacements. This is a concern in the development of any new flexible multibody formulation. A new set of generalized coordinates for the non-linear kinematic and dynamic analysis of multibody systems that consist of rigid bodies, flexible bodies that undergo small deformations, and very flexible bodies that undergo large deformations need to be introduced. This set, which consists of three different sets of coupled generalized coordinates, will be used in formulating the kinematic and dynamic equations as described in the following sections. The kinematic description used in this investigation leads to exact modeling of the rigid body dynamics. Continuity of the displacement gradients is also guaranteed for very flexible bodies. The nonlinear kinematic and algebraic constraint equations that describe the connections between different bodies of the system need to be developed in terms of the coupled sets of generalized coordinates. For very flexible bodies that experience large deformations, the kinematic constraint equations will formulated in terms of the absolute nodal coordinates and displacement gradients.

3. ***Sparse Matrix Structure*** A new augmented form of the multibody system equations of motion in terms of the three coupled sets of generalized coordinates and Lagrange multipliers can be developed. This new augmented form of the equations of motion can be used as the basis for developing a new computational algorithm for solving the *differential and algebraic equations* encountered in the large deformation multibody problems. Since the mass matrix associated with very flexible bodies in the system is constant, a Cholesky transformation can be defined and used to define an identity generalized inertia matrix associated with a new set of Cholesky coordinates. Using this new set of Cholesky coordinates, an optimum set of sparse matrix equations that govern the dynamics of the multibody system can be obtained and efficiently solved for the system accelerations.

4. ***Qualitative and Quantitative Evaluation of the Dynamic Coupling*** The dynamic inertia and elastic coupling between the absolute nodal coordinates used for the large deformations and the reference and elastic coordinates used to describe the rigid body motion and the small elastic deformations need to be examined analytically and numerically. Such a study will lead to a better understanding of the new structure of the dynamic equations of motion.

5. ***Contact Problem*** The formulation of the rigid body contact requires the use of set of parameters to describe the geometry of the contact surfaces. When the contact conditions are formulated using the constraint approach, the equations of motion in their augmented form are expressed in terms of the surface parameters. These parameters have no generalized inertia or forces associated

with them. As a result, the augmented form of the equations of motion can be formulated in terms of a coupled set of generalized coordinates and non-generalized surface parameters. The use of this approach allows determining the coordinates of the contact points on line and may require the numerical integration of the derivatives of the non-generalized surface parameters.

5. Generalized Coordinates

Most large displacement problems in structural mechanics are being solved using incremental solution procedures. One must resort to this solution strategy due to fundamental problems that arise due to the kinematic description of the large displacements of the existing finite elements. Furthermore, existing finite element solution procedures can lead to violation of the principle of work and energy when large displacements are considered unless specific measures are taken in the numerical integration scheme to prevent such a violation. It is the objective of this investigation to present a new non-incremental solution procedure that is well suited for the large displacement and deformation analysis of multibody system applications. Crucial to such a successful development is the use of a kinematic motion description that leads to accurate modeling of the dynamics of the finite element that undergoes large displacements.

The first step in developing a new formulation, is to select the system coordinates. While in most formulations, one coordinate type is used, in this research project it is proposed to develop a new methodology that utilizes different coordinate types in order to achieve efficiency and generality while maintaining accuracy. Different components in the system, depending on whether they are rigid, flexible or very flexible, are modeled using different types of coordinates. In this study, the following different sets of generalized coordinates are used; *reference coordinates*, *elastic coordinates* for small deformations, and *absolute nodal coordinates* for large deformations. In this investigation, the reference coordinates are used to define the location and orientation of a selected body reference. In the case of rigid bodies, a centroidal body coordinate system is used. In the three dimensional analysis, the reference coordinates consist of three translation coordinates and three or four orientation parameters depending on whether Euler angles or Euler parameters are used. This set of coordinates can be used to develop efficient models for bulky components that can be treated as rigid bodies. The elastic coordinates used for the small deformation problem define the body deformation with respect to the body coordinate system. Such a choice of a coordinate set allows using deformation modes in order to reduce the number of the system degrees of freedom. By using this set of elastic coordinates and the floating frame of reference formulation, an efficient algorithm for solving small deformation problems in multibody systems that takes into account the nonlinear coupling between the reference motion and the elastic deformation can be developed. In the floating frame of reference formulation, the global position vector of an arbitrary point on a body i in the system can be written, as shown in Fig.2, as follows:

$$\mathbf{r}^i = \mathbf{R}^i + \mathbf{A}^i(\bar{\mathbf{u}}_o^i + \mathbf{S}^i \mathbf{q}_f^i)$$

where $\mathbf{R}^i$ is the global position vector of the origin of the body coordinate system, $\mathbf{A}^i$ is the transformation matrix that defines the orientation of the body coordinate system, $\bar{\mathbf{u}}_o^i$ is the position vector of the arbitrary point in the undeformed state, $\mathbf{S}^i$ is the shape function matrix, and $\mathbf{q}_f^i$ is the vector of elastic coordinates that define the body deformation with respect to the body coordinate system. Note that the preceding equation can be used also to describe the rigid body kinematics by setting the last term equal to zero. In this special case, the rigid boy kinematics is described using the reference translation coordinates $\mathbf{R}^i$ and the orientation coordinates 2^i that define the transformation matrix $\mathbf{A}^i$.

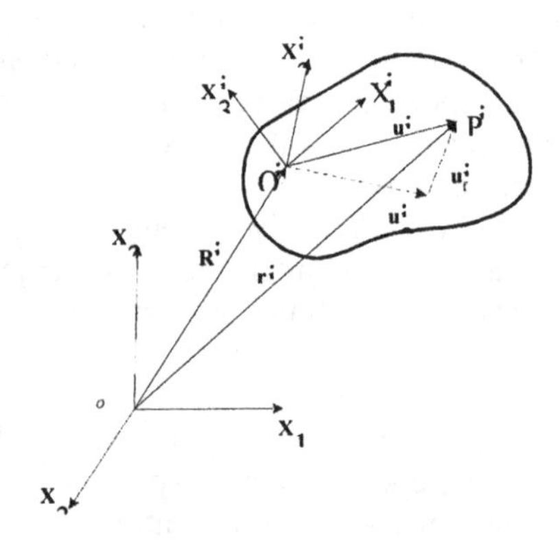

Figure 2 Floating Frame of Reference Formulation

The third set of generalized coordinates used to describe the large deformations is introduced using the absolute nodal coordinate formulation. This set of generalized coordinates consists of global displacement and slope coordinates and does not include any infinitesimal or finite rotations. By using the absolute nodal coordinates, exact modeling of the rigid body dynamics can be obtained. Furthermore, the absolute nodal coordinate formulation leads to a constant mass matrix in two- and three-dimensional problems. In this formulation, the global position vector of an arbitrary point on the flexible body is defined, as shown in Fig. 3, as

$$\mathbf{r}^i = \mathbf{S}_a^i \mathbf{q}_a^i$$

where $\mathbf{S}_a^i$ is a shape function matrix, and $\mathbf{q}_a^i$ is the vector of absolute coordinates that consist of global displacement and slope coordinates. It can be shown that the preceding equation can be used to describe an arbitrary rigid body motion and can also be effectively and efficiently used to describe the large deformation of beams, plates and shells.

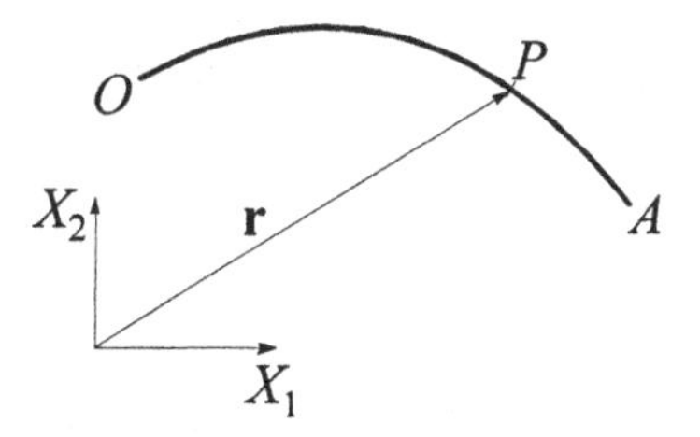

Figure 3 Absolute Nodal Coordinate Formulation

Using the different sets of coordinates above described, the vector of the generalized coordinates of the multibody system that contains rigid, flexible, and very flexible components can be described using the following set of system generalized coordinates:

$$\mathbf{q} = \begin{bmatrix} \mathbf{R}^{\mathrm{T}} & \theta^{\mathrm{T}} & \mathbf{q}_f^{\mathrm{T}} & \mathbf{q}_a^{\mathrm{T}} \end{bmatrix}^{\mathrm{T}}$$

where **R** and 2 are the reference coordinates of the rigid or flexible bodies in the system, $\mathbf{q}_f$ is the vector of elastic coordinates that define the small deformations of the bodies with respect to the body coordinate system, and $\mathbf{q}_a$ is the vector of absolute nodal coordinates used to describe the motion of bodies in the system that undergo large deformations. The preceding equation can be written in terms of three distinct sets of coordinates as follows:

$$\mathbf{q} = \begin{bmatrix} \mathbf{q}_r^{\mathrm{T}} & \mathbf{q}_f^{\mathrm{T}} & \mathbf{q}_a^{\mathrm{T}} \end{bmatrix}^{\mathrm{T}}$$

where $\mathbf{q}_r = [\mathbf{R}^{\mathrm{T}} \quad 2^{\mathrm{T}}]^{\mathrm{T}}$ is the vector of reference coordinates of the system.

6. Non-Generalized Coordinates

The general treatment of the rigid body contact problem requires introducing a set of parameters **s** that describe the geometry of the contact surfaces. In the augmented form of the equations of motion, there is no inertia or generalized forces associated with these parameters, and for this reason, they are referred to as non-generalized coordinates. Using these surface parameters, the tangents and normal to a surface at the contact point can be determined, and used to define the contact conditions. By using this approach, the contact points can be determined on line, and the coupling between the generalized and non-generalized surface parameters can be taken into consideration. By introducing the surface parameters for the rigid body contact, the vector of the system coordinates can be written as follows:

$$\mathbf{p}=\begin{bmatrix}\mathbf{q}_r^{\mathrm{T}} & \mathbf{q}_f^{\mathrm{T}} & \mathbf{q}_a^{\mathrm{T}} & \mathbf{s}^{\mathrm{T}}\end{bmatrix}^{\mathrm{T}}$$

In the following section, the formulation of the constrained dynamic equations of motion in terms of the generalized coordinates and non-generalized surface parameters is discussed.

7. Dynamic Equations

A new general flexible multibody algorithm must allow joint connectivity between rigid bodies, flexible bodies, and very flexible bodies as well as the general treatment of the contact problem. In this case, the joint constraints must be expressed in terms of the reference, elastic and absolute nodal coordinates. Considering also *driving constraints* that can depend explicitly on time, the vector of nonlinear constraint functions can be expressed in terms of the system reference, elastic and absolute nodal coordinates as well as the non-generalized surface parameters as follows:

$$\mathbf{C}(\mathbf{q}_r, \mathbf{q}_f, \mathbf{q}_a, \mathbf{s}, t)=\mathbf{0}$$

As demonstrated in the multibody dynamics literature, the floating frame of reference formulation leads to highly nonlinear expressions for the joint constraints as the result of using the body coordinate system that introduces geometric nonlinearities. The formulation of many of these joints, however, becomes much simpler when the absolute nodal coordinate formulation is used. Nonetheless, since slopes are used as coordinates in the absolute nodal coordinate formulation, the formulations of some joints require the development of new special techniques for defining the kinematics of selected reference frames at the joint definition points in terms of the absolute nodal coordinates.

The kinematic constraints that describe mechanical joints, contact conditions and specified motion trajectories can be adjoined to the system differential equations of motion using the technique of Lagrange multipliers. This leads to the following augmented form of the system equations of motion:

$$\begin{bmatrix}\mathbf{M}_{rr} & \mathbf{M}_{rf} & \mathbf{0} & \mathbf{0} & \mathbf{C}_{\mathbf{q}_r}^{\mathrm{T}} \\ \mathbf{M}_{fr} & \mathbf{M}_{ff} & \mathbf{0} & \mathbf{0} & \mathbf{C}_{\mathbf{q}_f}^{\mathrm{T}} \\ \mathbf{0} & \mathbf{0} & \mathbf{M}_{aa} & \mathbf{0} & \mathbf{C}_{\mathbf{q}_a}^{\mathrm{T}} \\ \mathbf{0} & \mathbf{0} & \mathbf{0} & \mathbf{0} & \mathbf{C}_{\mathbf{s}}^{\mathrm{T}} \\ \mathbf{C}_{\mathbf{q}_r} & \mathbf{C}_{\mathbf{q}_f} & \mathbf{C}_{\mathbf{q}_a} & \mathbf{C}_{\mathbf{s}} & \mathbf{0}\end{bmatrix}\begin{bmatrix}\ddot{\mathbf{q}}_r \\ \ddot{\mathbf{q}}_f \\ \ddot{\mathbf{q}}_a \\ \ddot{\mathbf{s}} \\ \boldsymbol{\lambda}\end{bmatrix}=\begin{bmatrix}\mathbf{Q}_r \\ \mathbf{Q}_f \\ \mathbf{Q}_a \\ \mathbf{0} \\ \mathbf{Q}_c\end{bmatrix}$$

where **M** refers to a mass sub-matrix, subscripts r, f, a, and s refer, respectively, to reference, elastic, absolute nodal coordinates, and non-generalized surface parameters, $\mathbf{C}_\mathbf{q}$ is the constraint Jacobian matrix associated with the generalized coordinates, $\mathbf{C}_\mathbf{s}$ is the constraint Jacobian matrix associated with the non-generalized surface parameters **s**, 8 is the vector of Lagrange multipliers, $\mathbf{Q}_r$, $\mathbf{Q}_f$, and $\mathbf{Q}_a$ are the generalized forces associated with reference, elastic, and absolute nodal coordinates, and $\mathbf{Q}_c$ is a quadratic velocity vector that results from the differentiation of the kinematic constraint equations twice with respect to time, that is

$$\mathbf{C}_{\mathbf{p}}\ddot{\mathbf{p}} = \mathbf{Q}_c$$

The augmented form of the equations of motion can be solved in order to obtain the second time derivative of the vectors of reference, elastic, absolute nodal coordinates and surface parameters as well as the vector of Lagrange multipliers. Lagrange multipliers can be used to determine the generalized constraint forces associated with the reference, elastic, and absolute nodal coordinates. The reference, elastic, absolute nodal accelerations and the time derivatives of the surface parameters can be integrated forward in time in order to determine the coordinates and velocities. The numerical algorithm proposed in this investigation ensures that the algebraic constraint equations are not violated.

The vector of Lagrange multipliers can also be used to determine the normal contact forces. These normal contact forces can be used to determine the creep forces required for accurate modeling of railroad vehicle system applications.

8. Cholesky Coordinates

Since the mass matrix $\mathbf{M}_{aa}$ associated with the absolute nodal coordinates is constant, a Cholesky transformation can be used to obtain a generalized identity mass matrix. This will lead to an optimum sparse matrix structure for the augmented form of the equations of motion of the system. The resulting augmented form of the equations of motion can be written as follows:

$$\begin{bmatrix} \mathbf{M}_{rr} & \mathbf{M}_{rf} & \mathbf{0} & \mathbf{0} & \mathbf{C}_{\mathbf{q}_r}^{\mathrm{T}} \\ \mathbf{M}_{fr} & \mathbf{M}_{ff} & \mathbf{0} & \mathbf{0} & \mathbf{C}_{\mathbf{q}_f}^{\mathrm{T}} \\ \mathbf{0} & \mathbf{0} & \mathbf{I} & \mathbf{0} & \mathbf{C}_{\mathbf{q}_{ch}}^{\mathrm{T}} \\ \mathbf{0} & \mathbf{0} & \mathbf{0} & \mathbf{0} & \mathbf{C}_{\mathbf{s}}^{\mathrm{T}} \\ \mathbf{C}_{\mathbf{q}_r} & \mathbf{C}_{\mathbf{q}_f} & \mathbf{C}_{\mathbf{q}_a} & \mathbf{C}_{\mathbf{s}} & \mathbf{0} \end{bmatrix} \begin{bmatrix} \ddot{\mathbf{q}}_r \\ \ddot{\mathbf{q}}_f \\ \ddot{\mathbf{q}}_{ch} \\ \ddot{\mathbf{s}} \\ \boldsymbol{\lambda} \end{bmatrix} = \begin{bmatrix} \mathbf{Q}_r \\ \mathbf{Q}_f \\ \mathbf{Q}_{ch} \\ \mathbf{0} \\ \mathbf{Q}_c \end{bmatrix}$$

where $\mathbf{I}$ is an identity matrix, and subscript *ch* refers to Cholesky coordinates.

9. Summary and Conclusions

The methodology proposed in this investigation has several special features that makes it suited for the analysis of complex multibody systems in which components undergo large displacements including large deformations. The proposed methodology does not require the use of special measures to satisfy the principle of work and energy, as demonstrated by the preliminary results presented in previous publications. Accurate results for highly nonlinear systems can be obtained using explicit integrators commonly used in the multibody simulation codes, as a consequence the linearity of the inertia forces become an important issue. As previously pointed out, the method does not require interpolation of rotations or slopes, automatically captures the effect of geometric centrifugal stiffening, and can be systematically applied to beam, plate and shell elements. Continuity of the displacement gradients is ensured, and as a consequence, the stress calculations are accurate. Another important feature of the proposed absolute nodal coordinate formulation is that it leads to a constant mass matrix in two- and three-dimensional applications in which many of the assumptions of Euler-Bernoulli, Timoshenko, and Mindlin beam and plate theories are relaxed. This important property allows for introducing Cholesky coordinates that lead to an identity generalized inertia matrix associated with the Cholesky coordinates. The result is an optimum sparse matrix structure for the augmented form of the multibody equations of motion.

In some mechanical system applications, the equations of motion are formulated in terms of a set of non-generalized coordinates such as the parameters used to describe the geometry of the surfaces in contact. These parameters have no generalized inertia or forces associated with them. The augmented form of the equations of motion can be formulated in terms of coupled set of generalized and non-generalized coordinates as demonstrated in this investigation. Lagrange multipliers associated with the contact constraints can be used to determine the normal contact forces required for the calculations of the tangential crepage forces and spin moments.

References

1. Greenwood, D.T. (1988) *Principles of Dynamics*, Second Edition, Prentice Hall.

2. Fowles, G.R. (1986) *Analytical Mechanics*, Fourth Edition, Saunders College Publishing.

3. Goldstein, H. (1950) *Classical Mechanics*, Addison-Wesley.

4. Nikravesh, P.E. (1988) *Computer Aided Analysis of Mechanical Systems*, Prentice Hall.

5. Shabana, A.A.. (2001) *Computational Dynamics*, Second Edition.

6. Agrawal, O.P., and Shabana, A.A. (1985) Dynamic analysis of multibody systems using component modes, *Computers and Structures*, **21**(6), 1301-1312.

7. Ashley, H. (1967) Observations on the dynamic behavior of large flexible bodies in orbit, *AIAA Journal*, **5**(3), 460-469.

8. Cavin, R.K., and Dusto, A.R. (1977) Hamilton's principle: finite element methods and flexible body dynamics, *AIAA Journal*, **15**(2), 1684-1690.

9. De Veubeke, B.F. (1976) The dynamics of flexible bodies, *Int. J. Eng. Sci.*, **14**, 895-913.

10. Hughes, P.C. (1979) Dynamics of chain of flexible bodies, *J. Astronaut. Sci.*, **27**(4), 359-380.

11. Cook, R.D. (1981) *Concepts and Applications of Finite Element Analysis*, Second Edition.

12. Huebner, K.H., Thornton, E.A., and Byrom, T.G. (1995) *The Finite Element Method for Engineers*, Third Edition, Wiley & Sons.

13. Tong, P., and Rossettos, J.N. (1977) *Finite Element Method*, The MIT Press.

14. Bathe, K.J. (1981) *Finite Element Procedures in Engineering Analysis*, Prentice Hall.

15. Zienkiewicz, O.C. (1979) *The Finite Element Method*, McGraw-Hill.

16. Shabana, A.A. (1982) Dynamics of large scale flexible mechanical systems", Ph.D. Thesis, University of Iowa, Iowa City.

17. Shabana, A.A., and Wehage, R.A. (1983) Coordinate reduction technique for transient analysis of spatial substructures with large angular rotations, *Journal of Structural mechanics*, **11**(3), 401-431.

18. Winfrey, R.C. (1971) Elastic link mechanism dynamics, *ASME J. Eng. Industry*, **93**, 268-272.

19. Winfrey, R.C. (1972) Dynamic analysis of elastic link mechanisms by reduction of coordinates, *ASME J. Eng. Industry*, **94**, 557-582.

20. Erdman, A.G., and Sandor, G.N. (1972) Kineto-Elastodynamics: A review of the state of the art and trends, *Mechanism and Machine Theory*, **7**, 19-33.

21. Bahgat, B., and Willmert, K.D. (1973) Finite element vibration analysis of planar mechanisms", *Mechanism and Machine Theory*, **8**, 497-516.

22. Lowen, G.G., and Chassapis, C. (1986) The elastic behavior of links: An update, *Mechanism and Machine Theory*, **21**(1), 33-42.

23. Rankin, C.C., and Brogan, F.A. (1986) An element independent co-rotational procedure for the treatment of large rotations, *ASME Journal Pressure Vessel Technology*, **108**, 165-174.

24. Benson, D.J., and Hallquist, J.D. (1986) A simple rigid body algorithm for structural dynamics programs, *International Journal for Numerical Methods in Engineering*, **22**, 723-749.

25. Wasfy, T.M., and Noor, A.K. (1996) Modeling and sensitivity analysis of multibody systems using new solid, shell, and beam elements, *Computer Methods in Applied Mechanics and Engineering*, **138**, 187-211.

26. Shabana, A. (1998) *Dynamics of Multibody Systems*, Second Edition, Cambridge University Press.

27. Omar, M.A., and Shabana, A.A. (2001) A two-dimensional shear deformable beam for large rotation and deformation problems, *Sound and Vibration*, **243**(3), 565-576.

28. Campanelli, M., Berzeri, M., and Shabana, A.A. (2000) Performance of the incremental and non-incremental finite element formulations in flexible multibody problems, *ASME Journal of Mechanical Design*, **122**(4), 498-507.

29. Shabana, A.A., and Yakoub, R.Y. (2001) Three-dimensional absolute nodal coordinate formulation for beam elements: Theory, *ASME Journal of Mechanical Design*, **123** (4), 606-613.

30. Yakoub, R.Y., and Shabana, A.A. (2001) Three-dimensional absolute nodal coordinate formulation for beam elements: implementation and applications, *ASME Journal of Mechanical Design*, **123** (4), 614-621.

31. Shampine, L., and Gordon, M. (1975) *Computer Solution of ODE: The Initial Value Problem*, Freeman.

GEOMETRICAL INTERPRETATION OF MULTIBODY DYNAMICS: THEORY AND IMPLEMENTATIONS

W. BLAJER
Technical University of Radom, Institute of Applied Mechanics
ul. Krasickiego 54, PL-26-600 Radom, Poland
wblajer@poczta.onet.pl

ABSTRACT. A unified geometrical interpretation of multibody dynamics codes is presented. The geometrical picture is built using the concepts of configuration manifolds, linear vector spaces, and projection techniques. The presented projection method leads to compact schemes for obtaining different types of equations of motion and for determination of constraint reactions, relevant to open-loop and closed-loop, and for holonomic and nonholonomic systems. The other useful implementations stimulated by the geometrical interpretation are improved schemes for constraint violation elimination, a novel approach to efficient determination of constraint reactions, a geometric interpretation of the augmented Lagrangian formulation, and an orthonormalization method.

1. Introduction

The observed rapid expansion of research on multibody dynamics stems from the current applications in many technological disciplines such as spacecraft, robotics, machine and vehicle dynamics and biomechanics, and is stimulated by the advances in computational techniques. Based on classical mechanics, numerous multibody formalisms has been developed to analyze the increasingly complex mechanical systems, and sophisticated numerical/symbolic methods are often involved. A legitimate means of presenting these problems in a systematic way is to illustrate them geometrically. The precise and powerful investigation tool of differential geometry can be stimulative, too.

The purpose of this paper is to associate the multibody dynamics procedures with the geometrical picture involving the concepts of configuration manifolds, linear vector spaces, and projection techniques. An unconstrained system is assigned a free configuration manifold, and is regarded as a generalized particle on the manifold. The system dynamics is then considered in the local tangent space to the manifold at the system representation point. Imposed constraints on the system, the tangent space splits into the velocity restricted and the velocity admissible subspaces, and the configuration confines to the holonomic constraint manifold. Followed these geometrical concepts, a uniform and compact treatment of both holonomic and nonholonomic systems can be demonstrated [3,4,11,13,15,16], which gives an intuitive geometrical insight into the problems solved and directly appeals to the geometry of constrained particle motion known from Newtonian dynamics. While exploiting the differential geometry formalism, the paper is written in standard matrix notation, well suited for computer implementations.

W. Schiehlen and M. Valášek (eds.), Virtual Nonlinear Multibody Systems, 17–36.

The presented projection method leads to compact schemes for obtaining different types of equations of motion and for the determination of constraint reactions, relevant to open-loop/closed-loop and holonomic/nonholonomic systems [3,4]. Uniform projective formulations and geometrical interpretations of many classical methods of multibody dynamics can also be demonstrated [4,6,11,13,15,16]. The other useful implementations are the improved schemes for constraint violation elimination [5,20], a geometrical interpretation and extended applications of the augmented Lagrangian formulation [6], and some novel multibody codes such as augmented joint coordinate method [7] and an orthonormalization method [2].

2. Unconstrained System Dynamics

Consider an n-degree-of-freedom autonomous system defined in generalized coordinates $\mathbf{p} = [p_1 \ \dots \ p_n]^T$ and velocities $\mathbf{v} = [v_1 \ \dots \ v_n]^T$, the latter being either generalized velocities or quasi-velocities. The system can be a collection of unconstrained rigid bodies or a Lagrangian system defined in independent variables. The generic matrix form of the initial governing equations is [3,4,12,14,17,18]:

$$\dot{\mathbf{p}} = \mathbf{A}(\mathbf{p})\,\mathbf{v} \tag{1}$$

$$\mathbf{M}(\mathbf{p})\,\dot{\mathbf{v}} + \mathbf{d}(\mathbf{p},\mathbf{v}) = \mathbf{f}(\mathbf{p},\mathbf{v},t) \tag{2}$$

where $\mathbf{A}$ is the $n \times n$ transformation matrix, $\mathbf{M}$ is the $n \times n$ symmetric and positive definite generalized mass matrix, $\mathbf{d} = [d_1 \ \dots \ d_n]^T$ represents in general the centrifugal, Coriolis, and gyroscopic dynamic terms, $\mathbf{f} = [f_1 \ \dots \ f_n]^T$ are the applied forces related to $\mathbf{v}$, and t is the time.

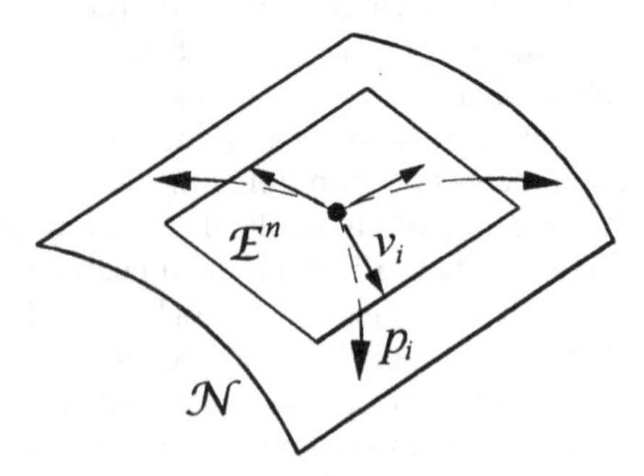

Figure 1. The manifold $\mathcal{N}$ and the space $\mathcal{E}^n$.

The system can be viewed as a point on the n-dimensional configuration manifold $\mathcal{N}$ of the system (Fig. 1). At each position $\mathbf{p} \in \mathcal{N}$, an n-dimensional tangent space to the manifold can then be defined, whose positive-definite quadratic form is expressed by the system kinetic energy $2T = \mathbf{v}^T\mathbf{M}\,\mathbf{v}$. So endowed with the metric, $\mathcal{N}$ becomes a Riemannian manifold [3,4,11,13,15,16], and the said local tangent space at $\mathbf{p} \in \mathcal{N}$ is an Euclidean (linear vector) space $\mathcal{E}^n$. The generalized mass matrix $\mathbf{M}$ is thus the metric tensor matrix of a basis referred to $\mathbf{v}$, and the dynamic equation (2) can be viewed as a matrix representation of vector formula $\vec{b} = \vec{f}$, where $\mathbf{b} = \mathbf{M}\dot{\mathbf{v}} + \mathbf{d}$ and $\mathbf{f}$ are the generalized dynamic and applied forces, respectively. According to the geometrical interpretation, the kinetic energy T defined above can be seen as the doubled dot product of the velocity vector $\vec{v}$ of the system, $2T = \vec{v} \circ \vec{v}$. Continuing with this geometrical setting, $\mathbf{v}$ and $\mathbf{M}\mathbf{v}$ are generalized velocities and generalized momenta of the system, and $\mathbf{a} = \dot{\mathbf{v}} + \mathbf{M}^{-1}\mathbf{d}$ and $\mathbf{b} = \mathbf{M}\dot{\mathbf{v}} + \mathbf{d}$ are the generalized accelerations and generalized dynamic forces of the system, respectively. The Appell's function [9] can then be defined as $2S = \mathbf{a}^T\mathbf{M}\mathbf{a} = (\dot{\mathbf{v}} + \mathbf{M}^{-1}\mathbf{d})^T\mathbf{M}\,(\dot{\mathbf{v}} + \mathbf{M}^{-1}\mathbf{d}) = \mathbf{b}^T\mathbf{M}^{-1}\mathbf{b}$, and $\mathbf{a}$ and $\mathbf{b}$ are respectively the contravariant and covariant representations of the same vector, denoted above as $\vec{b}$ ($2S = \vec{b} \circ \vec{b}$), in the linear vector space $\mathcal{E}^n$ (see [3,4,13,15,16] for more detailBy changing the velocity components from $\mathbf{v}$ to $\mathbf{v}'$, a new basis of $\mathcal{E}^n$ is introduced, and the transformation formula is $\mathbf{v} = \mathbf{B}\mathbf{v}'$, where $\mathbf{B}(\mathbf{p})$ is an $n \times n$ transformation matrix (for $\mathbf{v}' = \dot{\mathbf{p}}$, $\mathbf{B} = \mathbf{A}^{-1}$, and $\mathbf{A}$ is defined in (1)). By premultiplying (2) with $\mathbf{B}^T$, the dynamic equations (2) are firstly projected into the new basis [3,4]. Then, by using $\mathbf{v} = \mathbf{B}\mathbf{v}'$ and $\dot{\mathbf{v}} = \mathbf{B}\dot{\mathbf{v}}' + \boldsymbol{\gamma}$, where $\boldsymbol{\gamma} = \dot{\mathbf{B}}\mathbf{v}'$, the projected equations are expressed in the new velocity components,

$$\mathbf{B}^T[\mathbf{M}(\mathbf{B}\dot{\mathbf{v}}' + \dot{\mathbf{B}}\mathbf{v}') + \mathbf{d} = \mathbf{f}] \quad \Leftrightarrow \quad \mathbf{M}'(\mathbf{p})\dot{\mathbf{v}}' + \mathbf{d}'(\mathbf{p}, \mathbf{v}') = \mathbf{f}'(\mathbf{p}, \mathbf{v}', t) \tag{3}$$

where $\mathbf{M}' = \mathbf{B}^T\mathbf{M}\mathbf{B}$, $\mathbf{d}' = \mathbf{B}^T(\mathbf{M}\dot{\mathbf{B}}\mathbf{v}' + \mathbf{d})$ and $\mathbf{f}' = \mathbf{B}^T\mathbf{f}$ have the same meaning as in (2) but refer now to the new velocities $\mathbf{v}'$. This geometrically grounded projection technique constitutes an easy and automatic way for transformation of motion equations between different sets of coordinates and velocity components. Let us illustrate this process with the following simple example.

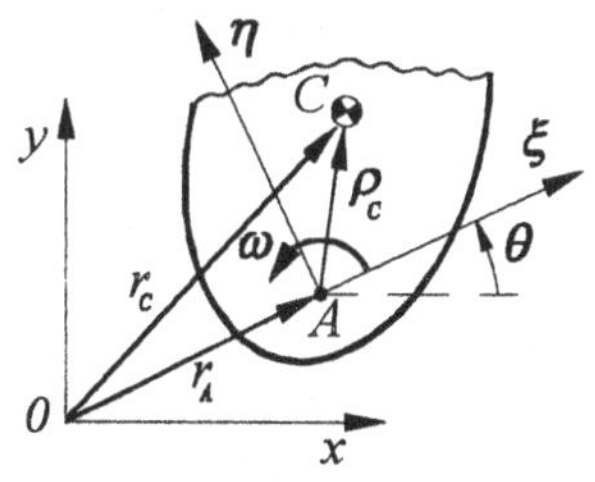

Figure 2. Planar motion of a rigid body: the reference frames Oxy and $A\xi\eta$.

Example 1: Dynamic equations of a rigid body in a body-fixed reference frame (planar case). The starting point are Newton-Euler equations of motion, which for the planar case simplify to $\mathbf{M}\dot{\mathbf{v}} = \mathbf{f}$, where $\mathbf{M} = \mathrm{diag}(m, m, J_C)$, m is the mass of the body and J_C is its moment of inertia with respect to the mass center C, $\mathbf{v} = [v_{Cx}\ \ v_{Cy}\ \ \omega]^T$ are the components of the velocity of C in the inertial frame Oxy and the body angular velocity, and $\mathbf{f} = [F_x\ \ F_y\ \ M_C]^T$ are the components in Oxy of the total of external forces on the body and their resultant moment with respect to C, respectively. Let derive the dynamic equations of both translational and rotational motions in any body-fixed reference frame $A\xi\eta$, i.e. in the velocity components $\mathbf{v}' = [v_{C\xi}\ \ v_{C\eta}\ \ \omega]^T$ (note that v_ξ and v_η are quasi-velocities). The velocity and acceleration transformation formulae are (Fig. 2):

$$\mathbf{v} = \begin{bmatrix} v_{Cx} \\ v_{Cy} \\ \omega \end{bmatrix} = \begin{bmatrix} \cos\theta & -\sin\theta & -\xi_C \sin\theta - \eta_c \cos\theta \\ \sin\theta & \cos\theta & \xi_C \cos\theta - \eta_c \sin\theta \\ 0 & 0 & 1 \end{bmatrix} \begin{bmatrix} v_{A\xi} \\ v_{A\eta} \\ \omega \end{bmatrix} = \mathbf{B}\,\mathbf{v}'$$

$$\dot{\mathbf{v}} = \begin{bmatrix} \dot{v}_{Cx} \\ \dot{v}_{Cy} \\ \dot{\omega} \end{bmatrix} = \mathbf{B} \begin{bmatrix} \dot{v}_{A\xi} \\ \dot{v}_{A\eta} \\ \dot{\omega} \end{bmatrix} + \begin{bmatrix} -\omega\,(v_{A\xi}\sin\theta + v_{A\eta}\cos\theta) - \omega^2(\xi_C\cos\theta - \eta_C\sin\theta) \\ \omega\,(v_{A\xi}\cos\theta - v_{A\eta}\sin\theta) - \omega^2(\xi_C\sin\theta + \eta_C\cos\theta) \\ 0 \end{bmatrix} = \mathbf{B}\,\dot{\mathbf{v}}' + \boldsymbol{\gamma}$$

where $\rho_C = [\xi_C \;\; \eta_C]^T$. The dynamic equations in $\mathbf{v}'$ can then be obtained from (3) by hand (for this simple case) or using computer symbolic manipulations, i.e.

$$\begin{bmatrix} m & 0 & -m\eta_C \\ 0 & m & m\xi_C \\ -m\eta_C & m\xi_C & J_A \end{bmatrix} \begin{bmatrix} \dot{v}_{A\xi} \\ \dot{v}_{A\eta} \\ \dot{\omega} \end{bmatrix} + \begin{bmatrix} -m(v_{A\eta} + \xi_C\omega)\omega \\ m(v_{A\xi} - \eta_C\omega)\omega \\ m(v_{A\xi}\xi_C + v_{A\eta}\eta_C)\omega \end{bmatrix} = \begin{bmatrix} F_\xi \\ F_\eta \\ M_A \end{bmatrix} \quad \Leftrightarrow \quad \mathbf{M}'\,\dot{\mathbf{v}}' + \mathbf{d}' = \mathbf{f}'$$

where $J_A = J_C + m(\xi_C^2 + \eta_C^2)$, $F_\xi = F_x \cos\theta + F_y \sin\theta$, $F_\eta = -F_x \sin\theta + F_y \cos\theta$, and $M_C = (-F_x \sin\theta + F_y \cos\theta)\xi_C - (F_x \cos\theta + F_y \sin\theta)\eta_C + M_C$.

3. Geometry of Constrained System Dynamics

Let the system described in (1) and (2) be subjected to m_h holonomic and m_{nh} non-holonomic constraints, $m = m_h + m_{nh}$, all scleronomic for simplicity (the rheonomic constraints are considered in e.g. [3,4]). The constraint equations, given implicitly, are:

$$\begin{cases} \boldsymbol{\Phi}(\mathbf{p}) = \mathbf{0} & \text{(4a)} \\ \mathbf{C}_{nh}(\mathbf{p})\,\mathbf{v} = \mathbf{0} & \text{(4b)} \end{cases}$$

The holonomic constraints (4a) define a k-dimensional submanifold $\mathcal{K}$ in $\mathcal{N}$, $k = n - m_h$, and the system configuration is confined to $\mathcal{K}$ on which k independent curvilinear coordinates $\mathbf{q} = [q_1 \;\cdots\; q_k]^T$ can be defined. By differentiating with respect to time the holonomic constraints one obtains $\dot{\boldsymbol{\Phi}} = \mathbf{C}_h(\mathbf{p})\,\mathbf{v} = \mathbf{0}$, where $\mathbf{C}_h = (\partial\boldsymbol{\Phi}/\partial\mathbf{p})\,\mathbf{A}$. The uniform constraint equations at the velocity and acceleration levels are then:

$$\boldsymbol{\Psi} = \mathbf{C}(\mathbf{p})\,\mathbf{v} = \mathbf{0} \tag{5}$$

$$\dot{\boldsymbol{\Psi}} = \mathbf{C}(\mathbf{p})\,\dot{\mathbf{v}} - \boldsymbol{\xi}(\mathbf{p},\mathbf{v}) = \mathbf{0} \tag{6}$$

where $\mathbf{C} = [\mathbf{C}_h^T \;\; \mathbf{C}_{nh}^T]^T$ is the $m \times n$ constraint matrix, and $\boldsymbol{\xi} = -\dot{\mathbf{C}}\,\mathbf{v}$ is the m-vector of constraint induced accelerations (see Figure 3 for the geometrical interpretation).

The m constraint vectors, represented in $\mathbf{C}$ as rows, span an m-dimensional *constrained subspace* $\mathcal{C}^m$ in $\mathcal{E}^n$. According to (5) the projection of the system velocity $\vec{v}$ into $\mathcal{C}^m$ vanishes (Fig. 3a), i.e. $\vec{v}$ is sunk in the *unconstrained subspace* $\mathcal{D}^r$, which complements $\mathcal{C}^m$ in $\mathcal{E}^n$, $\mathcal{D}^r \cup \mathcal{C}^m = \mathcal{E}^n$ and $\mathcal{D}^r \cap \mathcal{C}^m = 0$. For the considered case of systems with scleronomic constraints, $\mathcal{C}^m$ and $\mathcal{D}^r$ are then velocity restricted and velocity admissible subspaces, , respectively, called *orthogonal* and *tangent* subspaces in the following as well. The subspace $\mathcal{D}^r$ can be defined by $r = n - m$ vectors represented as columns of an $n \times r$ matrix **D** which satisfies the condition

$$\mathbf{C}\,\mathbf{D} = \mathbf{0} \qquad \Leftrightarrow \qquad \mathbf{D}^T\mathbf{C}^T = \mathbf{0} \tag{7}$$

i.e. **D** is an *orthogonal complement* matrix to the constraint matrix **C** [2-4,12,13,17,18].

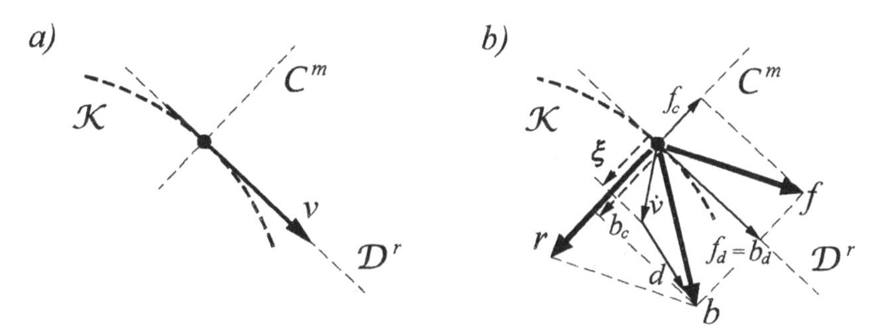

Figure 3. The geometry of constrained system dynamics.

The reactions of *ideal* constraints are by assumption collinear with the constraint vectors, and are represented in $\mathcal{C}^m$ by m Lagrange multipliers $\boldsymbol{\lambda} = [\lambda_1 \; \ldots \; \lambda_m]^T$. The representation of the generalized constraint reaction force $\vec{r}$ in $\mathcal{E}^n$ is $\mathbf{r} = -\mathbf{C}^T\boldsymbol{\lambda}$ [3,4,12,13,17,18], and the dynamic equations of the constrained system are

$$\mathbf{M}(\mathbf{p})\,\dot{\mathbf{v}} + \mathbf{d}(\mathbf{p},\mathbf{v}) = \mathbf{f}(\mathbf{p},\mathbf{v},t) - \mathbf{C}^T(\mathbf{p})\,\boldsymbol{\lambda} \tag{8}$$

often referred to as Lagrange's equations of type one. The vector form of (8) is $\vec{b} = \vec{f} + \vec{r}$ (Fig. 3b). By assuming that the constraints (4) express the prohibited relative translations/rotations and the vanishing relative translational/rotational velocities in the joints, the components of $\boldsymbol{\lambda}$ are the respective physical forces/torques. The governing equations of the constrained system are then the following $2n + m$ differential-algebraic equations (DAEs) in $2n$ differential variables **p** and **v**, and m algebraic variables $\boldsymbol{\lambda}$

$$\dot{\mathbf{p}} = \mathbf{A}\,\mathbf{v} \qquad\qquad \dot{\mathbf{p}} = \mathbf{A}(\mathbf{p})\,\mathbf{v} \tag{9a}$$

$$\begin{bmatrix} \mathbf{M} & \mathbf{C}^T \\ \mathbf{C} & \mathbf{0} \end{bmatrix} \begin{bmatrix} \dot{\mathbf{v}} \\ \boldsymbol{\lambda} \end{bmatrix} = \begin{bmatrix} \mathbf{f} - \mathbf{d} \\ \boldsymbol{\xi} \end{bmatrix} \qquad \Leftrightarrow \qquad \mathbf{H}(\mathbf{p}) \begin{bmatrix} \dot{\mathbf{v}} \\ \boldsymbol{\lambda} \end{bmatrix} = \mathbf{h}(\mathbf{p},\mathbf{v},t) \tag{9b}$$

and the initial values of the state variables must satisfy the lower-order constraint equations (4a) and (5), $\boldsymbol{\Phi}(\mathbf{p}_0) = \mathbf{0}$ and $\boldsymbol{\Psi}(\mathbf{p}_0, \mathbf{v}_0) = \mathbf{0}$. Since the leading matrix **H** in (9b) is invertible if only $\det(\mathbf{C}\,\mathbf{M}^{-1}\mathbf{C}^T) \neq 0$ [4], the equation can be solved for $\dot{\mathbf{v}}$ and standard ODE (ordinary differential equation) methods can be used to solve DAEs (9) for $\mathbf{p}(t)$ and $\mathbf{v}(t)$. Simultaneously $\boldsymbol{\lambda}(t)$ can be determined using the current state variables.

4. Constraint-Referred Projection Formulation

The formula (3) describes a projection of the dynamic equations (2) from a basis referred to $\mathbf{v}$ into a basis referred to $\mathbf{v}'$. In the case of a constrained system a similar projection of the dynamic constraint reaction-affected equations (8) can be performed with respect to the constrained and unconstrained directions, which is achieved by premultiplying (8) with $[\mathbf{D}^T \; (\mathbf{C}\mathbf{M}^{-1})^T]^T$. The premultiplication stands for the projection of (8) from $\mathcal{E}^n$ (basis referred to $\mathbf{v}$) into $\mathcal{D}^r$ and $\mathcal{C}^m$ [3,4,13], respectively, and yields:

$$\begin{bmatrix} \mathbf{D}^T \\ \mathbf{C}\mathbf{M}^{-1} \end{bmatrix} \left(\mathbf{M}\dot{\mathbf{v}} + \mathbf{d} = \mathbf{f} - \mathbf{C}^T\boldsymbol{\lambda} \right) \quad \begin{matrix} \Rightarrow \quad \mathbf{D}^T\mathbf{M}\dot{\mathbf{v}} + \mathbf{D}^T\mathbf{d} = \mathbf{D}^T\mathbf{f} \qquad\qquad\qquad\qquad (10a) \\ \Rightarrow \quad \mathbf{C}\dot{\mathbf{v}} + \mathbf{C}\mathbf{M}^{-1}\mathbf{d} = \mathbf{C}\mathbf{M}^{-1}\mathbf{f} - \mathbf{C}\mathbf{M}^{-1}\mathbf{C}^T\boldsymbol{\lambda} \quad (10b) \end{matrix}$$

The geometrical interpretation of (10a) and (10b) is $\vec{b}_d = \vec{f}_d$ and $\vec{b}_c = \vec{f}_c + \vec{r}$, respectively (Fig. 2b). Using the tangential projection (10a), $2n$ ODEs in $\mathbf{p}$ and $\mathbf{v}$ follow:

$$\begin{matrix} \dot{\mathbf{p}} = \mathbf{A}\,\mathbf{v} \\ \begin{bmatrix} \mathbf{D}^T\mathbf{M} \\ \mathbf{C} \end{bmatrix} \dot{\mathbf{v}} = \begin{bmatrix} \mathbf{D}^T(\mathbf{f} - \mathbf{d}) \\ \boldsymbol{\xi} \end{bmatrix} \end{matrix} \quad \Leftrightarrow \quad \begin{matrix} \dot{\mathbf{p}} = \mathbf{A}(\mathbf{p})\,\mathbf{v} \qquad (11a) \\ \mathbf{S}(\mathbf{p})\,\dot{\mathbf{v}} = \mathbf{s}(\mathbf{p}, \mathbf{v}, t) \qquad (11b) \end{matrix}$$

where the $n \times n$ matrix $\mathbf{S}$ in (11b) is by assumption invertible. The orthogonal projection (10b) provides then an algebraic equation for the determination of $\boldsymbol{\lambda}$

$$\boldsymbol{\lambda}(\mathbf{p}, \mathbf{v}, t) = (\mathbf{C}\mathbf{M}^{-1}\mathbf{C}^T)^{-1}[\mathbf{C}\mathbf{M}^{-1}(\mathbf{f} - \mathbf{d}) - \boldsymbol{\xi}] \qquad (12)$$

Example 2: *Nonholonomic system* (*absolute variable formulations*). Let us illustrate the contents of Sections 3 and 4 by considering a simple nonholonomic system shown in Figure 4a. A rigid bar moves on a horizontal plane. Its end A of moves along the y axis, and is connected to the origin by a spring whose stiffness coefficient and natural length are k and d, and thus $F_s = k(y_A - d)$. The other end B is equipped with a knife edge with the blade along the bar direction. The mass of the bar and its moment of inertia with respect to C (located in the middle of the bar of length $2b$) are m and J_C.

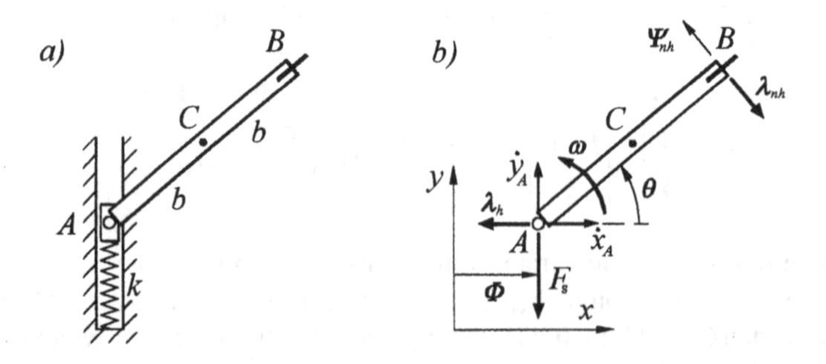

Figure 4. The sample nonholonomic system.

A convenient choice the absolute coordinates is $\mathbf{p} = [x_A \; y_A \; \theta]^T$. The dynamic equations of the unconstrained bar in these coordinates, $\mathbf{M}\ddot{\mathbf{p}} + \mathbf{d} = \mathbf{f}$, are:

$$\begin{bmatrix} m & 0 & -mb\sin\theta \\ 0 & m & mb\cos\theta \\ -mb\sin\theta & mb\cos\theta & J_A \end{bmatrix} \begin{bmatrix} \ddot{x}_A \\ \ddot{y}_A \\ \ddot{\theta} \end{bmatrix} + \begin{bmatrix} -mb\dot{\theta}^2\cos\theta \\ -mb\dot{\theta}^2\sin\theta \\ 0 \end{bmatrix} = \begin{bmatrix} 0 \\ -F_s \\ 0 \end{bmatrix}$$

where $J_A = J_C + mb^2$. The constraints on the bar, and the matrices $\mathbf{C}$ and ξ are:

$$\Phi = x_A = 0$$
$$\Psi_{nh} = -\dot{x}_A \sin\theta + \dot{y}_A \cos\theta + 2b\omega = 0$$

$$\mathbf{C} = \begin{bmatrix} 1 & 0 & 0 \\ -\sin\theta & \cos\theta & 2b \end{bmatrix}; \qquad \xi = \begin{bmatrix} 0 \\ (\dot{x}_A \cos\theta + \dot{y}_A \sin\theta)\dot{\theta} \end{bmatrix}$$

An orthogonal complement matrix $\mathbf{D}$ that satisfies the condition (7) can be guessed for this simple case. A reasonable proposition is

$$\mathbf{D}^T = \begin{bmatrix} 0 & -2b & \cos\theta \end{bmatrix}$$

Following the above definitions, two absolute variable formulations of the system, corresponding to (9) and (11), are possible.

The dependent variable formulations (9)/(13) and (11)/(14) involve the second-order differential constraint equations (6). The exact realization of only these constraint equations is thus assured by assumption. The lower-order constraint equations (4a) and (5) may be violated by the numerical solutions $\tilde{\mathbf{p}}(t)$ and $\tilde{\mathbf{v}}(t)$, burdened with the numerical error of integration, $\tilde{\mathbf{\Phi}} = \mathbf{\Phi}(\tilde{\mathbf{p}}) \neq \mathbf{0}$ and $\tilde{\mathbf{\Psi}} = \mathbf{\Psi}(\tilde{\mathbf{p}}, \tilde{\mathbf{v}}) \neq \mathbf{0}$. The numerical solution is free from the constraint violation problem when independent state variables are used. Most of the independent variable formulations can conveniently be interpreted geometrically and assisted with the projection technique.

5. Projective Formulation of Relevant Independent Variable Formulations

The dependent state variables $\mathbf{p} = [p_1 \; \dots \; p_n]^T$ and $\mathbf{v} = [v_1 \; \dots \; v_n]^T$ used in the previous sections are linked to the constraint equations *given implicitly*, m_h holonomic constraints (4a) imposed on $\mathbf{p}$ and m unified velocity (holonomic and nonholonomic) constraints (5) imposed on $\mathbf{v}$. There exist thus some $k = n - m_h$ independent coordinates $\mathbf{q} = [q_1 \; \dots \; q_k]^T$ which define the system position on the holonomic constraint manifold $\mathcal{K}$, and some $r = n - m$ independent speeds $\mathbf{u} = [u_1 \; \dots \; u_r]^T$ - the components of the system velocity in $\mathcal{D}^r$. The constraint equations *given explicitly* [4,18] are then relations between the dependent and independent state variables, i.e. $\mathbf{p} = \mathbf{g}(\mathbf{q})$ and $\mathbf{v} = \mathbf{D}\mathbf{u}$, if only $\mathbf{\Phi}(\mathbf{g}(\mathbf{q})) \equiv \mathbf{0}$ and $\mathbf{\Psi} = \mathbf{C}\mathbf{D}\mathbf{u} \equiv \mathbf{0} \Rightarrow \mathbf{C}\mathbf{D} = \mathbf{0}$. The explicit constraint equations play crucial role in the minimal-form formulations of multibody dynamics.

5.1 JOINT COORDINATE FORMULATION FOR OPEN-LOOP HOLONOMIC SYSTEMS

Let us assume for a while that the system described in (4) and (8) is an open-loop (tree structure) holonomic system, and $\mathbf{p}$ and $\mathbf{v}$ are the absolute variables that describe the positions and velocities of individual bodies with respect to a global non-moving reference frame. The constraints on the bodies due to the kinematical joints, given implicitly, are $\mathbf{\Phi}(\mathbf{p})=\mathbf{0}$, i.e. $m=m_h$ and $m_{nh}=0$ in (4). By choosing $k=n-m_h$ independent joint coordinates $\mathbf{q}=[q_1\ \dots\ q_k]^T$ to describe the relative configurations of the adjacent bodies in the joints, the holonomic constraint equations can be expressed explicitly as

$$\mathbf{p}=\mathbf{g}(\mathbf{q}) \tag{13}$$

and by assumption the implicit constraint equations are satisfied identically after substituting (13), $\mathbf{\Phi}(\mathbf{g}(\mathbf{q}))\equiv\mathbf{0}$. After differentiating (13) with respect to time one receives:

$$\mathbf{v}=\mathbf{D}(\mathbf{q})\,\dot{\mathbf{q}} \tag{14}$$

$$\dot{\mathbf{v}}=\mathbf{D}(\mathbf{q})\,\ddot{\mathbf{q}}+\boldsymbol{\gamma}(\mathbf{q},\dot{\mathbf{q}}) \tag{15}$$

where $\mathbf{D}=\mathbf{A}^{-1}(\partial\mathbf{g}/\partial\mathbf{q})$ and $\boldsymbol{\gamma}=\dot{\mathbf{D}}\dot{\mathbf{q}}$ are of dimensions $n\times k$ and $n\times 1$. After substituting (14) and (15) into (5) and (6) it follows that $\mathbf{D}$ defined above satisfies the conditions (7), i.e. $\mathbf{D}$ is an orthogonal complement to $\mathbf{C}$ (now $\mathbf{C}=\mathbf{C}_h$), and then $\mathbf{C}\boldsymbol{\gamma}=\boldsymbol{\xi}$.

By applying the explicit forms (13) – (15) of holonomic constraints to the projective scheme described in Section 4, from the tangential projection (10a) one obtains

$$\mathbf{D}^T\mathbf{M}(\mathbf{D}\ddot{\mathbf{q}}+\boldsymbol{\gamma})+\mathbf{D}^T\mathbf{d}=\mathbf{D}^T\mathbf{f} \quad\Leftrightarrow\quad \overline{\mathbf{M}}(\mathbf{q})\ddot{\mathbf{q}}+\overline{\mathbf{d}}(\mathbf{q},\dot{\mathbf{q}})=\overline{\mathbf{f}}(\mathbf{q},\dot{\mathbf{q}},t) \tag{16}$$

where $\overline{\mathbf{M}}=\mathbf{D}^T\mathbf{M}\mathbf{D}$ is the $k\times k$ generalized mass matrix, $\overline{\mathbf{d}}=\mathbf{D}^T(\mathbf{M}\boldsymbol{\gamma}+\mathbf{d})$ is the k vector of generalized dynamic forces due to the centrifugal, Coriolis and gyroscopic terms, and $\overline{\mathbf{f}}=\mathbf{D}^T\mathbf{f}$ is the k vector of generalized applied forces, all related to $\mathbf{q}$. The same result can be obtained by using different methods, see e.g. [12-14,16-18]. The formula (12) for determination of constraint reactions can then be expressed symbolically as $\boldsymbol{\lambda}(\mathbf{q},\dot{\mathbf{q}},t)$. By using $\mathbf{C}\boldsymbol{\gamma}=\boldsymbol{\xi}$ it can also be reformulated to

$$\boldsymbol{\lambda}(\mathbf{q},\dot{\mathbf{q}},t)=(\mathbf{C}\mathbf{M}^{-1}\mathbf{C}^T)^{-1}\mathbf{C}\,[\mathbf{M}^{-1}(\mathbf{f}-\mathbf{d})-\boldsymbol{\gamma}] \tag{17}$$

The scheme described above can also be treated as a general code for transforming the dynamic equations of multibody systems formulated initially in dependent variables $\mathbf{p}$ and $\mathbf{v}$ into a smaller set of equations in independent coordinates $\mathbf{q}$. By choosing $\mathbf{q}$ the holonomic constraint equations in the implicit form (4a) are eliminated, $\mathbf{\Phi}(\mathbf{g}(\mathbf{q}))\equiv\mathbf{0}$. However, the explicit formulation (13) of holonomic constraints is not always attainable in practice (e.g. for closed-loop systems).

Example 3: *Moving pendulum.* Consider a two-degree-of-freedom system shown in Figure 5. The relations (13) between the absolute coordinates $\mathbf{p}=[x_1\ y_1\ \theta_1\ x_2\ y_2\ \theta_2]^T$ and the joint coordinates $\mathbf{q}=[s\ \varphi]^T$, and $\mathbf{D}$ and $\boldsymbol{\gamma}$ as defined in (14) and (15), are:

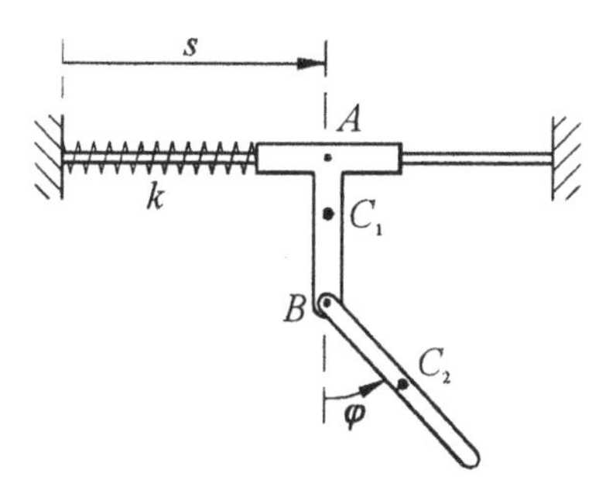

Figure 5. The moving pendulum.

$$\begin{aligned} x_1 &= s \\ y_1 &= a_1 \\ \theta_1 &= 0 \\ x_2 &= s + a_2 \sin\varphi \\ y_2 &= l_1 + a_2\cos\varphi \\ \theta_2 &= \varphi \end{aligned} \quad ; \quad \mathbf{D} = \begin{bmatrix} 1 & 0 \\ 0 & 0 \\ 0 & 0 \\ 1 & a_2\cos\varphi \\ 0 & -a_2\sin\varphi \\ 0 & 1 \end{bmatrix} \quad ; \quad \boldsymbol{\gamma} = \begin{bmatrix} 0 \\ 0 \\ 0 \\ -a_2\dot{\varphi}^2\sin\varphi \\ -a_2\dot{\varphi}^2\cos\varphi \\ 0 \end{bmatrix}$$

where $a_1 = AC_1$, $a_2 = BC_2$ and $l_1 = AB$. Using $\mathbf{M} = diag(m_1, m_1, J_1, m_2, m_2, J_2)$ and $\mathbf{f} = [-F_s \quad m_1 g \quad F_s a_1 \cos\theta_1 \quad 0 \quad m_2 g \quad 0]^T$ from the absolute ccordinate formulation, the final dynamic equations obtained using the scheme (16) are

$$\begin{bmatrix} m_1 + m_2 & m_2 a_2 \cos\varphi \\ m_2 a_2 \cos\varphi & J_2 + m_2 a_2^2 \end{bmatrix} \begin{bmatrix} \ddot{s} \\ \ddot{\varphi} \end{bmatrix} + \begin{bmatrix} -m_2 a_2 \dot{\varphi}^2 \sin\varphi \\ 0 \end{bmatrix} = \begin{bmatrix} -k(s-d) \\ -m_2 g\, a_2 \sin\varphi \end{bmatrix}$$

where m_i and J_i are the link masses and their central moments of inertia, g is the gravity acceleration, $F_s = k(s-d)$, and k and d are the stiffness coefficient and natural length of the spring.

5.2 VELOCITY PARTITIONING FORMULATION

The approach is usually referred to as *coordinate partitioning method* [19]. It is based on formulating the explicit constraint equations directly at the velocity and acceleration levels, in the form similar to (14) and (15). Thus, in a general sense of formulation of this paper, this is rather a *velocity partitioning method* [4].

If $\text{rank}(\mathbf{C}) = m = \max$, the implicit constraint equations (5) and (6) can always be resolved for some m dependent velocities/accelerations $\mathbf{w}/\dot{\mathbf{w}}$ in terms of the remaining r independent ones $\mathbf{u}/\dot{\mathbf{u}}$. The partition can be set symbolically as $\mathbf{v} = [\mathbf{u}^T \quad \mathbf{w}^T]^T$, which yields appropriate factorization of the constraint matrix $\mathbf{C} = [\mathbf{U} \quad \mathbf{W}]$, where $\mathbf{U}$ and $\mathbf{W}$ are the $m \times r$ and $m \times m$ matrices, respectively, and $\det(\mathbf{W}) \neq 0$ is assumed. By applying the partition to (5) and (6), the following explicit forms of the velocity and acceleration constraints can be obtained:

$$\mathbf{C}\mathbf{v} \equiv \mathbf{U}\mathbf{u}+\mathbf{W}\mathbf{w}=\mathbf{0} \quad \Rightarrow \quad \mathbf{v}=\begin{bmatrix}\mathbf{I}\\ -\mathbf{W}^{-1}\mathbf{U}\end{bmatrix}\mathbf{u} \equiv \mathbf{D}(\mathbf{p})\mathbf{u} \tag{18}$$

$$\mathbf{C}\dot{\mathbf{v}}+\boldsymbol{\xi} \equiv \mathbf{U}\dot{\mathbf{u}}+\mathbf{W}\dot{\mathbf{w}}-\boldsymbol{\xi}=\mathbf{0} \quad \Rightarrow \quad \dot{\mathbf{v}}=\mathbf{D}\dot{\mathbf{u}}+\begin{bmatrix}\mathbf{0}\\ \mathbf{W}^{-1}\boldsymbol{\xi}\end{bmatrix} \equiv \mathbf{D}(\mathbf{p})\dot{\mathbf{u}}+\boldsymbol{\gamma}(\mathbf{p},\mathbf{u}) \tag{19}$$

where **I** and **0** are the identity and null matrices of dimensions $r\times r$ and $r\times 1$, respectively. It is easy to check that the $n\times r$ matrix **D** is an orthogonal complement matrix to **C**, i.e. $\mathbf{C}\mathbf{D}=\mathbf{0}$. The relations (18) and (19) are thus the explicit forms of the constraints at the velocity and acceleration levels, and correspond to (14) and (15). Since **C** and as such **W** depends on the system position, in order to avoid possible singularities in the process of simulation, when $\det(\mathbf{W})\to 0$, variant formulations for different coordinate partitions can be prepared in advance to change between them when necessary. A projective criterion for best (optimal) coordinate partitioning was proposed in [8].

With the use of (18) and (19) and the projection technique, the governing equations of a constrained system can be transformed to $n+r$ ODEs whose symbolic form is:

$$\dot{\mathbf{p}}=\overline{\mathbf{A}}(\mathbf{p})\mathbf{u} \tag{20a}$$

$$\overline{\mathbf{M}}(\mathbf{p})\dot{\mathbf{u}}+\overline{\mathbf{d}}(\mathbf{p},\mathbf{u})=\overline{\mathbf{f}}(\mathbf{p},\mathbf{u},t) \tag{20b}$$

where $\overline{\mathbf{A}}=\mathbf{A}\mathbf{D}$, and $\overline{\mathbf{M}}$, $\overline{\mathbf{d}}$ and $\overline{\mathbf{f}}$ are as defined in (16). The constraint reactions can then be obtained from either (12) or (17), both represented symbolically as $\boldsymbol{\lambda}(\mathbf{p},\mathbf{u},t)$. The velocity partitioning method provides a general and useful code for ODE formulations of closed-loop holonomic systems and nonholonomic systems [4,12,14,17-19].

Example 4: *Nonholonomic system* (*velocity partitioning*). Let us reconsider the nonholonomic system of Example 2. From the analysis of **C** introduced there, it follows that $\dot{x}_A$ cannot be chosen as independent velocity (ill conditioned problem), and for $\dot{\theta}$ as independent velocity the partition is ill conditioned for $\theta\neq\pm\pi/2$. The best partition of $\dot{\mathbf{p}}=[\dot{x}_A\ \dot{y}_A\ \dot{\theta}]^T$ is $\mathbf{u}=[\dot{y}_A]$ and $\mathbf{w}=[\dot{x}_A\ \dot{\theta}]^T$, which is well conditioned for all possible positions of the bar. For this partition **C** is factorized into:

$$\mathbf{U}=\begin{bmatrix}0\\ \cos\theta\end{bmatrix}; \qquad \mathbf{W}=\begin{bmatrix}1 & 0\\ -\sin\theta & 2b\end{bmatrix}; \qquad \det(\mathbf{W})=2b$$

D and $\boldsymbol{\gamma}$ defined in (18) and (19) are then:

$$\mathbf{D}=\begin{bmatrix}0\\ 1\\ -\cos\theta/(2b)\end{bmatrix}; \qquad \boldsymbol{\gamma}=\begin{bmatrix}0\\ 0\\ -u^2\sin\theta\cos\theta/(4b^2)\end{bmatrix}$$

The kinematic equations (20a) are $\dot{\mathbf{p}}=\mathbf{D}\mathbf{u}$, and the single dynamic equation (20b) is

$$\left(m\sin^2\theta+\frac{J_A}{4b^2}\cos^2\theta\right)\dot{u}-\left(m-\frac{J_A}{4b^2}\right)\frac{u^2}{2b}\sin\theta\cos^2\theta=-F_s$$

In the above simple example all the expressions were introduced analytically. In more complicated cases the analytical derivations are unattainable in practice, however, and $\mathbf{D}$, γ and the final governing equations (20) are obtained numerically.

5.3 TREATMENT OF CLOSED-LOOP HOLONOMIC SYSTEMS

A multibody system may contain one ore more closed kinematic loops. The explicit forms (13) of constraints due to the kinematical joints in such systems, i.e. the relations between the dependent (say absolute) coordinates and some r independent coordinates, where r is equal to the number of degrees of freedom, are difficult to obtain in practice (if attainable at all). The minimal-form dynamic formulation described in Section 5.1 for open-loop systems cannot thus be directly extended to closed-loop systems.

Four basic treatments of closed-loop systems (Fig. 6) can be distinguished, which use n absolute coordinates $\mathbf{p}$ when all the joints in the system are "opened", and k joint coordinates $\mathbf{q}$ of an equivalent open-loop system obtained after cutting each closed loop at one of the kinematic joints, $k > r$. Both $\mathbf{p}$ and $\mathbf{q}$ are dependent. In each case the arising governing equations are thus DAEs, and can afterwards be transformed to ODEs by using the coordinate partitioning method. The DAE and ODE formulations in $\mathbf{p}$ have already been described, and are respectively: $2n+m$ DAEs (9) and $n+r$ ODEs (20).

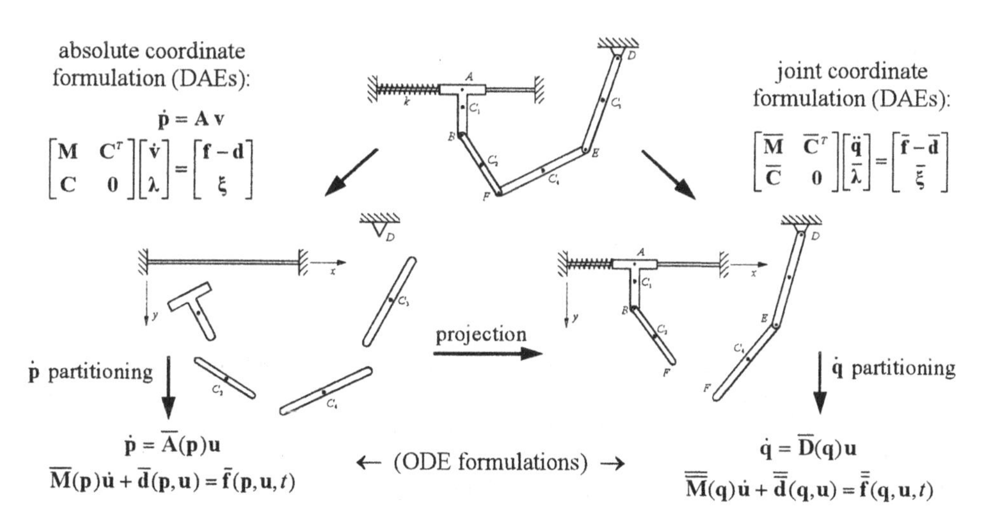

Figure 6. Treatment of closed-loop systems.

The formulations in $\mathbf{q}$ are as follows. The initial k dynamic equations for the equivalent open-loop system are formulated in the form (16) as $\overline{\mathbf{M}}(\mathbf{q})\ddot{\mathbf{q}} + \overline{\mathbf{d}}(\mathbf{q},\dot{\mathbf{q}}) = \overline{\mathbf{f}}(\mathbf{q},\dot{\mathbf{q}},t)$, obtained conveniently from the absolute coordinate formulation by using the scheme described in Section 5.1. The system described this way is then subjected to $l = m - r$ closing constraints due to the cut joints, $\overline{\boldsymbol{\Phi}}(\mathbf{q}) = \mathbf{0}$. Then, the following $2k+l$ DAEs in $\mathbf{q}$, $\dot{\mathbf{q}} = \overline{\mathbf{v}}$ and $\overline{\boldsymbol{\lambda}} = [\bar{\lambda}_1 \; \dots \; \bar{\lambda}_l]^T$ (closing constraint reactions) arise, whose structure corresponds to the structure of DAEs (9):

$$\dot{\mathbf{q}} = \overline{\mathbf{v}} \qquad\qquad \dot{\mathbf{q}} = \overline{\mathbf{v}} \tag{21a}$$

$$\begin{bmatrix} \overline{\mathbf{M}} & \overline{\mathbf{C}}^T \\ \overline{\mathbf{C}} & \mathbf{0} \end{bmatrix} \begin{bmatrix} \dot{\overline{\mathbf{v}}} \\ \overline{\boldsymbol{\lambda}} \end{bmatrix} = \begin{bmatrix} \overline{\mathbf{f}} - \overline{\mathbf{d}} \\ \overline{\boldsymbol{\xi}} \end{bmatrix} \quad \Leftrightarrow \quad \overline{\mathbf{H}}(\mathbf{q}) \begin{bmatrix} \dot{\overline{\mathbf{v}}} \\ \overline{\boldsymbol{\lambda}} \end{bmatrix} = \overline{\mathbf{h}}(\mathbf{q}, \overline{\mathbf{v}}, t) \tag{21b}$$

where $\overline{\mathbf{C}} = \partial \overline{\mathbf{\Phi}} / \partial \mathbf{q}$ and $\overline{\boldsymbol{\xi}} = -\dot{\overline{\mathbf{C}}} \dot{\mathbf{q}}$ are of dimensions $l \times k$ and $l \times 1$, respectively. Assumed $\overline{\mathbf{\Phi}}(\mathbf{q}_0) = \mathbf{0}$ and $\overline{\mathbf{C}}(\mathbf{q}_0)\dot{\mathbf{q}}_0 = \mathbf{0}$, the DAEs can be solved for $\mathbf{q}(t)$, $\dot{\mathbf{q}}(t)$ and $\overline{\boldsymbol{\lambda}}(t)$.

By using the coordinate partitioning method described in Section 5.2, $\dot{\mathbf{q}} = [\mathbf{u}^T \ \mathbf{w}^T]^T$ applied to $\overline{\mathbf{C}}\dot{\mathbf{q}} = \mathbf{0}$ and $\overline{\mathbf{C}}\ddot{\mathbf{q}} - \overline{\boldsymbol{\xi}} = \mathbf{0}$ yields the explicit forms of the closing constraints at the velocity and acceleration levels, $\dot{\mathbf{q}} = \overline{\mathbf{D}}(\mathbf{q})\mathbf{u}$ and $\ddot{\mathbf{q}} = \overline{\mathbf{D}}(\mathbf{q})\dot{\mathbf{u}} + \boldsymbol{\gamma}(\mathbf{q}, \mathbf{u})$, similar to those described in (18) and (19). The final $k + r$ ODEs in $\mathbf{q}$ and $\mathbf{u}$ are then:

$$\dot{\mathbf{q}} = \overline{\mathbf{D}}(\mathbf{q})\mathbf{u} \tag{22a}$$

$$\overline{\overline{\mathbf{M}}}(\mathbf{q})\dot{\mathbf{u}} + \overline{\overline{\mathbf{d}}}(\mathbf{q}, \mathbf{u}) = \overline{\overline{\mathbf{f}}}(\mathbf{q}, \mathbf{u}, t) \tag{22b}$$

where $\overline{\overline{\mathbf{M}}} = \overline{\mathbf{D}}^T \overline{\mathbf{M}} \overline{\mathbf{D}}$, $\overline{\overline{\mathbf{d}}} = \overline{\mathbf{D}}^T (\overline{\mathbf{M}} \overline{\boldsymbol{\gamma}} + \overline{\mathbf{d}})$, and $\overline{\overline{\mathbf{f}}} = \overline{\mathbf{D}}^T \overline{\mathbf{f}}$ The above ODEs can be considered as the minimal-form formulation for closed-loop systems.

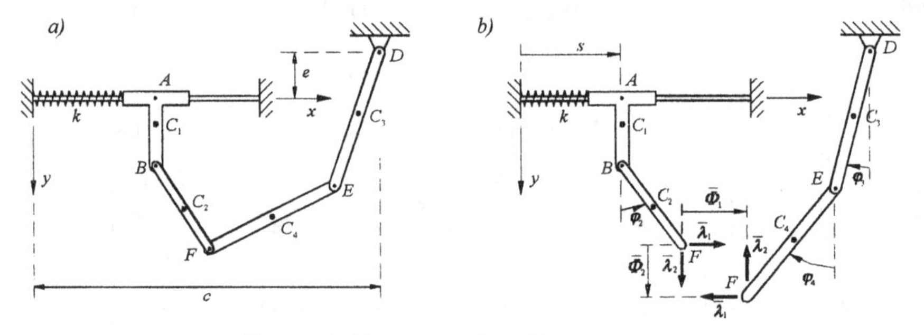

Figure 7. The sample closed-loop system.

Example 5: *Mechanism* (*the minimal-form formulation*). As an illustration let us consider the two-degree-of-freedom closed-loop system shown in Figure 7a. After cutting the closed loop at joint F two simple open-loop systems are obtained (Fig. 7b), with two-degrees-of-freedom each. The left one is the system analyzed in Example 3, and the other is a classical double pendulum. The dynamic equations $\overline{\mathbf{M}}\ddot{\mathbf{q}} + \overline{\mathbf{d}} = \overline{\mathbf{f}}$ in the joint coordinates $\mathbf{q} = [\mathbf{s} \ \varphi_2 \ \varphi_3 \ \varphi_4]^T$ of the cut system are easy to obtain by using the joint coordinate method described in Section 5.1. These equations will be not reported here for shortness. The closing constraint equations (given implicitly) are then

$$\overline{\Phi}_1 = c - l_3 \sin\varphi_3 - l_4 \sin\varphi_4 - s - l_2 \sin\varphi_2 = 0$$
$$\overline{\Phi}_2 = -e + l_3 \cos\varphi_3 + l_4 \cos\varphi_4 - l_1 - l_2 \cos\varphi_2 = 0$$

$$\overline{\mathbf{C}} = \begin{bmatrix} -1 & -l_2 \cos\varphi_2 & -l_3 \cos\varphi_3 & -l_4 \cos\varphi_4 \\ 0 & l_2 \sin\varphi_2 & -l_3 \sin\varphi_3 & -l_4 \sin\varphi_4 \end{bmatrix}; \qquad \overline{\boldsymbol{\xi}} = (\cdots)$$

There are six possible partitions of $\mathbf{q}$. Depending on the geometrical characteristics each of these partitions may lead, for certain system configurations, to singularities in the coordinate partitioning method formulation. For instance, the choice $\mathbf{u} = [\dot{s} \;\; \dot{\varphi}_2]^T$ is conditioned upon $\det(\mathbf{W}) = l_3 l_4 \sin(\varphi_4 - \varphi_3) \neq 0$. The matrix $\overline{\mathbf{D}}$ is then

$$\overline{\mathbf{D}}^T = \begin{bmatrix} 1 & 0 & \dfrac{-\sin\varphi_4}{l_3\sin(\varphi_4-\varphi_3)} & \dfrac{-l_2\sin(\varphi_4+\varphi_2)}{l_3\sin(\varphi_4-\varphi_3)} \\ 0 & 1 & \dfrac{\sin\varphi_3}{l_4\sin(\varphi_4-\varphi_3)} & \dfrac{l_2\sin(\varphi_3+\varphi_2)}{l_4\sin(\varphi_4-\varphi_3)} \end{bmatrix}; \qquad \overline{\gamma} = (\cdots)$$

The analytical form of the governing equations (22) is not reported for its complexity.

5.4 TREATMENT OF NONHOLONOMIC SYSTEMS

Seemingly to closed-loop holonomic systems, the treatment of nonholonomic systems can be grouped in four methods. In Section 3 a uniform treatment of holonomic and nonholonomic systems, respectively $m_{nh} = 0$ or $m_{nh} > 0$, are provided in the dependent (possible absolute) variables $\mathbf{p}$ and $\mathbf{v}$. These formulations are illustrated in Example 2. By applying the projection method, the initial $2n+m$ DAEs (9) can then be transformed to $n+r$ ODEs (20). The transformation is based on relations between the dependent and independent velocities/accelerations, $\mathbf{v} = \mathbf{D(p)u}$ and $\dot{\mathbf{v}} = \mathbf{D(p)}\dot{\mathbf{u}} + \boldsymbol{\gamma}(\mathbf{p},\mathbf{u})$. In general, these explicit forms of constraint equations, given implicitly in (5) and (6), can be determined by any method – by guess, inspection or, more usual, numerically [17], and the components of $\mathbf{u}$ may be either generalized velocities or quasi-velocities [3,4,13]. One effective techniques of the latter type is the velocity partitioning method described in Section 5.2, and illustrated in Example 4.

In DAEs (9) and ODEs (20) the m_h holonomic constraints (4a) and m_{nh} nonholonomic constraints (4b) are treated in a unified way, $m = m_h + m_{nh}$, and n dependent coordinates $\mathbf{p}$ are used. The other possibility is first to "eliminate" the holonomic constraints and obtain the dynamic equations (16) in $k = n - m_h$ independent coordinates $\mathbf{q}$, followed the scheme provided in Section 5.1. The system described this way will then be subjected to only nonholonomic constraints (4b), expressed in the new variables as $\overline{\mathbf{C}}_{nh}(\mathbf{q})\,\dot{\mathbf{q}} = \mathbf{0}$. The followed governing $2k + m_{nh}$ DAEs are then:

$$\dot{\mathbf{q}} = \overline{\mathbf{v}} \tag{23a}$$

$$\begin{bmatrix} \overline{\mathbf{M}} & \overline{\mathbf{C}}_{nh}^T \\ \overline{\mathbf{C}}_{nh} & \mathbf{0} \end{bmatrix} \begin{bmatrix} \dot{\overline{\mathbf{v}}} \\ \boldsymbol{\lambda}_{nh} \end{bmatrix} = \begin{bmatrix} \overline{\mathbf{f}} - \overline{\mathbf{d}} \\ \overline{\boldsymbol{\xi}}_{nh} \end{bmatrix} \tag{23b}$$

which are similar to (21). Further application of the projection method (possibly the $\dot{\mathbf{q}}$ partitioning scheme) will then lead to $k+r$ (minimal-form) ODEs (22) in the independent variables $\mathbf{q}$ and $\mathbf{u}$.

Example 6: *Nonholonomic system* (*the minimal-form formulation*). Let us come back once more to the nonholonomic system considered previously in Examples 2 and 4. By choosing the independent coordinates as $\mathbf{q} = [s \;\; \varphi]^T$, and starting from the initial dynamic equations in $\mathbf{p} = [x_A \;\; y_A \;\; \theta]^T$ reported in Example 2, the dynamic equations in $\mathbf{q}$ are easy to obtain by using the scheme provided in Section 5.1, i.e.

$$\begin{bmatrix} m & mb\cos\varphi \\ mb\cos\varphi & J_A \end{bmatrix}\begin{bmatrix} \ddot{s} \\ \ddot{\varphi} \end{bmatrix} + \begin{bmatrix} -m\dot{\varphi}^2 b\sin\varphi \\ 0 \end{bmatrix} = \begin{bmatrix} -F_s \\ 0 \end{bmatrix}$$

The nonholonomic constraint equation imposed on the system described this way is then $\overline{\Psi}_{nh} = \dot{s}\cos\varphi + 2b\dot{\varphi} = 0$. By applying the velocity partitioning method, for $\mathbf{u} = [\dot{s}]$, the minimal-form formulation in the independent state variables $\mathbf{q}$ and $\mathbf{u}$ is as follows:

$$\dot{s} = u$$

$$\dot{\varphi} = -\frac{\cos\varphi}{2b}u$$

$$\left(m\sin^2\varphi + \frac{J_C}{4b^2}\cos^2\varphi\right)\dot{u} - \left(m - \frac{J_C}{4b^2}\right)\frac{u^2}{2b}\sin\varphi\cos^2\varphi = -F_s$$

The dynamic equation is evidently the same as obtained previously in Example 4 (the same independent velocity was used in the two cases, and $\varphi = \theta$).

5.5 GIBBS'-APPELL'S EQUATIONS

The Gibbs'-Appell's method provides one with another general approach to holonomic and nonholonomic systems [4,9]. According to the geometrical interpretation of Section 2, the Appell's function, after using $\mathbf{v} = \mathbf{D}\mathbf{u}$ and $\dot{\mathbf{v}} = \mathbf{D}\dot{\mathbf{u}} + \boldsymbol{\gamma}$, can be written as

$$2S = (\dot{\mathbf{v}} + \mathbf{M}^{-1}\mathbf{d})^T \mathbf{M}\,(\dot{\mathbf{v}} + \mathbf{M}^{-1}\mathbf{d}) = (\mathbf{D}\dot{\mathbf{u}} + \boldsymbol{\gamma} + \mathbf{M}^{-1}\mathbf{d})^T \mathbf{M}\,(\mathbf{D}\dot{\mathbf{u}} + \boldsymbol{\gamma} + \mathbf{M}^{-1}\mathbf{d}) \tag{24}$$

The Gibbs'-Appell's equations [9], expressed respectively in $\mathbf{v}$ and $\mathbf{u}$, are then:

$$\left(\frac{\partial S}{\partial \dot{\mathbf{v}}}\right)^T = \mathbf{f} \quad \text{and} \quad \left(\frac{\partial S}{\partial \dot{\mathbf{u}}}\right)^T = \bar{\mathbf{f}} \tag{25}$$

where $\bar{\mathbf{f}} = \mathbf{D}^T\mathbf{f}$, and $(\partial S/\partial\dot{\mathbf{u}}) = \mathbf{D}^T(\partial S/\partial\dot{\mathbf{v}})$. Applying (24) one receives directly

$$\mathbf{M}\dot{\mathbf{v}} + \mathbf{d} = \mathbf{f} \quad \text{and} \quad \mathbf{D}^T\mathbf{M}(\mathbf{D}\dot{\mathbf{u}} + \boldsymbol{\gamma} + \mathbf{M}^{-1}\mathbf{d}) = \mathbf{D}^T\mathbf{f} \tag{26}$$

which are equivalent to (2) and (20), respectively. The matrix setting of Gibbs'-Appell's is thus equivalent to the projective formulation.

5.6 KANE'S METHOD

Kane's formalism [14] can also been interpreted geometrically [4,11,15] as the projection of the initial dynamic equations (2) (Newton-Euler equations) into the unconstrained subspace. More specifically, for an unconstrained system, the generalized active $\mathbf{K}$ and inertial $\mathbf{K}^*$ forces, which constitute Kane's equations, can be interpreted as:

$$\vec{f} - \vec{b} = 0 \quad \Leftrightarrow \quad \mathbf{K} + \mathbf{K}^* = \mathbf{0} \quad \Leftrightarrow \quad \mathbf{f} - (\mathbf{M}\dot{\mathbf{v}} + \mathbf{d}) = \mathbf{0} \tag{27}$$

For a constrained system, the relations between the dependent velocities $\mathbf{v}$ and independent generalized speeds $\mathbf{u}$ are used, $\mathbf{v} = \mathbf{D}\mathbf{u}$, and then $\dot{\mathbf{v}} = \mathbf{D}\dot{\mathbf{u}} + \gamma$, where $\mathbf{D}$ is named *partial velocity matrix* [14]. The projected Kane's equations are then (Fig. 3b):

$$\vec{f}_d - \vec{b}_d = 0 \quad \Leftrightarrow \quad \overline{\mathbf{K}} + \overline{\mathbf{K}}^* = \mathbf{0} \quad \Leftrightarrow \quad \mathbf{D}^T\mathbf{f} - \mathbf{D}^T\mathbf{M}[(\mathbf{D}\dot{\mathbf{u}} + \gamma) + \mathbf{d}] = \mathbf{0} \tag{28}$$

Evidently, Kane's formulation [14] entails an entire useful methodology for setting up motion equations in a practical way, impossible to discuss in this short comment.

6. Other Useful Implementations

6.1 GEOMETRIC ELIMINATION OF CONSTRAINT VIOLATION

For a system modeled in dependent state variables, when the governing equations involve the constraint equations differentiated with respect to time, one consequence of numerical truncation errors is possible violation of the original constraint equations by the numerical solutions. The geometrical scheme proposed in [20], and then developed in [3,5], consists in correcting the state variables so that to eliminate the constraint violations after each step of integration (or a sequence of steps). An important feature of these corrections is that they are performed in the constrained directions, and as such do not influence in practice the system motion (Fig. 8) along the constraints.

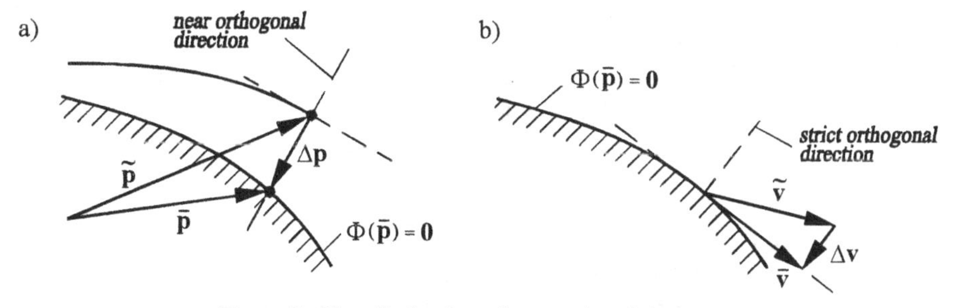

Figure 8. The elimination of constraint violations.

Let us consider the problem with reference to the dependent variable formulations (9) and (11). For a current numerical position $\tilde{\mathbf{p}}$ and velocity $\tilde{\mathbf{v}}$, the constraint violations are $\tilde{\boldsymbol{\Phi}} = \boldsymbol{\Phi}(\tilde{\mathbf{p}}) \neq \mathbf{0}$ and $\tilde{\boldsymbol{\Psi}} = \boldsymbol{\Psi}(\tilde{\mathbf{p}}, \tilde{\mathbf{v}}) \neq \mathbf{0}$, and the state corrections are [3,5]:

$$\Delta\mathbf{p} = -\mathbf{A}\mathbf{M}^{-1}\mathbf{C}^T(\mathbf{C}\mathbf{M}^{-1}\mathbf{C}^T)^{-1}\begin{bmatrix}\tilde{\boldsymbol{\Phi}} \\ \mathbf{0}\end{bmatrix} \quad \text{and} \quad \Delta\mathbf{v} = -\mathbf{M}^{-1}\mathbf{C}^T(\mathbf{C}\mathbf{M}^{-1}\mathbf{C}^T)^{-1}\tilde{\boldsymbol{\Psi}} \tag{29a,b}$$

where $\Delta \mathbf{p} = \overline{\mathbf{p}} - \widetilde{\mathbf{p}}$, $\Delta \mathbf{v} = \overline{\mathbf{v}} - \widetilde{\mathbf{v}}$, and $\overline{\mathbf{p}}$ and $\overline{\mathbf{v}}$ are the constraint consistent state variables, $\mathbf{\Phi}(\overline{\mathbf{p}}) = \mathbf{0}$ and $\mathbf{\Psi}(\overline{\mathbf{p}}, \overline{\mathbf{v}}) = \mathbf{0}$. The position correction scheme (29a) can possibly be replaced by $\Delta \mathbf{p} = -\mathbf{A}\mathbf{M}^{-1}\mathbf{C}_h^T(\mathbf{C}_h\mathbf{M}^{-1}\mathbf{C}_h^T)^{-1}\widetilde{\mathbf{\Phi}}$ as well [5].

The scheme (29b) "translates" the violation $\widetilde{\mathbf{\Psi}}$ from $\mathcal{C}^m$ into appropriate velocity correction $\Delta \mathbf{v}$ in $\mathcal{E}^n$, and the metric of the linear space is involved [3]. As stated in Section 2, $\mathcal{E}^n$ is a tangent space to the configuration manifold $\mathcal{N}$ on which the curvilinear coordinates $\mathbf{p}$ are defined (Fig. 1). This is why the position correction (29a) is somewhat "imprecise" – it is performed in $\mathcal{E}^n$ in lieu of on $\mathcal{N}$. That is why the position correction requires usually few iterations, and should be performed first. The velocity correction can then be done in one step (refer to [5] for numerical examples).

The numerical integration truncation errors yield inaccuracy in the constraint consistent solution to motion equations as well, and the problem relates to both the dependent and independent variable formulations. The total energy of a system, computed from the initial energy plus the energy input rate due to the external and dissipative forces, is a measure of this inaccuracy – an artificial constraint on the system [20]. The energy constraint equation and the improved scheme of velocity correction proposed in [5] are:

$$\Psi_E = E - E_0 - \int_{t_0}^{t} \dot{E}\, dt = 0 \; ; \qquad \hat{\mathbf{v}} = \sqrt{1 - \overline{\Psi}_E / \overline{T}}\; \overline{\mathbf{v}} \qquad (30),(31)$$

where $E = T + V$ is the sum of the kinetic T and potential V energies, $\dot{E}$ denotes the energy input rate to the system, $\overline{\Psi}_E = \Psi_E(\overline{\mathbf{p}}, \overline{\mathbf{v}}) \neq 0$ is the energy constraint violation by the constraint consistent state variables $\overline{\mathbf{p}}$ and $\overline{\mathbf{v}}$, and $\hat{\mathbf{v}}$ is the corrected velocity so that $\Psi_E(\overline{\mathbf{p}}, \hat{\mathbf{v}}) = 0$. Though the scheme (31) "works" numerically, it is argued in [5] that the energy constraint correction may be an inaccurate/defective means of improving the numerical simulation accuracy. Firstly, the evaluation of $\overline{\Psi}_E$ is usually far too imprecise for non-conservative systems ($\dot{E} \neq 0$). Then, we deal with only one condition (30), and for many-degree-of-freedom systems the "proportional" correction (31) of velocity components is questionable. Finally, the inaccuracies in T and V for $\overline{\mathbf{p}}$ and $\overline{\mathbf{v}}$, compared to their values for the "exact" state variables, may have opposite signs. The vanishing energy constraint violation may thus be achieved for "inexact" state variables as well. These assertions are justified by numerical experiments in [5].

6.2 AUGMENTED JOINT COORDINATE METHOD

The formulation of ODEs (16) involves only the explicit forms (13) – (15) of constraint equations. If the constraint reactions $\boldsymbol{\lambda}$ need to be determined according to (17), the constraint matrix $\mathbf{C}$ is required in addition, i.e. the implicit constraint equations (4) and (5) need to be introduced, and this means additional modeling effort. Another inconvenience of scheme (17) is that the $m \times m$ matrix $\mathbf{C}\mathbf{M}^{-1}\mathbf{C}^T$ must be inverted, which may be a numerically inefficient task. The novel scheme developed in [7] is released from these inconveniences. In the approach, the explicit constraint equations are modified to

$$\mathbf{p} = \mathbf{g}(\mathbf{q}, \mathbf{z}) \quad \Rightarrow \quad \dot{\mathbf{p}} = \left.\left(\frac{\partial \mathbf{g}}{\partial \mathbf{q}}\right)\right|_{\mathbf{z}=\mathbf{0}} \dot{\mathbf{q}} + \left.\left(\frac{\partial \mathbf{g}}{\partial \mathbf{z}}\right)\right|_{\mathbf{z}=\mathbf{0}} \dot{\mathbf{z}} \equiv \mathbf{D}(\mathbf{q})\dot{\mathbf{q}} + \mathbf{E}(\mathbf{q})\dot{\mathbf{z}} \quad (32),(33)$$

where the *k joint* coordinates $\mathbf{q}$ and *m open-constraint* coordinates $\mathbf{z}$ describe, respectively, the admissible and prohibited relative motions in the joints. The constraint equations (13) and (14) are then retrieved for $\mathbf{z}=\mathbf{0}$ and $\dot{\mathbf{z}}=\mathbf{0}$. In fact, the dependence on $\mathbf{z}$ in equations (33) is needed only to grasp the prohibited motion directions in the joints, represented as columns in the $n\times m$ matrix $\mathbf{E}$. An important characteristic of $\mathbf{E}$ is [7]

$$\mathbf{C}\mathbf{E}=\mathbf{I} \quad \Leftrightarrow \quad \mathbf{E}^T\mathbf{C}^T=\mathbf{I} \tag{34}$$

where $\mathbf{I}$ is the $m\times m$ identity matrix. In other words, $\mathbf{E}$ is a *pseudoinverse* of the rectangular matrix $\mathbf{C}$. Using $\mathbf{E}$, the new projection formula (10) ca be modified to:

$$\begin{bmatrix}\mathbf{D}^T\\ \mathbf{E}^T\end{bmatrix}\left(\mathbf{M}\dot{\mathbf{v}}+\mathbf{d}=\mathbf{f}-\mathbf{C}^T\boldsymbol{\lambda}\right) \quad \begin{matrix}\Rightarrow & \mathbf{D}^T\mathbf{M}(\mathbf{D}\ddot{\mathbf{q}}+\boldsymbol{\gamma})+\mathbf{D}^T\mathbf{d}=\mathbf{D}^T\mathbf{f} & \text{(35a)}\\ \Rightarrow & \mathbf{E}^T\mathbf{M}(\mathbf{D}\ddot{\mathbf{q}}+\boldsymbol{\gamma})+\mathbf{E}^T\mathbf{d}=\mathbf{E}^T\mathbf{f}-\boldsymbol{\lambda} & \text{(35b)}\end{matrix}$$

The tangential projection (35a) leads to ODEs (16), and (35b) results in

$$\boldsymbol{\lambda}(\mathbf{q},\dot{\mathbf{q}},\ddot{\mathbf{q}},t)=\mathbf{E}^T\left(\mathbf{f}-\mathbf{M}(\mathbf{D}\ddot{\mathbf{q}}+\boldsymbol{\gamma})\right) \tag{36}$$

In the scheme (36) the constraint reactions $\boldsymbol{\lambda}$ are then obtained directly in a "resolved" form – no matrix inversion is required. The implicit constraint equations (4) are not involved, either. Instead, their explicit forms are slightly modified to (33) with no much additional modeling effort. Moreover, the present scheme can conveniently be used to determine only some joint reactions – only the respective entries of $\mathbf{z}$ can be introduced, and the number of columns of $\mathbf{E}$ can appropriately be reduced. Finally, the method can also be extended to the velocity partitioning formulation (14), which modifies to

$$\dot{\mathbf{q}}=\begin{bmatrix}\mathbf{I}\\ -\mathbf{W}^{-1}\mathbf{U}\end{bmatrix}\mathbf{u}+\begin{bmatrix}\mathbf{0}\\ \mathbf{W}^{-1}\end{bmatrix}\dot{\mathbf{z}}\equiv\overline{\mathbf{D}}(\mathbf{q})\mathbf{u}+\overline{\mathbf{E}}(\mathbf{q})\dot{\mathbf{z}} \tag{37}$$

More details related to the above methodology for determination of constraint reactions can be found in [7], and the method corresponds to "bringing noncontributing forces into evidence" discussed by Kane [14], and then reconsidered e.g. in [10,15].

Example 7: *Moving pendulum* (*constraint reactions*). For the system analyzed previously in Example 3, we have $\mathbf{q}=[s\ \ \varphi]^T$ and $\mathbf{z}=[z_1\ z_2\ z_3\ z_4]^T$ (Fig. 9). The augmented formulation of explicit kinematic constraint equations and the matrix E are:

$$\begin{aligned}x_1&=s+a_1\sin z_2\\ y_1&=z_1+a_1\cos z_2\\ \theta_1&=z_2\\ x_2&=s+l_1\sin z_2+z_3+a_2\sin\varphi\\ y_2&=z_1+l_1\cos z_2+z_4+a_2\cos\varphi\\ \theta_2&=\varphi\end{aligned}\quad ; \quad \mathbf{E}=\begin{bmatrix}0&a_1&0&0\\1&0&0&0\\0&1&0&0\\0&l_1&1&0\\1&0&0&1\\0&0&0&0\end{bmatrix}$$

By rewriting equation (36) in the form $\boldsymbol{\lambda}=\mathbf{E}^T\mathbf{f}-\mathbf{E}^T\mathbf{M}\mathbf{D}\ddot{\mathbf{q}}-\mathbf{E}^T\mathbf{M}\boldsymbol{\gamma}$, we obtain then:

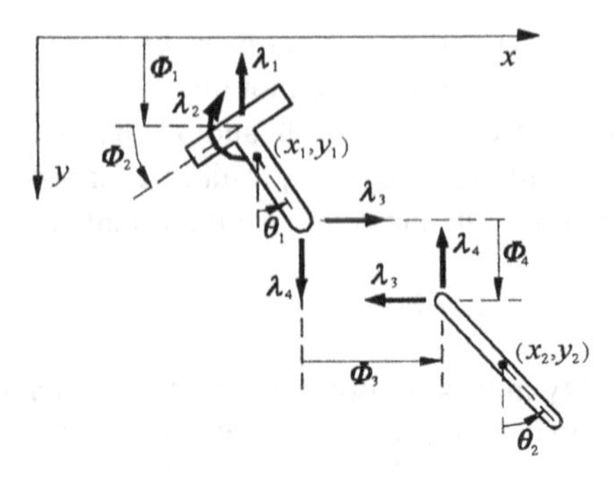

Figure 9. The joint and open-constraint coordinates, and the reaction forces.

$$\begin{bmatrix} \lambda_1 \\ \lambda_2 \\ \lambda_3 \\ \lambda_4 \end{bmatrix} = \begin{bmatrix} (m_1+m_2)g \\ 0 \\ 0 \\ m_2 g \end{bmatrix} - \begin{bmatrix} -m_2 a_2 \ddot{\varphi} \sin\varphi \\ (m_1 a_1 + m_2 l_1)\ddot{s} + m_2 l_1 a_2 \ddot{\varphi}\cos\varphi \\ m_2 \ddot{s} + m_2 a_2 \ddot{\varphi}\cos\varphi \\ -m_2 a_2 \ddot{\varphi}\sin\varphi \end{bmatrix} - \begin{bmatrix} -m_2 a_2 \dot{\varphi}^2 \cos\varphi \\ -m_2 l_1 a_2 \dot{\varphi}^2 \sin\varphi \\ -m_2 a_2 \dot{\varphi}^2 \sin\varphi \\ -m_2 a_2 \dot{\varphi}^2 \cos\varphi \end{bmatrix}$$

6.3 GEOMETRIC INTERPRETATION OF AUGMENTED LAGRANGIAN FORMULATION

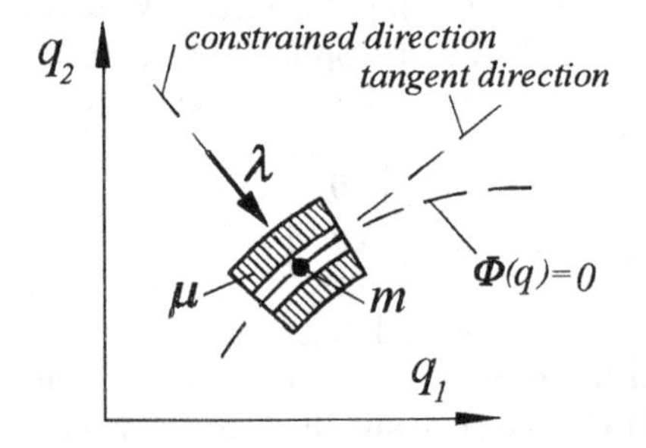

Figure 10. Geometric interpretation of the augmented Lagrangian formulation.

A variant method for handling constrained multibody systems, called *augmented Lagrangian formulation*, is discussed e.g. in [1,12]. A geometrical interpretation of the method is then provided in [6]. According to this interpretation (Fig. 10), instead of imposing constraints on a system in the traditional sense, large artificial masses μ resisting in the constrained directions are added, and the system motion is enforced to evolve primarily in the directions with smaller masses (unconstrained directions). The residual motion in the constrained directions is then removed by applying the constraint reactions to the system, estimated effectively in few iterations. The formulation is simple and leads to computationally efficient numerical codes. Applications of the formulation to the analysis of constrained multibody systems with possible singular configurations, massless links and redundant constraints can are shown in [6].

6.4 ORTHONORMALIZATION METHOD

The final example of implementations stimulated by the geometrical interpretation of multibody system dynamics is the orthonormalization method proposed in [2]. An effective scheme for converting the equations of motion expressed in terms of absolute variables, DAEs (9), into a convenient minimal-form set of equations in independent kinematical parameters $\mathbf{u}$ is developed. Exploiting the fact that $\mathcal{E}^n$ is a metric space, the Gram-Schmidt orthogonalization scheme is adopted to generate a genuine orthonormal basis of $\mathcal{D}^r$, $\mathbf{C} \xrightarrow{\text{orthonormalization}} \breve{\mathbf{D}}$, where $\mathbf{C}\breve{\mathbf{D}} = \mathbf{0}$. Using the orthonormal basis defined by $\breve{\mathbf{D}}$, r independent kinematic parameters $\breve{\mathbf{u}}$ can be introduced, $\mathbf{v} = \breve{\mathbf{D}}\breve{\mathbf{u}}$, and the dimension of $\breve{\mathbf{u}}$ is $[\sqrt{\text{kg}}\ \text{m/s}]$, just between the dimensions of momenta $\mathbf{M}\mathbf{v}$ and velocities $\mathbf{v}$ in the traditional settings. A useful peculiarity of this approach is that in the minimal-form dynamic equations (20b) produced this way one obtains $\breve{\mathbf{D}}^T\mathbf{M}\breve{\mathbf{D}} = \mathbf{I}$, i.e. the related mass matrix is the $r \times r$ identity matrix (the dynamic equations are obtained directly in a "resolved" form). The method is especially efficient for $\mathbf{M} = \text{const}$, it can thus be considered as an effective solver for absolute variable formulations [2].

7. Conclusion

The geometrical interpretation of multibody system dynamics can be both insightful and stimulative. Appealing directly to the geometry of particle dynamics known from Newtonian dynamics, it provides one with a precise and powerful investigation tool for handling constrained motion problems. The followed mathematical formulations resolve then themselves to compact matrix transformations suitable for computer implementations. Systems subject to holonomic and nonholonomic constraints can be treated, and the analysis can be performed in either generalized or quasi-velocities. Variant forms of equations of motions can be obtained, relevant to many other methods of classical mechanics. Finally, novel contributions and amendments in the theory of constrained systems can be stimulated.

8. References

1. Bayo, E. and Ledesma R. (1996) Augmented Lagrangian and mass-orthogonal projection methods for constrained multibody dynamics, *Nonlinear Dynamics* **9**, 113-130.
2. Blajer W. (1995) An effective solver for absolute variable formulation of multibody dynamics, *Computational Mechanics* **15**, 460-472.
3. Blajer, W. (1997) A geometric unification of constrained system dynamics, *Multibody System Dynamics* **1**, 3-21.
4. Blajer, W. (2001) A geometrical interpretation and uniform matrix formulation of multibody system dynamics, *ZAMM* **81**, 247-259.
5. Blajer, W. (2002) Elimination of constraint violation and accuracy aspects in numerical simulation of multibody systems, *Multibody System Dynamics* **7**, 265-284.
6. Blajer, W. (in press) Augmented Lagrangian formulation: geometrical interpretation and application to systems with singularities and redundancy, *Multibody System Dynamics*.

7. Blajer, W. and Schiehlen, W. (submitted) Efficient determination of constraint reactions in multibody systems", *Multibody System Dynamics.*
8. Blajer, W., Schiehlen, W., and Schirm, W. (1994) A projective criterion to the coordinate partitioning method for multibody dynamics, *Archive of Applied Mechanics* **64**, 86-98.
9. Desloge, E.A. (1988) The Gibbs-Appell equations of motion, *American Journal of Physics* **56**, 841-846.
10. Djerassi, S. (1997) Determination of noncontributing forces and noncontributing impulses in three-phase motions, *Journal of Applied Mechanics* **64**, 582-589.
11. Essén, H. (1994) On the geometry of nonholonomic dynamics, *Journal of Applied Mechanics* **61**, 689-694.
12. García de Jalón, J. and Bayo, E. (1994) *Kinematic and Dynamic Simulation of Multibody Systems: the Real-Time Challenge*, Springer-Verlag, New York.
13. Jungnickel, U. (1994) Differential-algebraic equations in Riemannian spaces and applications to multibody system dynamics, *ZAMM* **74**, 409-415.
14. Kane, T.R. and Levinson, D.A. (1985) *Dynamics: Theory and Applications*, McGraw-Hill, New York.
15. Lesser, M. (1992) A geometrical interpretation of Kane's equations, *Proceedings of the Royal Society in London* **A436**, 69-87.
16. Maißer, P. (1991) Analytical dynamics of multibody systems, *Computer Methods in Applied Mechanics and Engineering* **91**, 1391-1396.
17. Nikravesh, P.E. (1988) *Computer-Aided Analysis of Mechanical Systems*, Prince-Hall, Englewood Cliffs, New Jersey.
18. Schiehlen, W. (1997) Multibody system dynamics: roots and perspectives, *Multibody System Dynamics* **1**, 149-188.
19. Wehage, R.A. and Haug, E.J.)1982) Generalized coordinate partitioning for dimension reduction in analysis of constrained dynamic systems, *Journal of Mechanical Design* **104**, 247-255.
20. Yoon, S., Howe, R.M., and Greenwood, D.T. (1994) Geometric elimination of constraint violations in numerical simulation of Lagrangian equations, *Journal of Mechanical Design* **116**, 1058-1064.

VIRTUAL PROTOTYPING OF MULTIBODY SYSTEMS WITH LINEAR GRAPH THEORY AND SYMBOLIC COMPUTING

J.J. MCPHEE
University of Waterloo
Systems Design Engineering
Waterloo, Ontario, Canada N2L 3G1

Summary. The combined use of linear graph theory and symbolic computing leads to efficient dynamic models of multibody systems that facilitate real-time simulations. Graph theory allows the selection of coordinates that best suit a given problem, resulting in relatively simple and compact equations of motion. Special topologies, e.g. those for parallel robots, can be exploited using graph theory, which also facilitates the modelling of multibody and mechatronic systems using subsystem models.

Keywords. Multibody system dynamics, linear graph theory, symbolic computing, subsystem models, mechatronic systems, virtual prototyping.

1. Introduction

The most recent developments in applying linear graph theory to multibody system dynamics are presented, with particular emphasis on the symbolic generation of simple, compact equations that are well-suited for real-time simulation.

An overview of the modelling of flexible multibody systems is presented in the next sections, highlighting the unique features of linear graph theory and symbolic computer programming. Graph theory can be used to generate equations in terms of user-selected coordinates, including absolute, joint, or indirect coordinates, or some combination of the three. By selecting coordinates that are well-suited to a given problem, one can reduce the complexity of the kinematic and dynamic equations; one can also reduce the number of equations when compared to traditional absolute or joint coordinate formulations.

The use of symbolic computer programming leads to further efficiencies, because multiplications by 0 or 1 are eliminated and equation simplifications can be automated. The final equations can be exported as optimized C or Fortran

W. Schiehlen and M. Valášek (eds.), Virtual Nonlinear Multibody Systems, 37–56.

code for a subsequent real-time simulation; these equations do not have to be re-formulated at each time step, as is the case for numerical formulations. Symbolic programming was used to implement a dynamic formulation based on the principle of virtual work. The resulting Maple program (DynaFlex) for modelling flexible multibody systems is briefly described and demonstrated.

Graph theory can also be used to exploit special system topologies, such as that found in parallel robotic manipulators. A method is shown for automatically generating symbolic solutions for the inverse dynamics of parallel robots. The modelling of multibody systems is greatly facilitated by the use of subsystem models; a graph-theoretic generalization of the Norton and Thevenin theorems from electrical network theory is used to generate models of multibody and mechatronic subsystems. This subsystem modelling approach is demonstrated on a parallel robot and a controlled electro-mechanical system.

2. Flexible Multibody Systems: Linear Graph and Coordinate Selection

Consider the planar slider-crank mechanism shown in Figure 1, and its linear graph representation in Figure 2. Nodes in the graph represent body-fixed reference frames, while edges represent kinematic or dynamic transformations associated with physical components. Edges $m_1 - m_3$ represent the three bodies, which may be rigid or flexible. Newton's Laws require that the edges originate at an inertial frame (node) and terminate at a body-fixed frame (node) — the center of mass for a rigid body, and at one end of a flexible beam element. For clarity, the physical components are superimposed on the linear graph with dotted lines.

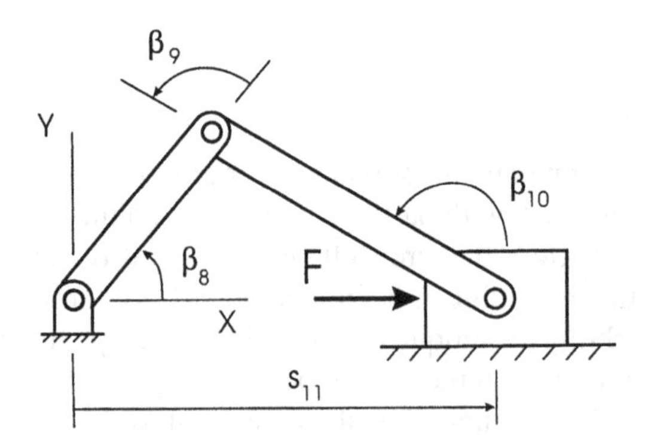

Figure 1. Slider-crank mechanism.

Edges $h_8 - h_{10}$ are the three revolute joints connecting the bodies, while s_{11} is the prismatic joint between the ground and the slider. The "arm elements" $r_4 - r_7$ represent transformations from the body frame to the body-fixed joint frames.

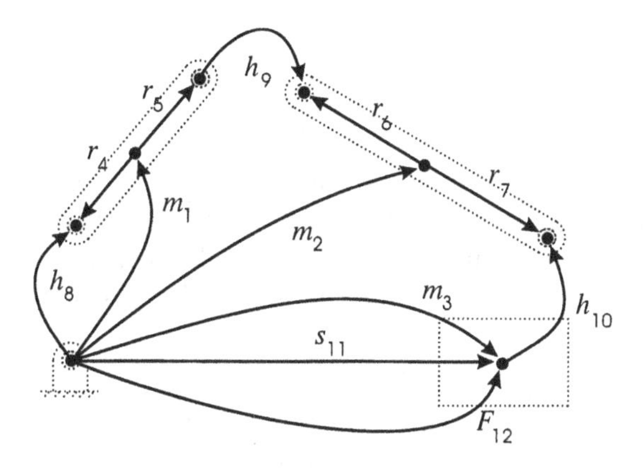

Figure 2. Linear graph of slider-crank mechanism.

These transformations are functions of the body rotation and, for flexible bodies, the elastic deflections. Finally, edge F_{12} is the external force acting on the slider. Additional physical effects, e.g. motor torques, springs, and weights, are easily added to the system graph.

Associated with each edge are through (force, torque) and across (translation, rotation) variables that satisfy the fundamental cutset and circuit equations [1]. For mechanical systems, the cutset equations provide dynamic equilibrium equations for any combination of components, while the circuit equations provide closure conditions around any loop. These cutset and circuit equations are systematically generated from an incidence matrix representation of the linear graph. Thus, the kinematic and dynamic equations can be automatically generated for a system with any topology, as demonstrated in the subsequent sections.

If a tree is selected for the graph, then the circuit equations can be used to express all kinematic variables as linear combinations of the coordinates associated with the tree edges ("branches") [2]. Thus, *by selecting a spanning tree, the branch coordinates* $\mathbf{q}$ *that appear in the kinematic and dynamic equations are defined.* This is an important feature of a graph-theoretic approach, since most other multibody formulations are restricted to a pre-defined set of coordinates, usually absolute [3] or joint [4].

As an example, consider the slider-crank mechanism from Figure 1. Rigid arm elements (i.e. $r_4 - r_7$) are always selected into the tree of a multibody system because they represent constant kinematic transformations; since there are no unknown coordinates associated with an arm element, they don't introduce any variables into $\mathbf{q}$. By also selecting h_8, h_{10}, and s_{11} into the tree, one will obtain equations in the corresponding branch coordinates $\mathbf{q} = [\beta_8, \beta_{10}, s_{11}]^T$. All cotree variables can be expressed in terms of $\mathbf{q}$ using the circuit equations, e.g. for the rotation of cotree rigid body m_2:

$$\underline{\theta}_2 = \underline{\theta}_{11} + \underline{\theta}_{10} - \underline{\theta}_7 \quad (1)$$

$$\underline{\omega}_2 = \underline{\omega}_{11} + \underline{\omega}_{10} - \underline{\omega}_7 \quad (2)$$

where $\underline{\theta}_i$ and $\underline{\omega}_i$ are the orientation and angular velocity for element i, as defined by the elemental constitutive equations, e.g. $\underline{\theta}_i = \beta_i \hat{k}$ for revolute joints 8, 9, and 10, and $\underline{\omega}_{11} = 0$ for the prismatic joint. For spatial systems, orientations cannot be represented by a vector $\underline{\theta}$ and so the circuit equation (1) is re-written in terms of rotation matrices $\mathbf{R}$:

$$\mathbf{R}_2 = \mathbf{R}_{11}\mathbf{R}_{10}\mathbf{R}_7^T \quad (3)$$

In this manner, all kinematic variables are systematically expressed in terms of $\mathbf{q}$.

Using graph theory, one can generate equations in absolute coordinates (select all bodies into the tree), joint coordinates (select joints into the tree), or some combination of the two. One can even select two different trees — one to generate rotational equations and one to generate translational equations. By doing this, one can reduce the total number of equations to be solved, and reduce their complexity at the same time [2]. By selecting coordinates that are tailored to the problem at hand, one can obtain equations that are well-suited to real-time simulation, in either a virtual reality environment or an operator-in-the-loop simulation.

3. Kinematics of Multibody Systems

For dependent branch coordinates, which is always the case for systems with closed kinematic chains, the kinematic constraint equations are generated by *projecting the circuit equations for cotree joints onto the reaction space for that joint.* For holonomic constraints, one obtains m nonlinear algebraic equations in terms of the n branch coordinates $\mathbf{q}$:

$$\mathbf{\Phi}(\mathbf{q}, t) = \mathbf{0} \quad (4)$$

where $n - m = f$, the degrees of freedom (DOF) of the system.

To demonstrate, consider again the slider-crank with the dependent branch coordinates $\mathbf{q} = [\beta_8, \beta_{10}, s_{11}]^T$, i.e. with h_8, h_{10}, and s_{11} selected into the tree. The reaction space for cotree joint h_9 is spanned by unit vectors $\hat{\imath}$ and $\hat{\jmath}$ (the directions of the joint reaction forces), onto which the circuit equation for h_9 is projected:

$$\mathbf{\Phi} = \underline{r}_9 \cdot \hat{\imath}, \hat{\jmath} \quad (5)$$

$$= (\underline{r}_6 - \underline{r}_7 + \underline{r}_{10} + \underline{r}_{11} - \underline{r}_8 + \underline{r}_4 - \underline{r}_5) \cdot \hat{\imath}, \hat{\jmath} \quad (6)$$

where $\underline{r}_i$ is the translational displacement vector for element i. Substituting the elemental constitutive equations, e.g. $\underline{r}_{10} = 0$, and evaluating,

$$\mathbf{\Phi} = \left\{ \begin{array}{c} L_{67}\cos\beta_{10} + s_{11} - L_{45}\cos\beta_8 \\ L_{67}\sin\beta_{10} - L_{45}\sin\beta_8 \end{array} \right\} = \mathbf{0} \quad (7)$$

where $L_{45} = L_4 + L_5$, etc., and $L_i = |\underline{r}_i|$. Thus, one obtains $m = 2$ constraint equations in terms of the $n = 3$ branch coordinates for this 1-DOF system.

To get a smaller set of coordinates and equations for the same example, one can select h_8, h_9, and h_{10} into the tree for translational equations, and h_8, h_{10}, and s_{11} into a second tree for generating rotational equations. This results in the $n = 2$ branch coordinates $\mathbf{q} = [\beta_8, \beta_{10}]^T$. The only non-null constraint equation is obtained from the circuit equation for cotree joint s_{11}, projected onto the unit vector $\hat{\jmath}$ normal to its axis of sliding (direction of normal reaction force):

$$\begin{aligned} \mathbf{\Phi} &= \underline{r}_{11} \cdot \hat{\jmath} && (8) \\ &= (\underline{r}_8 - \underline{r}_4 + \underline{r}_5 + \underline{r}_9 - \underline{r}_6 + \underline{r}_7 - \underline{r}_{10}) \cdot \hat{\jmath} && (9) \end{aligned}$$

Substituting constitutive equations and evaluating,

$$\mathbf{\Phi} = L_{45} \sin \beta_8 - L_{67} \sin \beta_{10} = \mathbf{0} \qquad (10)$$

the $m = n - f = 1$ kinematic constraint equation is obtained.

This procedure for generating kinematic equations is valid for any system topology, and for any selected set of branch coordinates.

4. Dynamics of Multibody Systems

Once the kinematic equations are obtained, the dynamic equations can be systematically generated by *projecting the cutset equations for branch components onto the motion space for that component* [2]:

$$\mathbf{M}\ddot{\mathbf{q}} + \mathbf{\Phi}_{\mathbf{q}}^T \lambda = \mathbf{F} \qquad (11)$$

where $\mathbf{M}$ is the mass matrix, $\mathbf{\Phi}_{\mathbf{q}}$ is the Jacobian of the constraint equations, the Lagrange multipliers λ correspond to reactions in cotree joints, and $\mathbf{F}$ contains external forces and quadratic velocity terms.

Alternatively, the principle of virtual work may be employed, especially when flexible bodies are included in the system model [5], to obtain the dynamic equations (11). To facilitate this approach, the first variation of the constraint equations is partitioned into f independent coordinates $\mathbf{q_i}$ and m dependent coordinates $\mathbf{q_d}$:

$$\delta \mathbf{\Phi} = \mathbf{\Phi}_{\mathbf{q_d}} \delta \mathbf{q_d} + \mathbf{\Phi}_{\mathbf{q_i}} \delta \mathbf{q_i} = \mathbf{0} \qquad (12)$$

where $\mathbf{\Phi}_{\mathbf{q_i}} = \partial \mathbf{\Phi} / \partial \mathbf{q_i}$ and $\mathbf{\Phi}_{\mathbf{q_d}} = \partial \mathbf{\Phi} / \partial \mathbf{q_d}$ is non-singular as long as the given physical constraints are not redundant. One can then solve these linear equations (12) for the transformation from dependent to independent variations:

$$\delta \mathbf{q_d} = -\mathbf{\Phi}_{\mathbf{q_d}}^{-1} \mathbf{\Phi}_{\mathbf{q_i}} \delta \mathbf{q_i} = \mathbf{J}\, \delta \mathbf{q_i} \qquad (13)$$

and for the transformation from all variations to independent variations:

$$\delta\mathbf{q} = \left\{ \begin{array}{c} \delta\mathbf{q_d} \\ \delta\mathbf{q_i} \end{array} \right\} = \left[\begin{array}{c} \mathbf{J} \\ \mathbf{1} \end{array} \right] \delta\mathbf{q_i} = \mathbf{G}\,\delta\mathbf{q_i} \tag{14}$$

where $\mathbf{G}$ is easily shown to be an orthogonal complement to the Jacobian $\mathbf{\Phi_q}$.

By summing the contributions of all working components to the system virtual work δW, and using the above transformation, one obtains:

$$\delta W = \mathbf{Q}^T\,\delta\mathbf{q} = \mathbf{Q}^T\mathbf{G}\,\delta\mathbf{q_i} = \mathbf{Q_i}^T\delta\mathbf{q_i} = 0 \tag{15}$$

where $\mathbf{Q}$ and $\mathbf{Q}_i = \mathbf{G}^T\mathbf{Q}$ are the generalized forces associated with $\delta\mathbf{q}$ and $\delta\mathbf{q_i}$, respectively. Since $\delta\mathbf{q}$ are not independent quantities, $\mathbf{Q}$ can only be set to zero if it is augmented by the Lagrange multipliers λ. The result is the set of dynamic equations (11), which are said to be written in "augmented" form.

Alternatively, the dynamic equations can be written in an "embedded" form, in which the constraint reactions λ do not appear. Since $\delta\mathbf{q_i}$ are independent, each generalized force $\mathbf{Q_i}$ can be set directly to zero to obtain one dynamic equation per degree of freedom:

$$\widetilde{\mathbf{M}}\,\ddot{\mathbf{q}} = \widetilde{\mathbf{F}} \tag{16}$$

where $\widetilde{\mathbf{M}} = \mathbf{G}^T\mathbf{M}$ is an unsymmetric $f \times n$ mass matrix and $\widetilde{\mathbf{F}} = \mathbf{G}^T\mathbf{F}$. Note that the embedded dynamic equations (16) can be directly obtained by pre-multiplying the augmented dynamic equations (11) by $\mathbf{G}^T$ to eliminate the Lagrange multipliers. In a later section, these embedded dynamic equations will be used to advantage in inverse dynamic analyses.

In a forward dynamic analysis, one has to solve a set of differential-algebraic equations (DAEs) to determine the time response $\mathbf{q}(t)$. Equations (4) and (11) comprise a set of $n + m$ index-3 DAEs that can be solved simultaneously for $\mathbf{q}(t)$ and $\lambda(t)$. An attractive alternative is to solve the smaller set of n index-2 DAEs (4) and (16) for only the branch coordinates $\mathbf{q}(t)$. There are quite a number of different numerical methods available for solving the DAEs that govern the forward dynamics of multibody systems [6].

5. Symbolic Computer Implementation: DynaFlex

Symbolic programming provided an effective computer implementation of our graph-theoretic virtual work formulation [7], especially when real-time simulations of complex mechanical systems were required. Multiplications by 0 and 1 are eliminated in a symbolic program, and trigonometric simplifications are automated. The kinematic and dynamic equations can be visually examined for physical insight, or exported as optimized C or Fortran code (in which repeated expressions are identified and evaluated only once) for subsequent simulation.

We have used the Maple symbolic programming language to develop our DynaFlex software[1] for modelling flexible multibody systems. A variety of modelling components is available, including: three-dimensional rigid bodies or elastic beams; revolute, prismatic, spherical, planar, universal, and other joints; kinematic drivers; springs, dampers, actuators, and other force components. Symbolic kinematic and dynamic equations can be generated in absolute or joint coordinates or a combination of both. Dynamic equations can be generated in either the augmented form (11) or the embedded form (16). The DynaFlex program requires an input file that describes the system's topology and parameters; this file can be easily created using the graphical user interface, DynaGUI.

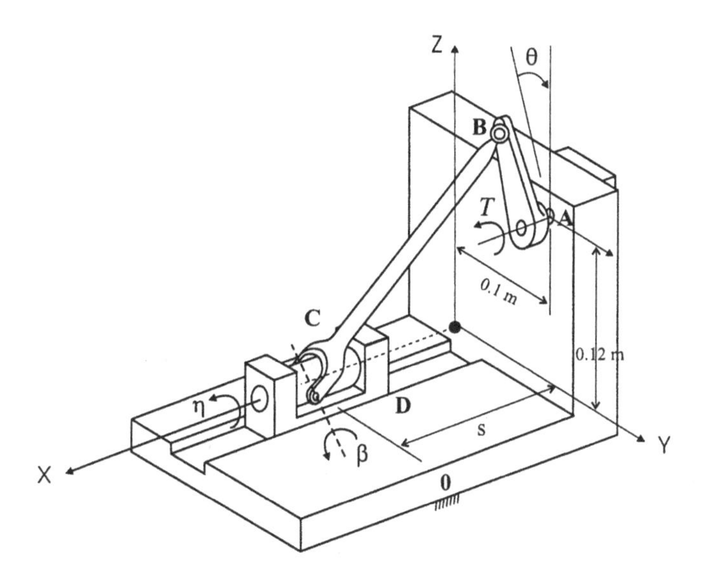

Figure 3. Spatial slider-crank mechanism.

To demonstrate the modelling process using DynaGUI and DynaFlex, consider the spatial slider-crank mechanism [3] shown in Figure 3. A crank is driven at a constant rotational speed by the torque T. The crank is connected to the ground by a revolute joint at A, and to a connecting rod by a spherical joint at B. A sliding block is attached to the ground by a prismatic joint at D, and to the connecting rod by a compound revolute-revolute joint at C. An inverse dynamic analysis is required to determine the driving torque T.

A screenshot of the DynaGUI model is shown in Figure 4. The three bodies are represented by rectangles; body-fixed frames are shown as attached circles, or as a square for the main body frame. The four joints are shown as lines connecting the three bodies and the ground. Each joint has a particular type and set of properties,

[1] Freely available for non-commercial research from http://real.uwaterloo.ca/~dynaflex

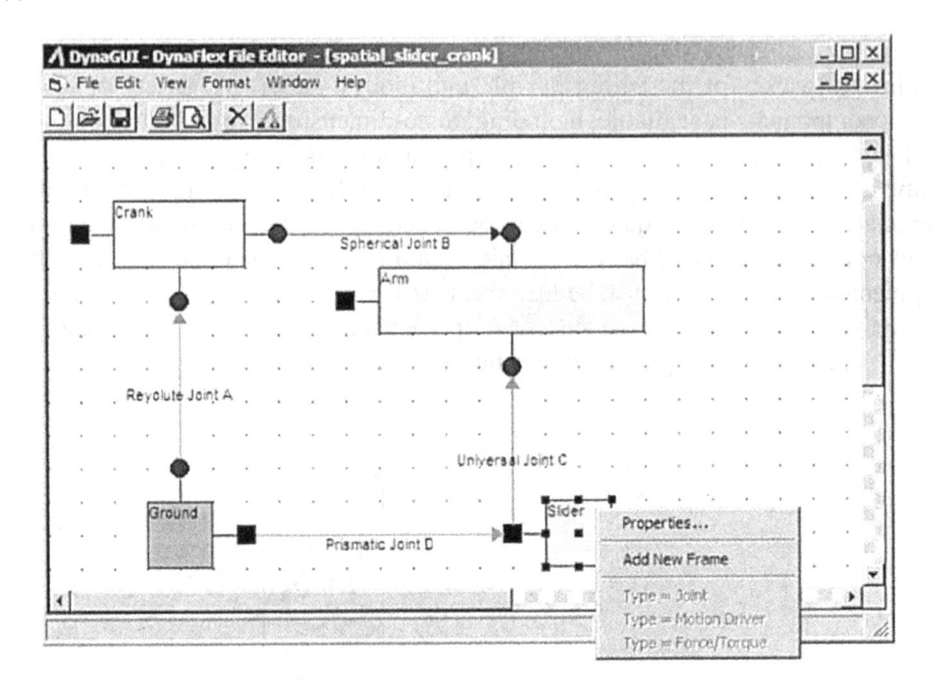

Figure 4. DynaGUI model of spatial slider-crank.

and may be in the tree or cotree. In this example, the spherical joint is placed in the cotree; the branch coordinates corresponding to the remaining joints in the tree are $\mathbf{q} = [s, \theta, \eta, \beta]^T$, which are shown in Figure 3.

The motion for the crank is prescribed as $\theta = 2\pi t$, and the corresponding torque T is automatically included in the virtual work computation. Combining the prescribed motion with the projected circuit equations for the cotree spherical joint, and substituting the link lengths given by [3], DynaFlex generates the kinematic equations:

$$\mathbf{\Phi} = \left\{ \begin{array}{c} 2\sin\theta + 6\sin\eta\sin\beta + 25/10 \\ 2\cos\theta + 6\cos\eta\sin\beta + 3 \\ 25s - 6\cos\beta \\ \theta - 2\pi t \end{array} \right\} = \mathbf{0} \tag{17}$$

which are easily solved for $\mathbf{q}(t)$. By requesting the dynamic equations in the embedded form of equation (16), DynaFlex produces a single dynamic equation for this 1-DOF system. The driving torque appears linearly in this equation, so a symbolic expression for T is easily obtained with Maple and plotted in Figure 5; the results are identical to those shown in [3], which were obtained by solving a large system of linear equations using numerical methods. Unlike the latter approach,

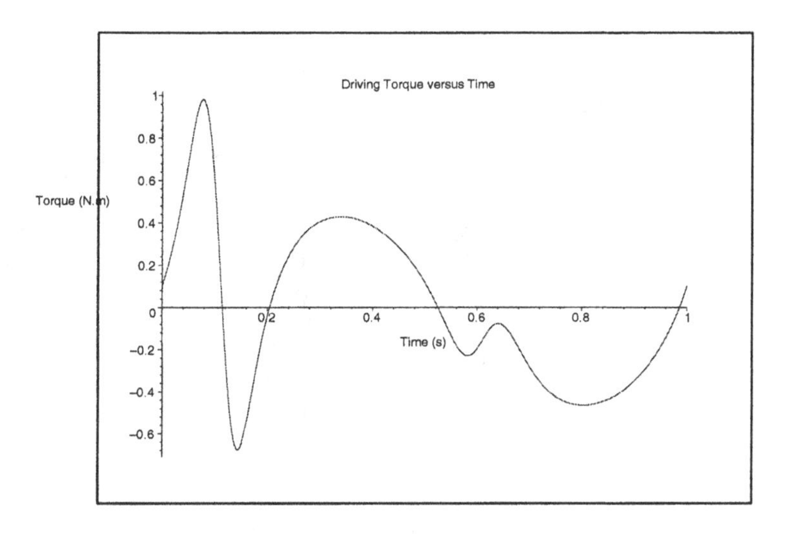

Figure 5. Driving torque on crank of spatial slider-crank.

our symbolic solution for T can be viewed, exported as optimized C or Fortran code, and used to facilitate the real-time control of a physical prototype.

6. Indirect Coordinates

By defining and selecting "virtual joints" into the tree [8], it is even possible to generate equations in "indirect coordinates" corresponding to bodies that are not necessarily adjacent [9]. Consider the open-loop multibody system taken from [9], and its corresponding linear graph in Figure 6.

Once again, bodies are represented by edges $m_1 - m_4$, body-fixed transformations by arm elements $r_5 - r_{11}$, revolute joints by edges $h_{12} - h_{14}$, and the spherical joint by b_{15}. In addition, two virtual joints are added: a virtual revolute joint vh_{16} representing the rotation of m_3 relative to m_1, and a virtual spherical joint vb_{17} to represent the rotation of m_4 relative to the ground. Fayet and Pfister [9] have shown that the corresponding indirect coordinates result in equations that are simpler in form than the more traditional joint coordinate equations. This is due to the fact that joint axes z_2 and z_3 are parallel, and absolute angular coordinates give simpler equations than relative angular coordinates [10]. With a graph-theoretic approach, one can generate the motion equations in indirect coordinates by simply selecting vh_{16} and vb_{17} into the tree in place of h_{14} and b_{15}.

Applying this approach to a serial manipulator comprised of bodies $m_1 - m_3$, the dynamic equations (11) were generated in terms of the joint coordinates $\mathbf{q}_\beta = [\beta_{12}, \beta_{13}, \beta_{14}]^T$ and the indirect coordinates $\mathbf{q}_\mathbf{i} = [\beta_{12}, \beta_{13}, \beta_{16}]^T$. Both sets of coordinates are independent for this serial manipulator, so there are no kinematic

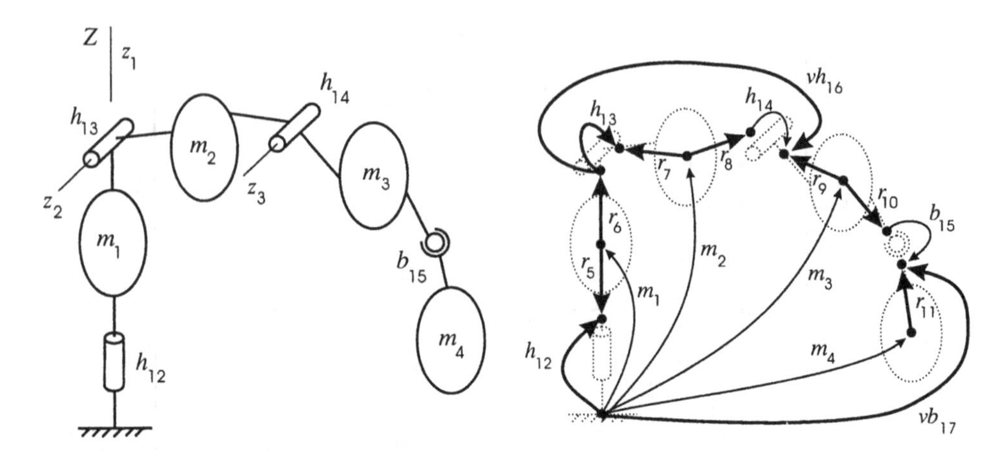

Figure 6. Open-loop multibody system and linear graph.

constraint equations and no constraint reactions λ appearing in the dynamic equations. However, some of the entries in these equations will differ for the two sets of coordinates, e.g. the second diagonal entry in the two mass matrices:

$$\begin{aligned} \mathbf{M}_\beta(2,2) &= I_{2z} + m_2 L_7^2 + I_{3z} + m_3[L_{78}^2 + 2L_{78}L_9\cos\beta_{14} + L_9^2] \\ \mathbf{M}_\mathbf{i}(2,2) &= I_{2z} + m_2 L_7^2 + m_3 L_{78}^2 \end{aligned}$$

The use of indirect coordinates results in a simplification of the mass matrix, for which the diagonal entries correspond to moments of inertia of "compound augmented bodies" [9]. In this case, $\mathbf{M}_\mathbf{i}(2,2)$ is the moment of inertia I_{2z} of m_2, plus the mass of m_3 concentrated at joint 14, about the z_2 axis. In contrast, the more complex $\mathbf{M}_\beta(2,2)$ is the combined moment of inertia of m_2 and m_3 about z_2.

TABLE 1. Number of operations.

	Adds.	Mults.	Functs.
$\mathbf{M}_\beta$:	27	53	12
$\mathbf{M}_\mathbf{i}$:	20	43	11
$\mathbf{F}_\beta$:	62	149	59
$\mathbf{F}_\mathbf{i}$:	60	138	57

The forces $\mathbf{F}$ are also simplified, albeit slightly, by the use of indirect coordinates. Table 1 provides the number of additions, multiplications, and function calls needed to evaluate $\mathbf{M}$ and $\mathbf{F}$ for both sets of coordinates. These are easily obtained using built-in routines from the Maple symbolic programming language.

7. Application to Parallel Robot Manipulators

For multibody systems with special topologies, such as the parallel robots shown in Figures 7 and 8 that have legs of identical configuration, linear graph theory can be used to exploit this topology during the equation generation process. In comparison with serial robots, parallel manipulators have very good performance in terms of rigidity, accuracy and dynamic characteristics. However, the closed chains in parallel manipulators lead to difficulties in obtaining the dynamic models needed for simulation, control, and design. Manual derivations may lead to very efficient equations, but they are tedious and prone to errors for complex parallel robots. Merlet [12] writes that one school of thought recommends that dynamic models should not be used because modelling errors are too numerous. Despite this, and the fact that dynamic equations can be generated automatically, many authors [13, 14] are still performing manual derivations for parallel manipulators.

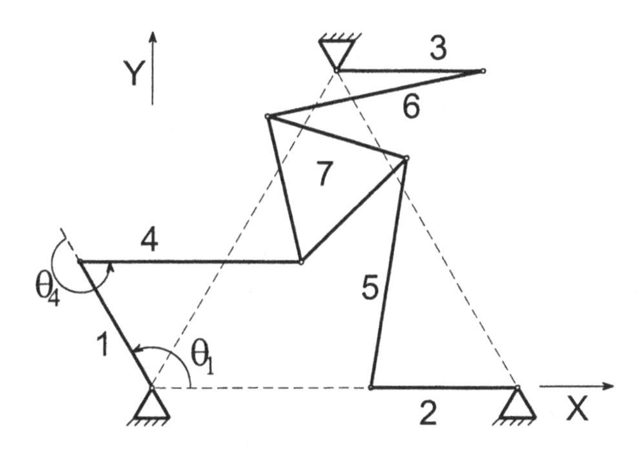

Figure 7. Planar 3-DOF parallel manipulator.

Symbolic solutions for the inverse dynamics are especially useful for real-time control of parallel manipulators and other multibody systems [11]. Given that parallel manipulators are frequently used as input devices for real-time simulations, or as platforms for flight simulators, the automatic generation of inverse dynamic solutions facilitates the development of these virtual reality environments.

To show how this can be effected using graph theory and symbolic computing, consider the 3-DOF planar parallel manipulator [15] shown in Figure 7. The end effector (labelled 7) is an equilateral triangle with sides of length L_7, links 1, 2, and 3 have a length L_1, links 4, 5, and 6 have a length of L_4, and the three ground-fixed revolute joints form an equilateral triangle with sides of length L_0. The three revolute joints connecting the end effector 7 to links 4, 5, and 6 are placed in the

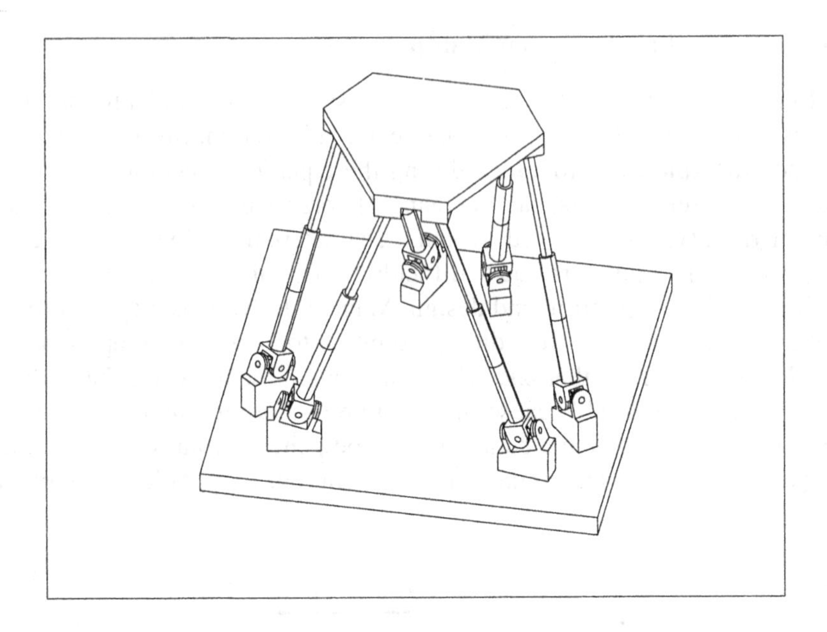

Figure 8. Spatial 6-DOF Stewart-Gough platform.

cotree of the linear graph representation; the corresponding joint coordinates are thereby eliminated from all equations. From the projected circuit equations for these three joints, one obtains the $m = 6$ nonlinear kinematic constraint equations:

$$\mathbf{\Phi} = \left\{ \begin{array}{c} -x_7 + \frac{1}{2}L_7\, c_7 - \frac{\sqrt{3}}{6}L_7\, s_7 + L_1 c_1 + L_4 c_{14} \\ -y_7 + \frac{1}{2}L_7 s_7 + \frac{\sqrt{3}}{6}L_7 c_7 + L_1 s_1 + L_4 s_{14} \\ L_0 - x_7 - \frac{1}{2}L_7\, c_7 - \frac{\sqrt{3}}{6}L_7\, s_7 + L_1 c_2 + L_4 c_{25} \\ -y_7 - \frac{1}{2}L_7 s_7 + \frac{\sqrt{3}}{6}L_7 c_7 + L_1 s_2 + L_4 s_{25} \\ -x_7 + \frac{\sqrt{3}}{3}L_7\, s_7 + L_1 c_3 + L_4 c_{36} + \frac{1}{2}L_0 \\ -y_7 - \frac{\sqrt{3}}{3}L_7 c_7 + L_1 s_3 + L_4 s_{36} + \frac{\sqrt{3}}{2}L_0 \end{array} \right\} = \mathbf{0} \qquad (18)$$

where $c_i \equiv \cos\theta_i$ and $s_{ij} \equiv \sin(\theta_i + \theta_j)$, etc, $\theta_i (i = 1...6)$, is the coordinate for the revolute joint on the proximal side of link i, and x_7, y_7, and θ_7 represent the motion of the end effector 7. Each of these 9 branch coordinates $\mathbf{q}$ correspond to a joint in the spanning tree of the graph, including a "virtual joint" between the ground and the end effector.

The constraint equations (18) can be used to obtain the transformation (13) from dependent variations to independent variations which, when combined with our symbolic virtual work routines, allows the dynamic equations to be written in the embedded form given by equation (16). In an inverse dynamic problem for which $\mathbf{q}(t)$ is known, equation (16) can be solved for the f actuator loads.

Note that the actuator loads appear linearly in $\widetilde{\mathbf{F}}$. Furthermore, by choosing the actuated joint variables as the independent coordinates $\mathbf{q}_i$, the actuator loads will be decoupled in equation (16), i.e. exactly one actuator load appears explicitly in each equation. Hence, no matrix inversion is needed, unlike the approach in [13].

The direct application of Maple algorithms to the solution of the linear equations (12) for the dependent variations will be called the *direct symbolical* approach. Due to memory limitations associated with symbolic programming, this approach will fail when the number of loop closure equations is large, which is the case for the 6-DOF Gough-Stewart platform [14] shown in Figure 8. In this situation, one can still obtain the dynamic equations (16) by using a *pseudo-variable* approach to solve equation (12) in a way that exploits the special topology [11].

To accomplish this, the set of coordinates are further partitioned as $\mathbf{q} = (\mathbf{q_i}, \mathbf{q_{dd}}, \mathbf{q_e})$, where $\mathbf{q_i}$ are the independent variables (associated with the actuators), $\mathbf{q_e}$ are the end effector variables, and $\mathbf{q_{dd}}$ are the variables associated with the unactuated joints. The so-called pseudo-independent variables are defined as $\mathbf{q_{pi}} = \mathbf{q_e}$, and the pseudo-dependent variables $\mathbf{q_{pd}} = (\mathbf{q_i}, \mathbf{q_{dd}})$. This re-partitioning of the joint coordinates leads to a largely decoupled linear system of equations (12); this is the essential feature of the pseudo-variable approach.

To demonstrate, consider selecting the actuated joint angles θ_1, θ_2, and θ_3 as independent coordinates $\mathbf{q_i}$ for the planar 3-DOF manipulator. The partial derivatives (Jacobian matrices) of the loop closure equations (18) with respect to the dependent and pseudo-dependent variables have the structures:

$$\mathbf{\Phi}_{\mathbf{q_d}} = \begin{bmatrix} * & & & * & & * \\ * & & & & * & * \\ & * & & * & & * \\ & * & & & * & * \\ & & * & * & & * \\ & & * & & * & * \end{bmatrix}, \quad \mathbf{\Phi}_{\mathbf{q_{pd}}} = \begin{bmatrix} * & * & & & & \\ * & * & & & & \\ & & * & * & & \\ & & * & * & & \\ & & & & * & * \\ & & & & * & * \end{bmatrix} \tag{19}$$

where an asterisk $*$ indicates a non-zero entry. One can clearly see the decoupling that is present in the Jacobian matrix $\mathbf{\Phi}_{\mathbf{q_{pd}}}$ for the pseudo-dependent variables. Once the pseudo-dependent variations are obtained, it is a simple matter [11] to recover the pseudo-independent variations and the symbolic transformation (13).

In addition to the direct symbolical and pseudo-variable approaches, three others have been implemented for comparison. In an *implicit symbolical* approach, the actuator loads are expressed in terms of the entries of the transformation matrix $\mathbf{J}$, which are calculated and saved symbolically. In a combined symbolic/numeric approach, the actuator loads are again expressed in terms of the entries in $\mathbf{J}$, which are obtained numerically by solving f linear systems $\mathbf{\Phi}_{\mathbf{q_d}}\mathbf{j}_k = -\varphi_k, k = 1 \ldots f$, where $\mathbf{j}_k$ and φ_k are the k-th column of the matrices $\mathbf{J}$ and $\mathbf{\Phi}_{\mathbf{q_i}}$, respectively. In another combined approach, the actuator loads are expressed in terms of $\mathbf{\Phi}_{\mathbf{q_d}}^{-1}$, which is also obtained using numerical methods during the evaluation process.

Symbolic expressions for the torques actuating the 3-DOF parallel manipulator are identical to those derived by hand in [15]. The explicit expressions are too lengthy to be displayed here[2].

TABLE 2. Computational efficiency for one inverse dynamic analysis (333-MHz Pentium II)

Approach	Flops	CPU [ms]
direct symbolic	967	6.0
pseudo-variable	871	6.1
implicit symbolic	879	7.6
symbolic/numeric (dummy $\mathbf{J}$)	923	15.4
symbolic/numeric (dummy $\mathbf{\Phi}_{\mathbf{q}_d}^{-1}$)	1235	19.8

The CPU time and number of flops required by Matlab for one inverse dynamic evaluation of the three driving torques are shown in Table 2. The kinematic solution, performed prior to the dynamic analysis, is not counted. It can be seen that the direct symbolic and pseudo-variable approaches are equivalent regarding the CPU time, while the implicit symbolic approach requires about 25% more CPU time. The combined symbolic/numeric approaches require more than twice the CPU time compared to the symbolic approaches.

We can draw the conclusion that, for problems of this complexity, the symbolic approaches are preferred because they are faster than the numerical approaches. Furthermore, they do not need matrix manipulation capabilities — this is advantageous in microprocessor-based control applications. For the more complex 6-DOF Gough-Stewart platform, the symbolic/numerical approaches outperform the purely symbolic approaches in terms of CPU time, especially if sparse matrix solution methods are employed.

8. Efficient Modelling via Subsystems

The modelling of multibody systems is greatly facilitated by the use of subsystem models. Models of entire sub-assemblies can be created and stored (symbolically) for future use. For systems with a repetitive structure, this can be very helpful. In a time-varying topological situation, e.g. a virtual reality application in which the user is adding or removing parts to their model, the system equations can be reformulated quickly by focusing attention only on the modified subsystems.

Subsystem models are particularly well-suited for the topologies that are typical of parallel manipulators. Each leg can be modelled as a subsystem that is kinematically decoupled from the other leg subsystems. This allows the equations

[2] Available upon request, for both the 3-DOF manipulator and the Gough-Stewart platform.

for kinematics and inverse dynamics to be generated and solved in parallel, either symbolically or numerically on a parallel computing platform. Graph theory is naturally suited to the modelling of systems via subsystem models. Although the subsystem modelling of electrical circuits was achieved decades ago using graph theory, it has only recently been extended to the nonlinear kinematics and dynamics of multibody systems [16].

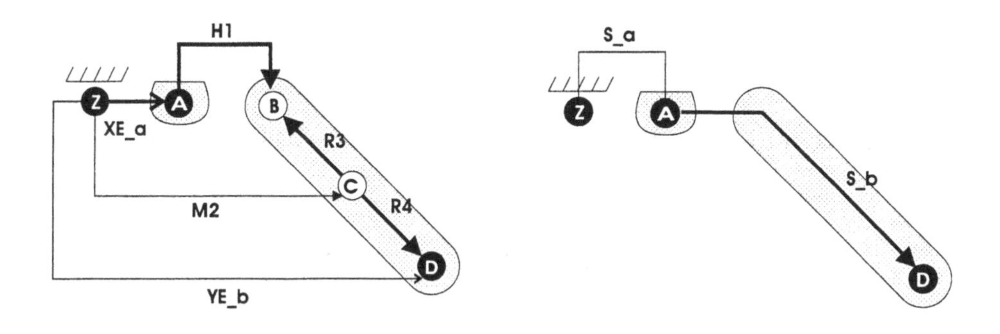

Figure 9. Single link plus revolute joint: augmented standard graph and subsystem model.

To demonstrate, consider the 3-DOF parallel manipulator again. The modelling process is started with a standard linear graph of a single link plus revolute joint, as shown on the left of Figure 9. External kinematic (XE_a) and dynamic (YE_b) excitations are applied to the boundary nodes A and D of this linear graph model. By matching the resulting response against that for the standard model, the constitutive equations are obtained for edges S_a and S_b in the subsystem model shown on the right of Figure 9. Essentially, this procedure represents a graph-theoretic generalization of the well-known Norton and Thevenin theorems from electrical network theory. Note that the internal nodes B and C have been eliminated from the subsystem model, which includes nodes A, D, and an inertial reference node Z. A simple electrical analogy would be the replacement of a series of linear resistors by a single equivalent resistance.

This same procedure is used to combine two of the single-link subsystems into a subsystem model of an entire leg, as shown in Figure 10. In this case, only the common internal node B is eliminated in the subsystem modelling process. Finally, by attaching three subsystem models of the legs to a standard linear graph model of the end effector (M0, R10, R11, R12) via three revolute joints (H4, H5, H6), the full system model in Figure 11 is obtained. The graph representation is simplified by the use of subsystem models, and the modelling process is made more efficient by the use and re-use of models for repeated subsystems.

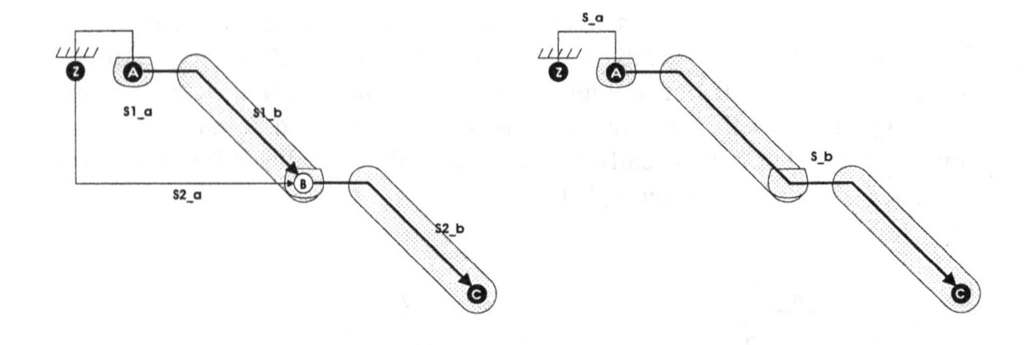

Figure 10. Two single-link subsystems combined to form a leg subsystem model.

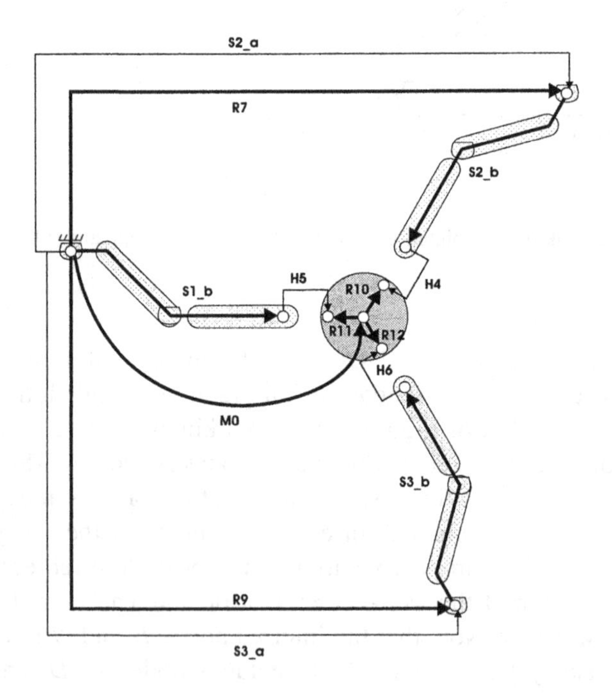

Figure 11. Parallel 3-DOF manipulator modelled with subsystem graphs.

9. Extension to Mechatronic Systems

Graph-theoretic modelling is not restricted to mechanical systems; it has long been applied to electrical, pneumatic, and hydraulic systems, and our graph-theoretic symbolic algorithms were easily extended to the modelling of electro-mechanical multibody systems [17].

To demonstrate, consider the two-link manipulator shown in Figure 12 with its

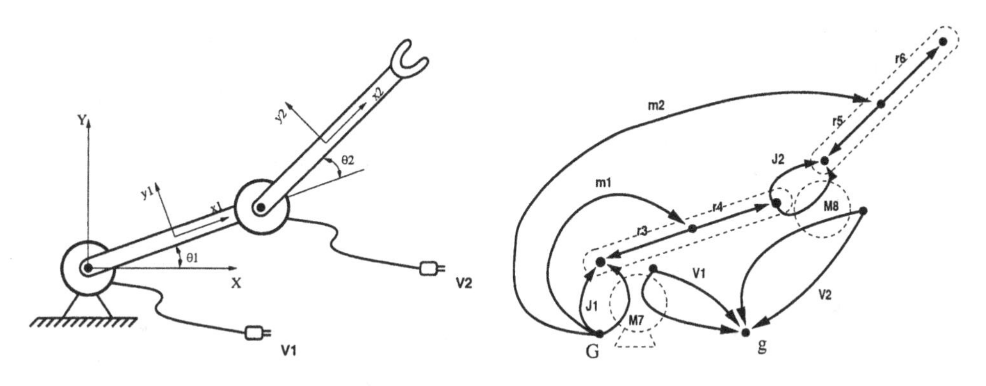

Figure 12. DC motor-driven robot manipulator and linear graph.

linear graph representation. The joint angles θ_1 and θ_2 are controlled by two DC motors which are powered by voltage sources V_1 and V_2. The linear graph of the system contains the standard mechanical components, the two voltage sources, and the two motors M_7 and M_8. Note that the graph consists of two parts, one for the mechanical domain and one for the electrical domain, and that the motors have edges in each domain. This is true for any transducer element that converts energy between different domains.

Four differential equations are generated for the two-link manipulator, two for each of the two domains, in terms of θ_1, θ_2, and the two motor currents. These system equations can be used to find symbolic inverse solutions for the motor currents required to drive a particular trajectory, or as the basis of a forward dynamic simulation that tests out different controller strategies [17].

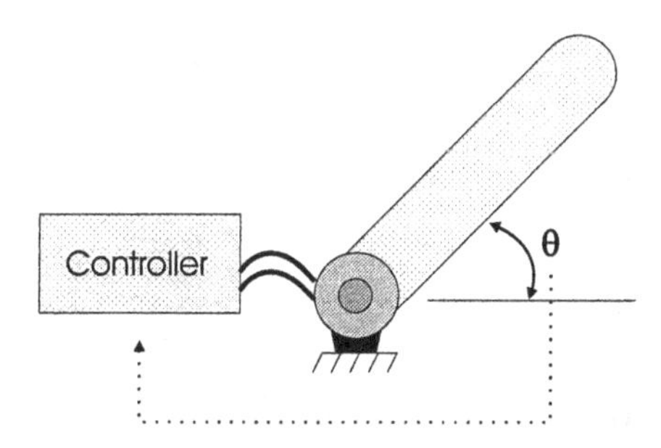

Figure 13. Mechatronic system: rotating link, angle sensor, motor, and controller.

The subsystem modelling algorithms from Section 8 are very well-suited to mechatronic systems. As an example, consider the simple electro-mechanical system shown in Figure 13, consisting of a rigid body, revolute joint and motor, sensor, and controller. A block diagram of this system is shown in Figure 14. The differencer component computes the difference between the voltage V_theta_desired corresponding to a desired angle and the voltage V_theta_actual that is output from the angle sensor. This voltage difference is used by the PD controller to send a control voltage V_control to the motor, which applies a corrective torque T_control to the link. Note that *Ge* is the common electrical ground while *Gm* is the inertial reference frame.

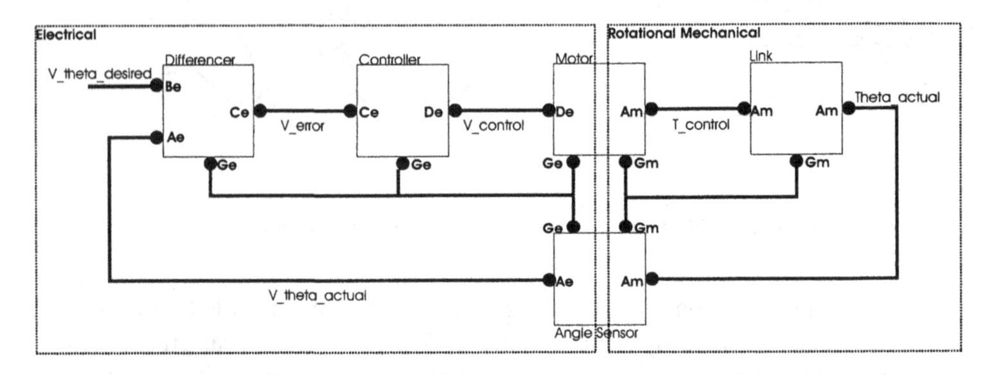

Figure 14. Block diagram representation of mechatronic system.

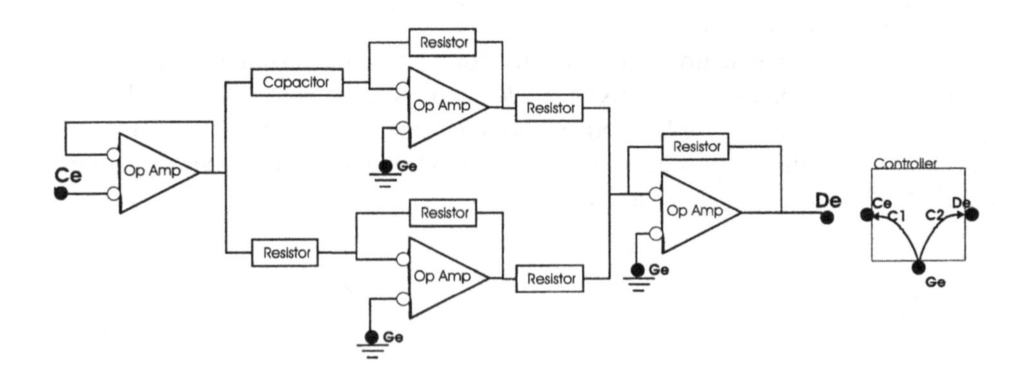

Figure 15. PD controller: block diagram representation and linear graph subsystem.

Note also that each component shown in Figure 14 is essentially a subsystem model. Shown in Figure 15 is a block diagram representation of the entire PD controller, consisting of 4 operational amplifiers (op amps), 6 resistors, and 1 capacitor. Solid circles correspond to boundary nodes of the graph-theoretic subsystem model shown on the right of Figure 15; the hollow circles are internal nodes that are eliminated from the subsystem representation. The constitutive equations for this subsystem model are directly obtained using the procedure described in Section 8, i.e. the graph-theoretic generalization of the Norton-Thevenin theorems.

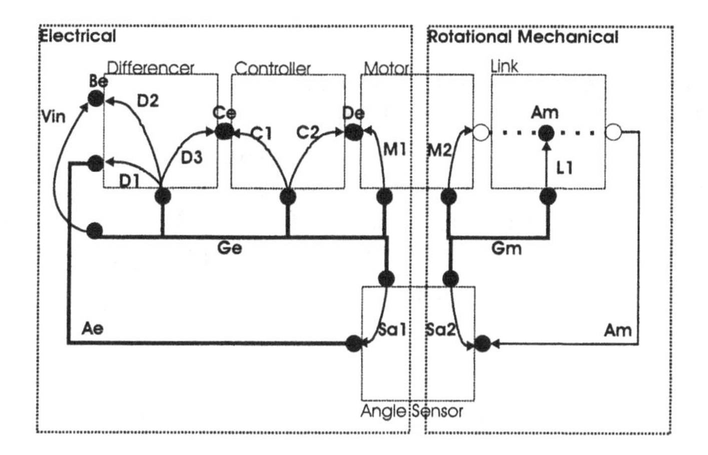

Figure 16. Combination of all graph-theoretic subsystem models.

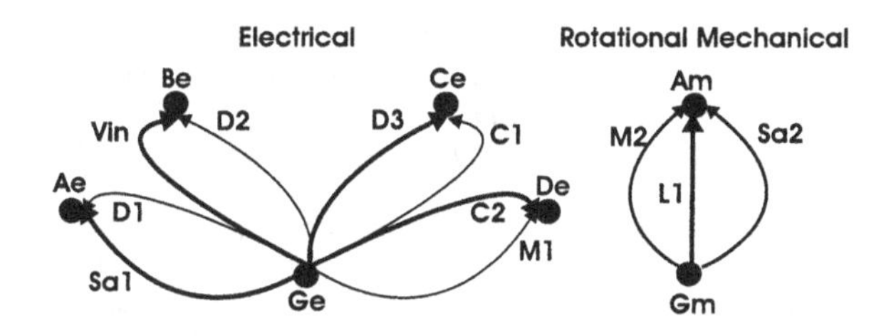

Figure 17. Standard condensed form of system-level graph.

Combining these graph-theoretic subsystem models gives the system-level linear graph shown in Figure 16. Note the similarity between this linear graph and the original block diagram representation, Figure 14. If one condenses the common nodes (e.g. *Ge*) in Figure 16, one obtains the standard linear graph representation of the mechatronic system shown in Figure 17, for which our symbolic algorithms will generate the governing system equations.

10. Conclusions

When combined with symbolic programming, linear graph theory provides a very efficient approach to modelling the kinematics and dynamics of flexible multibody systems. By selecting a spanning tree for a graph representation, one can define a set of coordinates that is well-suited to the problem at hand, including absolute, joint, or indirect coordinates. The special topologies of parallel manipulators can be exploited, especially if a subsystem modelling approach is adopted. Graph theory is equally applicable to electro-mechanical multibody systems.

11. Acknowledgements

Financial support for this research is provided by a Premier's Research Excellence Award and Waterloo Maple Inc., and by the Natural Sciences and Engineering Research Council of Canada. The author is grateful to Professor Clément Gosselin of Laval University, Canada, and Dr. Jiegao Wang of MD Robotics, Canada, for providing Figure 8. The author also thanks Mr. Scott Redmond (the author of DynaGUI) for providing Figure 4, and Mr. Chad Schmitke for providing material on subsystem modelling in Section 8.

References

1. Koenig, H.E., Tokad, Y., and Kesavan, H.K. (1967) *Analysis of Discrete Physical Systems*, McGraw-Hill, New York.
2. McPhee, J. (1998) Automatic generation of motion equations for planar mechanical systems using the new set of 'branch coordinates', *Mechanism and Machine Theory* **33**, 805-823.
3. Haug, E.J. (1989) *Computer-Aided Kinematics and Dynamics of Mechanical Systems*, Allyn and Bacon, Boston.
4. Wittenburg, J. (1977) *Dynamics of Systems of Rigid Bodies*, B.G. Teubner, Stuttgart.
5. Shi, P., and McPhee, J. (2000) Dynamics of flexible multibody systems using virtual work and linear graph theory, *Multibody System Dynamics* **4**, 355-381.
6. Eich-Soellner, E., and Führer, C. (1998) *Numerical Methods in Multibody Dynamics*, B.G. Teubner, Stuttgart.
7. Shi, P. and McPhee, J. (2002) Symbolic programming of a graph-theoretic approach to flexible multibody dynamics, *Mechanics of Structures and Machines* **30**, 123-154.
8. McPhee, J. (2001) A unified formulation of multibody kinematic equations in terms of absolute, joint, and indirect coordinates", *Proceedings* of the ASME Design Engineering Technical Conference, Pittsburgh U.S.A.
9. Fayet, M., and Pfister, F. (1994) Analysis of multibody systems with indirect coordinates and global inertia tensors, *European Journal of Mechanics, A/Solids* **13**, 431-457.
10. Huston, R.L., Liu, Y.S, and Liu, C. (1994) Use of absolute coordinates in computational multibody dynamics, *Computers and Structures* **52**, 17-25.
11. Geike, T. and McPhee, J. (2001) Inverse dynamic analysis of parallel manipulators with full mobility, *Mechanism and Machine Theory*, under review.
12. Merlet, J. (2000) *Parallel Robots*, Kluwer Academic Publishers, Dordrecht.
13. Tsai, L.-W. (2000) Solving the inverse dynamics of a Stewart-Gough manipulator by the principle of virtual work, *Journal of Mechanical Design* **122**, 3-9.
14. Wang, J. and Gosselin, C. (1998) New approach for the dynamic analysis of parallel manipulators, *Multibody System Dynamics* **2**, 317-334.
15. Ma, O. and Angeles, J. (1989) Direct kinematics and dynamics of a planar 3-DOF parallel manipulator, *Advances in Design Automation - 1989*, **3**, 313-320.
16. Schmitke, C. and McPhee, J. (2003) Effective use of subsystem models in multibody and mechatronic system dynamics, *International Journal of Computational Civil and Structural Engineering*, in press.
17. Scherrer, M. and McPhee, J. (2002) Dynamic modelling of electromechanical multibody systems, *Multibody System Dynamics*, in press.

IMPACT OF RIGID AND FLEXIBLE MULTIBODY SYSTEMS: DEFORMATION DESCRIPTION AND CONTACT MODELS

J.A.C. AMBRÓSIO
IDMEC/Instituto Superior Técnico
Av. Rovisco Pais, 1049-001 Lisboa, Portugal

Abstract

The use of multibody tools in a virtual environment, aiming to real-time, leads to models where a compromise between accuracy and computational efficiency has to be reached. Complex multibody models that undergo complex interactions often experience some level of deformations while contact between the system components or with external surfaces occurs. Depending on the objectives of the virtual environment different modeling assumptions, leading to various types of models, can be used. In this work several representations of the system flexibility able to represent the deformation of the components are presented. The plastic hinge and the finite segment approaches provide computationally efficient forms of representing large and small deformations but they lack the ability of describing the stresses and strains on the deformed components. The finite element approaches can describe with high level of accuracy the system deformation, including the stress and strain fields, but they have higher computational costs. In systems that experience contact or impact situations it is necessary to include proper contact models. In this work continuous force and unilateral contact models are presented. The choice between these contact models is discussed here taking into account the use of the different representations of the system flexibility. In the process, the equivalence between the different contact models is discussed. Applications to railway, automobile and biomechanical contact are used to demonstrate the use of the formulations proposed.

1. Introduction

The multibody dynamics methodologies present the necessary characteristics to model systems that experience contact and impact conditions. The study of the biomechanics of impact is an example of an area where the multibody biomechanical models have dominated the numerical procedures. The design of vehicles for crashworthiness has also relied for long in conceptual models, which are in fact multibody models. Other type of systems, such as roller chain drives, gear drives or mechanisms with joint clearances, also require the description of the contact and impact between the system components and, eventually, the deformation of those elements. In all applications of multibody dynamics formulations that involve contact and impact it is required that proper contact models are used, the numerical integration procedures can accommodate the sudden change of the system conditions and the deformations of the system components are represented such a way that the model deformation patterns can be reproduced by the model.

W. Schiehlen and M. Valášek (eds.), Virtual Nonlinear Multibody Systems, 57–81.

The design requirements of advanced mechanical and structural systems and the real-time simulation of complex systems exploit the ease of use of the powerful computational resources available today to create virtual prototyping environments. These simulation facilities play a fundamental role in the study of systems that undergo large rigid body motion while their components experience material or geometric nonlinear deformations. The standard methodologies are unable to deal with systems that undergo material nonlinear deformations and consequently are not useful for applications involving structural impact of multibody systems. To overcome these limitations, Nikravesh and Sung [1] suggested a rigid body description of the multibody system with the nonlinear deformations lumped in force elements acting the kinematic joints and referred to as plastic hinges. This technique, further developed by Ambrósio and Pereira [2] and by Dias and Pereira [3], is applicable to systems for which the pattern of deformation can be assumed beforehand.

Using reference frames fixed to planar flexible bodies, Song and Haug [4] suggest a finite element based methodology, which yields coupled gross rigid body motion and small elastic deformations. The idea behind Song and Haug's approach is further developed and generalized by Shabana and Wehage [5,6] that use substructuring and the mode component synthesis to reduce the number of generalized coordinates required to represent the flexible components. Yoo and Haug [7] account for the contribution of the truncated modes by introducing static correction modes.

The community studying space dynamics has been dealing with the dynamics of flexible bodies undergoing large rigid body motion. The orbiting space structural systems are characterized by the use of very flexible lightweight components. The need to characterize dynamically and control such systems, in particular, motivated valuable investigations on flexible multibody dynamic [8,9]. In the framework of the spinning spacecrafts modeling, Kane, Ryan and Banerjee [10] showed that though most of the flexible multibody methods, at the time, could capture the inertia coupling between the elastodynamics of the system components and their large motion but they would still produce incorrect results. Actually, the floating reference frame methods used in flexible multibody dynamics have the ability to lower the geometric nonlinearities of the flexible bodies, but they do not eliminate them because the moderate rotation assumption about the floating reference frame is still required [11]. Recognizing the problem posed and using some of the approaches well in line with those of the finite element community Cardona and Geradin proposed formulations for the nonlinear flexible bodies using either a geometrically exact model [12] or through substructuring [13]. Defining it as an absolute nodal coordinate formulation, Shabana [14] used finite rotations nodal coordinates enabling the capture of the geometric nonlinear deformations. Another approach taken by Ambrósio and Nikravesh [15] to model geometrically nonlinear flexible bodies is to relax the need for the structures to exhibit small moderate rotations about the floating frame by using an incremental finite element approach within the flexible body description. The approach is further extended to handle material nonlinearities of flexible multibody systems also [16].

Applications of rigid and flexible multibody systems to impact scenarios require an efficient description of the contact conditions [17,18]. Some of the characteristics required from these contact models, in a multibody dynamics framework, concern their contribution to the stable integration of the equations of motion, the description of the geometry and properties of the surfaces in contact and the correct representation energy dissipation due the local and global deformation effects not modeled by the flexible bodies. The continuous contact force model, proposed by Lankarani [19], fulfills the required characteristics to represent both nodal and rigid body contact. An alternative model, using the addition and deletion of kinematic constraints, is proposed for the nodal contact of the nonlinear finite elements representing the flexible components. In both cases of rigid and flexible body contact the dynamic friction forces are modeled.

Multibody formulations, able to describe material nonlinear deformations of the system components are reviewed here. Two methodologies able to describe efficiently contact in multibody systems are presented in this work. Finally, the methodologies discussed are used in several applications that involve vehicle and biomechanics impact.

2. Equations of Motion For Multibody Systems

The multibody system is a collection of rigid and flexible bodies joined together by kinematic joints and force elements, as shown in Fig. 1. For the ith body in the system $\mathbf{q}_i$ denotes a vector of coordinates which contains the Cartesian translation coordinates $\mathbf{r}_i$ and a set of rotational coordinates $\mathbf{p}_i$. A vector of velocities for a rigid body i is defined as $\mathbf{v}_i$, which contains the translation velocities $\mathbf{r}_i$ and the angular velocities ω_i. The vector of accelerations for the body is denoted by $\mathbf{v}_i$ and it is the time derivative of $\mathbf{v}_i$. For a multibody system with nb bodies, the vectors of coordinates, velocities, and accelerations are $\mathbf{q}$, $\mathbf{v}$ and $\mathbf{v}$ which contain the elements of $\mathbf{q}_i$, $\mathbf{v}_i$ and $\mathbf{v}_i$, respectively, for i=1, ...,nb.

The kinematic joints between rigid bodies are described by mr kinematic constraints

$$\boldsymbol{\Phi}\ \mathbf{q}) = \mathbf{0} \tag{1}$$

The time derivatives of the constraints result in the velocity and acceleration equations.

$$\dot{\boldsymbol{\Phi}}\quad \mathbf{Dv}\quad \mathbf{0} \tag{2}$$

$$\boldsymbol{\Phi} = \mathbf{Dv} + \mathbf{Dv} = \mathbf{0} \tag{3}$$

where $\mathbf{D}$ is the Jacobian matrix of the constraints. The constraint equations are included in the equations of motion using the Lagrange multiplier technique [20]

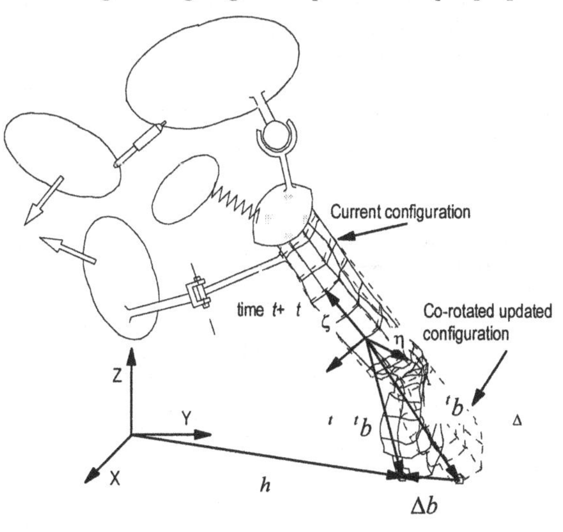

Figure 1. Generic multibody system in an impact fictitious scenario

$$\mathbf{M}\dot{\mathbf{v}} \quad \mathbf{D}^T\boldsymbol{\lambda} \quad \mathbf{g} \tag{4}$$

where $\mathbf{M}$ is the inertia matrix, $\boldsymbol{\lambda}$ is the vector of Lagrange multipliers, and $\mathbf{g} \quad \mathbf{g}(\mathbf{q},\mathbf{v})$ has the gyroscopic terms and forces and moments acting on the rigid and flexible bodies.

The solution of equation (4) together with equation (3) constitutes the set of differential-algebraic equations that represent the motion of the multibody system. These are written in a matrix form as

$$\begin{bmatrix} \mathbf{M} & \mathbf{D}^T \\ \mathbf{D} & \mathbf{0} \end{bmatrix} \begin{bmatrix} \dot{\mathbf{v}} \\ \boldsymbol{\lambda} \end{bmatrix} = \begin{bmatrix} \mathbf{g} \\ -\dot{\mathbf{D}}\mathbf{v} \end{bmatrix} \tag{5}$$

There is a wide range of multibody systems applications, involving impact that can use models made solely of rigid bodies but that still have a representation of deformations, using the plastic hinge concept for instance [1]. However, many applications of multibody systems require the description of the system deformation. The use of flexible multibody models that include the description of the large deformations of the system components by nonlinear finite elements is necessary for these applications. Before a methodology describing the distributed flexibility of multibody systems is presented the use of the plastic hinge concept in the context of multibody systems models to impact is briefly described.

2.1. PLASTIC HINGES FOR MULTIBODY NONLINEAR DEFORMATIONS

In many impact situations, the individual structural members are overloaded, principally in bending, giving rise to plastic deformations in highly localized regions, called plastic hinges. These deformations, presented in Fig. 2, develop at points where maximum bending moments occur, load application points, joints or locally weak areas [21] and, therefore, for most practical situations, their location is predicted well in advance. Multibody models obtained with this method are relatively simple. This methodology is known in the automotive, naval and aerospace industries as conceptual modeling [1,22,23].

The plastic hinge concept has been developed by using generalized spring elements to represent constitutive characteristics of localized plastic deformation of beams and kinematic joints [2], as illustrated in Fig. 3. The characteristics of the spring-damper that describe the properties of the plastic hinge are obtained by experimental component testing, finite element nonlinear analysis or simplified analytical methods. For a flexural plastic hinge the spring stiffness is expressed as a function of the change of the relative angle between two adjacent bodies connected by the plastic hinge, as shown in Fig. 4.

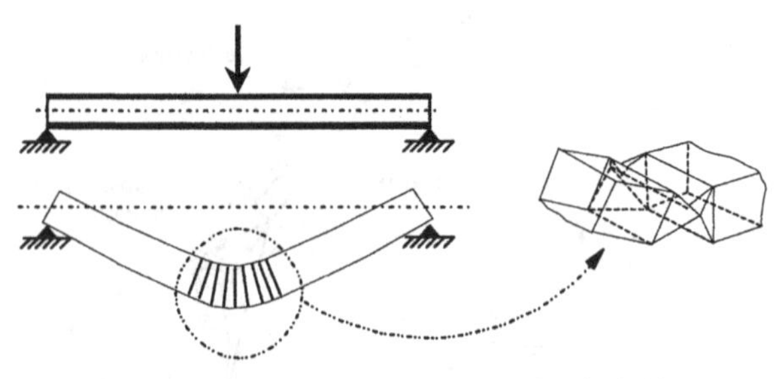

Figure 2. Localized deformations on a beam and a plastic hinge

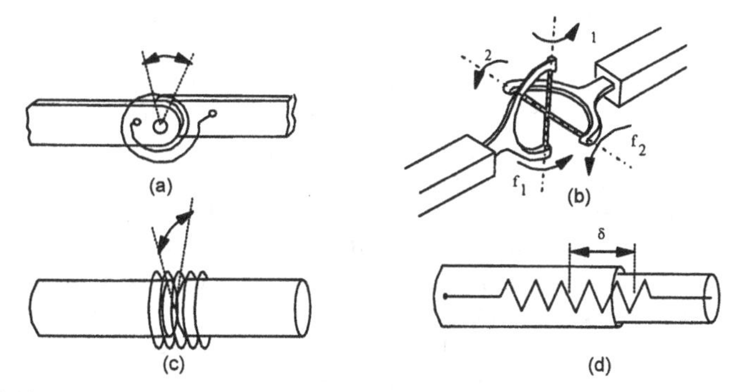

Figure 3. :Plastic hinge models for different loading conditions: a) one axis bending b) two axis bending; c) torsion; d) axial loading

For a bending plastic hinge the revolute joint axis must be perpendicular to the neutral axis of the beam and to the plastic hinge bending plane. The relative angle between the adjacent bodies measured in the bending plane is:

$$\,_{ij} \qquad \,_{i} - \,_{j} - \,_{ij}^{0} \tag{6}$$

where $\,_{ij}^{0}$ is the initial relative angle between the adjacent bodies.

The typical torque-angle constitutive relationship, as in Fig. 4, is found based on a kinematic folding model [24] for the case of a steel tubular cross section. This model is modified accounting for elastic-plastic material properties including strain hardening and strain rate sensitivity of some materials. A dynamic correction factor is used to account for the strain rate sensitivity [25].

$$P_d \,/\, P_s \qquad 1 \;+\; 0.07 \;\; V_0^{0.82} \tag{7}$$

Here P_d and P_s are the dynamic and static forces, respectively, and V_0 is the relative velocity between the adjacent bodies. The coefficients appearing in Eq. (7) are dependent on the type of cross section and material.

3. Flexible Multibody Systems

Though the use of crushable elements or plastic hinges can be a valuable approach they require that the pattern of deformation of the structural component is known in advance. In complex cases, eventually involving multiple impacts, it is not possible to predict the loading of the vehicle structural components. Therefore, the use of a nonlinear finite element description is irreplaceable [26].

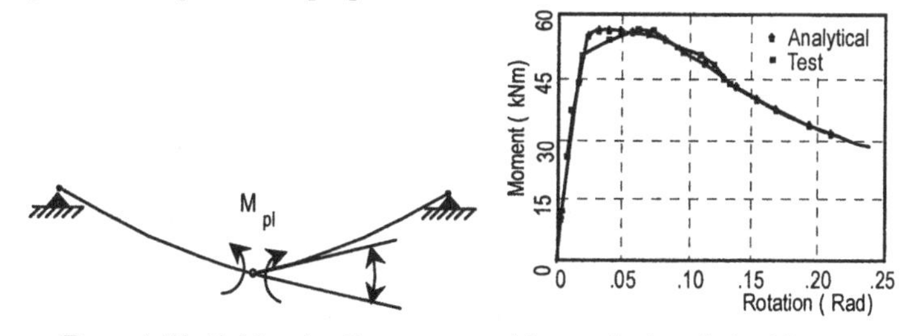

Figure 4. Plastic hinge bending moment and its constitutive relationship

3.1. EQUATIONS OF MOTION FOR A SINGLE FLEXIBLE BODY

The motion of a flexible body, depicted by Fig. 5, is characterized by a continuous change of its shape and by large displacements and rotations, associated to the gross rigid body motion. Let XYZ denote the inertial reference frame and ξηζ a body fixed coordinate frame. Let the principle of the virtual works be used to express the equilibrium of the flexible body in the current configuration $t+\ t$ and an updated Lagrangean formulation be used to obtain the equations of motion of the flexible body [26].

Let the finite element method be used to represent the equations of motion of the flexible body. The finite elements used in the discretization of the flexible body are assembled, leading to the body equations of motion. In order to improve the numerical efficiency of the solution of the equations obtained, a lumped mass formulation is used and the nodal accelerations $\ddot{\mathbf{u}}$, measured with respect to the body fixed frame, are substituted by the nodal accelerations relative to the inertial frame $\ddot{\mathbf{q}}_f$. Furthermore, it is assumed that the flexible body has a rigid and a flexible part and that the body fixed coordinate frame is attached to the center of mass of the rigid part, as shown in Fig. 6. The flexible and rigid parts are attached by the boundary nodes ψ. The procedure is described in Ambrósio and Nikravesh [26]. The resulting equations of motion for the flexible body are written as

$$\begin{bmatrix} m\mathbf{I}+\bar{\mathbf{A}}\underline{\mathbf{M}}^{*}\bar{\mathbf{A}}^{T} & -\bar{\mathbf{A}}\underline{\mathbf{M}}^{*}\mathbf{S} & \mathbf{0} \\ -\left(\bar{\mathbf{A}}\underline{\mathbf{M}}^{*}\mathbf{S}\right)^{T} & \mathbf{J}'+\mathbf{S}^{T}\underline{\mathbf{M}}^{*}\mathbf{S} & \mathbf{0} \\ \mathbf{0} & \mathbf{0} & \mathbf{M}_{ff} \end{bmatrix} \begin{bmatrix} \ddot{\mathbf{r}} \\ \dot{\boldsymbol{\omega}}' \\ \ddot{\mathbf{q}}'_{f} \end{bmatrix} = \begin{bmatrix} \mathbf{f}_{r}+\bar{\mathbf{A}}\mathbf{C}' \\ \mathbf{n}'-\tilde{\boldsymbol{\omega}}'\mathbf{J}'\boldsymbol{\omega}'-\mathbf{S}^{T}\mathbf{C}'-\bar{\mathbf{I}}^{T}\mathbf{C}'_{\theta} \\ \mathbf{g}'_{f}-\mathbf{f}-\left(\mathbf{K}_{L}+\mathbf{K}_{NL}\right)\mathbf{u}' \end{bmatrix} \tag{8}$$

where $\ddot{\mathbf{r}}$ and $\dot{\boldsymbol{\omega}}'$ are respectively the translational and angular accelerations of the body fixed reference frame, $\mathbf{J}'$ is the inertia tensor, expressed in body fixed coordinates, $\mathbf{f}_r$ is the vector of the external forces applied on the body and $\mathbf{n}'$ is the a vector with the force transport and external moments. Vector $\mathbf{u}'$ denotes the displacements increments from a previous configuration to the current configuration, measured in body fixed coordinates. The right-hand side of Eq. (8) contains, the vector generalized forces applied on the deformable body $\mathbf{g}$, matrices $\mathbf{K}_L$ and $\mathbf{K}_{NL}$, which are the linear and nonlinear stiffness matrices respectively, and $\mathbf{f}$ denotes the vector of equivalent nodal forces due to the state of stress. The reference to the linearity of the stiffness matrices $\mathbf{K}_L$ and $\mathbf{K}_{NL}$ is related to their relation with the displacements.

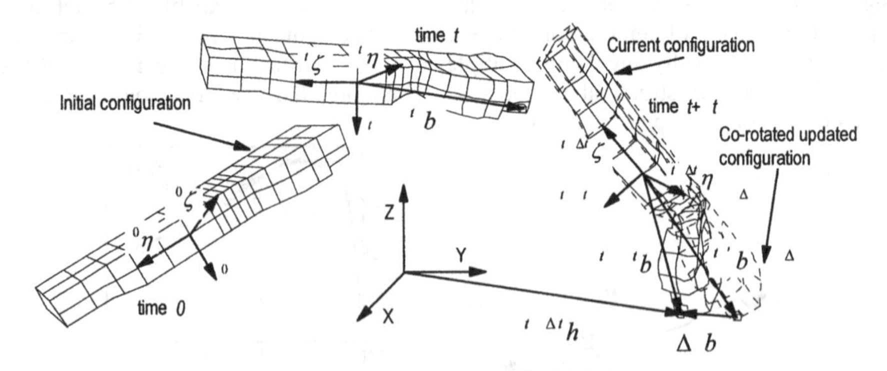

Figure 5. General motion of a flexible body

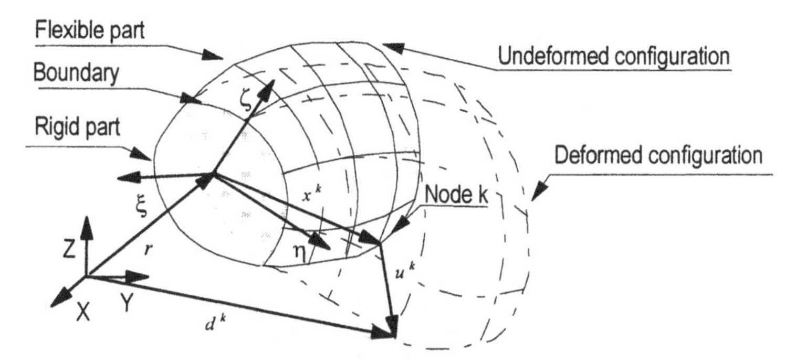

Figure 6. Flexible body with a rigid part

In Eq. (8) $\underline{\mathbf{M}}^*$ is a diagonal mass matrix containing the mass of the boundary nodes, and

$$\bar{\mathbf{A}}^T = \mathbf{A}\mathbf{A}\cdots\mathbf{A}^{\ T} ;\quad \mathbf{S} = \left\lfloor\right]\tilde{\mathbf{x}}_1' + \tilde{\boldsymbol{\delta}}_1'^{\ T}\ \tilde{\mathbf{x}}_2' + \tilde{\boldsymbol{\delta}}_2')^{\ T} \cdots\ \tilde{\mathbf{x}}_{\underline{n}}' + \tilde{\boldsymbol{\delta}}_{\underline{n}}')^T \Big]^T ;\quad \bar{\mathbf{I}} = \mathbf{I}\mathbf{I}\cdots\mathbf{I}]^T$$

Here $\mathbf{A}$ is the transformation matrix from the body fixed to global coordinate coordinates and $\mathbf{x}_k$ denotes the position of node k. Vectors $\mathbf{C}'$ and $\mathbf{C}$ are the reaction force and moment of the flexible part of the body over the rigid part respectively, given by

$$\mathbf{C}' = \mathbf{g}'_{\underline{\ }} - \mathbf{F}_{\underline{\ }} - \ \mathbf{K}_L + \mathbf{K}_{NL})_{\underline{\ }}\ \boldsymbol{\delta}' - \ \mathbf{K}_L + \mathbf{K}_{NL})_{\underline{\ }\theta}\ \boldsymbol{\theta}' \tag{9}$$

$$\mathbf{C}' = \mathbf{g}'_{\underline{\theta}} - \mathbf{F}_{\underline{\theta}} - \ \mathbf{K}_L + \mathbf{K}_{NL})_{\underline{\theta}}\ \boldsymbol{\delta}' - \ \mathbf{K}_L + \mathbf{K}_{NL})_{\underline{\theta}\theta}\ \boldsymbol{\theta}' \tag{10}$$

In these equations, the subscripts δ' and θ' refer to the partition of the vectors and matrices with respect to the translation and rotational nodal degrees of freedom. The underlined subscripts are referred to nodal displacements of the nodes fixed to the rigid part.

4. Contact Models

The description of any crash phenomena is strongly dependent on the contact/impact model used to describe the interaction between the system components. These contact models must not only be formulated in a form compatible with the multibody description used but also allow for the representation of the local deformations, friction forces, energy dissipation and still contribute for the stability of the numerical methods involved.

4.1. CONTACT DETECTION

Let the flexible body approach a surface during the motion of the multibody system, as represented in Fig. 7. Without lack of generality, let the impacting surface be described by a mesh of triangle patches. In particular, let the triangular patch, where node k of the flexible body will impact, be defined by points i, j and l. The normal to the outside surface of the contact patch is $\vec{n}\quad \vec{r}_{ij}\quad \vec{r}_{jl}$. $\times$

Let the position of the structural node k with respect to point i of the surface be

$$\mathbf{r}_{ik} = \mathbf{r}_k - \mathbf{r}_i \tag{11}$$

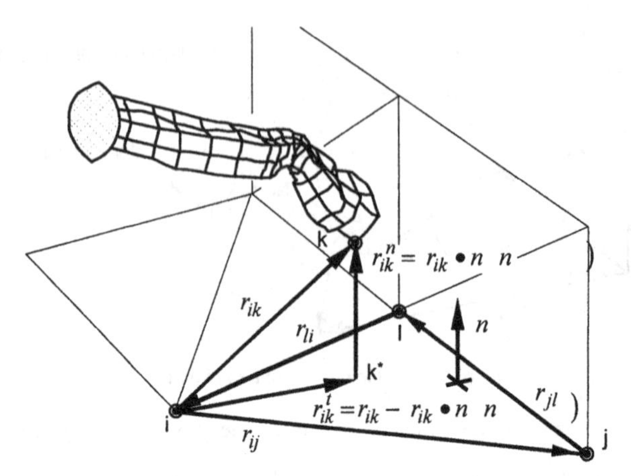

Figure 7. Contact detection between a finite element node and a surface

This vector is decomposed in its tangential component, which locates point k^* in the patch surface, and a normal component, given respectively by

$$\mathbf{r}_{ik}^{t} \quad \mathbf{r}_{ik} \quad \mathbf{r}_{ik}^{T}\,\mathbf{n}\big)\mathbf{n} \qquad - \tag{12}$$

$$\mathbf{r}_{ik}^{n} \quad \mathbf{r}_{ik}^{T}\,\mathbf{n}\big)\mathbf{n} \tag{13}$$

A necessary condition for contact is that node k penetrates the surface of the patch, i.e.

$$\mathbf{r}_{ik}^{T}\,\mathbf{n} \leq 0 \tag{14}$$

In order to ensure that a node does not penetrate the surface through its 'interior' face a thickness e must be associated to the patch. The thickness penetration condition is

$$-\mathbf{r}_{ik}^{T}\,\mathbf{n} \leq e \tag{15}$$

The condition described by equation (15) prevents that penetration is detected when the flexible body is far away, behind the contact surface. The remaining necessary conditions for contact results from the need for the node to be inside of the triangular patch, which are

$$\tilde{\mathbf{r}}_{ik}^{t}\mathbf{r}_{ij}\;^{T}\,\mathbf{n} \quad 0\big); \quad \tilde{\mathbf{r}}_{ik}^{t}\mathbf{r}_{jl}\big)^{T}\,\mathbf{n} \quad 0 \;\text{ and }\; \tilde{\mathbf{r}}_{ik}^{t}\mathbf{r}_{ki}\big)^{T}\,\mathbf{n} \geq 0 \tag{16}$$

Equations (14) through (16) are necessary conditions for contact. However, depending on the contact force model actually used, they may not be sufficient.

The contact detection algorithm is applicable to rigid body contact by using a rigid body point P instead of node k in equation (11). The global position of point P is given by $\mathbf{r}_i^P \quad \mathbf{r}_i + \mathbf{A}_i\mathbf{s}_i^P$, where $\mathbf{s}_i^P$ denotes the point location in body i frame.

4.2. CONTINUOUS FORCE MODEL

A model for the contact force must consider the material and geometric properties of the surfaces, contribute to a stable integration and account for some level of energy dissipation. Based on a Hertzian description of the contact forces between two solids, Lankarani and Nikravesh [19] propose a continuous force contact model that accounts for energy dissipation during impact.

Let the contact force between two bodies or a system component and an external object be a function of the pseudo-penetration and pseudo-velocity of penetration

$$\mathbf{f}_{s,i} \quad K \;^{n} + D \;\big)\; \mathbf{u} \tag{17}$$

where K is the equivalent stiffness, D is a damping coefficient and $\mathbf{u}$ is a unit vector normal to the impacting surfaces. The hysteresis dissipation is introduced in equation (17) by D , being the damping coefficient written as

$$D = \frac{3K \;1 \;\; e^2\big)}{4 \;\dot{}(\;)} \quad ^{n} \tag{18}$$

This coefficient is a function of the impact velocity $(\;)$, stiffness of the contacting surfaces and restitution coefficient e. The generalized stiffness coefficient K depends on the geometry material properties of the surfaces in contact. For the contact between a sphere and a flat surface the stiffness is [27]

$$K = 0.424\sqrt{r}\left(\frac{1-\;_{i}^{2}}{\pi E_i} + \frac{1-\;_{j}^{2}}{\pi E_j}\right)^{1} \qquad \Big) \tag{19}$$

where v_l and E_l are the Poisson's ratio and the Young's modulus associated to each surface and r is the radius of the impacting sphere.

The nonlinear contact force is obtained by substituting equation (18) into equation (17)

$$\mathbf{f}_{s,i} = K \quad ^{n}\left[1 + \frac{3 \;1-e^2\big)}{4}\frac{\dot{}}{\dot{}(\;)}\right]\mathbf{u} \tag{20}$$

This equation is valid for impact conditions, in which the contacting velocities are lower than the propagation speed of elastic waves, i.e., $\dot{}(\;) \leq 10^{\;5}\sqrt{\quad E/\rho}$.

4.3. FRICTION FORCES

The contact forces between the node and the surface include friction forces modeled using Coulomb friction. The dynamic friction forces in the presence of sliding are given by

$$\mathbf{f}^{friction} \;=\; -\mu_d\, f_d\left(\left|\mathbf{f}_k^{n}\right| / \left|\dot{\mathbf{q}}_k\right|\right)\dot{\mathbf{q}}_k \tag{21}$$

where μ_d is the dynamic friction coefficient, and $\dot{\mathbf{q}}_k$ is the velocity of node k. The dynamic correction coefficient f_d is expressed as

$$f_d \;=\; \begin{cases} 0 & \text{if} \quad \left|\dot{\mathbf{q}}_k\right| \leq v_0 \\ \left|\dot{\mathbf{q}}_k\right| - v_0 \;/\; v_1 - v_0\big) & \text{if} \quad v_0 \leq \left|\dot{\mathbf{q}}_k\right| \leq v_1 \\ 1 & \text{if} \quad \left|\dot{\mathbf{q}}_k\right| \geq v_1 \end{cases} \tag{22}$$

The dynamic correction factor prevents that the friction force changes direction for almost null values of the nodal tangential velocity, which is perceived by the integration algorithm as a dynamic response with high frequency contents, forcing it to reduce the time step size.

The friction model represented by equation (21) does not account for the adherence between the node and the contact surface. The interested reader is referred to the work of Wu, Yang and Haug [28] where stiction and sliding in multibody dynamics is discussed.

4.4. CONTACT MODELS USING UNILATERAL CONSTRAINTS

If contact between a node and a surface is detected, a kinematic constraint is imposed. Assuming fully plastic nodal contact, the normal components of the node k velocity and acceleration, with respect to the surface, are null during contact. Therefore the global nodal velocity and acceleration of node k, in the event of contact, become

$$\ddot{\mathbf{q}}_k = \ddot{\mathbf{q}}_k^{(\,)} - \ddot{\mathbf{q}}_k^{(\,)T}\mathbf{n}\big)\mathbf{n} \tag{23}$$

$$\dot{\mathbf{q}}_k = \dot{\mathbf{q}}_k^{(\,)} - \dot{\mathbf{q}}_k^{(\,)T}\mathbf{n}\big)\ \mathbf{n} \tag{24}$$

where $\dot{\mathbf{q}}_k^{(\,)}$ and $\ddot{\mathbf{q}}_k^{(\,)}$ represent the nodal velocity and acceleration immediately before impact.

The kinematic constraint implied by equations (23) and (24) is removed when the normal reaction force between the node and the surface becomes opposite to the surface normal. Representing by $\mathbf{f}_k$ the resultant of forces applied over node k, except for the surface reaction forces but including the internal structural forces due to the flexible body stiffness, the kinematic constraint over node k is removed when

$$\mathbf{f}_k^n = -\mathbf{f}_k^T\ \ \mathbf{n} > 0 \tag{25}$$

This model is not suitable to be used with rigid body contact. The sudden change of the rigid body velocity and acceleration would imply that the velocity and acceleration equations resulting from the kinematic constraints would not be fulfilled. Consequently constraint violations would develop throughout the integration of the state variables and the analysis would fail. The friction forces can still be included when using this contact model. However, as the normal contact reaction forces $\mathbf{f}_k^n$ are calculated using the Lagrange multipliers, the friction forces would be only approximate forces.

4.5. EXAMPLE OF IMPACT OF AN ELASTIC BEAM

The contact force models are exemplified with the oblique impact of a hyperelastic beam against a rigid wall. The impact scenario, proposed by Orden and Goicolea [29], is described in Fig 8. The beam, with a mass of 20 kg, length of 1 m and a circular cross-section, is made of a material with an elastic modulus of E=10^8 Pa, Poisson's ration of $\;$ = 0.27 and a mass density of ρ = 7850 kg/m^3. For this geometry and material properties the equivalent stiffness coefficient used in equation (20) is K=1.2 10^8 kg/m$^{2/3}$.

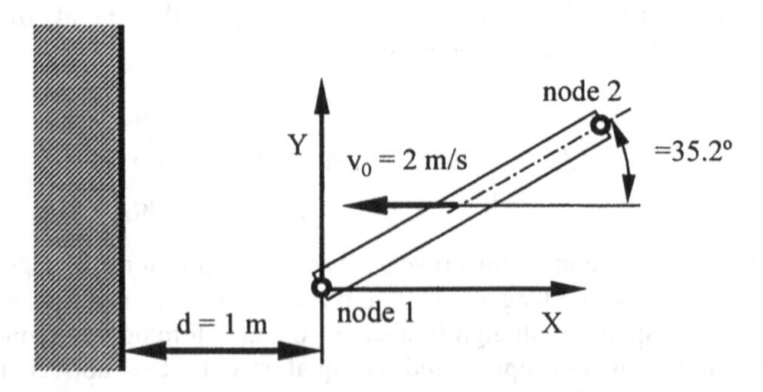

Figure 8. Impact scenario for an oblique elastic beam

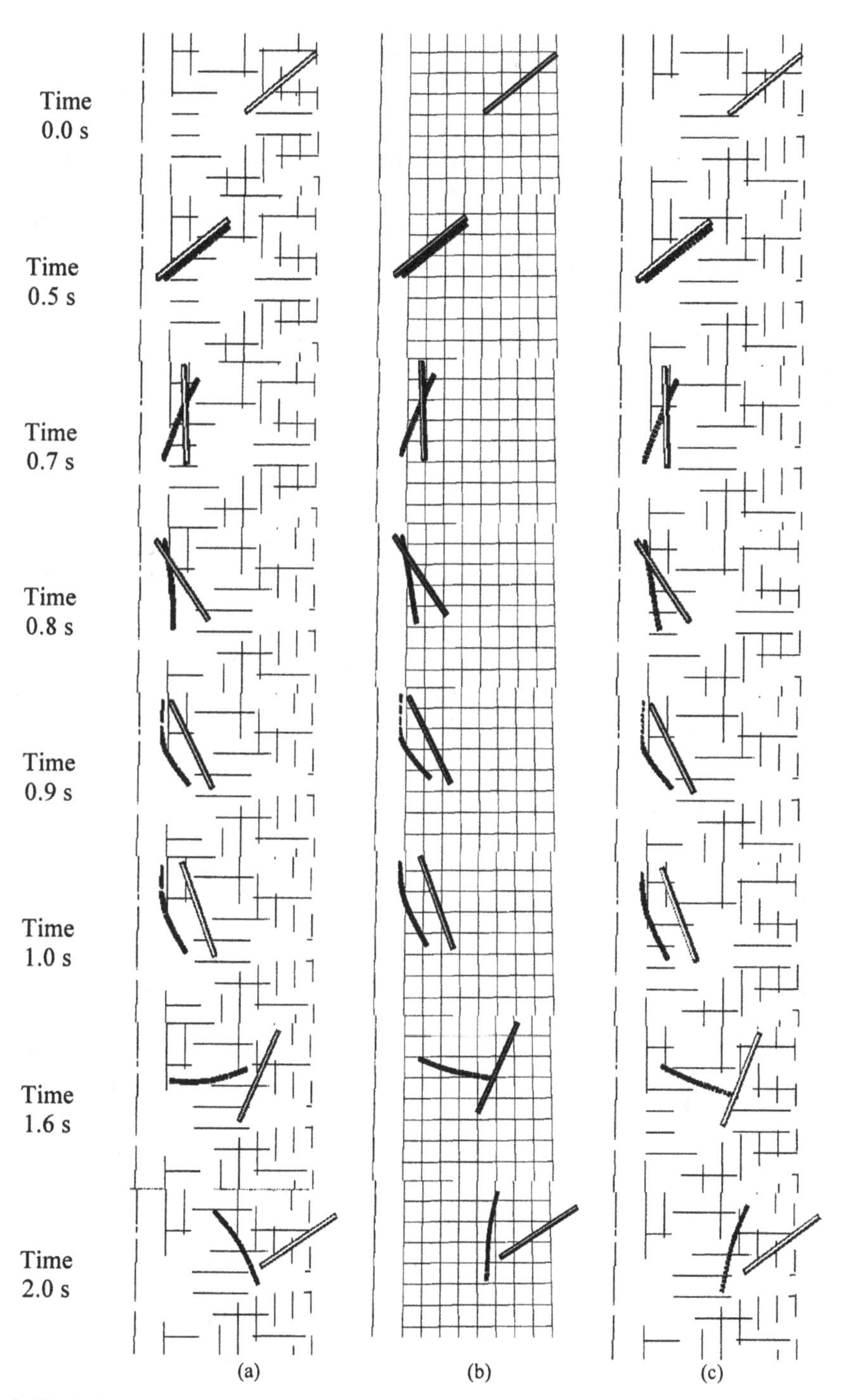

Figure 9. Frictionless impact of a hyperelastic beam in a rigid surface: (a) Beam model with 4 finite elements; (b) Model using 10 elements; (c) f.e.m. model with 20 elements

TABLE 1. Average time-step of the integration algorithm

Type of contact model	Beam model	Avg. time-step before contact	Avg. time-step during contact	Avg. time-step after contact
Continuous force	4 elements	$0.60\ 10^{-3}$	$0.20\ 10^{-3}$	$0.55\ 10^{-3}$
Continuous force	10 elements	$0.24\ 10^{-3}$	$0.15\ 10^{-3}$	$0.20\ 10^{-3}$
Continuous force	20 elements	$0.43\ 10^{-4}$	$0.34\ 10^{-4}$	$0.40\ 10^{-4}$
Kinematic constraint	4 elements	$0.60\ 10^{-3}$	$0.20\ 10^{-4}$	$0.55\ 10^{-3}$
Kinematic constraint	10 elements	$0.24\ 10^{-3}$	$0.74\ 10^{-5}$	$0.20\ 10^{-3}$

The motion of the impacting beam is shown in Fig. 9 together with that of a rigid bar with similar inertia. In Fig. 10, the contact forces developed during nodal impact are presented. The motion predicted for the beam model using 10 elements is similar to that presented by Orden and Goicolea [29] for a model made of 20 elements and using an energy-momentum formulation to describe contact.

A integration algorithm with variable time-step is used to integrate the equations of motion. The size of the integration time-step is controlled, during contact, by the time of travel of the elastic wave across one element, which is T_e=4.4 10^{-4} s for 20 elements [29]. Outside of contact, the integration time-step is controlled by the system response, generally associated to the flexible body higher frequencies. Table 1 shows the average time step taken by the algorithm in the various phases of the system motion.

Observing the contact forces, shown in Fig. 10, it is clear that the impact phenomena occurs with multiple contacts. Each of these contacts lasts 0.02 s, in average, which is similar to the contact duration of 0.018 s estimated by Orden and Goicolea [29] using the elastic wave travel time across the bar length.

5. Application To Train Impact

The application of the methodologies using only rigid multibody systems to train crashworthiness is exemplified by the study of a new anti-climber device for the interface between the cars of a train. This device, which ensures that train cars remain aligned during the crash, has been developed within the BRITE-EURAM project SAFETRAIN [30] and the analysis is described in detail by Milho, Ambrósio and Pereira in references [31,32].

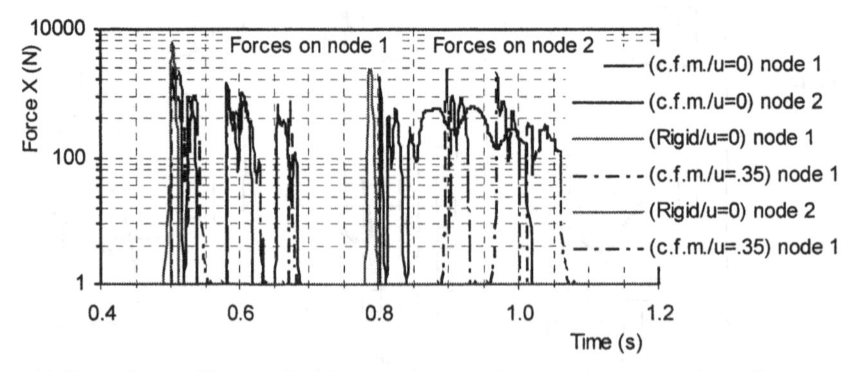

Figure 10. Forces developed between the (10 elements) beam end nodes and the rigid surface (c.f.m. – continuous force model; k.c. – kinematic constraint model)

5.1. RAILWAY CRASHWORTHINESS

The methodology developed here is applied to the simulation of a train collision. Table 2 presents the arrangement of a train set with eight individual car-bodies and it includes the length and the mass of each one of them while in Fig. 11 the topology of each individual car is shown. Five rigid bodies, B_1 through B_5, which represent the passenger compartment, boggie chassis and deformable end extremities, are used in the car-body model. The inertia and mechanical properties of the system components are described in reference [31].

HE LE LE LE LE LE LE LE HE

	1	2	3	4	5	6	7	8
Length (m)	20	26	26	26	26	26	26	26
Mass (10^3 Kg)	68	51	34	34	34	34	34	51

The high-energy zones (HE) are located in the extremities of the train set in the frontal zone of the motor car-body and in the opposing back zone, in the last car-body. The HE are potential impact extremities between two train sets. The low-energy zones (LE) are located in the remaining extremities of the train car-bodies and correspond to regions of contact between cars of the same train set.

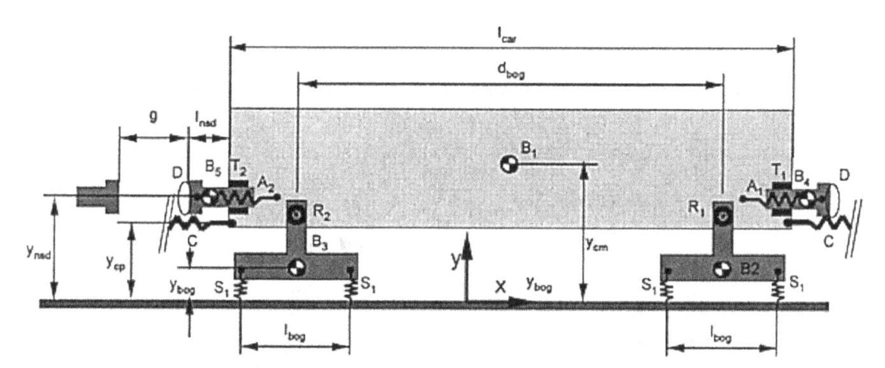

Figure 11. Car-body model for a single car

The objective is to identify, in the first design stage, the impact kinematics and the general forces developed during collision. In the second design stage the energy absorbing devices at the vehicle ends are defined. The first simulation scenarios are characterized by a moving train, with velocities of 30, 40 and 55 km/h, which collides with a parked train with brakes applied. Of importance to the anti-climber design are the simulation results for the contact forces and the relative displacements between car-bodies extremities [31].

The vertical relative displacement between the points of the contact surfaces defining the anti-climber devices is described by the distance g measured along the contact surface, see Fig. 12. This displacement, illustrated in Fig 13, is calculated when contact between the end extremities of the car-bodies occurs.

The tangential force in the anti-climber device, illustrated in Fig. 14, is described as the tangential component of the contact force between the end extremities of the car-bodies. It is observed that the tangential force at the interfaces, tend to increase both in magnitude and frequency in the final stage of the train impact.

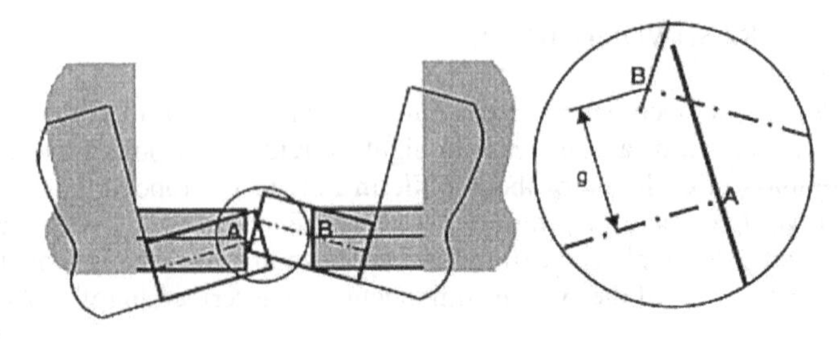

Figure 12. Anti-climber device contact geometry

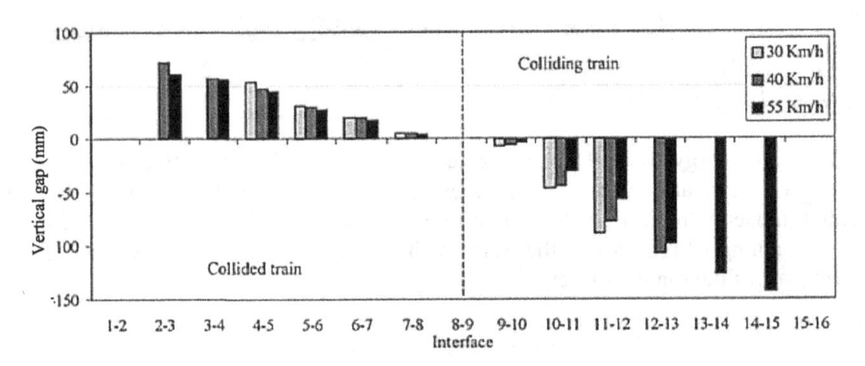

Figure 13. Vertical relative displacement between car-bodies in the interfaces

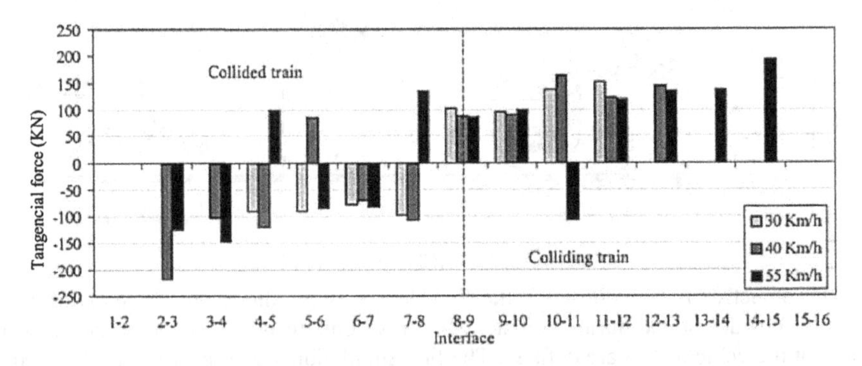

Figure 14. Maximum tangential force along the interfaces

In a second design stage the conceptual railway vehicle model presented herein is simulated in a train crash scenario similar to that of an experimental test performed to validate a low-energy end design developed within the framework of Brite/Euram III project SAFETRAIN [30]. The experimental test consists in having a vehicle moving with a velocity of 54 Km/h toward a composition with two vehicles stopped on the railroad, as depicted in Fig. 15. The two stopped vehicles are equipped with the low-energy ends and connected by a coupler device. Vehicle C is also equipped with a high-energy device in the colliding end. The vehicles were instrumented in order to measure accelerations, forces transmitted in the buffers and coupler and relative displacements between system components.

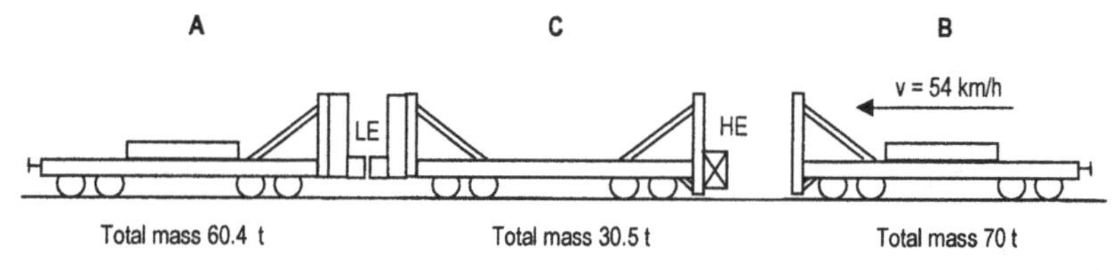

Figure 15. Collision scenario used in the numerical simulations and in the experimental test

The force time history for the buffers of wagon A is displayed in Fig. 16 for both the simulation and the experimental test. Note that the experimental test results are plotted for a single buffer while the expected force resulting from the simulation is shown for the cumulative effects of the two buffers of wagon A.

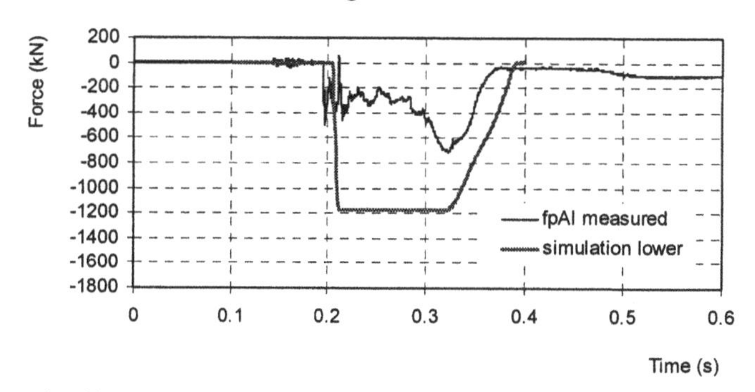

Figure 16. Force-time history of the buffers of wagon A (half of the force magnitude on the buffers is compared with the force measured in the left buffer of wagon A)

The force displacement history for the buffers of wagon A and wagon C are displayed in Fig. 17. There, it is observed that deformation predicted for the buffers of wagon C is a higher than that of wagon A. The buffer of wagon C is expected to be fully crushed.

The velocities of the three cars during the simulation are plotted in Fig. 18. There it can be observed that the velocities predicted by the model are very similar to those observed in the experimental test. The contact between wagon A and C is predicted to happen with no initial vertical gap between the buffers, as shown in Fig. 19. However, as the buffers approach each other, the vertical gap between the wagons increases, reaching a maximum of 15 mm. The vertical forces, that the buffers have to support in order to prevent overriding, presented in Fig. 19, oscillate with a maximum peak of 15 ton.

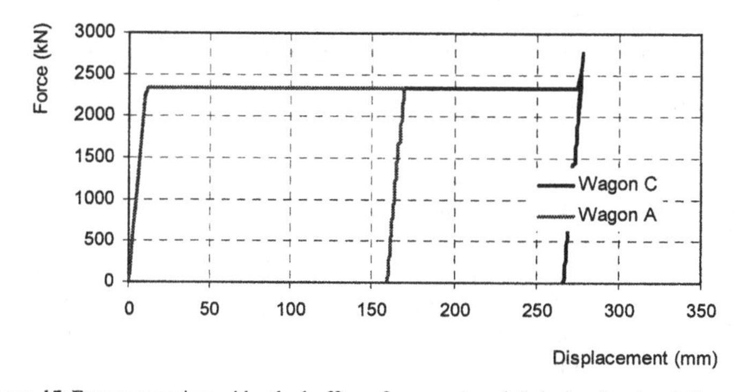

Figure 17. Forces experienced by the buffers of wagon A and C during the simulation

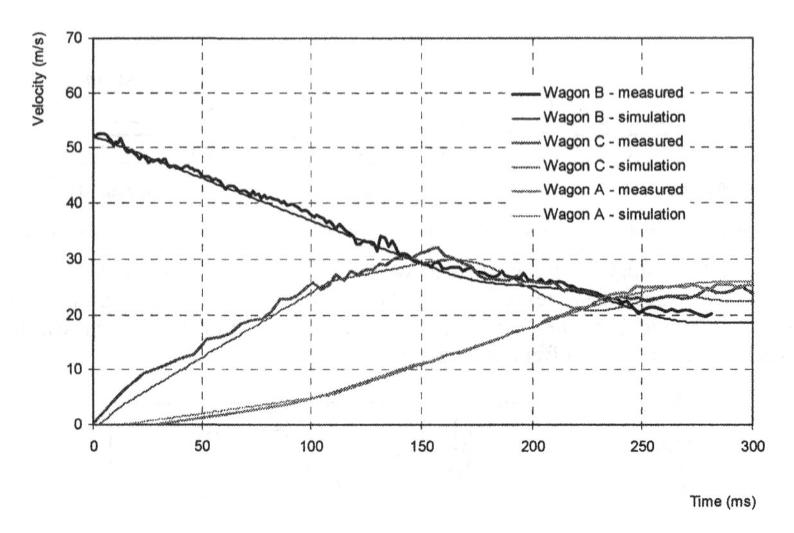

Figure 18. Time history for the wagons in the simulation and experimental test

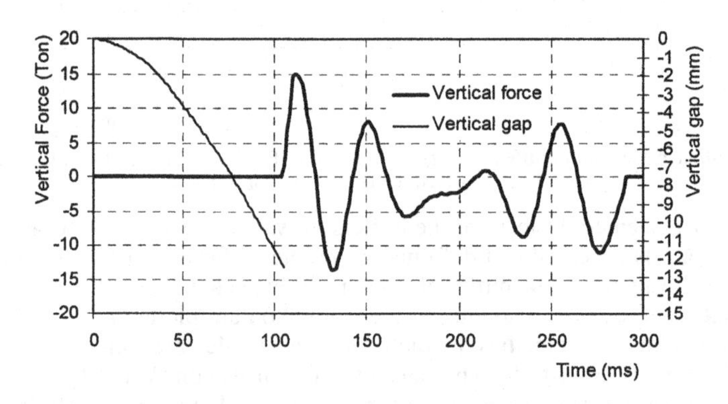

Figure 19. Time history of the gap and vertical contact forces in the buffers

The model presented for the preliminary design of the railway vehicle energy absorbing devices, is improved, based on the results obtained. The crushable structural elements mechanical characteristics are improved and the new force displacement curves implemented in the model. The procedure used is explained in detail in reference [32]

5.2. APPLICATION CASE : VEHICLE ROLLOVER WITH OCCUPANTS

The scenario in which the methodology described in this work is demonstrated is a rollover situation, depicted by Fig. 20, first proposed by Ambrósio, Nikravesh and Pereira [33]. This crash case is characterized by the existence of multiple impacts and by a complex interaction among the vehicle, occupants and ground that can hardly be represented by the more traditional approach of simulating the vehicle rollover first and then using the crash pulse as input for the occupants dynamic analysis. For more details on this application, the interested reader is referred to [34].

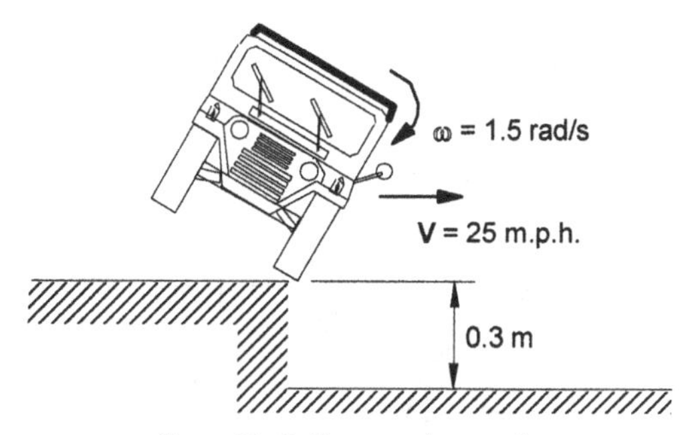

Figure 20. Rollover crash scenario.

5.2.1. *Biomechanical Model For The Vehicle Occupants*

For the representation of the vehicle occupants a suitable whole-body response biomechanical model is required. The biomechanical model, presented by Silva, Ambrósio and Pereira [35], includes 12 anatomic segments being the relative motion between adjacent bodies limited to be inside generalized cones of feasible motion. The model, pictured in Fig. 21, is general and accepts data for any individual.

In contact/impact simulations the relative kinematics of the head-neck and torso are important to the correct evaluation of the loads transmitted to the human body. Consequently, the head and neck are modeled as separate bodies and the torso is divided in two bodies. The hands and feet do not play a significant role in this type of problems and consequently are not modeled independently of the adjacent segments. In the biomechanical model no muscle activation is considered but the muscle passive behavior is represented by joint resistance torques. Physically unacceptable positions of the body segments are prevented by applying a set of penalty torques when adjacent segments of the biomechanical model reach the limit of their relative range of motion. The moment penalizing torque increases rapidly, from zero to a maximum value, when the two bodies interconnected by that joint, reach physically unacceptable positions. Details of this model can be found on reference [35].

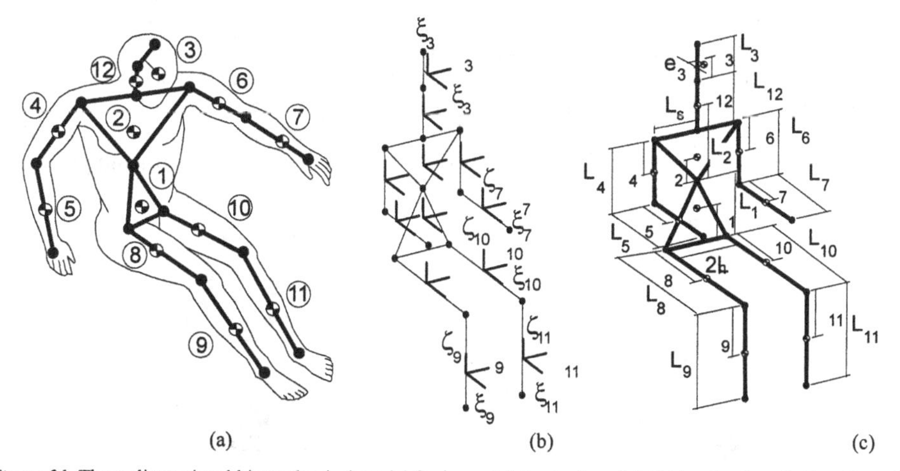

Figure 21. Three-dimensional biomechanical model for impact: (a) actual model; (b) local referential locations; (c) dimensions of the biomechanical segments.

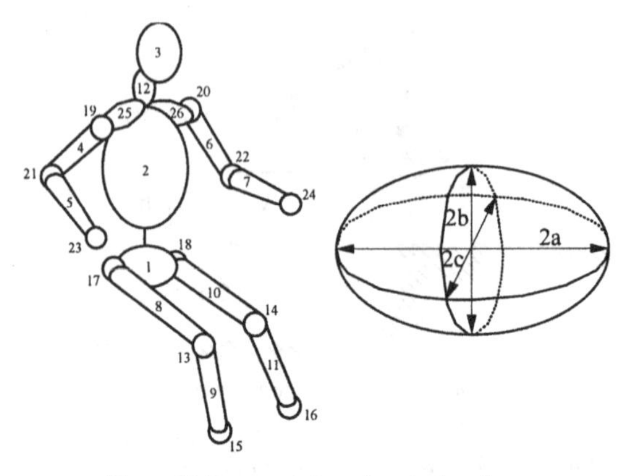

Figure 22. Representation of contact surfaces.

A set of contact surfaces is also defined for the calculation of the external forces exerted on the model when the surfaces of the bodies contact other objects or different body segments. These surfaces are ellipsoids and cylinders with the form depicted by Fig. 22. When contact between components of the biomechanical model is detected a contact force is applied in the point of contact and with a direction normal to the surface.

5.2.2. *Evaluation Of The Initial Positions Of The Vehicle Occupants*

The multibody dynamic analysis of the vehicle and occupant in a crash scenario requires that the initial positions and velocities of the system components are available. In order obtain these conditions for realistic occupant positions a process of recording the human body actual motion and to extract the position of its anatomical segments for every frame, designated by spatial reconstruction, is used here [36,37]. The most frequently used technique is photogrammetry, in which video cameras are used [37]. The laboratory apparatus of cameras is schematically represented in Fig. 23.

The images collected by a single camera are collections of two-dimensional information, resulting from the projection of a three-dimensional space into a two-dimensional one. Aziz and Karara [38], proposed a solution for the reconstruction process called Direct Linear Transformation. This method uses the two-dimensional information, collected by two or more cameras, to reconstruct the coordinates of the anatomical points.

The biomechanical model used in this work requires the reconstruction of the spatial position of the 23 anatomical points depicted in Fig. 24 for each frame of the analysis period. The spatial position and orientation of the anatomical segments of the biomechanical model are obtained from the spatial positions of these reconstructed points.

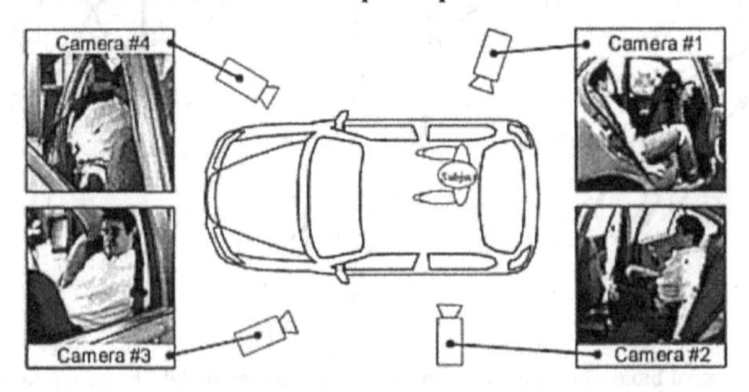

Figure 23. Vehicle and video cameras for the recording of the out-of-position occupants.

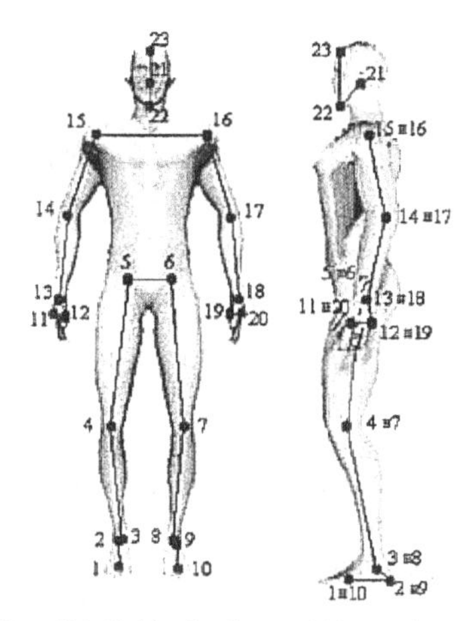

Figure 24. Set of 20 digitized points and kinematic structure.

With the setup described in Fig. 23 a seated occupant is asked to adopt positions similar to those that would be used when riding car. Among those, videotaped and reconstructed, the positions presented in Fig. 24 are used in the vehicle rollover.

5.2.3. Vehicle And Occupants Integrated Simulation

The vehicle rollover has been extensively analyzed with the purpose of studying the rollbar cage influence in the vehicle stability and its structural integrity. Here, the rollbar cage deformation is not included in the model though its deformations are implicitly described by the force contact model. Three 50%tile occupants, are also modeled and integrated with the vehicle. The two occupants in the front of the vehicle have shoulder and lap seat belts, described with the model by Laananen, Bolusbasi and Coltman [39], while the occupant seated in the back of the vehicle has no seatbelt. The initial positions of the occupants correspond to a normally seated driver, a front passenger that is bent and a rear occupant with a 'relaxed' position, all reconstructed and presented in Fig. 25. The vehicle and occupants are simulated here in a rollover situation described in Fig. 26.

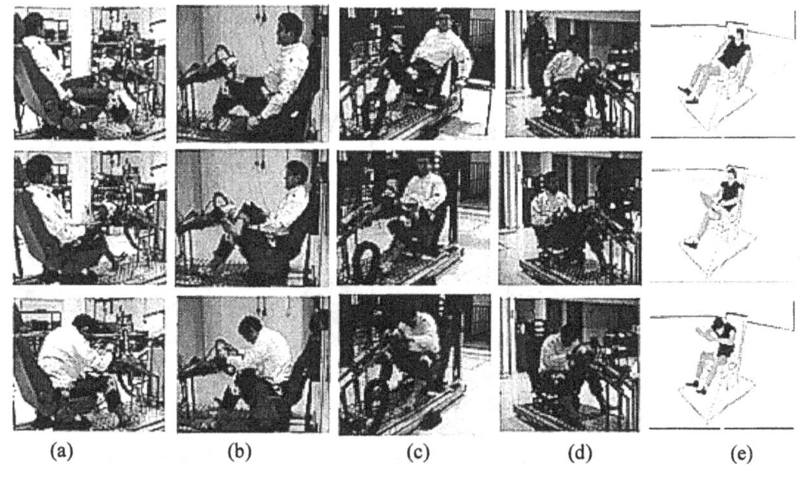

Figure 25. (a)-(d) Out-of-position occupants as viewed by the cameras and (e) spatial reconstructions.

Figure 26. Initial position of the vehicle and occupants for the rollover

The results of this simulation are pictured in Fig. 27, where several frames of the animation of the vehicle rollover with occupants are presented. It is noticeable in this sequence that the vehicle first impacts the ground with its left tires. At this point the rear occupant is ejected. The rollover motion of the vehicle proceeds with an increasing angular velocity, mainly due to the ground - tire contact friction forces. The occupants in the front of the vehicle are hold in place by the seat belts. Upon continuing its roll motion, the vehicle impacts the ground with its rollbar cage, while the ejection of the rear occupant is complete. Bouncing from the inverted position the vehicle completes another half turn and impacts the ground with the tires again.

The Severity Index, plotted in Fig. 28, indicates a very high probability of fatal injuries for the occupants under the conditions simulated. Notice that the model has rigid seats, rigid interior trimming for the dashboard, side and floor panels, and that the ground is also considered to be rigid. It is expected that the head accelerations are somehow lower if some compliance is included in the seats and side panels of the vehicle interior.

5.3. SPORTS VEHICLE MULTIBODY MODEL

The formulation presented here is applied to a replica of the original Lancia Stratos, shown in Fig. 29, in different frontal and oblique collision scenarios. The multibody model, described by rigid bodies and spring-damper elements, includes the suspension elements, steering system and chassis. The front crash-box, modeled by nonlinear finite elements, is also included [40].

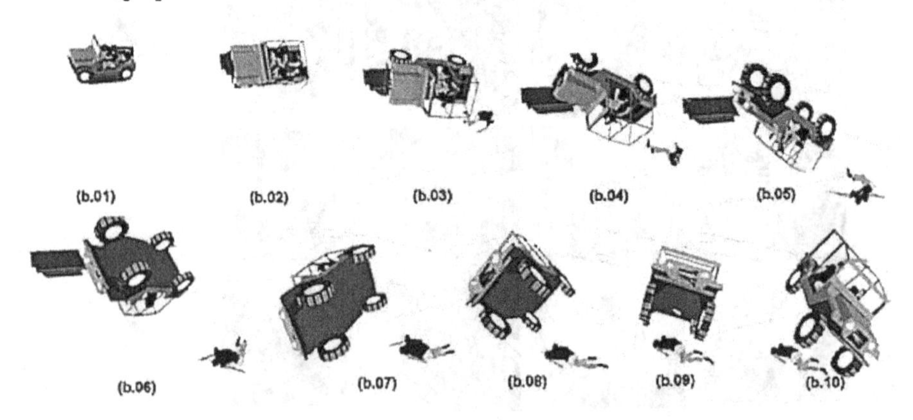

Figure 27. Front view of the rollover simulation of a vehicle with three occupants

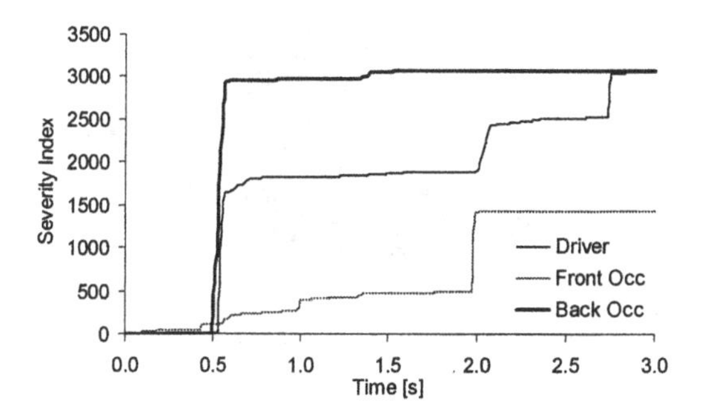

Figure 28. Severity Index for the vehicle occupants

Figure 29. Prototype of the sports car

The multibody model of the vehicle is composed of 16 rigid bodies. The system components include the front double A-arm suspension system, the rear McPherson suspension system, wheels and chassis as depicted by Fig. 30. To model the tire interaction with the ground an analytical tire model with comprehensive slip is used [41].

5.3.1. *Sports Car Crash-Box Model*

The sports car front crash box is depicted in Fig. 31. For the purpose of the frontal and oblique impacts at moderate velocities, the model for the chassis is considered rigid except for the front crash-box that is flexible.

The complete system has 15 degrees of freedom associated with the rigid bodies and 186 nodal degrees of freedom. This model is valid if plastic deformations are to occur in the parts of the vehicle modeled as flexible regions and no significant deformation takes place in the passenger compartment.

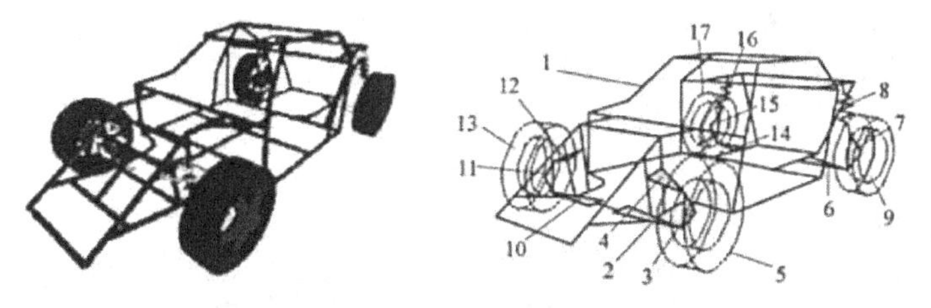

Figure 30. Multibody model of the sports vehicle

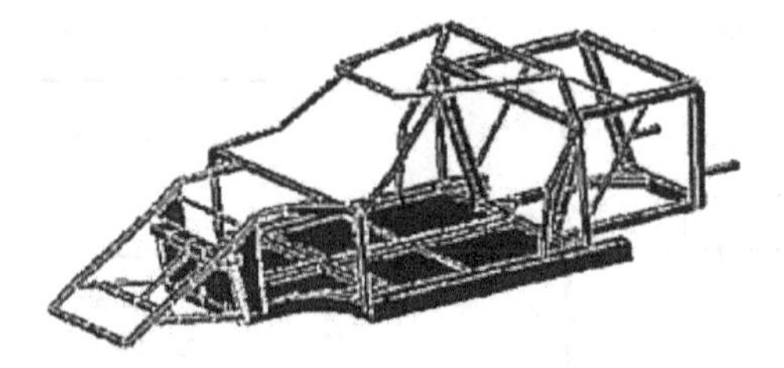

Figure 31. Model of sports car chassis with a nonlinear flexible front crash-box

5.3.2. Vehicle Frontal And Oblique Impact Simulations

The sports vehicle with the front crash-box is analyzed for various impact scenarios, represented in Fig. 32, where the angle of impact and the topology of the road are different. For the crash scenarios the contact models are applied and the existence of friction forces is considered. The simulations are carried until the vehicle reaches a full stop.

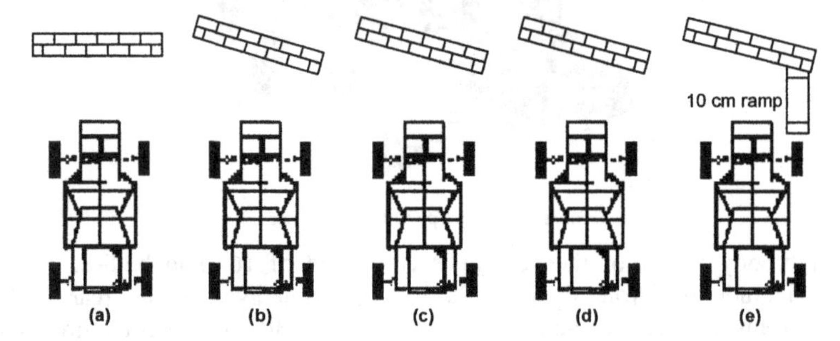

Figure 32. Impact with a) surface perpendicular to vehicle heading; b) 10° oblique surface, no friction; c) 20° oblique surface, no friction; d) 20° oblique surface, friction; e) Same as (d) but the vehicle goes over a 10 cm ramp.

The vehicle motion for different impact scenarios is presented in Fig. 33. For oblique contact with the obstacle there is a slight rotation of the vehicle during impact. This rotation is more visible in the case of frictionless impact. At the simulated impact speed the influence of the car suspension elements over the deformation mechanism is minimal. However, the suspension system plays an important role on the variation of the relative orientation between vehicle and obstacle before and after the crash.

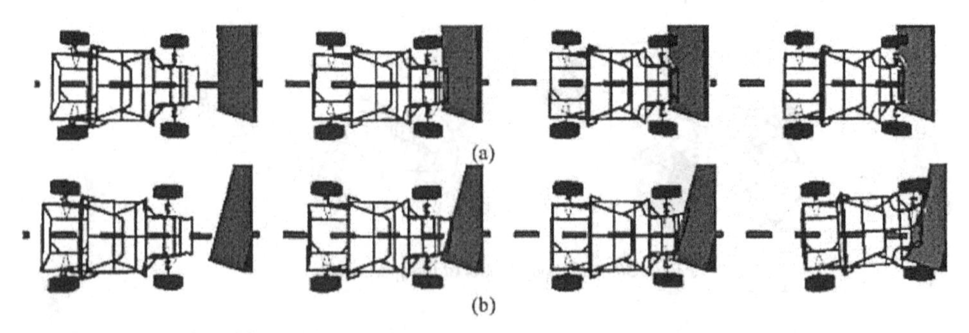

Figure 33. Motion of the vehicle for (a) Frontal impact; (b) 20° Oblique impact without contact friction

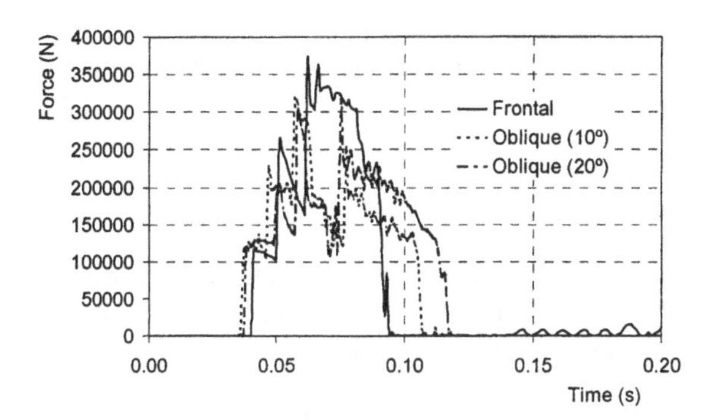

Figure 34. Forces developed between the front of the vehicle and impact surface

The efficiency of the crash-box to dissipate the kinetic energy of the vehicle decreases with the increase of the angle value between surface normal and vehicle heading. It is observed that for the frontal crash the front structure dissipates all kinetic energy. However, for oblique impacts with no surface friction, only part of that energy is absorbed by the crash-box. The vehicle motion is deflected, and would continue if no other component of the car entered in contact with the surface. In Fig. 34 the forces developed between vehicle and surface are plotted.

For the crash scenarios where friction is modeled, the deflection of the vehicle motion does not occur, enabling the structure to deform with a crushing mechanism similar to that of the frontal impact. Only a slight sideways translation of the vehicle is observed. This result clearly emphasizes the importance of a correct model for the friction forces developed during contact.

6. Conclusions

Deformation descriptions can be included in multibody system components for application in contact and impact situations. The deformations may be localized in a small region of the body, neighboring the point of contact, or they may occur in parts of the system away from contact. Therefore, the potential system deformation may be described by the contact model used, if these are localized at the contact point, or they may have to be included in the multibody formulation. Two representations of the system deformations have been proposed here.

The plastic hinge technique can be used when the deformation mechanisms are well known beforehand. Therefore, it is possible to model the structural components as a multibody sub-system having the kinematic joints located in the regions where the plastic hinges are expected. The second model reviewed uses nonlinear finite elements to represent the distributed deformations of the flexible body. This approach has the advantage of not requiring any particular knowledge of the deformation mechanisms beforehand but it is computationally more expensive.

The contact models used may also represent to a certain degree the structural deformations in the bodies in contact. A continuous force model has been described for both flexible and rigid multibody systems impact while a contact model based in unilateral constraints has been proposed for flexible multibody systems. It was shown that in the case of flexible multibody contact both models yield similar results, provided that the equivalent stiffness used for the continuous force model is based in the Hertzian elastic contact theory.

Through the application to vehicles and biomechanical impacts in diverse scenarios it was possible to appreciate the range of application of the different methodologies proposed and their combination.

Acknowledgements

The work reported was partially supported by Fundação para a Ciência e Tecnologia through the project PRAXIS/P/EME 14040/98, entitled *Human Locomotion Biomechanics Using Advanced Mathematical Models and Optimization Procedures.* Many co-workers contributed to material overviewed in this work. Among these, the author wants to acknowledge the contributions of Mr. Miguel Silva, Mr. João Gonçalves, Mr. João Milho, Prof. Manuel Seabra Pereira and Prof. João Abrantes.

7. References

1. Nikravesh, P., Chung, I. and Benedict, R. (1983) Plastic hinge approach to vehicle simulation using a plastic hinge technique, *J. Comp. Struct.* **16**, 385-400.
2. Ambrósio, J., Pereira, M. and Dias, J., (1996) Distributed and discrete nonlinear deformations on multibody systems, *Nonlinear Dynamics* **10**(4), 359-379.
3. Dias, J. and Pereira, M. (1997) Sensitivity analysis of rigid-flexible multibody systems, *Multibody System Dynamics* **1**(3), 303-322.
4. Song, J.O. and Haug, E.J. (1980) Dynamic analysis of planar flexible mechanisms, *Computer Methods in Applied Mechanics and Engineering*, **24**, 359-381.
5. Shabana, A.A. (1989) *Dynamics of Multibody Systems*, John Wiley & Sons, New York, New York.
6. Shabana A.A. and Wehage, R.A. (1989) A coordinate reduction technique for transient Analysis of spatial structures with large angular rotations, *Journal of Structural Mechanics* **11**, 401-431.
7. Yoo W.S.and Haug E.J., (1986) Dynamics of flexible mechanical systems using vibration and static correction modes, *ASME J. of Mechanisms, Transmissions and Automation in Design*, **108**, 315-322.
8. Modi, V.J. Suleman A. and Ng, A.C. (1991) An approach to dynamics and control of orbiting flexible structures, *Int. J. Nume. Methods Engng.* **32**, 1727-1748.
9. Banerjee, A.K. and Nagarajan, S. (1996) Efficient simulation of large overall motion of nonlinearly elastic beams, in Proceedings of ESA International Workshop on Advanced Mathematical Methods in the Dynamics of Flexible Bodies, ESA, Noordwijk, The Netherlands, June 3-5, pp. 3-23.
10. Kane, T., Ryan, R. and Banerjee, A. (1987) Dynamics of a cantilever beam attached to a moving base, *AIAA J of Guidance, Control and Dynamics* **10**, 139-151.
11. Geradin, M. (1996) Advanced methods in flexible multibody dynamics: review of element formulations and reduction methods, in Proceedings of ESA International Workshop on Advanced Mathematical Methods in the Dynamics of Flexible Bodies, ESA, Noordwijk, The Netherlands, June 3-5, pp. 83-106.
12. Cardona, A. and Geradin, M. (1988) A beam finite element non linear theory with finite rotations, *Int. J. Nume Methods in Engng.* **26**, 2403-2438.
13. Geradin, M. and Cardona, A. (1991)A Modelling of superelements in mechanism analysis, *Int. J. Nume Methods in Engng.* **32**, 1565-1594.
14. Shabana, A. (1997) Definition of the slopes and the finite element absolute nodal coordinate formulation, *Multibody System Dynamics* **1**, 339-348.
15. Ambrósio, J. and Nikravesh, P. (1992) Elastic-plastic deformations in multibody dynamics, *Nonlinear Dynamics* **3**, 85-104.
16. Ambrósio, J. and Seabra Pereira, M. (1994) Flexibility in multibody dynamics with applications to crashworthiness, In: M. Pereira and J. Ambrósio (ed.), Computer Aided Analysis of Rigid and Flexible Multibody Systems. Kluwer, Netherlands, pp. 199-232.
17. Pfeiffer, F., Glocker, C. (1996) *Multibody Dynamics With Unilateral Constraints*, John Wiley and Sons, New York, New York.
18. Martins, J.A.C., Barbarin, S., Raous, M., Pinto da Costa, A. (1999) Dynamic stability of finite dimensional linearly elastic systems with unilateral contact and Coulomb friction, *Comp. Meth. Appl. Mech. Engng.*, **177** (3-4), 289-328.

19. Lankarani H. M. and Nikravesh P. E. (1994) Continuous Contact Force Models for Impact Analysis in Multibody Systems, *Nonlinear Dynamics*, **5**, 193-207.
20. Nikravesh P. E. (1988) *Computer Aided Analysis Of Mechanical Systems*, Englewood Cliffs, New Jersey, Prentice-Hall.
21. Murray, N. (1983) The static approach to plastic collapse and energy dissipation in some thin-walled steel structures, In N. Jones & T. Wierzbicki Eds., *Structural Crashworthiness*, Butterworths: London, England, pp. 44-65.
22. Kindervater, C. (1997) Aircraft and helicopter crashworthiness: design and simulation, In J. Ambrósio, M. Pereira and F. Silva eds, *Crashworthiness Of Transportation Systems: Structural Impact And Occupant Protection*. NATO ASI Series E. V 332, Kluwer Academic Publishers, Dordrecht, Netherlands, pp. 525-577.
23. Matolcsy, M. (1997) Crashworthiness of bus structures and rollover protection. In J. Ambrósio, M. Pereira and F. Silva eds, *Crashworthiness Of Transportation Systems: Structural Impact And Occupant Protection*, NATO ASI Series E. Vol. 332, Kluwer Academic Publishers: Dordrecht, Netherlands, pp. 321-360.
24. Kecman, D. (1983) Bending collapse of rectangular and square section tubes, *Int. J. of Mech. Sci*, **25**(9-10), 623-636.
25. Winmer, A. (1977) Einfluß der belastungsgeschwindigkeit auf das festigkeits- und verformungsverhalten am beispiel von kraftfarhzeugen, *ATZ*, **77**(10), 281-286.
26. Ambrósio, J.A.C. and Nikravesh, P. (1992) Elastic-plastic deformations in multibody dynamics, *Nonlinear Dynamics*, **3**, 85-104.
27. Lankarani, H., Ma, D. and Menon, R., (1995) Impact dynamics of multibody mechanical systems and application to crash responses of aircraft occupant/structure, In M. Pereira and J. Ambrósio eds., *Computer Aided Analysis Of Rigid And Flexible Mechanical Systems*, NATO ASI Series E. Vol. 268, Kluwer Academic Publishers: Dordrecht, Netherlands, pp. 239-265.
28. Wu, S., Yang, S. and Haug, E., (1984) *Dynamics of Mechanical Systems with Coulomb Friction, Stiction, Impact and Constraint Addition-Deletion*, Technical Report 84-19, Center for Computer Aided Design, University of Iowa, Iowa City, Iowa.
29. Orden, J.C.G. and Goicolea, J.M. (2000) Conserving properties in constrained dynamics of flexible multibody systems, *Multibody System Dynamics*, **4**, 221-240.
30. SAFETRAIN, (2001) *Dynamic Tests*, BRITE/EURAM Project BE-309, SAFETRAIN Technical Report T8.2-F, Deutsche Bann, Berlin, Germany.
31. Milho, J., Ambrósio, J., and Pereira, M. (2002) A multibody methodology for the design of anti-climber devices for train crashworthiness simulation, *International Journal of Crashworthiness*, **7**(1), 7-20.
32. Milho, J., Ambrósio, J. and Pereira, M. (2002) Validated multibody model for train crash analysis, In E. Chirwa ed., *Proceedings of ICRASH2002 International Crashworthiness Conference*, Melbourne, Australia, February 25-27.
33. Ambrósio, J.A.C., Nikravesh, P.E. and Pereira, M.S. (1990) Crashworthiness Analysis of a Truck, *Mathematical Computer Modelling*, **14**, 959-964.
34. Silva, M.P.T. and Ambrósio, J.A.C. (2002) Out-of-position vehicle occupants models in a multibody integrated simulation environment, In *Proceedings of the 2002 IRCOBI Conference on the Biomechanics of Impact*, Munich, Germany, September 18-20.
35. Silva, M., Ambrósio, J. and Pereira, M., (1997) Biomechanical Model with Joint Resistance for Impact Simulation, *Multibody System Dynamics*, **1**(1), 65-84.
36. Allard, P., Stokes, I. and Blanchi, J. (1995) *Three-Dimensional Analysis of Human Movement*, Human Kinetics, Urbana-Champagne, Illinois.
37. Nigg, B. and Herzog, W. (1999) *Biomechanics of the Musculo-skeletal System*, John Wiley & Sons, New York.
38. Addel-Aziz, Y. and Karara, H., (1971) Direct linear transformation from comparator coordinates into object space coordinates in close-range photogrammetry, *Proceedings of the Symposium on Close-range Photogrammetry*, Falls Church, Virginia, pp. 1-18.
39. Laananen, D.H., Bolukbasi, A.O. and Coltman, J.W. (1983) *Computer Simulation of an Aircraft Seat and Occupant in a Crash Environment: Volume 1*, Technical Report DOT/FAA/CT-82/33-I, US Department of Transportation, Federal Aviation Administration, Atlantic City, New Jersey.
40. Ambrósio, J.A.C. and Gonçalves, J.P.C. (2001) Vehicle crashworthiness design and analysis by means of nonlinear flexible multibody dynamics, *Int. J. of Vehicle Design*, **26**(4), 309-330.
41. Gim G. (1988) *Vehicle Dynamic Simulation With a Comprehensive Model for Pneumatic Tires*, Ph.D. Dissertation, University of Arizona, Tucson, Arizona.

MODEL REDUCTION TECHNIQUES IN FLEXIBLE MULTIBODY DYNAMICS

PARVIZ E. NIKRAVESH
Department of Aerospace and Mechanical Engineering
University of Arizona, Tucson, AZ 85721 U.S.A.
pen@email.arizona.edu

In most applications of multibody dynamics, it is desirable to reduce the number of equations of motion in order to improve computational efficiency. This issue becomes even more important when the model contains deformable bodies. Therefore it would be desirable to reduce the number of the deformable body degrees-of-freedom without too much loss in the accuracy of the results. In this paper we present the equations of motion for a simple but general rigid-flexible multibody system. The equations are presented in a semi-abstract form in order to keep our focus on reduction techniques. We first discuss three different methods to attach a body frame to a moving deformable body. Then several model reduction techniques are reviewed and the advantages and disadvantages of each technique are briefly discussed.

1. Introduction

One interesting issue in multibody dynamics is computational efficiency and real-time simulation. This issue becomes even more serious when a multibody model contains deformable bodies. Although a deformable body may represent only small linear deflections, the overall multibody equations must be treated as a nonlinear system. Therefore, it is highly desirable to reduce the number of degrees-of-freedom in such a system.

Before we discuss model reduction techniques, we review another interesting topic in flexible multibody dynamics—how to attach a reference frame to a moving deformable body. In literature this is known as the floating reference frame. The most common practice is to attach the reference frame to a few nodes as it is a common practice in structural finite element modeling. In addition to this method we discuss the use of another relatively well-known method called the mean-axis conditions. We also discuss a completely new method that ensures the reference frame to be the principal-axes of the deformable body.

In the past decades researchers have proposed a variety of model reduction techniques for deformable bodies that undergo rigid-body motions. Each technique is based on certain assumptions and, consequently, approximation error is introduced into the model. Therefore, it is important to understand the differences among these methods and apply them where appropriate. In this paper we review several model reduction methods.

It is assumed that the reader has some familiarity with the equations of motion for a deformable body. These equations have been stated in this paper without proof. In order to concentrate on concepts and not to be distracted by the complex form of these equations, we assume that a node in a finite-element model exhibits only translational degrees-of-freedom. Elimination of the rotational degrees-of-freedom does not make the discussion any less general. The rotational degrees-of-freedom can be added to these formulations if necessary. Most of the topics discussed in this paper could be referenced to more than one source. However, in order to save space the list of references has been kept to a minimum.

2. Notation

In this paper matrix notation is used in order to keep our attention on concepts without any loss of details. The reader should find the notation to be effective in multibody formulation of the equations of motion. The following nomenclature is used:

W. Schiehlen and M. Valášek (eds.), Virtual Nonlinear Multibody Systems, 83–102.

Vectors and arrays:
Lower-case bold-face characters
Matrices:
Upper-case bold-face characters
Scalars:
Lower-case light-face characters
Right superscripts:
- a nodal-fixed *axes*
- b boundary node
- d deleted (truncated) mode
- k kept mode
- m master (kept) node
- s slave (deleted) node
- u unconstrained (free) node

Right subscripts:
- c constrained
- f free-free (unconstrained)
- G Guyan
- r index of a rigid body

Left superscripts:
- $`$ (prime) components are described in the body frame

Over-scores:
- $\sim$ (tilde) transforms a 3-vector to a skew-symmetric matrix
- $\wedge$ (hat) stacks vertically 3-vectors or 3×3 skew-symmetric matrices
- $-$ (bar) repeats a 3×3 matrix to form a diagonal or block-diagonal matrix

The right superscripts and subscripts are self-explanatory. The left superscript "prime" indicates that an entity is described in terms of its body components. Therefore, an entity without a "prime" left superscript refers to the global components. Assume that $\mathbf{b}^i$ is a 3-vectorand $\tilde{\mathbf{b}}^i$ is a 3×3 skew-symmetric matrix for $i=1, \ldots, n$. Further assume that $\mathbf{I}$ is a 3×3 identity matrix and $\mathbf{A}$ is a 3×3 rotational transformation matrix. The following examples show how we construct an array from n 3-vectors and how we use over-scores "^" and "$-$":

$$\mathbf{b} = \begin{Bmatrix} \mathbf{b}^1 \\ \vdots \\ \mathbf{b}^n \end{Bmatrix}, \quad \hat{\tilde{\mathbf{b}}} = \begin{bmatrix} \tilde{\mathbf{b}}^1 \\ \vdots \\ \tilde{\mathbf{b}}^n \end{bmatrix}, \quad \hat{\mathbf{I}} = \begin{bmatrix} \mathbf{I} \\ \vdots \\ \mathbf{I} \end{bmatrix}, \quad \bar{\mathbf{I}} = \begin{bmatrix} \mathbf{I} & & \\ & \ddots & \\ & & \mathbf{I} \end{bmatrix}, \quad \bar{\mathbf{A}} = \begin{bmatrix} \mathbf{A} & & \\ & \ddots & \\ & & \mathbf{A} \end{bmatrix}$$

3. Equations of Motion for A Deformable Body

In this section we state the equations of motion for a deformable body without considering how a reference frame is attached to that body. The equations of motion are initially stated for a structure without any rigid-body motion. Then, the structure is allowed to deform and undergo gross translation and rotation.

3.1. EQUATIONS OF MOTION IN NODAL SPACE

Assume that a deformable body is described by n_{nodes} nodes. The positions of the nodes are defined with respect to a non-moving body-frame $\xi - \eta - \zeta$ as shown in Fig. 1. The mass and stiffness matrices, $`\mathbf{M}$ and $`\mathbf{K}$, are $n_{dof} \times n_{dof}$ each where $n_{dof} = 3 \times n_{nodes}$. The equations of motion for this body are written as

$$`\mathbf{M}\,`\ddot{\delta} = `\mathbf{f} - `\mathbf{K}\,`\delta \tag{3.1}$$

where $`\mathbf{f}$ and $`\delta$ are each n_{dof} arrays containing external forces and nodal translational deflections respectively. We are reminded that the left superscript "prime" indicates that a vector, an array, or a matrix is described in terms of its components in the $\xi-\eta-\zeta$ frame. Since we have not yet imposed any boundary conditions, the stiffness matrix $`\mathbf{K}$ has a rank deficiency of six.

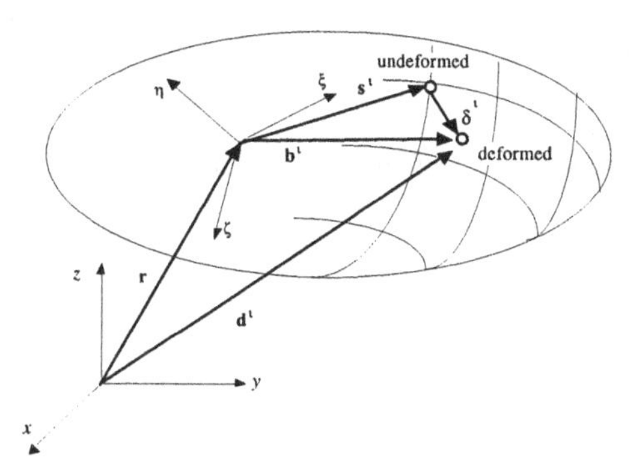

Figure 1. A deformable body and its body and global reference frames.

We introduce a second non-moving coordinate frame, x-y-z, that does not coincide with $\xi-\eta-\zeta$. A 3×3 rotational transformation matrix, $\mathbf{A}$, between the two non-moving frames is defined. Components of any vector in the $\xi-\eta-\zeta$ frame can be transformed to the x-y-z frame; e.g., $\delta^i = \mathbf{A}\ `\delta^i$. Using the block-diagonal matrix $\overline{\mathbf{A}}$, and the transformations $`\delta = \overline{\mathbf{A}}^{\mathrm{T}}\delta$ and $`\ddot{\delta} = \overline{\mathbf{A}}^{\mathrm{T}}\ddot{\delta}$, Eq. 3.1 is expressed in terms of the x-y-z components of its entities as

$$\mathbf{M}\ddot{\delta} = \mathbf{f} - \mathbf{K}\delta \tag{3.2}$$

where,

$$\mathbf{M} = \overline{\mathbf{A}}\ `\mathbf{M}\,\overline{\mathbf{A}}^{\mathrm{T}},\quad \mathbf{K} = \overline{\mathbf{A}}\ `\mathbf{K}\,\overline{\mathbf{A}}^{\mathrm{T}},\quad \mathbf{f} = \overline{\mathbf{A}}\ `\mathbf{f} \tag{3.3}$$

Most mass matrices have the characteristics to yield $\mathbf{M} = `\mathbf{M}$, but in general this may not be the case[1]. Equations 3.1 or 3.2 are valid only if there is no rigid body motion; i.e., the nodal accelerations are only due to structural deformation.

Now we allow the flexible body and its $\xi-\eta-\zeta$ frame to move with respect to the x-y-z frame. For a flexible body with n_{nodes} nodes, as shown in Fig. 1, we can write n_{dof} equations in n_{dof} accelerations as

$$\mathbf{M}\ddot{\mathbf{d}} = \mathbf{f} - \mathbf{K}\delta \tag{3.4}$$

The array of accelerations, $\ddot{\mathbf{d}}$, contains the absolute nodal accelerations described with respect to the x-y-z frame. Hence, the external and structural forces are also expressed in terms of their components in the same frame. The mass matrices and the force arrays in Eqs. 3.2 and 3.4 are the same—the only difference is in the acceleration arrays.

The absolute position of a typical node i, as shown in Fig. 1, is expressed as:

$$\mathbf{d}^i = \mathbf{r} + \mathbf{b}^i = \mathbf{r} + \mathbf{s}^i + \delta^i \tag{3.5}$$

[1] For most lumped and consistent mass matrices with only translational nodal degrees-of-freedom, $\mathbf{M} = `\mathbf{M}$. In certain cases, for example when the mass matrix is condensed through Guyan reduction, $\mathbf{M} \neq `\mathbf{M}$.

The absolute velocity and acceleration of the node are expressed as:

$$\dot{\mathbf{d}}^i = \dot{\mathbf{r}} - \tilde{\mathbf{b}}^i \omega + \dot{\delta}^i \tag{3.6}$$

$$\ddot{\mathbf{d}}^i = \ddot{\mathbf{r}} - \tilde{\mathbf{b}}^i \dot{\omega} + \ddot{\delta}^i + \mathbf{w}^i \tag{3.7}$$

where ω and $\dot{\omega}$ are the angular velocity and acceleration of the body-frame, and

$$\mathbf{w}^i = \tilde{\omega}\tilde{\omega}\mathbf{b}^i + 2\tilde{\omega}\dot{\delta}^i \tag{3.8}$$

The position, velocity, and acceleration of all the nodes are written in stack form as:

$$\mathbf{d} = \hat{\mathbf{I}}\mathbf{r} + \mathbf{b} \tag{3.9}$$

$$\dot{\mathbf{d}} = \hat{\mathbf{I}}\dot{\mathbf{r}} - \hat{\hat{\mathbf{b}}}\omega + \dot{\delta} = \begin{bmatrix} \hat{\mathbf{I}} & -\hat{\hat{\mathbf{b}}} & \bar{\mathbf{I}} \end{bmatrix} \begin{Bmatrix} \dot{\mathbf{r}} \\ \omega \\ \dot{\delta} \end{Bmatrix} \tag{3.10}$$

$$\ddot{\mathbf{d}} = \hat{\mathbf{I}}\ddot{\mathbf{r}} - \hat{\hat{\mathbf{b}}}\dot{\omega} + \ddot{\delta} + \mathbf{w} = \begin{bmatrix} \hat{\mathbf{I}} & -\hat{\hat{\mathbf{b}}} & \bar{\mathbf{I}} \end{bmatrix} \begin{Bmatrix} \ddot{\mathbf{r}} \\ \dot{\omega} \\ \ddot{\delta} \end{Bmatrix} + \bar{\bar{\omega}}\bar{\bar{\omega}}\mathbf{b} + 2\bar{\bar{\omega}}\dot{\delta} \tag{3.11}$$

We substitute Eq. 3.11 into Eq. 3.4 to get

$$\mathbf{M}\begin{bmatrix} \hat{\mathbf{I}} & -\hat{\hat{\mathbf{b}}} & \bar{\mathbf{I}} \end{bmatrix} \begin{Bmatrix} \ddot{\mathbf{r}} \\ \dot{\omega} \\ \ddot{\delta} \end{Bmatrix} = \mathbf{f} - \mathbf{M}\mathbf{w} - \mathbf{K}\delta \tag{3.12}$$

Equation 3.12 can also be written in a different form if we pre-multiply it by the transpose of the coefficient matrix of Eq. 3.10 (or Eq. 3.12):

$$\begin{bmatrix} m\mathbf{I} & -\hat{\mathbf{I}}^{\mathrm{T}}\mathbf{M}\hat{\hat{\mathbf{b}}} & \hat{\mathbf{I}}^{\mathrm{T}}\mathbf{M} \\ -\hat{\hat{\mathbf{b}}}^{\mathrm{T}}\mathbf{M}\hat{\mathbf{I}} & \mathbf{J} & -\hat{\hat{\mathbf{b}}}^{\mathrm{T}}\mathbf{M} \\ \mathbf{M}\hat{\mathbf{I}} & -\mathbf{M}\hat{\hat{\mathbf{b}}} & \mathbf{M} \end{bmatrix} \begin{Bmatrix} \ddot{\mathbf{r}} \\ \dot{\omega} \\ \ddot{\delta} \end{Bmatrix} = \begin{Bmatrix} \hat{\mathbf{I}}^{\mathrm{T}}(\mathbf{f} - \mathbf{M}\mathbf{w}) \\ -\hat{\hat{\mathbf{b}}}^{\mathrm{T}}(\mathbf{f} - \mathbf{M}\mathbf{w}) \\ \mathbf{f} - \mathbf{M}\mathbf{w} - \mathbf{K}\delta \end{Bmatrix} \tag{3.13}$$

where,

$$\hat{\mathbf{I}}^{\mathrm{T}}\mathbf{M}\hat{\mathbf{I}} = m\mathbf{I} \qquad (\hat{\mathbf{I}}^{\mathrm{T}}\,\text{`}\mathbf{M}\hat{\mathbf{I}} = m\mathbf{I}) \tag{3.14.a}$$

$$\hat{\hat{\mathbf{b}}}^{\mathrm{T}}\mathbf{M}\hat{\hat{\mathbf{b}}} = \mathbf{J} \qquad (\text{`}\hat{\hat{\mathbf{b}}}^{\mathrm{T}}\,\text{`}\mathbf{M}\,\text{`}\hat{\hat{\mathbf{b}}} = \text{`}\mathbf{J}) \tag{3.14.b}$$

$$\hat{\mathbf{I}}^{\mathrm{T}}\mathbf{K}\delta = \mathbf{0} \qquad (\hat{\mathbf{I}}^{\mathrm{T}}\,\text{`}\mathbf{K}\,\text{`}\delta = \mathbf{0}) \tag{3.14.c}$$

$$\hat{\hat{\mathbf{b}}}^{\mathrm{T}}\mathbf{K}\delta = \mathbf{0} \qquad (\text{`}\hat{\hat{\mathbf{b}}}^{\mathrm{T}}\,\text{`}\mathbf{K}\,\text{`}\delta = \mathbf{0}) \tag{3.14.d}$$

In Eqs. 3.14.a and 3.14.b, m and $\mathbf{J}$ represent the total mass and the instantaneous rotational inertia matrix of the body. In Eq. 3.13 we have also used the two identities given in Eqs. 3.14.c and 3.14.d. These identities state that the sum of internal structural forces and the sum of internal moments are equal to zero. Also note that these and some other identities can be expressed either in the non-moving or in the moving reference frames.

Equation 3.13 can also be transformed to other forms. For example, we can express the nodal deformation array with respect to the body-frame:

$$\begin{bmatrix} m\mathbf{I} & -\hat{\mathbf{I}}^{\mathrm{T}}\mathbf{M}\hat{\tilde{\mathbf{b}}} & \hat{\mathbf{I}}^{\mathrm{T}}\mathbf{M}\overline{\mathbf{A}} \\ -\hat{\tilde{\mathbf{b}}}^{\mathrm{T}}\mathbf{M}\hat{\mathbf{I}} & \mathbf{J} & -\hat{\tilde{\mathbf{b}}}^{\mathrm{T}}\mathbf{M}\overline{\mathbf{A}} \\ \overline{\mathbf{A}}^{\mathrm{T}}\mathbf{M}\hat{\mathbf{I}} & -\overline{\mathbf{A}}^{\mathrm{T}}\mathbf{M}\hat{\tilde{\mathbf{b}}} & {}^{\backprime}\mathbf{M} \end{bmatrix} \begin{Bmatrix} \ddot{\mathbf{r}} \\ \dot{\omega} \\ {}^{\backprime}\ddot{\delta} \end{Bmatrix} = \begin{Bmatrix} \hat{\mathbf{I}}^{\mathrm{T}}(\mathbf{f}-\mathbf{M}\mathbf{w}) \\ -\hat{\tilde{\mathbf{b}}}^{\mathrm{T}}(\mathbf{f}-\mathbf{M}\mathbf{w}) \\ {}^{\backprime}\mathbf{f}-{}^{\backprime}\mathbf{M}{}^{\backprime}\mathbf{w}-{}^{\backprime}\mathbf{K}{}^{\backprime}\delta \end{Bmatrix} \tag{3.15}$$

Or, we may express the angular velocity and acceleration vectors in terms of their body components as:

$$\begin{bmatrix} m\mathbf{I} & -\hat{\mathbf{I}}^{\mathrm{T}}{}^{\backprime}\mathbf{M}\hat{\tilde{\mathbf{b}}}\mathbf{A} & \hat{\mathbf{I}}^{\mathrm{T}}\mathbf{M}\overline{\mathbf{A}} \\ -\mathbf{A}^{\mathrm{T}}\hat{\tilde{\mathbf{b}}}^{\mathrm{T}}\mathbf{M}\hat{\mathbf{I}} & {}^{\backprime}\mathbf{J} & -{}^{\backprime}\hat{\tilde{\mathbf{b}}}^{\mathrm{T}}{}^{\backprime}\mathbf{M} \\ \overline{\mathbf{A}}^{\mathrm{T}}\mathbf{M}\hat{\mathbf{I}} & -{}^{\backprime}\mathbf{M}{}^{\backprime}\hat{\tilde{\mathbf{b}}} & {}^{\backprime}\mathbf{M} \end{bmatrix} \begin{Bmatrix} \ddot{\mathbf{r}} \\ {}^{\backprime}\dot{\omega} \\ {}^{\backprime}\ddot{\delta} \end{Bmatrix} = \begin{Bmatrix} \hat{\mathbf{I}}^{\mathrm{T}}(\mathbf{f}-\mathbf{M}\mathbf{w}) \\ -{}^{\backprime}\hat{\tilde{\mathbf{b}}}^{\mathrm{T}}({}^{\backprime}\mathbf{f}-{}^{\backprime}\mathbf{M}{}^{\backprime}\mathbf{w}) \\ {}^{\backprime}\mathbf{f}-{}^{\backprime}\mathbf{M}{}^{\backprime}\mathbf{w}-{}^{\backprime}\mathbf{K}{}^{\backprime}\delta \end{Bmatrix} \tag{3.16}$$

Equations 3.12, 3.13, 3.15, and 3.16 are different representations of Eq. 3.4. Although so far we have not discussed how to attach a reference frame to the body, let us assume that the modal deflections are known; therefore, the array of forces at the right-hand side of these equations can be determined. Equation 3.4 contains n_{dof} equations that can be solved to determine n_{dof} unknown accelerations. Equation 3.12 contains n_{dof} equations in $n_{dof}+6$ unknown accelerations and therefore they cannot be solved for the accelerations. Equation 3.13 (3.15 or 3.16) contains $n_{dof}+6$ equations in $n_{dof}+6$ unknown accelerations. However, since in each set of equations there are six redundant equations, they cannot be solved for the accelerations either.

3.1.1. *Boundary and Unconstrained Nodes*

In a multibody system, a flexible body is often connected to other rigid or flexible bodies. The connections are through kinematic joints, springs, dampers, or other elements. For this purpose we split the nodes of a flexible body into two sets as shown in Fig. 2: the *boundary* nodes having superscript "*b*", and the *unconstrained* nodes with superscript "*u*". The boundary nodes will be used to connect the deformable body to other bodies, and the unconstrained nodes remain free. Based on this categorization, all entities associated with a flexible body will be split into subsets, for example:

$$\mathbf{K}=\begin{bmatrix}\mathbf{K}^{bb} & \mathbf{K}^{bu}\\ \mathbf{K}^{ub} & \mathbf{K}^{uu}\end{bmatrix}=\begin{bmatrix}\mathbf{K}^{b,}\\ \mathbf{K}^{u,}\end{bmatrix}=\begin{bmatrix}\mathbf{K}^{,b} & \mathbf{K}^{,u}\end{bmatrix},\quad \mathbf{M}=\begin{bmatrix}\mathbf{M}^{bb} & \mathbf{M}^{bu}\\ \mathbf{M}^{ub} & \mathbf{M}^{uu}\end{bmatrix}=\begin{bmatrix}\mathbf{M}^{b,}\\ \mathbf{M}^{u,}\end{bmatrix}=\begin{bmatrix}\mathbf{M}^{,b} & \mathbf{M}^{,u}\end{bmatrix}$$

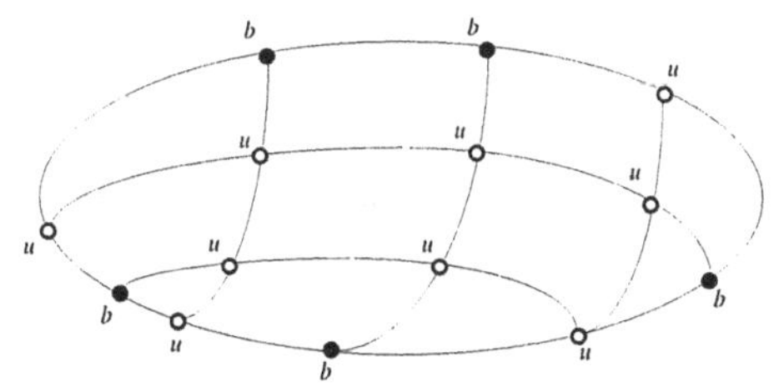

Figure 2. Boundary and unconstrained nodes.

4. Body Frames

In order to attach a reference frame to a deformable body, we need to define six boundary conditions. With these boundary conditions, the number of nodal deflection degrees-of-freedom becomes $n_{dof}=3\times n_{nodes}-6$; i.e., the rigid-body degrees-of-freedom are removed.

The most common procedure in the finite element method to set six of the nodal deflections to zero. Another method that is not widely used is to define six equations called the mean-axis conditions. A third method defines the body-frame to coincide with the instantaneous principal-axes of the deformable body. In some literatures, the body-frame is referred to as a floating frame.

4.1 NODAL-FIXED FRAME

In this method, when there are no nodal rotational degrees-of-freedom, we choose three nodes of the deformable body for the frame to be attached to. As shown in Fig. 3, we assume that the origin of the frame is always at node o, that node j always remains along the ξ-axis, while node k remains on the $\xi-\eta$ plane. This yields:

$$\begin{gathered} \grave{\delta}^{o}_{(\xi)} = \grave{\delta}^{o}_{(\eta)} = \grave{\delta}^{o}_{(\zeta)} = 0 \\ \grave{\delta}^{j}_{(\eta)} = \grave{\delta}^{j}_{(\zeta)} = 0 \\ \grave{\delta}^{k}_{(\zeta)} = 0 \end{gathered} \tag{4.1}$$

We split the array of nodal displacement and any associated entity into two sets. A set that defines the reference frame, denoted by superscript "a" for *axis*, and another set for the remaining deflections denoted by superscript "u" for *unconstrained* as:

$$\grave{\delta} = \begin{Bmatrix} \grave{\delta}^{a} \\ \grave{\delta}^{u} \end{Bmatrix} = \begin{Bmatrix} \mathbf{0} \\ \grave{\delta}^{u} \end{Bmatrix}, \quad \ddot{\grave{\delta}} = \begin{Bmatrix} \ddot{\grave{\delta}}^{a} \\ \ddot{\grave{\delta}}^{u} \end{Bmatrix} = \begin{Bmatrix} \mathbf{0} \\ \ddot{\grave{\delta}}^{u} \end{Bmatrix} \tag{4.2}$$

The a-set may or may not be part of the boundary set that was described in section 3.1.1.

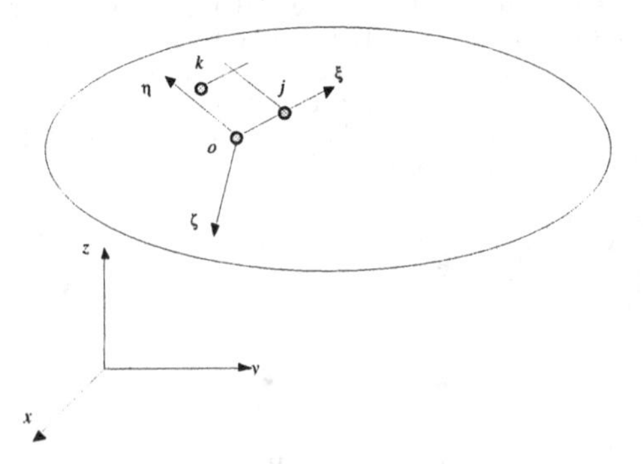

Figure 3. A local frame is attached to a body using nodal-fixed conditions.

With this partitioning of the nodal deflections, Eq. 3.12 becomes:

$$\begin{bmatrix} \mathbf{M}\hat{\mathbf{I}} & -\mathbf{M}\hat{\bar{\mathbf{b}}} & \bar{\mathbf{A}}\grave{\mathbf{M}}^{u} \end{bmatrix} \begin{Bmatrix} \ddot{\mathbf{r}} \\ \dot{\omega} \\ \ddot{\grave{\delta}}^{u} \end{Bmatrix} = \bar{\mathbf{A}}(\grave{\mathbf{f}} - \grave{\mathbf{K}}^{u}\grave{\delta}^{u} - \grave{\mathbf{M}}^{u}\grave{\mathbf{w}}^{u}) \tag{4.3}$$

where we have n_{dof} equations in n_{dof} accelerations.

The nodal-fixed boundary conditions can also be implemented in Eq. 3.16 (or 3.15) to obtain:

$$\begin{bmatrix} m\mathbf{I} & -\hat{\mathbf{I}}^{T}`\mathbf{M}`\hat{\tilde{\mathbf{b}}}\overline{\mathbf{A}} & \hat{\mathbf{I}}^{T}`\mathbf{M}`''\overline{\mathbf{A}} \\ -\overline{\mathbf{A}}^{T}`\hat{\tilde{\mathbf{b}}}^{T}`\mathbf{M}\hat{\mathbf{I}} & `\mathbf{J} & -`\hat{\tilde{\mathbf{b}}}^{T}`\mathbf{M}'' \\ \overline{\mathbf{A}}^{T}`\mathbf{M}''`\hat{\mathbf{I}} & -`\mathbf{M}''`\hat{\tilde{\mathbf{b}}} & `\mathbf{M}'''' \end{bmatrix} \begin{Bmatrix} \ddot{\mathbf{r}} \\ `\dot{\omega} \\ `\ddot{\delta}'' \end{Bmatrix} = \begin{Bmatrix} \mathbf{A}\hat{\mathbf{I}}^{T}(`\mathbf{f} - `\mathbf{M}'''`\mathbf{w}'') \\ -`\hat{\tilde{\mathbf{b}}}^{T}(`\mathbf{f} - `\mathbf{M}'''`\mathbf{w}'') \\ `\mathbf{f}'' - `\mathbf{K}''''`\delta'' - `\mathbf{M}''''`\mathbf{w}'' \end{Bmatrix} \tag{4.4}$$

We can use either Eq. 4.3 or Eq. 4.4 as our equations of motion.

4.2 MEAN-AXIS CONDITIONS

Another method of attaching a reference frame to a deformable body is known as the *mean-axis*[2]. The *translational* and *rotational* mean axis conditions are defined respectively at the velocity level as[3] [1]

$$\hat{\mathbf{I}}^{T}\mathbf{M}\dot{\delta} = \mathbf{0} \qquad (\hat{\mathbf{I}}^{T}`\mathbf{M}`\dot{\delta} = \mathbf{0}) \tag{4.5}$$

$$\hat{\tilde{\mathbf{b}}}^{T}\mathbf{M}\dot{\delta} = \mathbf{0} \qquad (`\hat{\tilde{\mathbf{b}}}^{T}`\mathbf{M}`\dot{\delta} = \mathbf{0}) \tag{4.6}$$

The time derivative of these equations, following some simplifications, yields

$$\hat{\mathbf{I}}^{T}\mathbf{M}\ddot{\delta} = \mathbf{0} \qquad (\hat{\mathbf{I}}^{T}`\mathbf{M}`\ddot{\delta} = \mathbf{0}) \tag{4.7}$$

$$\hat{\tilde{\mathbf{b}}}^{T}\mathbf{M}\ddot{\delta} = -\dot{\tilde{\delta}}^{T}\mathbf{M}\dot{\delta} \qquad (`\hat{\tilde{\mathbf{b}}}^{T}`\mathbf{M}`\ddot{\delta} = -`\dot{\tilde{\delta}}^{T}`\mathbf{M}`\dot{\delta}) \tag{4.8.a}$$

The right-hand side of Eq. 4.8.a could be zero depending on the characteristics of the mass matrix (refer to footnote 1). Therefore, the simpler form of Eq. 4.8.a is:

$$\hat{\tilde{\mathbf{b}}}^{T}\mathbf{M}\ddot{\delta} = \mathbf{0} \qquad (`\hat{\tilde{\mathbf{b}}}^{T}`\mathbf{M}`\ddot{\delta} = \mathbf{0}) \tag{4.8.b}$$

Other identities can be derived based on Eqs. 4.5 and 4.6. For example, if the origin of the frame is positioned at the mass center when the body is in its undeformed state; i.e., $\hat{\mathbf{I}}^{T}\mathbf{M}\mathbf{s} = \mathbf{0}$, then the integral of Eq. 4.5 yields

$$\hat{\mathbf{I}}^{T}\mathbf{M}\delta = \mathbf{0} \qquad (\hat{\mathbf{I}}^{T}`\mathbf{M}`\delta = \mathbf{0}) \tag{4.9.a}$$

and

$$\hat{\mathbf{I}}^{T}\mathbf{M}\mathbf{b} = \mathbf{0} \qquad (\hat{\mathbf{I}}^{T}`\mathbf{M}`\mathbf{b} = \mathbf{0}) \tag{4.9.b}$$

In turn we can show that:

$$\hat{\mathbf{I}}^{T}\mathbf{M}\hat{\tilde{\mathbf{b}}} = \hat{\tilde{\mathbf{b}}}^{T}\mathbf{M}\hat{\mathbf{I}} = \mathbf{0} \qquad (\hat{\mathbf{I}}^{T}`\mathbf{M}`\hat{\tilde{\mathbf{b}}} = `\hat{\tilde{\mathbf{b}}}^{T}`\mathbf{M}\hat{\mathbf{I}} = \mathbf{0}) \tag{4.9.c}$$

Any of the three forms of Eq. 4.9 keeps the origin of the reference frame at the instantaneous mass center of the body.

---

[2] The mean-axis conditions can be obtained from the minimization of the deformation kinetic energy. The kinetic energy associated with the deformation is expressed as:

$$T = \tfrac{1}{2}\dot{\delta}^{T}\mathbf{M}\dot{\delta} = \tfrac{1}{2}(\dot{\mathbf{d}} - \hat{\mathbf{I}}\dot{\mathbf{r}} + \hat{\tilde{\mathbf{b}}}\omega)^{T}\mathbf{M}(\dot{\mathbf{d}} - \hat{\mathbf{I}}\dot{\mathbf{r}} + \hat{\tilde{\mathbf{b}}}\omega)$$

The partial derivatives of the kinetic energy with respect to the translational and angular velocities of the body-frame are:

$$T_{\dot{r}} = -(\dot{\mathbf{d}} - \hat{\mathbf{I}}\dot{\mathbf{r}} + \hat{\tilde{\mathbf{b}}}\omega)^{T}\mathbf{M}\hat{\mathbf{I}} = -\dot{\delta}^{T}\mathbf{M}\hat{\mathbf{I}}$$

$$T_{\omega} = \tfrac{1}{2}(\dot{\mathbf{d}} - \hat{\mathbf{I}}\dot{\mathbf{r}} + \hat{\tilde{\mathbf{b}}}\omega)^{T}\mathbf{M}\tilde{\mathbf{b}} = \dot{\delta}^{T}\mathbf{M}\tilde{\mathbf{b}}$$

The minimum kinetic energy is achieved if these derivatives are set to zero.

[3] If nodal rotational degrees-of-freedom are included, the rotational mean axis condition finds a slightly different form.

Substituting Eqs. 4.7, 4.8.a, and 4.9.c into Eq. 3.13 yields:

$$\begin{bmatrix} m\mathbf{I} & \mathbf{0} & \mathbf{0} \\ \mathbf{0} & \mathbf{J} & \mathbf{0} \\ \mathbf{M}\hat{\mathbf{I}} & -\mathbf{M}\hat{\tilde{\mathbf{b}}} & \mathbf{M} \end{bmatrix} \begin{Bmatrix} \ddot{\mathbf{r}} \\ \dot{\boldsymbol{\omega}} \\ \ddot{\boldsymbol{\delta}} \end{Bmatrix} = \begin{Bmatrix} \hat{\mathbf{I}}^{\mathrm{T}}(\mathbf{f}-\mathbf{M}\mathbf{w}) \\ -\hat{\tilde{\mathbf{b}}}^{\mathrm{T}}(\mathbf{f}-\mathbf{M}\mathbf{w}) + \tilde{\dot{\boldsymbol{\delta}}}^{\mathrm{T}}\mathbf{M}\dot{\boldsymbol{\delta}} \\ \mathbf{f}-\mathbf{K}\boldsymbol{\delta}-\mathbf{M}\mathbf{w} \end{Bmatrix} \tag{4.10.a}$$

The coefficient matrix in Eq. 4.10.a is non-singular. Therefore, if all of the forces are known, the equations can be solved for the accelerations. If we use Eq. 4.8.b in the above process instead of Eq. 4.8.a, Eq. 4.10.a finds the following form:

$$\begin{bmatrix} m\mathbf{I} & \mathbf{0} & \mathbf{0} \\ \mathbf{0} & \mathbf{J} & \mathbf{0} \\ \mathbf{M}\hat{\mathbf{I}} & -\mathbf{M}\hat{\tilde{\mathbf{b}}} & \mathbf{M} \end{bmatrix} \begin{Bmatrix} \ddot{\mathbf{r}} \\ \dot{\boldsymbol{\omega}} \\ \ddot{\boldsymbol{\delta}} \end{Bmatrix} = \begin{Bmatrix} \hat{\mathbf{I}}^{\mathrm{T}}(\mathbf{f}-\mathbf{M}\mathbf{w}) \\ -\hat{\tilde{\mathbf{b}}}^{\mathrm{T}}(\mathbf{f}-\mathbf{M}\mathbf{w}) \\ \mathbf{f}-\mathbf{K}\boldsymbol{\delta}-\mathbf{M}\mathbf{w} \end{Bmatrix} \tag{4.10.b}$$

For notational simplicity, in the remaining parts of this paper we use Eq. 4.10.b instead of its more general form of Eq. 4.10.a.

Depending on the characteristics of the mass matrix, some of the terms on the right-hand side of Eq. 4.10 can further be simplified. For example, for lumped or consistent mass matrices, the term $\hat{\mathbf{I}}^{\mathrm{T}}\mathbf{M}\mathbf{w} = \mathbf{0}$. Similarly the term $\hat{\tilde{\mathbf{b}}}^{\mathrm{T}}\mathbf{M}\mathbf{w}$ could be simplified as well. With such simplifications it becomes clear that the first row in Eq. 4.10 represents the *Newton's equation* of motion for the instantaneous mass center of the deformable body, where its right-hand side consists only of the sum of external forces. The second row is the *Euler's equation* for the rotation of the deformable body based on its instantaneous rotational inertia matrix.

Equations 4.7 and 4.8.b can be incorporated into Eq. 3.13 with the aid of Lagrange multipliers to obtain:

$$\left[\begin{array}{ccc|c} m\mathbf{I} & -\hat{\mathbf{I}}^{\mathrm{T}}\mathbf{M}\hat{\tilde{\mathbf{b}}} & \hat{\mathbf{I}}^{\mathrm{T}}\mathbf{M} & \mathbf{0} \\ -\hat{\tilde{\mathbf{b}}}^{\mathrm{T}}\mathbf{M}\hat{\mathbf{I}} & \mathbf{J} & -\hat{\tilde{\mathbf{b}}}^{\mathrm{T}}\mathbf{M} & \mathbf{0} \\ \mathbf{M}\hat{\mathbf{I}} & -\mathbf{M}\hat{\tilde{\mathbf{b}}} & \mathbf{M} & \mathbf{D}_{m-a}^{\mathrm{T}} \\ \hline \mathbf{0} & \mathbf{0} & \mathbf{D}_{m-a} & \mathbf{0} \end{array}\right] \begin{Bmatrix} \ddot{\mathbf{r}} \\ \dot{\boldsymbol{\omega}} \\ \ddot{\boldsymbol{\delta}} \\ \hline \boldsymbol{\lambda}_{m-a} \end{Bmatrix} = \begin{Bmatrix} \hat{\mathbf{I}}^{\mathrm{T}}(\mathbf{f}-\mathbf{M}\mathbf{w}) \\ -\hat{\tilde{\mathbf{b}}}^{\mathrm{T}}(\mathbf{f}-\mathbf{M}\mathbf{w}) \\ \mathbf{f}-\mathbf{K}\boldsymbol{\delta}-\mathbf{M}\mathbf{w} \\ \hline \mathbf{0} \end{Bmatrix} \tag{4.11}$$

where,

$$\mathbf{D}_{m-a} = \begin{bmatrix} \hat{\mathbf{I}}^{\mathrm{T}}\mathbf{M} \\ \hat{\tilde{\mathbf{b}}}^{\mathrm{T}}\mathbf{M} \end{bmatrix} \tag{4.11.a}$$

and $\boldsymbol{\lambda}_{m-a}$ contains six Lagrange multipliers. In the process of numerical integration of the equations of motion, Eq. 4.11 exhibits more stable characteristics than Eq. 4.10.b. In Eq. 4.11 the mean-axis conditions at the acceleration level are explicitly present and therefore they are definitely enforced. In Eq. 4.10.b these conditions are not explicitly present. This means that when we solve Eq. 4.10.b for the accelerations, the result may not exactly satisfy Eqs. 4.7 and 4.8.b. Although the amount or error could be extremely small, in the process of forward integration this error may accumulate and it may cause numerical instability.

4.3 PRINCIPAL-AXES

It is possible to enforce the body-frame to coincide with the principal-axes of the body as a body deforms [2]. The first condition is that the origin of the $\xi-\eta-\zeta$ frame to be at the mass center of the body. This is the same as the translational mean-axis condition as expressed in

Eqs. 4.5, 4.7 and 4.9 assuming that the origin is initially positioned at the mass center when the body is in its undeformed state.

The rotational condition is that the three products of inertia to be equal to zero. The rotational inertia matrix of Eq. 3.14.b in terms of the $\xi-\eta-\zeta$ components of its entities can be written as

$$`\mathbf{J} = `\hat{\tilde{\mathbf{b}}}^{\mathrm{T}}\, `\mathbf{M}\, `\hat{\tilde{\mathbf{b}}} = \begin{bmatrix} j_{\xi\xi} & j_{\xi\eta} & j_{\xi\zeta} \\ j_{\eta\xi} & j_{\eta\eta} & j_{\eta\zeta} \\ j_{\zeta\xi} & j_{\zeta\eta} & j_{\zeta\zeta} \end{bmatrix} \tag{4.12}$$

It will be shown in the next sub-section that the products of inertia can be computed as

$$j_{\xi\eta} = `\mathbf{b}^{\mathrm{T}}\, `\mathbf{N}_{\xi\eta}\, `\mathbf{b}, \quad j_{\xi\zeta} = `\mathbf{b}^{\mathrm{T}}\, `\mathbf{N}_{\xi\zeta}\, `\mathbf{b}, \quad j_{\eta\zeta} = `\mathbf{b}^{\mathrm{T}}\, `\mathbf{N}_{\eta\zeta}\, `\mathbf{b} \tag{4.13}$$

These three products of inertia are set to zero and enforced as constraints:

$$`\mathbf{b}^{\mathrm{T}}\, `\mathbf{N}_{\xi\eta}\, `\mathbf{b} = 0, \quad `\mathbf{b}^{\mathrm{T}}\, `\mathbf{N}_{\xi\zeta}\, `\mathbf{b} = 0, \quad `\mathbf{b}^{\mathrm{T}}\, `\mathbf{N}_{\eta\zeta}\, `\mathbf{b} = 0 \tag{4.14}$$

The velocity constraints are obtained from the time derivative of Eq. 4.14 as:

$$`\mathbf{b}^{\mathrm{T}} `\mathbf{M}_{\xi\eta}\, `\dot{\delta} = 0, \quad `\mathbf{b}^{\mathrm{T}} `\mathbf{M}_{\xi\zeta}\, `\dot{\delta} = 0, \quad `\mathbf{b}^{\mathrm{T}} `\mathbf{M}_{\eta\zeta}\, `\dot{\delta} = 0 \tag{4.15}$$

where the matrices $`\mathbf{M}_{\xi\eta}$, $`\mathbf{M}_{\xi\zeta}$, and $`\mathbf{M}_{\eta\zeta}$ are discussed in the following sub-section. The acceleration constraints are obtained from the time derivative of Eq. 4.15 as,

$$\begin{aligned} `\mathbf{b}^{\mathrm{T}}\, `\mathbf{M}_{\xi\eta}\, `\ddot{\delta} &= -`\dot{\delta}^{\mathrm{T}}\, `\mathbf{M}_{\xi\eta}\, `\dot{\delta} \\ `\mathbf{b}^{\mathrm{T}}\, `\mathbf{M}_{\xi\zeta}\, `\ddot{\delta} &= -`\dot{\delta}^{\mathrm{T}}\, `\mathbf{M}_{\xi\zeta}\, `\dot{\delta} \\ `\mathbf{b}^{\mathrm{T}}\, `\mathbf{M}_{\eta\zeta}\, `\ddot{\delta} &= -`\dot{\delta}^{\mathrm{T}}\, `\mathbf{M}_{\eta\zeta}\, `\dot{\delta} \end{aligned} \tag{4.16}$$

The translational and rotational conditions at the acceleration level; i.e., Eqs. 4.7 and 4.16, can be written in matrix form as

$$\mathbf{D}_{p-a}\, `\ddot{\delta} = \gamma_{p-a} \tag{4.17}$$

where,

$$\mathbf{D}_{p-a} = \begin{bmatrix} \hat{\mathbf{I}}^{\mathrm{T}}\, `\mathbf{M} \\ `\mathbf{b}^{\mathrm{T}}\, `\mathbf{M}_{\xi\eta} \\ `\mathbf{b}^{\mathrm{T}}\, `\mathbf{M}_{\xi\zeta} \\ `\mathbf{b}^{\mathrm{T}}\, `\mathbf{M}_{\eta\zeta} \end{bmatrix}, \quad \gamma_{p-a} = -\begin{Bmatrix} \mathbf{0} \\ `\dot{\delta}^{\mathrm{T}}\, `\mathbf{M}_{\xi\eta}\, `\dot{\delta} \\ `\dot{\delta}^{\mathrm{T}}\, `\mathbf{M}_{\xi\zeta}\, `\dot{\delta} \\ `\dot{\delta}^{\mathrm{T}}\, `\mathbf{M}_{\eta\zeta}\, `\dot{\delta} \end{Bmatrix} \tag{4.17.a}$$

Equation 3.16, with the aid of six Lagrange multipliers λ_{p-a}, is modified to include the acceleration constraints of Eq. 4.17 as:

$$\left[\begin{array}{ccc|c} m\mathbf{I} & \mathbf{0} & \mathbf{0} & \mathbf{0} \\ \mathbf{0} & `\mathbf{J} & -`\hat{\tilde{\mathbf{b}}}^{\mathrm{T}} `\mathbf{M} & \mathbf{0} \\ `\mathbf{M}\hat{\mathbf{I}}\mathbf{A} & -`\mathbf{M}`\hat{\tilde{\mathbf{b}}} & `\mathbf{M} & \mathbf{D}_{p-a}^{\mathrm{T}} \\ \hline \mathbf{0} & \mathbf{0} & \mathbf{D}_{p-a} & \mathbf{0} \end{array}\right] \begin{Bmatrix} \ddot{\mathbf{r}} \\ `\dot{\omega} \\ `\ddot{\delta} \\ \hline \lambda_{p-a} \end{Bmatrix} = \begin{Bmatrix} \hat{\mathbf{I}}^{\mathrm{T}}(\mathbf{f} - \mathbf{M}\mathbf{w}) \\ -`\hat{\tilde{\mathbf{b}}}^{\mathrm{T}}(`\mathbf{f} - `\mathbf{M}`\mathbf{w}) \\ `\mathbf{f} - `\mathbf{K}`\delta - `\mathbf{M}`\mathbf{w} \\ \hline \gamma_{p-a} \end{Bmatrix} \tag{4.18}$$

4.3.1 *Auxiliary Mass Matrices*

In this sub-section we show how to construct the auxiliary mass matrices in Eqs. 4.14-4.16. The symmetric mass matrix for the deformable body with n_{nodes} is expressed in its most general form as

$$
{}^{\backprime}\mathbf{M} = \begin{bmatrix} {}^{\backprime}\mathbf{M}^{1,1} & {}^{\backprime}\mathbf{M}^{1,2} & \cdots & {}^{\backprime}\mathbf{M}^{1,n_{nodes}} \\ {}^{\backprime}\mathbf{M}^{2,1} & {}^{\backprime}\mathbf{M}^{2,2} & \cdots & {}^{\backprime}\mathbf{M}^{2,n_{nodes}} \\ \vdots & \vdots & \ddots & \vdots \\ {}^{\backprime}\mathbf{M}^{n_{nodes},1} & {}^{\backprime}\mathbf{M}^{n_{nodes},2} & \cdots & {}^{\backprime}\mathbf{M}^{n_{nodes},n_{nodes}} \end{bmatrix}
$$

where ${}^{\backprime}\mathbf{M}^{i,j} = {}^{\backprime}\mathbf{M}^{j,i\mathrm{T}}$. We define the following *selector* matrices:

$$
\mathbf{C}_{\xi} \equiv \begin{bmatrix} 0 & 0 & 0 \\ 0 & 0 & +1 \\ 0 & -1 & 0 \end{bmatrix},\quad \mathbf{C}_{\eta} \equiv \begin{bmatrix} 0 & 0 & -1 \\ 0 & 0 & 0 \\ +1 & 0 & 0 \end{bmatrix},\quad \mathbf{C}_{\zeta} \equiv \begin{bmatrix} 0 & +1 & 0 \\ -1 & 0 & 0 \\ 0 & 0 & 0 \end{bmatrix}
$$

The following matrices can systematically be constructed:

$$
{}^{\backprime}\mathbf{N}_{\xi\eta} = \overline{\mathbf{C}}_{\xi}^{\mathrm{T}}\, {}^{\backprime}\mathbf{M}\, \overline{\mathbf{C}}_{\eta},\quad {}^{\backprime}\mathbf{N}_{\xi\zeta} = \overline{\mathbf{C}}_{\xi}^{\mathrm{T}}\, {}^{\backprime}\mathbf{M}\, \overline{\mathbf{C}}_{\zeta},\quad {}^{\backprime}\mathbf{N}_{\eta\zeta} = \overline{\mathbf{C}}_{\eta}^{\mathrm{T}}\, {}^{\backprime}\mathbf{M}\, \overline{\mathbf{C}}_{\zeta} \tag{4.19}
$$

And,

$$
{}^{\backprime}\mathbf{M}_{\xi\eta} = {}^{\backprime}\mathbf{N}_{\xi\eta}^{\mathrm{T}} + {}^{\backprime}\mathbf{N}_{\xi\eta},\quad {}^{\backprime}\mathbf{M}_{\xi\zeta} = {}^{\backprime}\mathbf{N}_{\xi\zeta}^{\mathrm{T}} + {}^{\backprime}\mathbf{N}_{\xi\zeta},\quad {}^{\backprime}\mathbf{M}_{\eta\zeta} = {}^{\backprime}\mathbf{N}_{\eta\zeta}^{\mathrm{T}} + {}^{\backprime}\mathbf{N}_{\eta\zeta} \tag{4.20}
$$

For a deformable body these matrices are constructed only once. If the original mass matrix is either banded or diagonal, the process in Eqs. 4.19 and 4.20 can greatly be simplified.

5. Nodal and Modal Transformations

The number of nodes and consequently the number of degrees-of-freedom associated with a deformable body may be too large for any realistic numerical simulation. Therefore, various processes have been developed to reduce the number of degrees-of-freedom in such systems. The reduction process may be performed either in the nodal or modal space.

5.1. STATIC CONDENSATION

In the so-called static or Guyan condensation, it is assumed that some of the nodes of a finite element model are kept and the rest are deleted [3]. For this purpose, we use superscripts m and s for *master* (kept) and *slave* (deleted) nodes[4]. The structural equations of motion, Eq. 3.1, are re-written as

$$
\begin{bmatrix} {}^{\backprime}\mathbf{M}^{mm} & {}^{\backprime}\mathbf{M}^{ms} \\ {}^{\backprime}\mathbf{M}^{sm} & {}^{\backprime}\mathbf{M}^{ss} \end{bmatrix} \begin{Bmatrix} {}^{\backprime}\ddot{\boldsymbol{\delta}}^{m} \\ {}^{\backprime}\ddot{\boldsymbol{\delta}}^{s} \end{Bmatrix} + \begin{bmatrix} {}^{\backprime}\mathbf{K}^{mm} & {}^{\backprime}\mathbf{K}^{ms} \\ {}^{\backprime}\mathbf{K}^{sm} & {}^{\backprime}\mathbf{K}^{ss} \end{bmatrix} \begin{Bmatrix} {}^{\backprime}\boldsymbol{\delta}^{m} \\ {}^{\backprime}\boldsymbol{\delta}^{s} \end{Bmatrix} = \begin{Bmatrix} {}^{\backprime}\mathbf{f}^{m} \\ {}^{\backprime}\mathbf{f}^{s} \end{Bmatrix} \tag{5.1}
$$

We assume that the inertia of the entire structure is allocated to the master nodes and no external forces are applied to the slave nodes; i.e.,

$$
\begin{bmatrix} {}^{\backprime}\mathbf{M}_{\mathrm{G}} & \mathbf{0} \\ \mathbf{0} & \mathbf{0} \end{bmatrix} \begin{Bmatrix} {}^{\backprime}\ddot{\boldsymbol{\delta}}^{m} \\ {}^{\backprime}\ddot{\boldsymbol{\delta}}^{s} \end{Bmatrix} + \begin{bmatrix} {}^{\backprime}\mathbf{K}^{mm} & {}^{\backprime}\mathbf{K}^{ms} \\ {}^{\backprime}\mathbf{K}^{sm} & {}^{\backprime}\mathbf{K}^{ss} \end{bmatrix} \begin{Bmatrix} {}^{\backprime}\boldsymbol{\delta}^{m} \\ {}^{\backprime}\boldsymbol{\delta}^{s} \end{Bmatrix} = \begin{Bmatrix} {}^{\backprime}\mathbf{f}^{m} \\ \mathbf{0} \end{Bmatrix} \tag{5.2}
$$

---

[4] The *boundary* nodes are a sub-set of the *master* nodes. The *slave* nodes contain some or all of the *unconstrained* nodes.

where $`\mathbf{M}_G$ is the mass matrix associated with the master nodes containing the entire structural inertia. This matrix will be determined as a function of the original sub-matrices. The second row of Eq. 5.2 yields:

$$`\mathbf{K}^{sm}\,`\delta^{m} + `\mathbf{K}^{ss}\,`\bar{\delta}^{s} = \mathbf{0}$$

From this equation we get

$$`\delta^{s} = `\mathbf{G}\,`\delta^{m} \tag{5.3}$$

where

$$`\mathbf{G} = -\,`\mathbf{K}^{ss^{-1}}\,`\mathbf{K}^{sm} \tag{5.4}$$

is known as the static condensation matrix. This matrix is used to determine the reduced mass and stiffness matrices as:

$$`\mathbf{M}_G = `\mathbf{M}^{mm} + `\mathbf{M}^{ms}`\mathbf{G} + `\mathbf{G}^{\mathrm{T}}\,`\mathbf{M}^{sm} + `\mathbf{G}^{\mathrm{T}}\,`\mathbf{M}^{ss}`\mathbf{G} \tag{5.5}$$

$$`\mathbf{K}_G = `\mathbf{K}^{mm} - `\mathbf{K}^{ms}\,`\mathbf{K}^{ss^{-1}}\,`\mathbf{K}^{sm} \tag{5.6}$$

Matrices $`\mathbf{M}_G$ and $`\mathbf{K}_G$ can now be used in most of the equations of motion instead of $`\mathbf{M}$ and $`\mathbf{K}$. However, some of the simplifications that were discussed for the mean-axis conditions may no longer be applicable when we use a condensed mass matrix.

5.2. MODAL MATRICES

In the following sub-sections we use modal data for different mass and stiffness matrices. In general, it is assumed that the matrix of mode shapes, $`\Psi$, provides the following transformation between the modal and nodal coordinates:

$$`\delta = `\Psi\,\mathbf{z} \tag{5.7}$$

If we define another modal matrix as

$$\Psi = \bar{\mathbf{A}}\,`\Psi \tag{5.8}$$

then we obtain another form of the transformation equation as

$$\delta = \Psi\,\mathbf{z} \tag{5.9}$$

Equation 5.8 is also applicable to matrix $`\mathbf{G}$.

5.2.1. *Free-Free Modes*

The matrices of the mode shapes and modal frequencies obtained from $`\mathbf{M}$ and $`\mathbf{K}$ are:

$$\left[\hat{\mathbf{I}} \;\; -`\hat{\tilde{\mathbf{s}}} \;\; `\Psi_f\right], \;\; diag(\mathbf{0},\; \mathbf{0},\; \Lambda_f) \tag{5.10}$$

In the matrix of mode shapes, $\hat{\mathbf{I}}$ and $-`\hat{\tilde{\mathbf{s}}}$ represent the translational and rotational rigid-body mode shapes (each is an $n_{dof} \times 3$ matrix), and $`\Psi_f$ which is an $n_{dof} \times (n_{dof} - 6)$ matrix represents the deformation mode shapes. The diagonal matrix of eigenvalues contains six zero eigenvalues associated with the rigid-body modes and the natural frequencies, Λ_f. Note that the subscript *"f"* emphasizes that these entities correspond to a *free-free* structure.

The modal-to-nodal transformation equation is written as

$$`\delta = `\Psi_f\,\mathbf{z} \tag{5.11}$$

The modal stiffness and mass matrices are determined as

$$M_f = \Psi_f^{\mathrm{T}}\,\grave{}\mathbf{M}\Psi_f, \qquad K_f = \Psi_f^{\mathrm{T}}\,\grave{}\mathbf{K}\Psi_f \tag{5.12}$$

Considering the partitioning of the nodes based on the boundary and unconstrained nodes, Eq. 5.11 becomes

$$\begin{Bmatrix} \grave{}\delta^b \\ \grave{}\delta^u \end{Bmatrix} = \begin{bmatrix} \Psi_f^b \\ \Psi_f^u \end{bmatrix} \mathbf{z} \tag{5.13}$$

5.2.2. *Constrained Modes*

Assume that the boundary nodes are constrained not to have any deflections; i.e., $\grave{}\delta^b = \grave{}\ddot{\delta}^b = \mathbf{0}$. In this case sub-matrices $\grave{}\mathbf{M}^{uu}$ and $\grave{}\mathbf{K}^{uu}$ are used to obtain modal matrices Ψ_c and Λ_c. If the boundary nodes eliminate at least six degrees-of-freedom, then there will be no zero eigenvalues in Λ_c. The subscript "*c*" emphasizes that these entities correspond to a *constrained* structure. The transformation equation is

$$\grave{}\delta^u = \Psi_c \mathbf{z} \tag{5.14}$$

The modal stiffness and mass matrices are obtained as

$$M_c = \Psi_c^{\mathrm{T}}\,\grave{}\mathbf{M}^{uu}\Psi_c, \qquad K_c = \Psi_c^{\mathrm{T}}\,\grave{}\mathbf{K}^{uu}\Psi_c \tag{5.15}$$

5.2.3. *Constrained Modes and Static Condensation*

We can assume that the deflections of the unconstrained (free) nodes are expressed in terms of the deflection of the boundary nodes and the constrained modal matrix Ψ_c as

$$\grave{}\delta^u = \mathbf{E}\,\grave{}\delta^b + \Psi_c \mathbf{z}$$

where $\mathbf{E}$ is a coefficient matrix to be determined. For the complete set of deflections this equation is written as

$$\begin{Bmatrix} \grave{}\delta^b \\ \grave{}\delta^u \end{Bmatrix} = \begin{bmatrix} \mathbf{I} & \mathbf{0} \\ \mathbf{E} & \Psi_c \end{bmatrix} \begin{Bmatrix} \grave{}\delta^b \\ \mathbf{z} \end{Bmatrix} \tag{5.16}$$

Using the coefficient matrix from Eq. 5.16, the stiffness matrix is transformed to modal space as

$$\begin{bmatrix} \mathbf{I} & \mathbf{E}^{\mathrm{T}} \\ \mathbf{0} & \Psi_c^{\mathrm{T}} \end{bmatrix} \begin{bmatrix} \grave{}\mathbf{K}^{bb} & \grave{}\mathbf{K}^{bu} \\ \grave{}\mathbf{K}^{ub} & \grave{}\mathbf{K}^{uu} \end{bmatrix} \begin{bmatrix} \mathbf{I} & \mathbf{0} \\ \mathbf{E} & \Psi_c \end{bmatrix} \Rightarrow$$
$$\begin{bmatrix} \grave{}\mathbf{K}^{bb} + \grave{}\mathbf{K}^{bu}\mathbf{E} + \mathbf{E}^{\mathrm{T}}(\grave{}\mathbf{K}^{ub} + \grave{}\mathbf{K}^{uu}\mathbf{E}) & (\grave{}\mathbf{K}^{bu} + \mathbf{E}^{\mathrm{T}}\grave{}\mathbf{K}^{uu})\Psi_c \\ \Psi_c^{\mathrm{T}}(\grave{}\mathbf{K}^{ub} + \grave{}\mathbf{K}^{uu}\mathbf{E}) & \Psi_c^{\mathrm{T}}\grave{}\mathbf{K}^{uu}\Psi_c \end{bmatrix} \tag{5.17}$$

In order to uncouple the revised stiffness matrix between the boundary and the unconstrained nodes, we set

$$\mathbf{E} = -\,\grave{}\mathbf{K}^{uu^{-1}}\,\grave{}\mathbf{K}^{ub} \tag{5.18}$$

We observe that $\mathbf{E}$ is identical to the static condensation matrix in Guyan reduction as stated in Eq. 5.4. Here, the master and slave nodes are replaced by the boundary and unconstrained nodes. Then, Eq. 5.16 becomes [4, 5]

$$\begin{Bmatrix} \grave{}\delta^b \\ \grave{}\delta^u \end{Bmatrix} = \begin{bmatrix} \mathbf{I} & \mathbf{0} \\ \grave{}\mathbf{G} & \Psi_c \end{bmatrix} \begin{Bmatrix} \grave{}\delta^b \\ \mathbf{z} \end{Bmatrix} \tag{5.19}$$

Now Eq. 5.17 is written as

$$\begin{bmatrix} \mathbf{I} & {}^{\backprime}\mathbf{G}^{\mathrm{T}} \\ \mathbf{0} & \Psi_c^{\mathrm{T}} \end{bmatrix} \begin{bmatrix} {}^{\backprime}\mathbf{K}^{bb} & {}^{\backprime}\mathbf{K}^{bu} \\ {}^{\backprime}\mathbf{K}^{ub} & {}^{\backprime}\mathbf{K}^{uu} \end{bmatrix} \begin{bmatrix} \mathbf{I} & \mathbf{0} \\ {}^{\backprime}\mathbf{G} & \Psi_c \end{bmatrix} \Rightarrow \begin{bmatrix} {}^{\backprime}\mathbf{K}_G & \mathbf{0} \\ \mathbf{0} & \boldsymbol{K}_c \end{bmatrix} \tag{5.20}$$

Similarly, for the mass matrix we have

$$\begin{bmatrix} \mathbf{I} & {}^{\backprime}\mathbf{G}^{\mathrm{T}} \\ \mathbf{0} & \Psi_c^{\mathrm{T}} \end{bmatrix} \begin{bmatrix} {}^{\backprime}\mathbf{M}^{bb} & {}^{\backprime}\mathbf{M}^{bu} \\ {}^{\backprime}\mathbf{M}^{ub} & {}^{\backprime}\mathbf{M}^{uu} \end{bmatrix} \begin{bmatrix} \mathbf{I} & \mathbf{0} \\ {}^{\backprime}\mathbf{G} & \Psi_c \end{bmatrix} \Rightarrow \begin{bmatrix} {}^{\backprime}\mathbf{M}_G & \mathbf{M}_{Gc}^{\mathrm{T}} \\ \mathbf{M}_{Gc} & \boldsymbol{M}_c \end{bmatrix} \tag{5.21}$$

where $\mathbf{M}_{Gc} \equiv \Psi_c^{\mathrm{T}}({}^{\backprime}\mathbf{M}^{ub} + {}^{\backprime}\mathbf{M}^{uu}\,{}^{\backprime}\mathbf{G})$.

5.2.4. *Constrained Modes and Modal Condensation*

In section 7, the equations of motion will be truncated to obtain smaller sets of equations. When this process is performed in modal space, the mode shapes and the corresponding entities are split into the kept and deleted sets. The process that is discussed in this sub-section will be applicable to the condensation techniques used in sub-section 7.2.4.

Assume that the equations of motion for a structure, Eq. 3.1, are transformed to modal space using the constrained transformations; i.e., Eqs. 5.14 and 5.15. We then split the modal coordinates into the kept and deleted modes to get:

$$\begin{bmatrix} {}^{\backprime}\mathbf{M}^{bb} & {}^{\backprime}\mathbf{M}^{bu}\Psi_c^{k} & {}^{\backprime}\mathbf{M}^{bu}\Psi_c^{d} \\ \Psi_c^{k\mathrm{T}}\,{}^{\backprime}\mathbf{M}^{ub} & \boldsymbol{M}_c^{k} & \mathbf{0} \\ \Psi_d^{k\mathrm{T}}\,{}^{\backprime}\mathbf{M}^{ub} & \mathbf{0} & \boldsymbol{M}_c^{d} \end{bmatrix} \begin{Bmatrix} {}^{\backprime}\ddot{\boldsymbol{\delta}}^{b} \\ \ddot{\mathbf{z}}^{k} \\ \ddot{\mathbf{z}}^{d} \end{Bmatrix} + \begin{bmatrix} {}^{\backprime}\mathbf{K}^{bb} & {}^{\backprime}\mathbf{K}^{bu}\Psi_c^{k} & {}^{\backprime}\mathbf{K}^{bu}\Psi_c^{d} \\ \Psi_c^{k\mathrm{T}}\,{}^{\backprime}\mathbf{K}^{ub} & \boldsymbol{K}_c^{k} & \mathbf{0} \\ \Psi_d^{k\mathrm{T}}\,{}^{\backprime}\mathbf{K}^{ub} & \mathbf{0} & \boldsymbol{K}_c^{d} \end{bmatrix} \begin{Bmatrix} {}^{\backprime}\boldsymbol{\delta}^{b} \\ \mathbf{z}^{k} \\ \mathbf{z}^{d} \end{Bmatrix} = \begin{Bmatrix} {}^{\backprime}\mathbf{f}^{b} \\ \Psi_c^{k\mathrm{T}}\,{}^{\backprime}\mathbf{f}^{u} \\ \Psi_c^{d\mathrm{T}}\,{}^{\backprime}\mathbf{f}^{u} \end{Bmatrix} \tag{5.22}$$

In a process similar to that of the static condensation, we assume that the inertia associated with the deleted modes are allocated to the boundary nodes, and furthermore, there are no forces associated with the deleted modes [6]. With this assumption, the third row of Eq. 5.22 yields

$$\mathbf{z}^{d} = \Theta\,{}^{\backprime}\boldsymbol{\delta}^{b} \tag{5.23}$$

where

$$\Theta = -\boldsymbol{K}^{d^{-1}}\Psi_c^{d\mathrm{T}}\,{}^{\backprime}\mathbf{K}^{ub} \tag{5.24}$$

Note that Θ transforms a set of nodal coordinates to modal coordinates.

6. Equations of Motion for A Rigid-Flexible Multibody System

A multibody system may contain both flexible and rigid bodies. The Newton-Euler equations for a typical free rigid body are written as

$$\begin{bmatrix} \mathbf{M}_r & \mathbf{0} \\ \mathbf{0} & \mathbf{J}_r \end{bmatrix} \begin{Bmatrix} \ddot{\mathbf{r}}_r \\ \dot{\boldsymbol{\omega}}_r \end{Bmatrix} = \begin{Bmatrix} \mathbf{f}_r \\ \mathbf{n}_r \end{Bmatrix} \tag{6.1}$$

where $\mathbf{M}_r = m_r\mathbf{I}$ is the mass of the body, $\mathbf{J}_r$ is the rotational inertia matrix, and $\mathbf{f}_r$ is the external force vector acting on the body. The term $\mathbf{n}_r = \mathbf{n} + \tilde{\mathbf{s}}_r^{P}\,\mathbf{f}_r - \tilde{\boldsymbol{\omega}}_r\mathbf{J}_r\boldsymbol{\omega}_r$ is the sum of applied moments, $\mathbf{n}$, the moment of the force $\mathbf{f}_r$, and the gyroscopic moment. It is assumed that the origin of the body-frame is at the mass center. The Euler equations are normally written in terms of the body components, however, in order to be consistent with some of our flexible body formulations, we have expressed the equations in terms of the global components.

In order to discuss different formulations for a rigid-flexible multibody system, without any loss of generality, we concentrate on a simple multibody system containing only one rigid and one flexible body. We further assume that the connection between the two bodies, as shown in Fig. 4, is at point *b* representing one spherical joint.

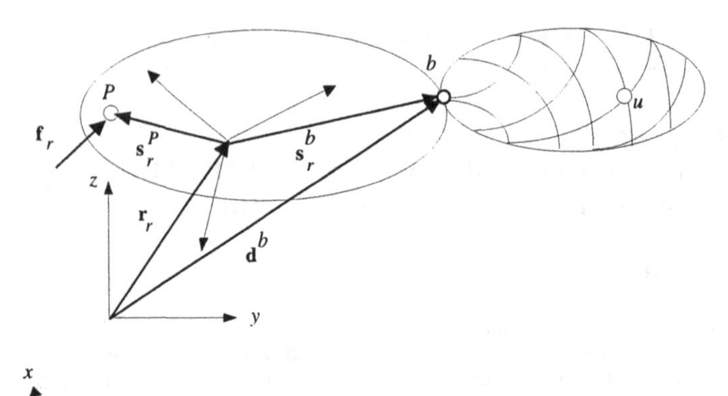

Figure 4. A simple rigid-flexible multibody system.

The kinematic constraint equations for a typical spherical joint is

$$\mathbf{r}_r + \mathbf{s}_r^b - \mathbf{d}^b = \mathbf{0} \tag{6.2}$$

The velocity and acceleration constraints are:

$$\dot{\mathbf{r}}_r - \tilde{\mathbf{s}}_r^b \omega_r - \dot{\mathbf{d}}^b = \mathbf{0} \tag{6.3}$$

$$\ddot{\mathbf{r}}_r - \tilde{\mathbf{s}}_r^b \dot{\omega}_r - \ddot{\mathbf{d}}^b + \tilde{\omega}_r \tilde{\omega}_r \mathbf{s}_r^b = \mathbf{0} \tag{6.4}$$

Using Eq. 3.7, Eq. 6.4 can also be expressed as

$$\ddot{\mathbf{r}}_r - \tilde{\mathbf{s}}_r^b \dot{\omega}_r - \ddot{\mathbf{r}} + \tilde{\mathbf{b}}^b \dot{\omega} - \ddot{\delta}^b = \gamma^b \tag{6.5}$$

where

$$\gamma^b = -\tilde{\omega}_r \tilde{\omega}_r \mathbf{s}_r^b + \mathbf{w}^b \tag{6.6}$$

If there are several spherical joints between the bodies, Eq. 6.4 or 6.5 is repeated for every constrained node.

In the following sub-sections the equations of motion for the system of Fig. 4 are derived in several forms. For the deformable body in these equations we only consider the mean-axis conditions. For the nodal-fixed and the principal-axes conditions, the equations of motion will have similar forms with slight differences.

6.1. ABSOLUTE NODAL MOTION

With the aid of Lagrange multipliers [7], we append Eqs. 6.1, 3.4, and 6.4 to form the equations of motion for the multibody system as

$$\begin{bmatrix} \mathbf{M}_r & \mathbf{0} & \mathbf{0} & \mathbf{0} & \mathbf{I} \\ \mathbf{0} & \mathbf{J}_r & \mathbf{0} & \mathbf{0} & -\tilde{\mathbf{s}}_r^{b\,\mathrm{T}} \\ \mathbf{0} & \mathbf{0} & \mathbf{M}^{bb} & \mathbf{M}^{bu} & -\mathbf{I} \\ \mathbf{0} & \mathbf{0} & \mathbf{M}^{ub} & \mathbf{M}^{uu} & \mathbf{0} \\ \mathbf{I} & -\tilde{\mathbf{s}}_r^b & -\mathbf{I} & \mathbf{0} & \mathbf{0} \end{bmatrix} \begin{Bmatrix} \ddot{\mathbf{r}}_r \\ \dot{\omega}_r \\ \ddot{\mathbf{d}}^b \\ \ddot{\mathbf{d}}^u \\ \lambda \end{Bmatrix} = \begin{Bmatrix} \mathbf{f}_r \\ \mathbf{n}_r \\ \mathbf{f}^b - \mathbf{K}^{b,}\delta \\ \mathbf{f}^u - \mathbf{K}^{u,}\delta \\ -\tilde{\omega}_r \tilde{\omega}_r \mathbf{s}_r^b \end{Bmatrix} \tag{6.7}$$

where λ contains three Lagrange multipliers.

6.2. NODAL DEFLECTIONS

We now transform Eq. 6.7 to another form where, instead of the absolute accelerations, accelerations associated with the nodal deformation are used. For this purpose we use Eqs. 4.10.b, 6.1 and 6.5 to get

$$
\left[\begin{array}{cc|cccc|c}
\mathbf{M}_r & \mathbf{0} & \mathbf{0} & \mathbf{0} & \mathbf{0} & \mathbf{0} & \mathbf{I} \\
\mathbf{0} & \mathbf{J}_r & \mathbf{0} & \mathbf{0} & \mathbf{0} & \mathbf{0} & -\tilde{\mathbf{s}}_r^{b\mathrm{T}} \\
\hline
\mathbf{0} & \mathbf{0} & m\mathbf{I} & \mathbf{0} & \mathbf{0} & \mathbf{0} & -\mathbf{I} \\
\mathbf{0} & \mathbf{0} & \mathbf{0} & \mathbf{J} & \mathbf{0} & \mathbf{0} & \tilde{\mathbf{b}}^{b\mathrm{T}} \\
\mathbf{0} & \mathbf{0} & \mathbf{M}^{b\prime}\hat{\mathbf{I}} & -\mathbf{M}^{b\prime}\hat{\tilde{\mathbf{b}}} & \mathbf{M}^{bb} & \mathbf{M}^{bu} & -\mathbf{I} \\
\mathbf{0} & \mathbf{0} & \mathbf{M}^{u\prime}\hat{\mathbf{I}} & -\mathbf{M}^{u\prime}\hat{\tilde{\mathbf{b}}} & \mathbf{M}^{ub} & \mathbf{M}^{uu} & \mathbf{0} \\
\hline
\hat{\mathbf{I}} & -\tilde{\mathbf{s}}_r^{b} & -\mathbf{I} & \tilde{\mathbf{b}}^{b} & -\mathbf{I} & \mathbf{0} & \mathbf{0}
\end{array}\right]
\left\{\begin{array}{c}
\ddot{\mathbf{r}}_r \\ \dot{\omega}_r \\ \hline \ddot{\mathbf{r}} \\ \dot{\omega} \\ \ddot{\delta}^b \\ \ddot{\delta}^u \\ \hline \lambda
\end{array}\right\}
=
\left\{\begin{array}{c}
\mathbf{f}_r \\ \mathbf{n}_r \\ \hline \mathbf{f}_o \\ \mathbf{n}_o \\ \mathbf{f}_t^b \\ \mathbf{f}_t^u \\ \hline \gamma^b
\end{array}\right\}
\tag{6.8}
$$

where $\mathbf{f}_o = \hat{\mathbf{I}}^{\mathrm{T}}(\mathbf{f} - \mathbf{M}\mathbf{w})$, $\mathbf{n}_o = -\hat{\tilde{\mathbf{b}}}^{\mathrm{T}}(\mathbf{f} - \mathbf{M}\mathbf{w})$, and $\mathbf{f}_t^j \equiv \mathbf{f}^j - \mathbf{M}^{j\prime}\mathbf{w} - \mathbf{K}^{j\prime}\delta\,; j = b,\ u.$

6.3. FREE-FREE MODES

We can transform Eq. 6.8 to modal coordinates using the free-free modal data. Using Eqs. 5.11-5.13, we get

$$
\left[\begin{array}{cc|ccc|c}
\mathbf{M}_r & \mathbf{0} & \mathbf{0} & \mathbf{0} & \mathbf{0} & \mathbf{I} \\
\mathbf{0} & \mathbf{J}_r & \mathbf{0} & \mathbf{0} & \mathbf{0} & -\tilde{\mathbf{s}}_r^{b\mathrm{T}} \\
\hline
\mathbf{0} & \mathbf{0} & m\mathbf{I} & \mathbf{0} & \mathbf{0} & -\mathbf{I} \\
\mathbf{0} & \mathbf{0} & \mathbf{0} & \mathbf{J} & \mathbf{0} & \tilde{\mathbf{b}}^{b\mathrm{T}} \\
\mathbf{0} & \mathbf{0} & \Psi_f^{\mathrm{T}}\mathbf{M}\hat{\mathbf{I}} & -\Psi_f^{\mathrm{T}}\mathbf{M}\hat{\tilde{\mathbf{b}}} & \boldsymbol{M}_f & -\Psi_f^{\mathrm{T}} \\
\hline
\mathbf{I} & -\tilde{\mathbf{s}}_r^{b} & -\mathbf{I} & \tilde{\mathbf{b}}^{b} & -\Psi_f & \mathbf{0}
\end{array}\right]
\left\{\begin{array}{c}
\ddot{\mathbf{r}}_r \\ \dot{\omega}_r \\ \hline \ddot{\mathbf{r}} \\ \dot{\omega} \\ \ddot{\mathbf{z}} \\ \hline \lambda
\end{array}\right\}
=
\left\{\begin{array}{c}
\mathbf{f}_r \\ \mathbf{n}_r \\ \hline \mathbf{f}_o \\ \mathbf{n}_o \\ \boldsymbol{f}_f \\ \hline \gamma^b
\end{array}\right\}
\tag{6.9}
$$

where $\boldsymbol{f}_f \equiv \Psi_f^{\mathrm{T}}(`\mathbf{f} - `\mathbf{M}`\mathbf{w}) - \boldsymbol{K}_f\mathbf{z}.$

6.4. CONSTRAINED MODES

Equation 6.8 can also be transformed to modal coordinates using the constrained modal data from Eqs. 5.14 and 5.15:

$$
\left[\begin{array}{cc|cccc|c}
\mathbf{M}_r & \mathbf{0} & \mathbf{0} & \mathbf{0} & \mathbf{0} & \mathbf{0} & \mathbf{I} \\
\mathbf{0} & \mathbf{J}_r & \mathbf{0} & \mathbf{0} & \mathbf{0} & \mathbf{0} & -\tilde{\mathbf{s}}_r^{b\mathrm{T}} \\
\hline
\mathbf{0} & \mathbf{0} & m\mathbf{I} & \mathbf{0} & \mathbf{0} & \mathbf{0} & -\mathbf{I} \\
\mathbf{0} & \mathbf{0} & \mathbf{0} & \mathbf{J} & \mathbf{0} & \mathbf{0} & \tilde{\mathbf{b}}^{b\mathrm{T}} \\
\mathbf{0} & \mathbf{0} & \mathbf{M}_1\hat{\mathbf{I}} & -\mathbf{M}_1\hat{\tilde{\mathbf{b}}} & `\mathbf{M}^{bb} & \mathbf{M}_3^{\mathrm{T}} & -\mathbf{A}^{\mathrm{T}} \\
\mathbf{0} & \mathbf{0} & \mathbf{M}_2\hat{\mathbf{I}} & -\mathbf{M}_2\hat{\tilde{\mathbf{b}}} & \mathbf{M}_3 & \boldsymbol{M}_c & \mathbf{0} \\
\hline
\mathbf{I} & -\tilde{\mathbf{s}}_r^{b} & -\mathbf{I} & \tilde{\mathbf{b}}^{b} & -\mathbf{A} & \mathbf{0} & \mathbf{0}
\end{array}\right]
\left\{\begin{array}{c}
\ddot{\mathbf{r}}_r \\ \dot{\omega}_r \\ \hline \ddot{\mathbf{r}} \\ \dot{\omega} \\ `\ddot{\delta}^b \\ \ddot{\mathbf{z}} \\ \hline \lambda
\end{array}\right\}
=
\left\{\begin{array}{c}
\mathbf{f}_r \\ \mathbf{n}_r \\ \hline \mathbf{f}_o \\ \mathbf{n}_o \\ `\mathbf{f}_t^b \\ \boldsymbol{f}_c \\ \hline \gamma^b
\end{array}\right\}
\tag{6.10}
$$

where $\mathbf{f}_t^b \equiv \mathbf{f}^b - \mathbf{M}^{b\prime}\mathbf{w} - \mathbf{K}^{b\prime}\delta$, $\boldsymbol{f}_c \equiv \Psi_c^{\mathrm{T}}(`\mathbf{f}^u - `\mathbf{M}^{u\prime}`\mathbf{w} - `\mathbf{K}^{ub}`\delta^b) - \boldsymbol{K}_c\mathbf{z}$, $\mathbf{M}_1 \equiv \bar{\mathbf{A}}^{\mathrm{T}}\mathbf{M}^{b\prime}$, $\mathbf{M}_2 \equiv \Psi_c^{\mathrm{T}}\mathbf{M}^{u\prime}$, and $\mathbf{M}_3 \equiv \Psi_c^{\mathrm{T}}`\mathbf{M}^{ub}$.

6.5. CONSTRAINED MODES AND STATIC CONDENSATION

Equations 5.19-5.21 transform Eq. 6.8 to the following form:

$$\left[\begin{array}{cc|cccc|c} \mathbf{M}_r & \mathbf{0} & \mathbf{0} & \mathbf{0} & \mathbf{0} & \mathbf{0} & \mathbf{I} \\ \mathbf{0} & \mathbf{J}_r & \mathbf{0} & \mathbf{0} & \mathbf{0} & \mathbf{0} & -\tilde{\mathbf{s}}_r^{b\,\mathrm{T}} \\ \hline \mathbf{0} & \mathbf{0} & m\mathbf{I} & \mathbf{0} & \mathbf{0} & \mathbf{0} & -\mathbf{I} \\ \mathbf{0} & \mathbf{0} & \mathbf{0} & \mathbf{J} & \mathbf{0} & \mathbf{0} & \tilde{\mathbf{b}}^{b\mathrm{T}} \\ \mathbf{0} & \mathbf{0} & \mathbf{M}_1\hat{\mathbf{I}} & -\mathbf{M}_1\hat{\tilde{\mathbf{b}}} & `\mathbf{M}_G & \mathbf{M}_{Gc}^{\mathrm{T}} & -\mathbf{A}^{\mathrm{T}} \\ \mathbf{0} & \mathbf{0} & \mathbf{M}_2\hat{\mathbf{I}} & -\mathbf{M}_2\hat{\tilde{\mathbf{b}}} & \mathbf{M}_{Gc} & \mathit{M}_c & \mathbf{0} \\ \hline \mathbf{I} & -\tilde{\mathbf{s}}_r^b & -\mathbf{I} & \tilde{\mathbf{b}}^b & -\mathbf{A} & \mathbf{0} & \mathbf{0} \end{array}\right] \left\{\begin{array}{c} \ddot{\mathbf{r}}_r \\ \dot{\omega}_r \\ \hline \ddot{\mathbf{r}} \\ \dot{\omega} \\ `\ddot{\delta}^b \\ \ddot{\mathbf{z}} \\ \hline \lambda \end{array}\right\} = \left\{\begin{array}{c} \mathbf{f}_r \\ \mathbf{n}_r \\ \hline \mathbf{f}_o \\ \mathbf{n}_o \\ `\mathbf{f}_t^b \\ f_c \\ \hline \gamma^b \end{array}\right\} \qquad (6.11)$$

where $`\mathbf{f}_t^b \equiv `\mathbf{f}^b - `\mathbf{M}^{b,`}\mathbf{w} + `\mathbf{G}^{\mathrm{T}}(`\mathbf{f}^u - `\mathbf{M}^{u,`}\mathbf{w}) - `\mathbf{K}_G `\delta^b$, $\quad f_c \equiv \Psi_c^{\mathrm{T}}(`\mathbf{f}^u - `\mathbf{M}^{u,`}\mathbf{w}) - \mathit{K}_c\mathbf{z}$, $\mathbf{M}_1 \equiv \overline{\mathbf{A}}^{\mathrm{T}}\mathbf{M}^{b,} + \mathbf{G}^{\mathrm{T}}\mathbf{M}^{u,}$, and $\mathbf{M}_2 \equiv \Psi_c^{\mathrm{T}}\mathbf{M}^{u,}$.

At any instant in time if the coordinates, deflections, velocities, and forces are known, Eqs. 6.7, 6.8, 6.9, 6.10, or 6.11 can be solved to determine the accelerations and Lagrange multipliers.

7. Model Reduction

In section 6 the equations of motion for a multibody system were stated in a variety of forms. Due to possible large number of deformable degrees-of-freedom, numerical integration of these equations may be too costly. In order to reduce these equations into a smaller set, some of the deformable body degrees-of-freedom and the associated equations must be eliminated. This, obviously, introduces approximation errors into the results. One contributing factor to this error is the degree of coupling between the kept and the deleted equations. Ideally, the kept and the deleted equations should be completely uncoupled. However, an inspection of the equations of motion reveals that the coupling between these two sets of equations exists in different forms, both in the coefficient matrix and also in the force array.

The degree of coupling in the mass matrix can be lowered if the mass matrix of the deformable body is diagonal (lumped mass matrix). For a non-diagonal mass matrix, if the elements around the boundary nodes are small relative to other elements, then the sub-matrices $\mathbf{M}^{bu}$ and $\mathbf{M}^{ub}$ could be neglected, in order to remove some of the coupling terms. In the following sub-sections we discuss some critical issues associated with the coupling terms in different formulations.

7.1. NODAL COORDINATES

In order to reduce the number of equations in the nodal space formulation, the static condensation process of section 5.1 can be applied to Eq. 6.8. With this process, some or all of the unconstrained nodes are removed and a set of condensed mass and stiffness matrices are obtained. It should be noted that with these condensed matrices, some of the simplifications that we considered in the equations of motion may no longer be valid. For example, we may need to use Eq. 4.10.a instead of 4.10.b in the derivation of Eq 6.8.

7.2. MODAL COORDINATES

The modal equations of motion for a single deformable body can be truncated to a smaller set since these equations are uncoupled between the kept and the deleted modes. However, when the deformable body is part of a multibody system, the reduction process is not as straight forward. As it will be seen in the following sub-sections, the coefficient matrix (mass matrix and/or the Jacobian matrix) contains coupling terms between the kept and the deleted modes.

For truncation purposes, the modal matrices are split into a kept set (superscript k) and a deleted set (superscript d). In the following sub-sections, only that portion of the modal equations of motion that are important in this process are shown. In the truncation process the main idea is to eliminate the modal accelerations $\ddot{\mathbf{z}}^d$ and its corresponding equations from the equations of motion.

7.2.1. *Free-Free Modes*
Equation 6.9 is split into the kept and deleted parts as

$$\begin{bmatrix} \cdots & -\Psi_f^{k\mathrm{T}}\mathbf{M}\hat{\mathbf{I}} & -\Psi_f^{k\mathrm{T}}\mathbf{M}\hat{\tilde{\mathbf{b}}} & M_f^k & \mathbf{0} & -\Psi_f^{k\mathrm{T}} \\ \cdots & -\Psi_f^{d\mathrm{T}}\mathbf{M}\hat{\mathbf{I}} & -\Psi_f^{d\mathrm{T}}\mathbf{M}\hat{\tilde{\mathbf{b}}} & \mathbf{0} & M_f^d & -\Psi_f^{d\mathrm{T}} \\ \cdots & -\mathbf{I} & \tilde{\mathbf{b}}^b & -\Psi_f^k & -\Psi_f^d & \mathbf{0} \end{bmatrix} \begin{Bmatrix} \vdots \\ \ddot{\mathbf{z}}^k \\ \ddot{\mathbf{z}}^d \\ \lambda \end{Bmatrix} = \begin{Bmatrix} \vdots \\ f_f^k \\ f_f^d \\ \gamma^b \end{Bmatrix} \tag{7.1}$$

where $f_f^j \equiv \Psi_f^{j\mathrm{T}}(\grave{\mathbf{f}} - \grave{\mathbf{M}}\grave{\mathbf{w}}) - K_f^j\mathbf{z}$; $j = k, d$. If we eliminate $\ddot{\mathbf{z}}^d$ and the corresponding equation, we observe that the term $-\Psi_f^d\ddot{\mathbf{z}}^d$ in the acceleration constraints is also eliminated. This elimination causes the constraints to be violated and to cause erroneous results.

7.2.2. *Constrained Modes*
Equation 6.10 is re-written in the following form for the reduction process:

$$\begin{bmatrix} \cdots & \grave{\mathbf{M}}^{bb} & \mathbf{M}_3^{k\mathrm{T}} & \mathbf{M}_3^{d\mathrm{T}} & -\mathbf{A}^\mathrm{T} \\ \cdots & \mathbf{M}_3^k & M_c^k & \mathbf{0} & \mathbf{0} \\ \cdots & \mathbf{M}_3^d & \mathbf{0} & M_c^d & \mathbf{0} \\ \cdots & -\mathbf{A} & \mathbf{0} & \mathbf{0} & \mathbf{0} \end{bmatrix} \begin{Bmatrix} \vdots \\ \grave{\ddot{\delta}}^b \\ \ddot{\mathbf{z}}^k \\ \ddot{\mathbf{z}}^d \\ \lambda \end{Bmatrix} = \begin{Bmatrix} \vdots \\ \grave{\mathbf{f}}^b \\ f_c^k \\ f_c^d \\ \gamma^b \end{Bmatrix} \tag{7.2}$$

where $f_c^j \equiv \Psi_c^{j\mathrm{T}}(\grave{\mathbf{f}}^u - \grave{\mathbf{M}}^{u,}\grave{\mathbf{w}} - \grave{\mathbf{K}}^{ub}\grave{\delta}^b) - K_c^j\mathbf{z}$; $j = k, d$. If we eliminate $\ddot{\mathbf{z}}^d$ and the corresponding equation, we observe that unlike Eq. 7.1, we do not eliminate anything from the constraint equation. Therefore there will be no violation in the constraints due to this process. However, the term $\mathbf{M}_3^{d\mathrm{T}}\ddot{\mathbf{z}}^d$ gets eliminated from the equation associated with the boundary node. This truncation process can produce enough error to make the results unacceptable.

7.2.3. *Constrained Modes and Static Condensation*
For the reduction process, Eq. 6.11 is re-written as

$$\begin{bmatrix} \cdots & \grave{\mathbf{M}}_G & \mathbf{M}_{Gc}^{k\ \mathrm{T}} & \mathbf{M}_{Gc}^{d\ \mathrm{T}} & -\mathbf{A}^\mathrm{T} \\ \cdots & \mathbf{M}_{Gc}^k & M_c^k & \mathbf{0} & \mathbf{0} \\ \cdots & \mathbf{M}_{Gc}^d & \mathbf{0} & M_c^d & \mathbf{0} \\ \cdots & -\mathbf{A} & \mathbf{0} & \mathbf{0} & \mathbf{0} \end{bmatrix} \begin{Bmatrix} \vdots \\ \grave{\ddot{\delta}}^b \\ \ddot{\mathbf{z}}^k \\ \ddot{\mathbf{z}}^d \\ \lambda \end{Bmatrix} = \begin{Bmatrix} \vdots \\ \grave{\mathbf{f}}_t^b \\ f_c^k \\ f_c^d \\ \gamma^b \end{Bmatrix} \tag{7.3}$$

where $\mathbf{M}_{Gc}^j \equiv \Psi_c^{j\mathrm{T}}(\grave{\mathbf{M}}^{ub} + \grave{\mathbf{M}}^{uu}\grave{\mathbf{G}})$ and $f_c^j \equiv \Psi_c^{j\mathrm{T}}(\grave{\mathbf{f}}^u - \grave{\mathbf{M}}^{u,}\grave{\mathbf{w}}) - K_c^j\mathbf{z}; j = k, d$. Eliminating $\ddot{\mathbf{z}}^d$ from this equation does not effect the constraint equation. However, we do eliminate the term $\mathbf{M}_{Gc}^{d\ \mathrm{T}}\ddot{\mathbf{z}}^d$ from the equation associated with the boundary node. The amount of error associated with this elimination is much smaller (almost negligible) compared to the elimination of $\mathbf{M}_3^{d^{\mathrm{T}}}\ddot{\mathbf{z}}^d$ from Eq. 7.2. This difference is due to the fact that in Eq. 7.2, $\boldsymbol{\delta}^u$ is purely dependent on the modal coordinates. However, in Eq. 7.3, $\boldsymbol{\delta}^u$ is a function of both $\boldsymbol{\delta}^b$ and the modal coordinates (refer to Eq 5.19). Therefore, truncating some of the modes from Eq. 7.3 is not as severe as that in Eq. 7.2.

7.2.4. *Constrained Modes and Modal Condensation*

The transformation matrix of Eq. 5.23 can be substituted into Eq. 7.2 in order to express the deleted modes in terms of the boundary nodal deflections. Then the equations of motion associated with the boundary nodes are pre-multiplied by Θ^{T} and the equations associated with the deleted modes are removed to obtain

$$\begin{bmatrix} \cdots & \mathbf{M}_{m1}\hat{\mathbf{I}} & -\mathbf{M}_{m1}\hat{\tilde{\mathbf{b}}} & \mathbf{M}_4 & \mathbf{M}_{m3}^{k\ \mathrm{T}} & -\Theta^{\mathrm{T}}\mathbf{A}^{\mathrm{T}} \\ \cdots & \mathbf{M}_2^k\hat{\mathbf{I}} & -\mathbf{M}_2^k\hat{\tilde{\mathbf{b}}} & \mathbf{M}_3^k & M_c^k & \mathbf{0} \\ \cdots & -\mathbf{I} & \tilde{\mathbf{b}}^b & -\mathbf{A} & \mathbf{0} & \mathbf{0} \end{bmatrix} \begin{Bmatrix} \vdots \\ {}'\ddot{\boldsymbol{\delta}}^b \\ \ddot{\mathbf{z}}^k \\ \boldsymbol{\lambda} \end{Bmatrix} = \begin{Bmatrix} \vdots \\ \Theta^{\mathrm{T}}\,{}'\mathbf{f}_t^b \\ f_c^k \\ \gamma^b \end{Bmatrix} \tag{7.4}$$

where $\mathbf{M}_{m1} \equiv \Theta^{\mathrm{T}}\mathbf{M}_1$, $\mathbf{M}_4 \equiv \Theta^{\mathrm{T}}({}'\mathbf{M}^{bb} + \mathbf{M}_3^{d^{\mathrm{T}}}\Theta)$, and $\mathbf{M}_{m3}^k \equiv \mathbf{M}_3^k\Theta$. Simulation with Eq. 7.4 provides almost the same results as those obtained from the truncated Eq. 7.3. The process that led us to Eq. 7.4 has restored some of the information associated with the deleted modes into the equations of motion. This is analogous to the restoration of information in Eq. 7.3 through the Guyan matrix. The main difference is that in Eq. 7.4 the information is restored in the modal space, whereas in Eq. 7.3 it is done in the nodal space.

8. Discussion

The two main issues that were discussed in this paper were how to attach a reference frame to a moving deformable body, and how to reduce the size of the degrees-of-freedom and the corresponding equations of motion for a deformable body within a multibody system environment. In this section we briefly discuss some of the important features, advantages and disadvantages of the techniques corresponding to each issue.

In order to show some of the features associated with the three choices of defining a body-frame for a deformable body, a simple multibody system is provided. The example consists of a deformable three-dimensional rectangular shaped plank suspended from the two ends by rigid links as shown in Fig. 5. The plank is modeled with solid finite elements and the links are modeled as rigid bodies. All six joints are assumed to be spherical. The system is released from an initial configuration shown in Fig. 5(a). The only external force acting on this system is the gravity. The plank drops to its extreme lowest position as shown in Fig. 5(b), it deforms, and then it bounces upward as expected. When the plank reaches its lowest position, large reaction forces are developed in the spherical joints. We expect the system to bounce upward symmetrically right after the plank hits the bottom. This is exactly the response we observe, as shown in Fig. 5(c), when we use the mean-axis or the principal-axes frames. However, the response from the nodal-fixed frame is non-symmetric as shown in Fig. 5(d).

Assume that with the nodal-fixed frame we have positioned the origin of the body-frame at node *i*. Node *j* is symmetrically positioned from node *i*. However, in the equations of motion with the nodal-fixed frame, nodes *i* and *j* are treated differently since node *i* is the origin of the body-frame and is not allowed to have any displacements with respect to the reference frame. In contrast node *j* is allowed to have displacement with respect to the body-frame. Due to the algorithmic, truncation, and round-off errors, after a while, specifically following an impulsive type reaction (plank hitting the bottom), the response can lose its symmetricity. This is not the case for the mean-axis and the principal-axes formulations since all the nodes are treated uniformly and there is no loss of symmetricity due to the accumulation of error. With the nodal-fixed frame, even if we position the origin of the body-frame at a centrally located node, such as node k in our example, the symmetricity issue will not be completely resolved. The reason is that the other three nodal-fixed conditions for the ξ and η axes will treat some of the symmetrically positioned nodes differently.

Possible advantages of using a nodal-fixed frame are: (a) having a slightly smaller number of equations and variables; (b) for most users familiar with the finite element modeling this is a natural way to define a body-frame. For the mean-axis and the principal-axes frames, the slight increase in the number of equations and variables should not be a major issue if we consider the overall large number of degrees-of-freedom of the system. The advantages gained from treating all of the nodes identically may be important when we deal with high-precision systems. The advantages or disadvantages between the mean-axis and the principal-axes could be a matter of preference. The rotational mean-axis condition is a non-holonomic equation where all of the principal-axes conditions are holonomic. This could be an issue to consider when we deal with monitoring of error and constraint violations at the coordinate level.

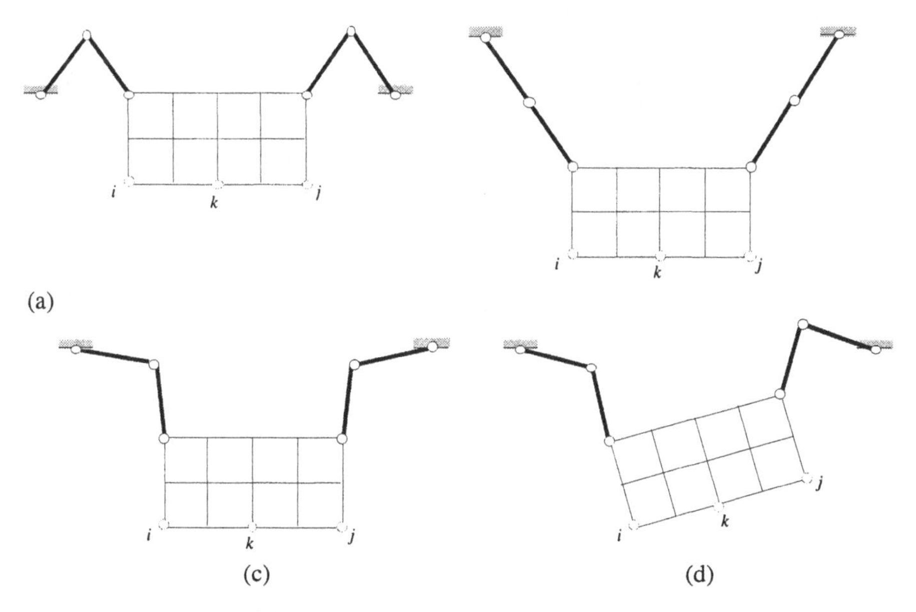

Figure 5. A rigid-flexible multibody system in its (a) initial configuration, (b) reaching the lowest configuration, (c) bouncing upward symmetrically, and (d) bouncing upward non-symmetrically.

Any of the choices of reference frames could be used with any of the model reduction techniques that were discussed in the paper. Again matching our choice of reference frame with our choice of model reduction is more or less a matter pr individual preference.

The simplest model reduction technique is the static condensation in nodal space. The mass and stiffness matrices are condensed once by eliminating some of the nodal degrees-of-freedom from a deformable body. This process normally causes the mass matrix to loose certain characteristics that causes some simplifications in the equations of motion not to be valid. Furthermore, it is possible that condensed mass and stiffness matrices to cause numerical instability. Therefore this reduction technique should be used with caution.

Model reduction in modal space without any compensation terms, either from a free-free or a constrained structure (such as in Eqs. 7.1 or 7.2), is not recommended. Even if we keep a large number of modes, these processes yield severe constraint violation and inaccurate results. In contrast, when compensation terms are incorporated in the condensation process (such as in Eqs. 7.3 and 7.4), reasonably accurate results should be expected with no constraint violation. The accuracy of the results is proportional to the degree of mode truncation. Determining compensation terms, either from a static condensation or a mode condensation process, is a simple task. Several variations of these compensation terms can be found in published literature.

9. References

1. Agrawal, O.P. and Shabana, A.A. (1986) Application of Deformable-Body Mean Axis to Flexible Multibody System Dynamics, *Computer methods in Applied Mechanics and Engineering* **56**, 217-245.

2. Nikravesh, P.E. (2001) The Use of Principal Axes in Moving Deformable Bodies for Multibody Dynamics, Unpublished class notes, University of Arizona.

3. Guyan, R.J. (1965) Reduction of Stiffness and Mass Matrices, *AIAA Journal*, Vol. **3**, No. 2, 380.

4. Hurty, W.C. (1965) Dynamic Analysis of Structural Systems Using Component Modes, *AIAA Journal*, Vol. **3**, No. 4, 678-685.

5. Craig, R.Jr.R. and Bampton, M.C.C. (1968) Coupling of Substructures for Dynamic Analyses, *AIAA Journal*, Vol. **6**, No. 7, 1313-1319.

6. Nikravesh, P.E. (1998) Model reduction with mode condensation, Unpublished class notes, University of Arizona.

7. Nikravesh, P.E. (1988) *Computer-Aided Analysis of Mechanical Systems*, Prentice-Hall.

UNILATERAL PROBLEMS OF DYNAMICS

F. Pfeiffer
Lehrstuhl für Angewandte Mechanik
Technische Universität München
85747 Garching
Germany

Abstract. Contact processes may be represented by local discretization, by a rigid approach or by a mixed method using both ideas. For the dynamics of mechanical systems a rigid body approach is described achieving good results also for multiple contact problems. This paper considers mainly contacts in multibody systems where the corresponding contact constraints vary with time thus generating structure-variant systems. The equations of motion for dynamical systems with such an unilateral behavior are discussed, solution methods and applications are presented.

Keywords: unilateral contacts, multibody dynamics, complementarity, friction, impacts

1. Introduction

A large variety of contact configurations exists in mechanical and biomechanical systems. Contacts may be closed, and the colliding bodies may detach again. Within a closed contact we might have sliding or stiction, both features connected with local friction. If two bodies come into contact, they usually penetrate into each other leading to local deformations. If contacts are accompanied by tangential forces and by tangential relative velocities within the contact plane, we get in addition to normal also tangential deformations. Depending on the dynamical (or statical) environment contacts may change their state, from closure to detachment, from sliding to stiction, and vice versa. We call a contact active, if it is closed or if we have stiction, otherwise we call a contact passive.

Active contacts always exhibit contact forces which in the general case of normal and tangential deformations follow from the local material properties of the colliding bodies, from the external (with respect to the contact) dynamics (statics) and from external forces. Considering contacts that way leads to complicated

W. Schiehlen and M. Valášek (eds.), Virtual Nonlinear Multibody Systems, 103–140.

problems of continuum mechanics which as a rule must be solved by numerical algorithms like FEM or BEM. For the treatment of dynamical problems this approach is too costly and in many cases also not adequate. The rigid body approach gives quicker results and applies better to large dynamical systems.

Under rigid body approach we understand a contact behavior characterized at least in the local contact zone by no deformations and thus by rigid body properties. The contact process is then governed by certain contact laws like those by Newton and Poisson and appropriate extensions of them. A fundamental law with respect to rigid body models is the complementarity rule, sometimes called corner's law or Signorinis's law. It states that in contact dynamics either relative kinematical quantities are zero and the accompanying constraint forces or constraint force combinations are not zero, or vice versa. For a closed contact the relative normal distance and normal velocity of the colliding bodies are zero and the constraint force in normal direction is not zero, or vice versa. For stiction the tangential relative velocity is zero, and the constraint force is located within the friction cone, which means that the difference of the static friction force and constraint force is not zero, or vice versa. The resulting inequalities are indispensable for an evaluation of the transitions between the various contact states.

From these properties follows a well defined indicator behavior giving the transition phases. For normal passive contacts the normal relative distance (or velocity) of the colliding bodies indicates the contact state. If it becomes zero the contact will be active, the indicator "relative distance" becomes a constraint accompanied by a constraint force. The end of an active contact is then indicated by the constraint force. If it changes sign, indicating a change from pressure to tension, we get detachment, and the contact again transits to a passive state.

In the case of friction we have in the passive state a non-vanishing tangential relative velocity as an indicator. If it becomes zero there might be a transition from sliding to stiction depending on the force balance within or on the friction cone. The indicator "relative tangential velocity" then becomes a constraint leading to tangential constraint forces. The contact remains active as long as the maximum static friction force is larger than the constraint force, which means that there is a force balance within the friction cone. If this "friction saturation" becomes zero the contact again might go into a passive state with non-zero tangential relative velocity.

If we deal with multibody systems including unilateral constraints the problem of multiple contacts and their interdependences arises. A straightforward solution of these processes would come out with a combinatiorial problem of huge dimension. Therefore a formulation applying complementarity rules and the resulting inequalities is a must. The mathematical methods developed in that area the last twenty years assure nearly always an unambiguous solution even in the cases of mutually dependent contacts.

The field of multibody dynamics with unilateral contacts must be seen before the background of various research activities in the area of variational inequalities in connection with convex energy functions on the one side [20], [21], [27], [36] and in the area of hemivariational inequalities in connection with non-convex energy functions [28]. Both areas are young, the first one being around 35 years and the second only 15 years old. The two scientists who are the most connected with this development are Moreau in Montpellier, France, and Panagiotopoulos in Thessaloniki, Greece. Moreau started already in the seventies to formulate the non-continuous properties of non-smooth mechanics [21], [22], [23] and not much later Panagiotopoulos considered inequality problems [27] which led him consistently to the development of hemivariational inequalities [28]. Most non-smooth mechanical problems possess non-convex features. Nearly all applications at that time were of statical or quasi-statical nature.

In Sweden Lötstedt considered a series of practical examples [15], [16] and established a school, which in more recent time has been continued very successfully by Klarbring and his group [11], [12]. Klarbring also treats statical and quasi-statical problems with applications to FEM.

The sphere of influence of these scientists increases since the significance of the area has been noticed. Especially [10] gives an excellent contribution to dry friction, also in connection with frictional impacts. The book [20] presents a proof with respect to the existence of Jean's – Moreau's impact theory. A very good survey on this topic may be found in [5]. In the meantime literature is increasing considerably, therefore only a few examples are given.

The papers [13], [14] deal with a two-dimensional contact including dry friction. The resulting linear complementarity problem is solved by a modified Simplex-algorithm. The authors of [18], [25] study the self-excitation of frictional vibrations. Frictional constraints are described by variational inequalities and evaluated on the basis of measured friction characteristics. The paper [1] discusses among other topics also the existence of unambiguous solutions for the generalized accelerations concerning frictional contact problems, a question by the way, which also has been regarded in [15], [16].

Non-smooth mechanics allows a general theoretical description but relevant problems must be solved numerically. Inspite of many valueable contributions to the numerics of non-continuous systems [24] the existing algorithms are still extremely time-consuming. As was indicated above we get for plane constacts a linear and for spatial contacts a nonlinear complementarity problem. For linear complementarity problems exist solution procedures [24] on the basis of extended linear programming theories. The existence of a solution is assured. This is not the case for nonlinear complementarity. At the time being various approximations have been suggested [11], [12], [17] and applied to practical problems ([38] to [44]). The algorithms work in most cases convincingly, but the computing times explode because of the iterative character of nearly all solution methods. Therefore the numerical solution of all kinds of complementarity problems is still a topic of current research.

The author's institute prosecutes applied and engineering mechanics and is involved in unilateral problems since more than fifteen years exclusively in connection with multibody dynamics. The fundaments of that area might be found in [3], [4] which present multibody theory on the basis of the projection method yielding the most efficient theory for multibody applications. In the course of the years and very much due to practical requirements this type of multibody theory has been combined with inequality formulations to describe the unilateral behavior of multiple contacts [29], [30], [31]. Especially the already mentioned complementarity properties in combination with the kinematics and kinetics of the appropriate indicators were the main keys for a consistent description of multibody systems with many unilateral contacts [31]. An extension to impacts before the background of Moreau's work was then straightforward [8], [9]. The corresponding theory has recently and successfully been confirmed by extensive measurements [2].

Based on these fundaments a large variety of industrial applications has been considered. The paper [6] and the dissertation [7] deal with roller chains very often used in automotive industry. Roller chains typically possess some hundred degrees of freedom and hundreds of contacts, where also stick-slip processes take place in the chain-wheel contacts. CVT-chains for Continuous Variable Transmissions do not have such a large amount of degrees of freedom but particularly difficult spatial contacts within the conic wheel discs. Even efficiencies have been evaluated with an excellent agreement of theory and measurements [33], [39], [40]. Another area of application is the field of assembly processes in manufacturing, where during mating processes contacts are closed and detached, and stick-slip phenomena (up to jamming) occur. Large efforts have been made to describe in a very general way the relative kinematics of colliding bodies in terms of differential geometry [19]. The results were convincing. They are now used in all other applications.

Another example of manufacturing concerns vibration conveyors which are very often used to transport, to orient and to select small parts like screws, bolts, electronic components and the like. Without the theory on multibody systems with unilateral contacts a treatment of such a machine would have been impossible. The dissertation [44] solves this problem theoretically and proves it experimentally.

In the following we shall give a survey of the theory including impacts with friction, and we shall present some typical applications.

2. Models

With models we mean the mechanical especially the dynamical behavior of our system under consideration expressed by the laws of kinematics and kinetics. This leads to a mathematical description in the form of nonlinear differential equations with inequality constraints, which as a rule must be solved numerically. As we treat multibody systems with unilateral contacts we start with a short representation of elastic multibody dynamics, proceed to contact kinematics in terms of differential geometry and explain the special features of unilateral contact behavior. We then formulate the equations of motion including the non-smoothness of our system and give finally a survey on impacts with friction described by a newly developed theory.

2.1 MULTIBODY DYNAMICS

The theory of multibody dynamics including interconnected rigid and/or elastic bodies is in the meantime well established and also commercialized to a large extend. Most of the existing computer codes are based on the so-called projection method, which considers the free motion within the hyper-spaces tangential to the constraint surfaces. The Jacobians performing these projections are derivations with respect to the constraint equations usually formulated on a velocity level [3], [4], [31]. As the equations of motion for multibody systems represent the basis for all further unilateral considerations we shall give a short survey.

We consider a multibody system with f degrees of freedom which we later shall subdivide in "rigid" and "elastic" degrees of freedom. Applying the principle of virtual power (Jourdain) we get

$$\sum_{m_i} \int_{m_i} \left(\frac{\partial_R \boldsymbol{v}}{\partial \dot{\boldsymbol{q}}} \right)_i^T ({}_R\boldsymbol{a} - {}_R\boldsymbol{f})_i \, dm_i = 0 \ , \tag{1}$$

where ${}_R\boldsymbol{v}_i \in \mathbb{R}^3$, ${}_R\boldsymbol{a}_i \in \mathbb{R}^3$ are the absolute velocity, acceleration of body i in a body-fixed reference system, ${}_R\boldsymbol{f}_i \in \mathbb{R}^3$ are all forces applied to body i and $\dot{\boldsymbol{q}} \in \mathbb{R}^f$ are the generalized velocities. m_i is the mass of body i. The velocity and acceleration vectors must be evaluated in a body-fixed frame which yields (see Figure 1)

$$\begin{aligned} {}_R\boldsymbol{v} &= {}_R\boldsymbol{v}_A + {}_R\tilde{\boldsymbol{\omega}}_{IR} \left({}_R\boldsymbol{x}_0 + {}_R\overline{\boldsymbol{r}}_{el} \right) + {}_R\dot{\overline{\boldsymbol{r}}}_{el} \ , \\ {}_R\boldsymbol{a} &= {}_R\boldsymbol{a}_A + {}_R\tilde{\boldsymbol{\omega}}_{IR} \, {}_R\tilde{\boldsymbol{\omega}}_{IR} \, {}_R\boldsymbol{x}_0 + {}_R\dot{\tilde{\boldsymbol{\omega}}}_{IR} \, {}_R\boldsymbol{x}_0 + {}_R\tilde{\boldsymbol{\omega}}_{IR} \, {}_R\tilde{\boldsymbol{\omega}}_{IR} \, {}_R\overline{\boldsymbol{r}}_{el} + \\ & \quad + {}_R\dot{\tilde{\boldsymbol{\omega}}}_{IR} \, {}_R\overline{\boldsymbol{r}}_{el} + 2 {}_R\tilde{\boldsymbol{\omega}}_{IR} \, {}_R\dot{\overline{\boldsymbol{r}}}_{el} + {}_R\ddot{\overline{\boldsymbol{r}}}_{el} . \end{aligned} \tag{2}$$

The following abbreviations are used (index i always omitted):

$\boldsymbol{v} = {}_R\boldsymbol{v}_A$	absolute velocity, acceleration of the reference base R,
$\boldsymbol{a} = {}_R\boldsymbol{a}_A$	expressed in the R-system,
$\boldsymbol{\omega} = {}_R\boldsymbol{\omega}_{IR}$	angular velocity of the reference system expressed in the R-system,
$\boldsymbol{x}_0 = {}_R\boldsymbol{x}_0$	vector from R to the mass element dm, in the undeformed configuration,
$\overline{\boldsymbol{r}} = {}_R\overline{\boldsymbol{r}}_{el}$	displacement vector,
$\boldsymbol{J}_T = \left(\frac{\partial_R \boldsymbol{v}_A}{\partial \dot{\boldsymbol{q}}} \right)$	Jacobian of translation, $\boldsymbol{J}_T \in \mathbb{R}^{3,f}$,
$\boldsymbol{J}_R = \left(\frac{\partial_R \boldsymbol{\omega}_{IR}}{\partial \dot{\boldsymbol{q}}} \right)$	Jacobian of rotation, $\boldsymbol{J}_R \in \mathbb{R}^{3,f}$,
$\tilde{\boldsymbol{\omega}} \boldsymbol{r} = \boldsymbol{\omega} \times \boldsymbol{r}$	definition of tilde tensor.

Combining the equations (1), (2) and the above abbreviations we come out with

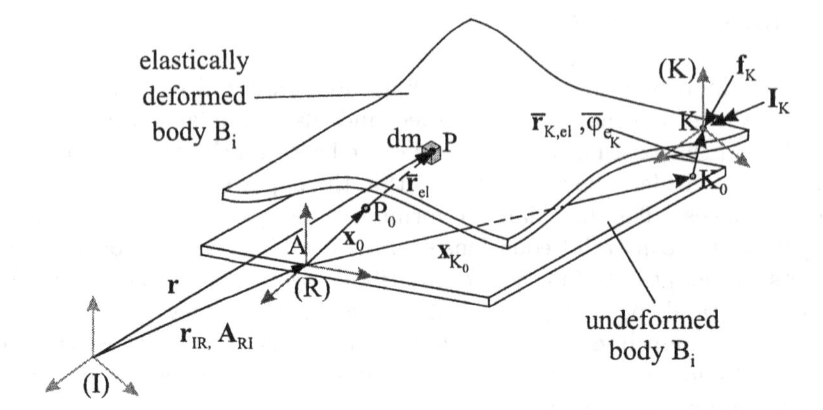

Figure 1: Coordinates for a deformed body B_i

the following set of equations of motion

$$\sum \int_{m_i} \left[\boldsymbol{J}_T^T + \boldsymbol{J}_R^T \left(\tilde{\boldsymbol{x}}_0 + \tilde{\overline{\boldsymbol{r}}} \right) + \left(\frac{\partial \dot{\overline{\boldsymbol{r}}}}{\partial \dot{\boldsymbol{q}}} \right)^T \right] \cdot \\ \cdot \left(\boldsymbol{a} + \tilde{\boldsymbol{\omega}}\tilde{\boldsymbol{\omega}}\boldsymbol{x}_0 + \dot{\tilde{\boldsymbol{\omega}}}\boldsymbol{x}_0 + \tilde{\boldsymbol{\omega}}\tilde{\boldsymbol{\omega}}\overline{\boldsymbol{r}} + \dot{\tilde{\boldsymbol{\omega}}}\overline{\boldsymbol{r}} + 2\tilde{\boldsymbol{\omega}}\dot{\overline{\boldsymbol{r}}} + \ddot{\overline{\boldsymbol{r}}} - \boldsymbol{f} \right) dm = 0, \tag{3}$$

where elastic influences are regarded as shown in Figure 1. The radius vector $\boldsymbol{r}$ from an inertial frame (I) to a mass element dm_i of the deformed body B_i can be written as

$$\boldsymbol{r} = \boldsymbol{r}_{IR} + \boldsymbol{x}_0 + \overline{\boldsymbol{r}}_{el} \ , \tag{4}$$

where $\boldsymbol{r}_{IR}$ is the vector from (I) to (R), $\boldsymbol{x}_0$ the vector to the mass element in the undeformed configuration and $\overline{\boldsymbol{r}}_{el}$ the displacement vector. As in most cases of technical relevancy we assume that the elastic displacements of the components are very small compared to their geometric dimensions.

This allows the introduction of a Ritz-approach for the elastic deformation [4]

$$\overline{\boldsymbol{r}}_i(\boldsymbol{x}_0, t) = \boldsymbol{W}_i(\boldsymbol{x}_o)\, \boldsymbol{q}_{el,i}(t) \ , \\ \boldsymbol{W}_i = (\boldsymbol{w}_1, \ldots, \boldsymbol{w}_j, \ldots)_i = \begin{pmatrix} \overline{\overline{\boldsymbol{r}}}_x^T \\ \overline{\overline{\boldsymbol{r}}}_y^T \\ \overline{\overline{\boldsymbol{r}}}_z^T \end{pmatrix}_i \ , \tag{5}$$

with $(\boldsymbol{q}_{el} \in \mathbb{R}^{n_{el}})_i$ and $(\boldsymbol{W} \in \mathbb{R}^{3,n_{el}})_i$. This well-known superposition of Ansatz- or shape-functions $\boldsymbol{w}_{ji}(\boldsymbol{x}_0)$ requires their completeness property [4]. These shape functions might be evaluated by measurements, by FEM-calculations, by analytic approaches, or they might be approximated by cubic spline systems. Anyway, they must be in accordance with the elastic behavior of the system.

Combining equations (3), (4), (5) we must consider the dependencies $\boldsymbol{a}(\ddot{\boldsymbol{q}})$, $\dot{\boldsymbol{\omega}}(\ddot{\boldsymbol{q}})$, $\ddot{\boldsymbol{q}}_{el,i}(\ddot{\boldsymbol{q}})$ which says that all absolute and elastic accelerations depend on the generalized accelerations $\ddot{\boldsymbol{q}}$. This property can be expressed by

$$\begin{aligned} \boldsymbol{a} &= \boldsymbol{J}_T\ddot{\boldsymbol{q}} + \underline{\boldsymbol{a}}(\boldsymbol{q},\dot{\boldsymbol{q}},t) \ , \\ \dot{\boldsymbol{\omega}} &= \boldsymbol{J}_R\ddot{\boldsymbol{q}} + \underline{\dot{\boldsymbol{\omega}}}(\boldsymbol{q},\dot{\boldsymbol{q}},t) \ . \end{aligned} \tag{6}$$

$$\ddot{\boldsymbol{q}}_{el,i} = \frac{\partial \boldsymbol{q}_{el,i}}{\partial \boldsymbol{q}}\ddot{\boldsymbol{q}} = \frac{\partial \dot{\boldsymbol{q}}_{el,i}}{\partial \dot{\boldsymbol{q}}}\ddot{\boldsymbol{q}} = \boldsymbol{J}_E\ddot{\boldsymbol{q}} \ . \tag{7}$$

The equations (3) to (7) can now be put into the form

$$\begin{aligned} \sum_i \Bigg\{ \int_{m_i} & \boldsymbol{J}_T^T \left[\boldsymbol{J}_T - (\tilde{\boldsymbol{x}}\tilde{\bar{\boldsymbol{r}}})\boldsymbol{J}_R + \boldsymbol{W}\boldsymbol{J}_E\right] dm + \\ & + \int_{m_i} \boldsymbol{J}_R^T(\tilde{\boldsymbol{x}}_0 + \tilde{\bar{\boldsymbol{r}}}) \left[\boldsymbol{J}_T - (\tilde{\boldsymbol{x}}_0 + \tilde{\bar{\boldsymbol{r}}})\boldsymbol{J}_R + \boldsymbol{W}\boldsymbol{J}_E\right] dm + \\ & + \int_{m_i} \boldsymbol{J}_E^T\boldsymbol{W}^T \left[\boldsymbol{J}_T - (\tilde{\boldsymbol{x}}_0 + \tilde{\bar{\boldsymbol{r}}})\boldsymbol{J}_R + \boldsymbol{W}\boldsymbol{J}_E\right] dm \Bigg\} \ddot{\boldsymbol{q}} + \\ & + \int_{m_i} \left[\boldsymbol{J}_T^T + \boldsymbol{J}_R^T(\tilde{\boldsymbol{x}}_0 + \tilde{\bar{\boldsymbol{r}}}) + \boldsymbol{J}_E^T\boldsymbol{W}^T\right] \times \\ & \times \left[\underline{\boldsymbol{a}} + \tilde{\omega}\tilde{\omega}(\tilde{\boldsymbol{x}}_0 + \bar{\boldsymbol{r}}) + \underline{\dot{\boldsymbol{\omega}}}(\tilde{\boldsymbol{x}}_0 + \bar{\boldsymbol{r}}) + 2\tilde{\omega}\dot{\bar{\boldsymbol{r}}} - \boldsymbol{f}\right] dm = 0 \ . \end{aligned} \tag{8}$$

After lengthy and tedious calculations it is always possible to bring equations (8) into a standard form, which we need for all further considerations. It writes

$$\begin{aligned} & \boldsymbol{M}(\boldsymbol{q},t)\ddot{\boldsymbol{q}} - \boldsymbol{h}(\boldsymbol{q},\dot{\boldsymbol{q}},t) = 0 \\ & (\boldsymbol{q} \in \mathbb{R}^f, \ \ \boldsymbol{h} \in \mathbb{R}^f, \ \ \boldsymbol{M} \in \mathbb{R}^{f,f}) \end{aligned} \tag{9}$$

Equations (9) include all bilateral constraints. They represent the maximum number of minimal (generalized) coordinates, and from this all unilateral contacts are in a state that they do not block any further degree of freedom. If some of the unilateral constraints become active, then the remaining number of degrees of freedom is smaller than f.

2.2 CONTACT KINEMATICS

Geometry and kinematics are the fundaments in establishing models of dynamical systems. In the case of unilateral contacts this is especially important, because magnitudes of relative kinematics serve as indicators for passive contacts and as constraints for active contacts (see chapter 1). As most of the applications require more or less arbitrary body contours it makes sense to derive the kinematical contact equations in a general form applying well-known rules of the differential geometry of surfaces [8], [19], [31].

In practice we find two types of contacts, two- and three-dimensional ones. For the two-dimensional case the contacting bodies lie in a plane thus defining a contact line with given direction. Only the sense of direction has to be determined. Problems of that kind are connected with linear complementarity.

For the three-dimensional case the contacting bodies have a spatial form, and the contact takes place in a plane allowing two tangential directions. The resulting direction for the contact process is not known beforehand and usually must be determined iteratively. Problems of that kind are connected with nonlinear complementarity.

2.2.1 *Plane Contact Kinematics*

Plane contact kinematics has been presented in the dissertation [8] and since then applied to many practical problems (see also [31]). We start with the geometry of a single body a motion as indicated in Figure 2. We assume a convex contour and describe it by a parameter s. Connecting with s a moving trihedral $(\boldsymbol{t}, \boldsymbol{n}, \boldsymbol{b})$ and introducing a body-fixed frame B we write

$$\begin{aligned}
&{}_B\boldsymbol{t} = {}_B\boldsymbol{r}'_{P\Sigma}; \ \kappa_B\boldsymbol{n} = {}_B\boldsymbol{r}''_{P\Sigma}; \ (\ \cdot\ )' = \tfrac{d}{ds} \ ; \\
&{}_B\boldsymbol{n} = {}_B\boldsymbol{b} \times {}_B\boldsymbol{t}; \ {}_B\boldsymbol{b} = {}_B\boldsymbol{t} \times {}_B\boldsymbol{n} \ ; \\
&{}_B\boldsymbol{t} = {}_B\boldsymbol{n} \times_B \boldsymbol{b} \ . \\
&{}_B\boldsymbol{n}' = {}_B\boldsymbol{b} \times {}_B\boldsymbol{t}' = {}_B\boldsymbol{b} \times {}_B\boldsymbol{n}\kappa = -\kappa_B\boldsymbol{t} \\
&{}_B\boldsymbol{t}' = \kappa_B\boldsymbol{n} \ .
\end{aligned} \tag{10}$$

For planar contours the binormal ${}_B\boldsymbol{b}$ is constant. Therefore

$$\begin{aligned}
{}_B\dot{\boldsymbol{n}} &= {}_B\boldsymbol{n}'\dot{s} = -\kappa\dot{s}\, {}_B\boldsymbol{t} \ , \\
{}_B\dot{\boldsymbol{t}} &= {}_B\boldsymbol{t}'\dot{s} = +\kappa\dot{s}\, {}_B\boldsymbol{n} \ .
\end{aligned} \tag{11}$$

On the other hand, the absolute changes of $\boldsymbol{n}$ and $\boldsymbol{t}$ are given by the Coriolis equation

$$\begin{aligned}
{}_B(\dot{\boldsymbol{n}}) &= {}_B\dot{\boldsymbol{n}} + {}_B\tilde{\boldsymbol{\Omega}}\, {}_B\boldsymbol{n} \ , \\
{}_B(\dot{\boldsymbol{t}}) &= {}_B\dot{\boldsymbol{t}} + {}_B\tilde{\boldsymbol{\Omega}}\, {}_B\boldsymbol{t} \ ,
\end{aligned} \tag{12}$$

where we must keep in mind that ${}_B\boldsymbol{\omega}_{IB} = {}_B\boldsymbol{\Omega}$ for body-fixed frames B. Putting (11) into (12) we get a coordinate-free representation of the overall changes $\dot{\boldsymbol{n}}, \dot{\boldsymbol{t}}$.

$$\begin{aligned}
\dot{\boldsymbol{n}} &= \tilde{\boldsymbol{\Omega}}\boldsymbol{n} - \kappa\dot{s}\boldsymbol{t} \ , \\
\dot{\boldsymbol{t}} &= \tilde{\boldsymbol{\Omega}}\boldsymbol{t} + \kappa\dot{s}\boldsymbol{n} \ ,
\end{aligned} \tag{13}$$

which we can evaluate in any basis. The main advantage of (13) consists of the eliminated, frame-dependent differentiations ${}_B\dot{\boldsymbol{n}}$ and ${}_B\dot{\boldsymbol{t}}$.

In the same manner we proceed with the contour vector $\boldsymbol{r}_{P\Sigma}$. According to (11), (12) we write

$$\begin{aligned}
{}_B\dot{\boldsymbol{r}}_{P\Sigma} &= {}_B\boldsymbol{r}'_{P\Sigma}\dot{s} = \dot{s}\, {}_B\boldsymbol{t} \ , \\
{}_B(\dot{\boldsymbol{r}}_{P\Sigma}) &= {}_B\dot{\boldsymbol{r}}_{P\Sigma} + {}_B\tilde{\boldsymbol{\Omega}}_B\boldsymbol{r}_{P\Sigma} \ ,
\end{aligned} \tag{14}$$

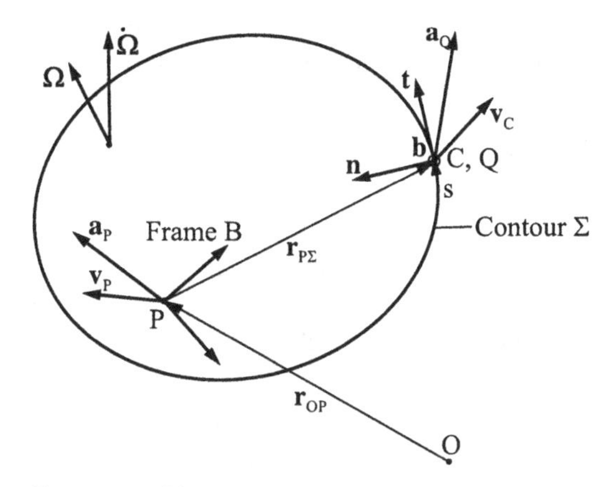

Figure 2: Planar contour geometry

and eliminate ${}_B\dot{\boldsymbol{r}}_{P\Sigma}$. Then we get the absolute changes of $\boldsymbol{r}_{P\Sigma}$,

$$\dot{\boldsymbol{r}}_{P\Sigma} = \tilde{\boldsymbol{\Omega}}\boldsymbol{r}_{P\Sigma} + \dot{s}\boldsymbol{t} \ . \tag{15}$$

Due to $\boldsymbol{v}_\Sigma = \boldsymbol{v}_P + \dot{\boldsymbol{r}}_{P\Sigma}$, the absolute velocity of the moving contour point is given by

$$\boldsymbol{v}_\Sigma = \boldsymbol{v}_P + \tilde{\boldsymbol{\Omega}}\boldsymbol{r}_{P\Sigma} + \dot{s}\boldsymbol{t} \ , \tag{16}$$

where

$$\boldsymbol{v}_C := \boldsymbol{v}_P + \tilde{\boldsymbol{\Omega}}\boldsymbol{r}_{P\Sigma} \ . \tag{17}$$

The velocity $\boldsymbol{v}_C$ results from rigid body kinematics and corresponds to the velocity of a body-fixed point at the contour. From (16) and (17) we see that

$$\boldsymbol{v}_\Sigma = \boldsymbol{v}_C + \dot{s}\boldsymbol{t} \ . \tag{18}$$

Next, we want to derive the absolute acceleration of C by differentiating (17) with respect to time:

$$\dot{\boldsymbol{v}}_C = \dot{\boldsymbol{v}}_P + \dot{\tilde{\boldsymbol{\Omega}}}\boldsymbol{r}_{P\Sigma} + \tilde{\boldsymbol{\Omega}}\dot{\boldsymbol{r}}_{P\Sigma} \ . \tag{19}$$

With $\dot{\boldsymbol{v}}_C = \boldsymbol{a}_C, \dot{\boldsymbol{v}}_P = \boldsymbol{a}_P$ and $\dot{\boldsymbol{r}}_{P\Sigma}$ from eq. (14) we get

$$\boldsymbol{a}_C = \boldsymbol{a}_P + \dot{\tilde{\boldsymbol{\Omega}}}\boldsymbol{r}_{P\Sigma} + \tilde{\boldsymbol{\Omega}}\tilde{\boldsymbol{\Omega}}\boldsymbol{r}_{P\Sigma} + \tilde{\boldsymbol{\Omega}}\boldsymbol{t}\dot{s} \tag{20}$$

which is *not* the acceleration of a body-fixed point on the contour. Only the part

$$\boldsymbol{a}_Q := \boldsymbol{a}_P + \dot{\tilde{\boldsymbol{\Omega}}}\boldsymbol{r}_{P\Sigma} + \tilde{\boldsymbol{\Omega}}\tilde{\boldsymbol{\Omega}}\boldsymbol{r}_{P\Sigma} \tag{21}$$

corresponds to such an acceleration, so we can write

$$\boldsymbol{a}_C = \boldsymbol{a}_Q + \tilde{\boldsymbol{\Omega}}\boldsymbol{t}\dot{s} \ . \tag{22}$$

Later we have to determine the relative velocities of contact points in the normal and tangential directions and their time derivatives. For this purpose we introduce the scalars

$$v_n = \boldsymbol{n}^T \boldsymbol{v}_C; \;\; v_t = \boldsymbol{t}^T \boldsymbol{v}_C \tag{23}$$

and state their derivatives as

$$\begin{aligned} \dot{v}_n &= \dot{\boldsymbol{n}}^T \boldsymbol{v}_C + \boldsymbol{n}^T \dot{\boldsymbol{v}}_C \;, \\ \dot{v}_t &= \dot{\boldsymbol{t}}^T \boldsymbol{v}_C + \boldsymbol{t}^T \dot{\boldsymbol{v}}_C \;. \end{aligned} \tag{24}$$

With $\dot{\boldsymbol{n}}, \dot{\boldsymbol{t}}$ from (13), $\dot{\boldsymbol{v}}_C = \boldsymbol{a}_C$ from (22), and noting $\boldsymbol{n}^T \tilde{\boldsymbol{\Omega}} \boldsymbol{t} = \boldsymbol{b}^T \boldsymbol{\Omega}$, $\boldsymbol{t}^T \tilde{\boldsymbol{\Omega}} \boldsymbol{t} = 0$ we derive

$$\begin{aligned} \dot{v}_n &= \boldsymbol{n}^T \left(\boldsymbol{a}_Q - \tilde{\boldsymbol{\Omega}} \boldsymbol{v}_C \right) - \kappa \dot{s} \boldsymbol{t}^T \boldsymbol{v}_C + \dot{s} \boldsymbol{b}^T \boldsymbol{\Omega} \\ \dot{v}_t &= \boldsymbol{t}^T \left(\boldsymbol{a}_Q - \tilde{\boldsymbol{\Omega}} \boldsymbol{v}_C \right) + \kappa \dot{s} \boldsymbol{n}^T \boldsymbol{v}_C \;. \end{aligned} \tag{25}$$

With this basis we are able to derive the relative kinematics of two bodies such as relative distances, relative velocities and accelerations. Without going into details and refering especially to [31] we summarize the following relationships (see Fig. 3):

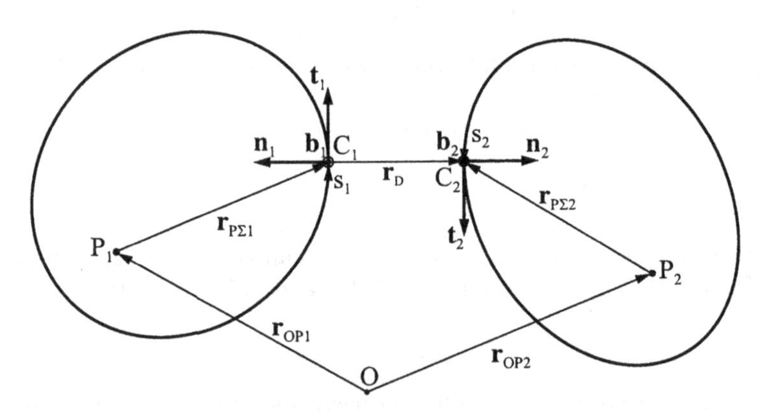

Figure 3: General orientation of two bodies

Potential contact points can be characterized by

$$\boldsymbol{n}_1^T(s_1) \cdot \boldsymbol{t}_2(s_2) = 0 \;\; \Leftrightarrow \;\; \boldsymbol{n}_2^T(s_2) \cdot \boldsymbol{t}_1(s_1) = 0 \;. \tag{26}$$

$$\boldsymbol{r}_D^T(s_1, s_2) \cdot \boldsymbol{t}_1(s_1) = 0, \;\; \boldsymbol{r}_D^T(s_1, s_2) \cdot \boldsymbol{t}_2(s_2) = 0 \;. \tag{27}$$

From each set we need only one equation. The relative distance g_N is

$$g_N(\boldsymbol{q}, t) = \boldsymbol{r}_D^T \boldsymbol{n}_2 = -\boldsymbol{r}_D^T \boldsymbol{n}_1. \tag{28}$$

Since the normal vectors always point inward, g_N is positive for separation and negative for overlapping. Therefore, a changing sign of g_N from positive to negative indicates a transition from initially separated bodies to contact.

With these equations and considering Fig. 3 we derive the relative velocites in normal and tangential direction

$$\dot{g}_N = \boldsymbol{n}_1^T \boldsymbol{v}_{C1} + \boldsymbol{n}_2^T \boldsymbol{v}_{C2}; \;\; \dot{g}_T = \boldsymbol{t}_1^T \boldsymbol{v}_{C1} + \boldsymbol{t}_2^T \boldsymbol{v}_{C2}, \tag{29}$$

where $\boldsymbol{v}_{C1}, \boldsymbol{v}_{C2}$ are the absolute velocities of the potential contact points C_1, C_2. These velocities might be expressed by the generalized velocities $\dot{\boldsymbol{q}}$ and some Jacobians $\boldsymbol{J}_{C1}, \boldsymbol{J}_{C2}$ to give

$$\boldsymbol{v}_{C1} = \boldsymbol{J}_{C1}\dot{\boldsymbol{q}} + \tilde{\boldsymbol{j}}_{C1}; \;\; \boldsymbol{v}_{C2} = \boldsymbol{J}_{C2}\dot{\boldsymbol{q}} + \tilde{\boldsymbol{j}}_{C2}. \tag{30}$$

Putting (29) into (30) yields

$$\dot{g}_N = \boldsymbol{w}_N^T \dot{\boldsymbol{q}} + \tilde{w}_N; \;\; \dot{g}_T = \boldsymbol{w}_T^T \dot{\boldsymbol{q}} + \tilde{w}_T \tag{31}$$

with

$$\begin{aligned} \boldsymbol{w}_N &= \boldsymbol{J}_{C1}^T \boldsymbol{n}_1 + \boldsymbol{J}_{C2}^T \boldsymbol{n}_2; \\ \boldsymbol{w}_T &= \boldsymbol{J}_{C1}^T \boldsymbol{t}_1 + \boldsymbol{J}_{C2}^T \boldsymbol{t}_2, \\ \tilde{w}_N &= \tilde{\boldsymbol{j}}_{C1}^T \boldsymbol{n}_1 + \tilde{\boldsymbol{j}}_{C2}^T \boldsymbol{n}_2; \\ \tilde{w}_T &= \tilde{\boldsymbol{j}}_{C1}^T \boldsymbol{t}_1 + \tilde{\boldsymbol{j}}_{C2}^T \boldsymbol{t}_2, \end{aligned} \tag{32}$$

which we use in the following as a representation of the relative velocities. It may be noticed here that a negative value of $\dot{g}_N$ corresponds to an approaching process of the bodies and coincides at vanishing distance $g_N = 0$ with the relative velocity in the normal direction before an impact. In the case of a continual contact ($g_N = \dot{g}_N = 0$) the term $\dot{g}_T$ shows the relative sliding velocity of the bodies, which we can use to determine the time points of transitions from sliding ($\dot{g}_T \neq 0$) to sticking or rolling ($\dot{g}_T = 0$).

The relevant accelerations follow from a further time differentiation. We get

$$\ddot{g}_N = \boldsymbol{w}_N^T \ddot{\boldsymbol{q}} + \overline{w}_N; \;\; \ddot{g}_T = \boldsymbol{w}_T^T \ddot{\boldsymbol{q}} + \overline{w}_T \tag{33}$$

where $\boldsymbol{w}_N, \boldsymbol{w}_T$ are given by (32), and $\overline{w}_n, \overline{w}_T$ are

$$\begin{aligned} \overline{w}_N &= \boldsymbol{n}_1^T (\bar{\boldsymbol{j}}_{Q1} - \tilde{\boldsymbol{\Omega}}_1 \boldsymbol{v}_{C1}) - \kappa_1 \dot{s}_1 \boldsymbol{t}_1^T \boldsymbol{v}_{C1} + \dot{s}_1 \boldsymbol{b}_{12}^T \boldsymbol{\Omega}_1 \\ &\quad + \boldsymbol{n}_2^T (\bar{\boldsymbol{j}}_{Q2} - \tilde{\boldsymbol{\Omega}}_2 \boldsymbol{v}_{C2}) - \kappa_2 \dot{s}_2 \boldsymbol{t}_2^T \boldsymbol{v}_{C2} + \dot{s}_2 \boldsymbol{b}_{12}^T \boldsymbol{\Omega}_2, \\ \overline{w}_T &= \boldsymbol{t}_1^T (\bar{\boldsymbol{j}}_{Q1} - \tilde{\boldsymbol{\Omega}}_1 \boldsymbol{v}_{C1}) + \kappa_1 \dot{s}_1 \boldsymbol{n}_1^T \boldsymbol{v}_{C1} \\ &\quad + \boldsymbol{t}_2^T (\bar{\boldsymbol{j}}_{Q2} - \tilde{\boldsymbol{\Omega}}_2 \boldsymbol{v}_{C2}) + \kappa_2 \dot{s}_2 \boldsymbol{n}_2^T \boldsymbol{v}_{C2}. \end{aligned} \tag{34}$$

with

$$\begin{aligned} \dot{s}_1 &= \frac{\kappa_2 \boldsymbol{t}_1^T (\boldsymbol{v}_{C2} - \boldsymbol{v}_{C1}) - \kappa_2 g_N \boldsymbol{b}_{12}^T \boldsymbol{\Omega}_1 + \boldsymbol{b}_{12}^T (\boldsymbol{\Omega}_2 - \boldsymbol{\Omega}_1)}{\kappa_1 + \kappa_2 + g_N \kappa_1 \kappa_2} \\ \dot{s}_2 &= \frac{\kappa_1 \boldsymbol{t}_1^T (\boldsymbol{v}_{C2} - \boldsymbol{v}_{C1}) - \kappa_1 g_N \boldsymbol{b}_{12}^T \boldsymbol{\Omega}_2 - \boldsymbol{b}_{12}^T (\boldsymbol{\Omega}_2 - \boldsymbol{\Omega}_1)}{\kappa_1 + \kappa_2 + g_N \kappa_1 \kappa_2} \end{aligned} \tag{35}$$

The angular velocities $\boldsymbol{\Omega}_1, \boldsymbol{\Omega}_2$ relate to the two contacting bodies (Fig. 2, 3).

2.2.2 *Spatial Contact Kinematics*

Spatial contact kinematics has been presented in the dissertation [19] and since then also has been applied to many practical problems. The situation in this case is of course more complex. We still assume that the two approaching bodies are convex (Figure 4) at least in that area where contact points might occur. The two bodies are moving with $\boldsymbol{v}_i, \boldsymbol{\Omega}_i (i = 1, 2)$. For the description of a surface Σ we need two parameters s and t : $\boldsymbol{r}_\Sigma = \boldsymbol{r}_\Sigma(s, t)$. The tangents $\boldsymbol{s}$ and $\boldsymbol{t}$, which span the tangent plane at a point of the surface, are defined as:

$$\boldsymbol{s} = \frac{\partial \boldsymbol{r}_\Sigma}{\partial s}\,, \quad \boldsymbol{t} = \frac{\partial \boldsymbol{r}_\Sigma}{\partial t}\;. \tag{36}$$

From these basic vectors the fundamental magnitudes of the first order are calculated:

$$E = \boldsymbol{s}^T\boldsymbol{s}\,, \quad F = \boldsymbol{s}^T\boldsymbol{t}\,, \quad G = \boldsymbol{t}^T\boldsymbol{t}\,. \tag{37}$$

The normalized normal vector $\boldsymbol{n}$ is perpendicular to the tangent plane and pointing outwards:

$$\boldsymbol{n} = \frac{\boldsymbol{s} \times \boldsymbol{t}}{\sqrt{EG - F^2}}\,. \tag{38}$$

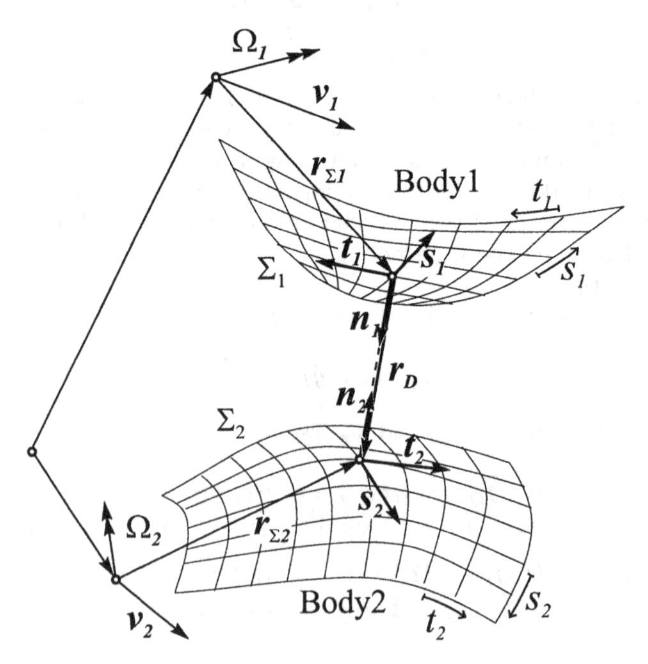

Figure 4: Contact geomety of two surfaces

We further need the fundamental magnitudes of the second order:

$$L = \boldsymbol{n}^T \frac{\partial^2 \boldsymbol{r}_\Sigma}{\partial s^2}, \quad M = \boldsymbol{n}^T \frac{\partial^2 \boldsymbol{r}_\Sigma}{\partial s \partial t}, \quad N = \boldsymbol{n}^T \frac{\partial^2 \boldsymbol{r}_\Sigma}{\partial t^2}. \tag{39}$$

For a contact point we demand, that the normal vector of body 1 ($\boldsymbol{n}_1$) and the distance vector $\boldsymbol{r}_D$ are perpendicular to the tangent vectors of body 2 ($\boldsymbol{s}_2$ and $\boldsymbol{t}_2$). Thus we obtain four nonlinear equations:

$$\begin{aligned} \boldsymbol{n}_1^T \boldsymbol{s}_2 &= 0\,, & \boldsymbol{r}_D^T \boldsymbol{s}_2 &= 0\,, \\ \boldsymbol{n}_1^T \boldsymbol{t}_2 &= 0\,, & \boldsymbol{r}_D^T \boldsymbol{t}_2 &= 0\,. \end{aligned} \tag{40}$$

This nonlinear problem has to be solved at every time step of the numerical integration. After the solution is found the distance g_N between the possible contact points can be calculated as

$$g_N = \boldsymbol{n}_1^T \boldsymbol{r}_D = -\boldsymbol{n}_2^T \boldsymbol{r}_D\,. \tag{41}$$

g_N is used as an indicator for the contact state. Its value is positive for 'no contact' and negative for penetration. The constraints are again formulated on velocity level, where in the spatial case we have three of them, one in normal direction $\dot{g}_N$ and two in the tangential directions $\dot{g}_S, \dot{g}_T$:

$$\begin{aligned} \dot{g}_N(\boldsymbol{q},\dot{\boldsymbol{q}},t) &= \boldsymbol{n}_1^T(\boldsymbol{v}_{\Sigma 2} - \boldsymbol{v}_{\Sigma 1})\,, \\ \dot{g}_S(\boldsymbol{q},\dot{\boldsymbol{q}},t) &= \boldsymbol{s}_1^T(\boldsymbol{v}_{\Sigma 2} - \boldsymbol{v}_{\Sigma 1})\,, \\ \dot{g}_T(\boldsymbol{q},\dot{\boldsymbol{q}},t) &= \boldsymbol{t}_1^T(\boldsymbol{v}_{\Sigma 2} - \boldsymbol{v}_{\Sigma 1})\,, \end{aligned} \tag{42}$$

with $\boldsymbol{v}_{\Sigma 1}$ and $\boldsymbol{v}_{\Sigma 2}$ being defined in an analoguous way as in (30). Differentiating these equations with respect to time leads to the constraints on acceleration level:

$$\begin{aligned} \ddot{g}_N &= \boldsymbol{n}_1^T(\dot{\boldsymbol{v}}_{\Sigma 2} - \dot{\boldsymbol{v}}_{\Sigma 1}) + \dot{\boldsymbol{n}}_1^T(\boldsymbol{v}_{\Sigma 2} - \boldsymbol{v}_{\Sigma 1})\,, \\ \ddot{g}_S &= \boldsymbol{s}_1^T(\dot{\boldsymbol{v}}_{\Sigma 2} - \dot{\boldsymbol{v}}_{\Sigma 1}) + \dot{\boldsymbol{s}}_1^T(\boldsymbol{v}_{\Sigma 2} - \boldsymbol{v}_{\Sigma 1})\,, \\ \ddot{g}_T &= \boldsymbol{t}_1^T(\dot{\boldsymbol{v}}_{\Sigma 2} - \dot{\boldsymbol{v}}_{\Sigma 1}) + \dot{\boldsymbol{t}}_1^T(\boldsymbol{v}_{\Sigma 2} - \boldsymbol{v}_{\Sigma 1})\,. \end{aligned} \tag{43}$$

The time derivatives of the contact point velocities $\boldsymbol{v}_{\Sigma 1}$ and $\boldsymbol{v}_{\Sigma 2}$ can be written in the form:

$$\begin{aligned} \dot{\boldsymbol{v}}_{\Sigma 1} &= \boldsymbol{J}_{\Sigma 1}(\boldsymbol{q},t)\ddot{\boldsymbol{q}} + \bar{\boldsymbol{j}}_{\Sigma 1}(\dot{\boldsymbol{q}},\boldsymbol{q},t)\,, \\ \dot{\boldsymbol{v}}_{\Sigma 2} &= \boldsymbol{J}_{\Sigma 2}(\boldsymbol{q},t)\ddot{\boldsymbol{q}} + \bar{\boldsymbol{j}}_{\Sigma 2}(\dot{\boldsymbol{q}},\boldsymbol{q},t)\,. \end{aligned} \tag{44}$$

The vectors $\dot{\boldsymbol{n}}_1, \dot{\boldsymbol{s}}_1$ and $\dot{\boldsymbol{t}}_1$ are determined by the formulas of Weingarten and Gauss, which express the derivatives of the normal vector and of the tangent vectors in terms of the basic vectors:

$$\begin{aligned} \dot{\boldsymbol{n}}_1 &= \boldsymbol{\Omega}_1 \times \boldsymbol{n}_1 + \frac{\partial \boldsymbol{n}_1}{\partial s_1}\dot{s}_1 + \frac{\partial \boldsymbol{n}_1}{\partial t_1}\dot{t}_1\,, \qquad \text{where:} \\ \frac{\partial \boldsymbol{n}_1}{\partial s_1} &= \underbrace{\frac{M_1 F_1 - L_1 G_1}{E_1 G_1 - F_1^2}}_{\alpha_1} \boldsymbol{s}_1 + \underbrace{\frac{L_1 F_1 - M_1 E_1}{E_1 G_1 - F_1^2}}_{\beta_1} \boldsymbol{t}_1\,, \end{aligned} \tag{45}$$

$$\frac{\partial \boldsymbol{n}_1}{\partial t_1} = \underbrace{\frac{N_1F_1 - M_1G_1}{E_1G_1 - F_1^2}}_{\alpha_1'}\boldsymbol{s}_1 + \underbrace{\frac{M_1F_1 - N_1E_1}{E_1G_1 - F_1^2}}_{\beta_1'}\boldsymbol{t}_1 ,$$

$$\begin{aligned}
\dot{\boldsymbol{s}}_1 &= \boldsymbol{\Omega}_1 \times \boldsymbol{s}_1 + \frac{\partial \boldsymbol{s}_1}{\partial s_1}\dot{s}_1 + \frac{\partial \boldsymbol{s}_1}{\partial t_1}\dot{t}_1 , \qquad \text{where:} \\
\frac{\partial \boldsymbol{s}_1}{\partial s_1} &= \Gamma^1_{11,1}\boldsymbol{s}_1 + \Gamma^2_{11,1}\boldsymbol{t}_1 + L_1\boldsymbol{n}_1 , \\
\frac{\partial \boldsymbol{s}_1}{\partial t_1} &= \Gamma^1_{12,1}\boldsymbol{s}_1 + \Gamma^2_{12,1}\boldsymbol{t}_1 + M_1\boldsymbol{n}_1 ,
\end{aligned} \tag{46}$$

$$\begin{aligned}
\dot{\boldsymbol{t}}_1 &= \boldsymbol{\Omega}_1 \times \boldsymbol{t}_1 + \frac{\partial \boldsymbol{t}_1}{\partial s_1}\dot{s}_1 + \frac{\partial \boldsymbol{t}_1}{\partial t_1}\dot{t}_1 , \qquad \text{where:} \\
\frac{\partial \boldsymbol{t}_1}{\partial s_1} &= \Gamma^1_{12,1}\boldsymbol{s} + \Gamma^2_{12,1}\boldsymbol{t} + M_1\boldsymbol{n}_1 , \\
\frac{\partial \boldsymbol{t}_1}{\partial t_1} &= \Gamma^1_{22,1}\boldsymbol{s} + \Gamma^2_{22,1}\boldsymbol{t} + N_1\boldsymbol{n}_1 .
\end{aligned} \tag{47}$$

The definition of the Christoffel symbols $\Gamma^\sigma_{\alpha\beta}$, $\alpha, \beta, \sigma = 1, 2$, can be found in standard textbooks. Inserting equations (45), (46), (47) in (43) yields the constraint equations:

$$\begin{aligned}
\ddot{g}_N = {} & \boldsymbol{n}_1^T(\boldsymbol{J}_{\Sigma 2} - \boldsymbol{J}_{\Sigma 1})\ddot{q} + \boldsymbol{n}_1^T\left(\bar{\boldsymbol{j}}_{\Sigma 2} - \bar{\boldsymbol{j}}_{\Sigma 1}\right) + (\boldsymbol{v}_{\Sigma 2} - \boldsymbol{v}_{\Sigma 1})^T(\boldsymbol{\Omega}_1 \times \boldsymbol{n}_1) + (\boldsymbol{v}_{\Sigma 2} - \boldsymbol{v}_{\Sigma 1})^T \cdot \\
& \left((\alpha_1\boldsymbol{s}_1 + \beta_1\boldsymbol{t}_1)\dot{s}_1 + (\alpha_1'\boldsymbol{s}_1 + \beta_1'\boldsymbol{t}_1)\dot{t}_1\right) , \\
\ddot{g}_S = {} & \boldsymbol{s}_1^T(\boldsymbol{J}_{\Sigma 2} - \boldsymbol{J}_{\Sigma 1})\ddot{q} + \boldsymbol{s}_1^T\left(\bar{\boldsymbol{j}}_{\Sigma 2} - \bar{\boldsymbol{j}}_{\Sigma 1}\right) + (\boldsymbol{v}_{\Sigma 2} - \boldsymbol{v}_{\Sigma 1})^T(\boldsymbol{\Omega}_1 \times \boldsymbol{s}_1) + (\boldsymbol{v}_{\Sigma 2} - \boldsymbol{v}_{\Sigma 1})^T \cdot \\
& \left((\Gamma^1_{11,1}\boldsymbol{s}_1 + \Gamma^2_{11,1}\boldsymbol{t}_1 + L_1\boldsymbol{n}_1)\dot{s}_1 + (\Gamma^1_{12,1}\boldsymbol{s}_1 + \Gamma^2_{12,1}\boldsymbol{t}_1 + M_1\boldsymbol{n}_1)\dot{t}_1\right) , \\
\ddot{g}_T = {} & \boldsymbol{t}_1^T(\boldsymbol{J}_{\Sigma 2} - \boldsymbol{J}_{\Sigma 1})\ddot{q} + \boldsymbol{t}_1^T\left(\bar{\boldsymbol{j}}_{\Sigma 2} - \bar{\boldsymbol{j}}_{\Sigma 1}\right) + (\boldsymbol{v}_{\Sigma 2} - \boldsymbol{v}_{\Sigma 1})^T(\boldsymbol{\Omega}_1 \times \boldsymbol{t}_1) + (\boldsymbol{v}_{\Sigma 2} - \boldsymbol{v}_{\Sigma 1})^T \cdot \\
& \left((\Gamma^1_{12,1}\boldsymbol{s}_1 + \Gamma^2_{12,1}\boldsymbol{t}_1 + M_1\boldsymbol{n}_1)\dot{s}_1 + (\Gamma^1_{22,1}\boldsymbol{s}_1 + \Gamma^2_{22,1}\boldsymbol{t}_1 + N_1\boldsymbol{n}_1)\dot{t}_1\right) .
\end{aligned} \tag{48}$$

As can be seen, these equations depend only on the Jacobians with respect to the contact points $\boldsymbol{J}_{\Sigma 1}$, $\boldsymbol{J}_{\Sigma 2}$, the basic vectors of the surfaces and the time derivatives of the contour parameters $\dot{s}_1, \dot{t}_1, \dot{s}_2, \dot{t}_2$. The Jacobians are known from the rigid body algorithm, the basic vectors from the surface description. The time derivatives of the contour parameters can be calculated by deriving equation (40) with respect to time:

$$\begin{aligned}
\left(\boldsymbol{n}_1^T\boldsymbol{s}_2\right)^\bullet &= 0 , \qquad & \left(\boldsymbol{r}_D^T\boldsymbol{s}_2\right)^\bullet &= 0 , \\
\left(\boldsymbol{n}_1^T\boldsymbol{t}_2\right)^\bullet &= 0 , \qquad & \left(\boldsymbol{r}_D^T\boldsymbol{t}_2\right)^\bullet &= 0 ,
\end{aligned} \tag{49}$$

which means, that the conditions for the contact point should not change while the two bodies are moving. Evaluating eq. (49) we obtain a system of equations which are linear in the derivatives of the contour parameters

$$\begin{pmatrix} \boldsymbol{s}_2^T(\alpha_1\boldsymbol{s}_1+\beta_1\boldsymbol{t}_1) & \boldsymbol{s}_2^T(\alpha_1'\boldsymbol{s}_1+\beta_1'\boldsymbol{t}_1) & L_2 & M_2 \\ \boldsymbol{t}_2^T(\alpha_1\boldsymbol{s}_1+\beta_1\boldsymbol{t}_1) & \boldsymbol{t}_2^T(\alpha_1'\boldsymbol{s}_1+\beta_1'\boldsymbol{t}_1) & M_2 & N_2 \\ -\boldsymbol{s}_1^T\boldsymbol{s}_2 & -\boldsymbol{s}_1^T\boldsymbol{s}_2 & \boldsymbol{s}_2^T\boldsymbol{s}_2 & \boldsymbol{s}_2^T\boldsymbol{t}_2 \\ -\boldsymbol{s}_1^T\boldsymbol{t}_2 & -\boldsymbol{s}_1^T\boldsymbol{t}_2 & \boldsymbol{s}_2^T\boldsymbol{t}_2 & \boldsymbol{t}_2^T\boldsymbol{t}_2 \end{pmatrix} \cdot \begin{pmatrix} \dot{s}_1 \\ \dot{t}_1 \\ \dot{s}_2 \\ \dot{t}_2 \end{pmatrix} = \begin{pmatrix} (\boldsymbol{s}_2\times\boldsymbol{n}_1)^T(\boldsymbol{\Omega}_2-\boldsymbol{\Omega}_1) \\ (\boldsymbol{t}_2\times\boldsymbol{n}_1)^T(\boldsymbol{\Omega}_2-\boldsymbol{\Omega}_1) \\ \boldsymbol{s}_2^T(\boldsymbol{v}_{\Sigma 1}-\boldsymbol{v}_{\Sigma 2}) \\ \boldsymbol{t}_2^T(\boldsymbol{v}_{\Sigma 1}-\boldsymbol{v}_{\sigma 2}) \end{pmatrix} \tag{50}$$

This linear problem has to be solved at every time step of numerical integration. Let us summarize the constraint equations in the well known form, by rewriting eq. (48):

$$\begin{aligned} \ddot{g}_N &= \boldsymbol{w}_N^T\ddot{\boldsymbol{q}}+\overline{w}_N \, , \\ \ddot{g}_S &= \boldsymbol{w}_S^T\ddot{\boldsymbol{q}}+\overline{w}_S \, , \\ \ddot{g}_T &= \boldsymbol{w}_T^T\ddot{\boldsymbol{q}}+\overline{w}_T \, . \end{aligned} \tag{51}$$

The terms in eq. (48), which are linearly dependent on $\ddot{\boldsymbol{q}}$, are collected in the constraint vectors $\boldsymbol{w}_N, \boldsymbol{w}_S$ and $\boldsymbol{w}_T$, all the rest is included in the scalars $\overline{w}_N, \overline{w}_S, \overline{w}_T$.

2.3 MULTIPLE UNILATERAL CONTACTS

Multiple contacts in multibody systems include a combinatorial problem of large dimensions. If the state in one contact changes, for example from contact to detachment or from slip to stick, all other contacts are also influenced which makes a search for a new set of contact configurations necessary. To not prosecuting the combinatorial process we need extended contact laws which describe unambiguously the transitions for the possible contact states and which generate only consistent contact configurations. In a first step we define all contact sets, which can be found in a multibody system [31]:

$$\begin{aligned} I_A &= \{1,2,\ldots,n_A\} && \text{with } n_A \text{ elements} \\ I_C &= \{i \in I_A : g_{Ni}=0\} && \text{with } n_C \text{ elements} \\ I_N &= \{i \in I_C : \dot{g}_{Ni}=0\} && \text{with } n_N \text{ elements} \\ I_T &= \{i \in I_N : |\dot{\boldsymbol{g}}_{Ti}|=0\} && \text{with } n_T \text{ elements} \end{aligned} \tag{52}$$

These sets describe the kinematic state of each contact point. The set I_A consists of the n_A indices of all contact points. As an example we consider eq. 9. It belongs to the set combination $I_A \setminus \{I_C, I_N, I_T\}$. The elements of the set I_C are the n_C indices of the unilateral constraints with vanishing normal distance $g_{Ni}=0$, but arbitrary relative velocity in the normal direction. In the index set I_N are the n_N indices of the potentially active normal constraints which fulfill the necessary conditions for continuous contact (vanishing normal distance $g_{Ni}=0$ and no relative velocity $\dot{g}_{Ni}$ in the normal direction). The index set I_N includes for example all contact states with slipping. The n_T elements of the set I_T are the indices of the potentially active tangential constraints. The corresponding normal constraints are closed and the relative velocities $\dot{g}_{Ti}$ in the tangential direction are zero. The numbers of elements of the index sets I_C, I_N and I_T are not constant because there are variable states of constraints due to separation and stick-slip phenomena.

As a next step we must organize all transitions from contact to detachment and from stick to slip and the corresponding reversed transitions. In **normal direction** of a contact we find the following situation [31]:

- Passive contact i
  $g_{Ni}(\boldsymbol{g}, t) \geq 0, \;\; \lambda_{Ni} = 0, \;\;$ indicator g_{Ni}
- Transition to contact
  $g_{Ni}(\boldsymbol{g}, t) = 0, \;\; \lambda_{Ni} \geq 0$
- Active contact i
  $g_{Ni}(\boldsymbol{g}, t) = 0, \;\; \lambda_{Ni} > 0$, indicator λ_{Ni}, constraint $g_{Ni} = 0$
- Transition to detachment
  $g_{Ni}(\boldsymbol{q}, t) \geq 0, \;\; \lambda_{Ni} = 0$ (53)

The kinematical magnitudes $g_{Ni}, \dot{g}_{Ni}, \ddot{g}_{Ni}$ are given with the equations (28), (31), (33) and (41), (42), (51), where $\dot{g}_{Ni}, \ddot{g}_{Ni}$ are needed when we go to a velocity or an acceleration level. The constraint forces λ_{Ni} must be compressive forces. If they change sign, we get separation. The properties eq. (53) establish a complementarity behavior which might be expressed by n_N (set I_N) complementarity conditions (put on an acceleration level)

$$\ddot{\boldsymbol{g}}_N \geq 0 \; ; \; \boldsymbol{\lambda}_N \geq 0 \; ; \; \ddot{\boldsymbol{g}}_N^T \boldsymbol{\lambda}_N = 0 \; . \tag{54}$$

The variational inequality

$$-\ddot{\boldsymbol{g}}_N^T(\boldsymbol{\lambda}_N^* - \boldsymbol{\lambda}_N) \leq 0 \; ; \; \boldsymbol{\lambda}_N \in C_N \; ; \; \forall \boldsymbol{\lambda}_N^* \in C_N \; , \tag{55}$$

is equivalent to the complementary conditions (54). The convex set

$$C_N = \{\boldsymbol{\lambda}_N^* : \boldsymbol{\lambda}_N^* \geq 0\} \tag{56}$$

contains all admissible contact forces λ_{Ni}^* in the normal direction [31], [41].

The complementarity problem defined in eq. (54) might be interpreted as a corner law which requires for each contact $\ddot{g}_{Ni} \geq 0, \;\; \lambda_{Ni} \geq 0, \;\; \ddot{g}_{Ni}\lambda_{Ni} = 0$. Figure 5 illustrates this property.

With respect to the **tangential direction** of a contact we shall confine our considerations to the application of Coulomb's friction law which in no way means a loss of generality. The complementary behavior is a characteristic feature of all

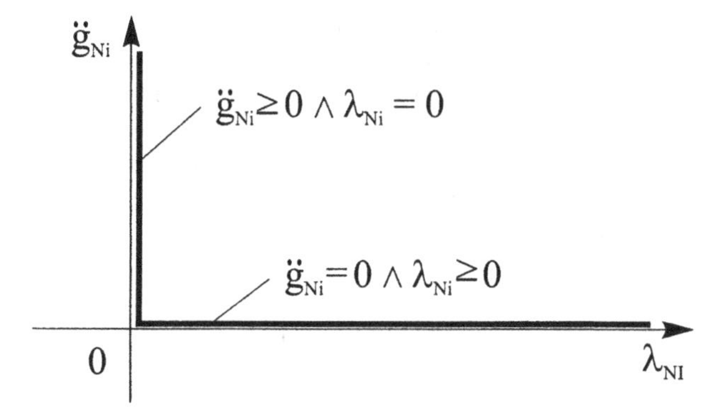

Figure 5: Corner law for normal contacts (Signorini's law)

contact phenomena independent of the specific physical law of contact. Furthermore we assume that within the infinitesimal small time step for a transition from stick to slip and vice versa the coefficients of static and sliding friction are the same, which may be expressed by

$$\lim_{\dot{g}_{Ti} \to 0} \mu_i(\dot{g}_{Ti}) = \mu_{0i} \tag{57}$$

For $\dot{g}_{Ti} \neq 0$ any friction law may be applied (see Fig. 6). With this property Coulomb's friction law distinguishes between the two cases

$$\begin{array}{llcll} \text{stiction:} & |\boldsymbol{\lambda}_{Ti}| < \mu_{0i}\lambda_{Ni} & \Rightarrow & |\dot{\boldsymbol{g}}_{Ti}| = 0 & (\text{Set } I_T) \\ \text{sliding:} & |\boldsymbol{\lambda}_{Ti}| = \mu_{0i}\lambda_{Ni} & \Rightarrow & |\dot{\boldsymbol{g}}_{Ti}| > 0 & (\text{Set } I_N \backslash I_T) \end{array} \tag{58}$$

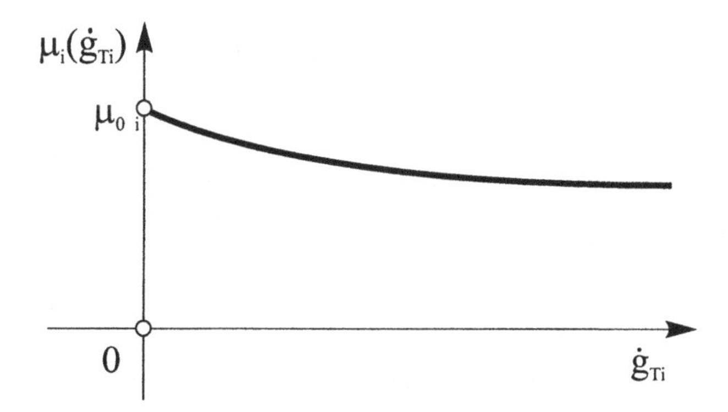

Figure 6: Typical friction characteristic

Equation (58) formulates the mechanical property, that we are for a frictional contact within the friction cone if the relative tangential velocity is zero and the tangential constraint force $|\boldsymbol{\lambda}_{Ti}|$ is smaller than the maximum static friction force $(\mu_{0i}\lambda_{Ni})$. Then we have stiction. We are on the friction cone if we slide with $|\dot{g}_{Ti}| > 0$. At a transition point the friction force is then $(\mu_{0i}\lambda_{Ni})$ (see eq. 57). In addition we must regard the fact that in the tangential contact plane we might get one or two directions according to a plane or a spatial contact. From this we summarize in the following way:

- Passive contact i (Sliding, Set $I_N \backslash I_T$)
  $|\dot{g}_{Ti}| \geq 0, \ |\mu_{0i}\lambda_{Ni}| - |\boldsymbol{\lambda}_{Ti}| = 0$, indicator $|\dot{g}_{Ti}|$
- Transition Slip to Stick
  $|\dot{g}_{Ti}| = 0, \ |\mu_{0i}\lambda_{Ni}| - |\boldsymbol{\lambda}_{Ti}| \geq 0$
- Active contact i (Sticking, Set I_T)
  $|\dot{g}_{Ti}| = 0, \ |\mu_{0i}\lambda_{Ni}| - |\boldsymbol{\lambda}_{Ti}| > 0$, indicator $|\mu_{0i}\lambda_{Ni}| - |\boldsymbol{\lambda}_{Ti}|$, constraint $|\dot{g}_{Ti}| = 0$
- Transition Stick to Slip
  $|\dot{g}_{Ti}| \geq 0, \ |\mu_{0i}\lambda_{Ni}| - |\boldsymbol{\lambda}_{Ti}| = 0$ (59)

From a numerical standpoint of view we have to check the indicator for a change of sign, which then requires a subsequent interpolation. For a transition from stick to slip one must examine the possible development of a non-zero relative tangential acceleration as a start for sliding.

Equation (58) put on an acceleration level can then be written in a more detailed form

$$\begin{aligned} |\lambda_{Ti}| &< \mu_{0i}\lambda_{Ni} \wedge \ddot{g}_{Ti} = 0 \quad (i \in I_T \text{ sticking}) \\ \lambda_{Ti} &= +\mu_{0i}\lambda_{Ni} \wedge \ddot{g}_{Ti} \leq 0 \quad (i \in I_N \backslash I_T \text{ negative sliding}) \\ \lambda_{Ti} &= -\mu_{0i}\lambda_{Ni} \wedge \ddot{g}_{Ti} \geq 0 \quad (i \in In \backslash I_T \text{ positive sliding}) \end{aligned} \tag{60}$$

This contact law may be represented by a double corner law as indicated in Fig. 7. To transform the law (60) for tangential constraints into a complementarity condition we must decompose the double corner into single ones. A decomposition into four elementary laws is given in [8], a decomposition into two elements in [37], [38]. In any case we come out with a complementarity problem of the form [31]

$$\boldsymbol{y} = \boldsymbol{A}\boldsymbol{x} + \boldsymbol{b} \ , \ \ \boldsymbol{y} \geq 0 \ , \ \ \boldsymbol{x} \geq 0 \ , \ \ \boldsymbol{y}^T\boldsymbol{x} = 0 \ , \qquad \boldsymbol{y}, \boldsymbol{x} \in \mathbb{R}^{n*} \tag{61}$$

where $n^* = n_N + 4n_T$ in the case of decomposition into four and $n^* = n_N + 2n_T$ for a decomposition into two elementary corners. The quantity $\boldsymbol{x}$ includes the contact forces and one part of the decomposed accelerations, the quantity $\boldsymbol{y}$ the relative accelerations and in addition the friction saturation defined as the difference of static friction force and tangential constraint force $(\mu_{0i}\lambda_{Ni} - |\lambda_{Ti}|)$. Equation (61) describes a linear complementarity problem thus being adequate for plane contacts.

In the case of spatial contacts the friction saturation contains the geometric sum of two possible friction directions leading to a nonlinearity which cannot be solved in a straightforward way. Solution procedures are discussed in [11], [12], [38], [42], [44].

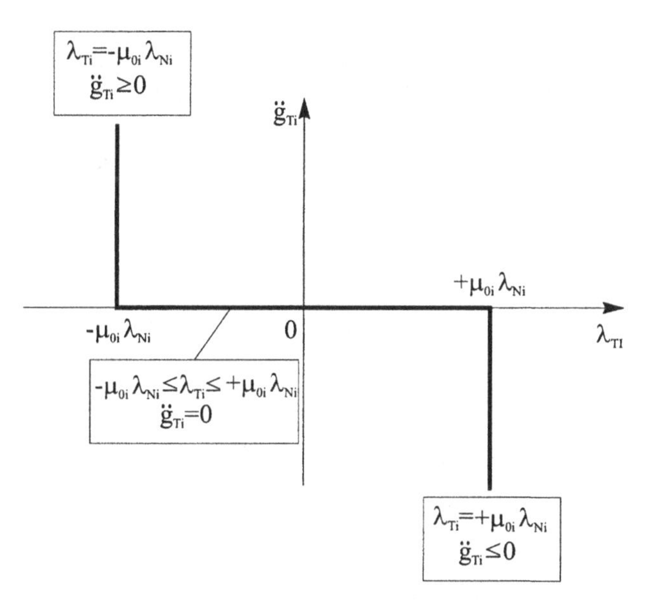

Figure 7: Corner law for tangential constraints

Similar as in the normal case we can represent the contact law eq. (60) by a variational inequality of the form [8], [42]

$$\ddot{g}_{Ti}^{T}(\boldsymbol{\lambda}_{Ti}^{*} - \boldsymbol{\lambda}_{Ti}) \geq 0 \ ; \ \boldsymbol{\lambda}_{Ti} \in C_{Ti} \ , \ \forall \boldsymbol{\lambda}_{Ti}^{*} \in C_{Ti} \ . \tag{62}$$

The convex set C_{Ti} contains all admissible contact forces $\boldsymbol{\lambda}_{Ti}^{*}$ in tangential direction

$$C_{Ti} = \{\boldsymbol{\lambda}_{Ti}^{*} | \, |\boldsymbol{\lambda}_{Ti}| \leq \mu_0 \boldsymbol{\lambda}_{Ni} \ ; \ \forall i \in I_T\} \tag{63}$$

For a derivation of the **equations of motion** including unilateral effects we must combine the multibody equations (9) with the unilateral constraints (33) or (51) and (60). In a first step we include the constraint forces into eq. (9) keeping in mind that in a system with additional unilateral constraints, the number of degrees of freedom is variable. To avoid difficulties with many different sets of minimal coordinates, we take one set of generalized coordinates (eq. (9) for the combination $(I_A \backslash I_c)$, $(I_A \backslash I_N)$, $(I_A \backslash I_T)$, see equation (52)) and consider the active unilateral constraints as additional constraints which necessarily are accompanied by constraint forces. We include these constraint forces, which are in fact the contact forces, into the equations of motion (9) by a Lagrange multiplier technique.

The constraint vectors $\boldsymbol{w}_{Ni}$ and $\boldsymbol{w}_{Ti}$ in the equations (33), (51) are arranged as columns in the constraint matrices

$$\begin{aligned} \boldsymbol{W}_N &= [\ldots, \boldsymbol{w}_{Ni}, \ldots] \in \mathbb{R}^{f,n_N} \quad ; \quad i \in I_N \\ \boldsymbol{W}_T &= [\ldots, \boldsymbol{w}_{Ti}, \ldots] \in \mathbb{R}^{f,2n_T} \quad ; \quad i \in I_T \end{aligned} \tag{64}$$

for all active constraints. The constraint matrices are transformation matrices from the space of constraints to the configuration space. The transposed matrices are used for the transition from the configuration space to the space of constraints. The contact forces have the amounts λ_{Ni} (normal forces) and the components λ_{Ti1} and λ_{Ti2} (tangential forces). These elements are combined in the vectors of constraint forces

$$\begin{aligned} \boldsymbol{\lambda}_N(t) &= \begin{pmatrix} \vdots \\ \lambda_{Ni}(t) \\ \vdots \end{pmatrix} \in \mathbb{R}^{n_N} \quad ; \quad i \in I_N \\ \boldsymbol{\lambda}_T(t) &= \begin{pmatrix} \vdots \\ \boldsymbol{\lambda}_{Ti}(t) \\ \vdots \end{pmatrix} \in \mathbb{R}^{2n_T} \quad ; \quad i \in I_T \end{aligned} \tag{65}$$

with $\boldsymbol{\lambda}_{Ti}(t) = [\lambda_{Ti1}(t), \lambda_{Ti2}(t)]^T$. In general, the contact forces are time-varying quantities. By the constraint vectors and matrices in equation (51), the contact forces can be expressed in the configuration space. These forces are then added to equation (9) to give

$$\boldsymbol{M}\ddot{\boldsymbol{q}} - \boldsymbol{h} - \sum_{i \in I_N} (\boldsymbol{w}_{Ni} + \boldsymbol{W}_{Ti}\boldsymbol{\lambda}_{Ti}) = 0 \ . \tag{66}$$

For the index sets see eq. (52). The contact forces $\boldsymbol{\lambda}_{Ti}$ in equation (66) can be passive forces of sticking contacts or active forces of sliding contacts. We express the tangential forces of the $n_N - n_T$ sliding contacts by the corresponding normal forces using Coulomb's friction law by

$$\boldsymbol{\lambda}_{Ti} = -\mu_i\left(|\dot{\boldsymbol{g}}_{Ti}|\right) \frac{\dot{\boldsymbol{g}}_{Ti}}{|\dot{\boldsymbol{g}}_{Ti}|} \lambda_{Ni} \ ; \quad i \in I_N \backslash I_T \tag{67}$$

where the coefficients $\mu_i(|\dot{\boldsymbol{g}}_{Ti}|)$ of sliding friction may depend on time. The negative sign relates to the opposite direction of relative velocity and friction force. The sliding forces of equation (66) in the configuration space are then

$$\boldsymbol{W}_{Ti}\boldsymbol{\lambda}_{Ti} = -\mu_i(|\dot{\boldsymbol{g}}_{Ti}|)\boldsymbol{W}_{Ti}\frac{\dot{\boldsymbol{g}}_{Ti}}{|\dot{\boldsymbol{g}}_{Ti}|}\lambda_{Ni} \ ; \quad i \in I_N \backslash I_T \ . \tag{68}$$

A substitution of these forces into equation (66) yields the equations of motion

$$\boldsymbol{M}(\boldsymbol{q},t)\ddot{\boldsymbol{q}}(t) - \boldsymbol{h}(\boldsymbol{q},\dot{\boldsymbol{q}},t) - [\boldsymbol{W}_N + \boldsymbol{H}_R,\ \boldsymbol{W}_T]\begin{pmatrix} \boldsymbol{\lambda}_N(t) \\ \boldsymbol{\lambda}_T(t) \end{pmatrix} = 0 \ , \tag{69}$$

with the additional contact forces as Lagrange multipliers. The matrices $\boldsymbol{W}_N$ and $\boldsymbol{W}_T$ contain components from the eqs. (33) or (51). The matrix $\boldsymbol{H}_R \in \mathbb{R}^{f,n_N}$ of the sliding contacts has the same dimension as the constraint matrix $\boldsymbol{W}_N$. For $n_T \leq n_N, \boldsymbol{H}_R$ consists of the $n_N - n_T$ columns

$$-\mu_i \boldsymbol{W}_{Ti} \frac{\dot{g}_{Ti}}{|\dot{g}_{Ti}|} \ ; \quad i \in I_N \backslash I_T \ ,$$

while the other n_T columns contain only zero-elements.

The relative accelerations of the active normal and tangential constraints in the equations (33) or (51) can be combined by means of the constraint matrices (64) in the matrix notation. Together with equation (69) we get the system of equations

$$\begin{aligned} &\boldsymbol{M}\ddot{\boldsymbol{q}} - \boldsymbol{h} - [\boldsymbol{W}_N + \boldsymbol{H}_R,\ \boldsymbol{W}_T]\begin{pmatrix} \boldsymbol{\lambda}_N \\ \boldsymbol{\lambda}_T \end{pmatrix} = 0 && \in \mathbb{R}^f \\ &\ddot{\boldsymbol{g}}_N = \boldsymbol{W}_N^T \ddot{\boldsymbol{q}} + \overline{\boldsymbol{w}}_N && \in \mathbb{R}^{n_N} \\ &\ddot{\boldsymbol{g}}_T = \boldsymbol{W}_T^T \ddot{\boldsymbol{q}} + \overline{\boldsymbol{w}}_T && \in \mathbb{R}^{n_T} \end{aligned} \tag{70}$$

The unknown quantities are the generalized accelerations $\ddot{\boldsymbol{q}} \in \mathbb{R}^f$, the contact forces in the normal direction $\boldsymbol{\lambda}_N \in \mathbb{R}^{n_N}$ and in the tangential direction $\boldsymbol{\lambda}_T \in \mathbb{R}^{2n_T}$, as well as the corresponding relative accelerations $\ddot{\boldsymbol{g}}_N \in \mathbb{R}^{n_N}$ and $\ddot{\boldsymbol{g}}_T \in \mathbb{R}^{2n_T}$. For the determination of the $f + 2(n_N + 2n_T)$ quantities, we have up to now $f + n_N + 2n_T$ equations. In the following the system of equations (70) will be completed by including the missing $n_N + 2n_T$ contact laws. In general, the kinematic equations are dependent on each other if there is more than one contact point per rigid body. The situation results in linearly dependent columns of the constraint matrices $\boldsymbol{W}_N, \boldsymbol{W}_T$ in equation (70). Such constraints are called dependent constraints.

Before including the contact laws we shall evaluate the equations (70) a bit further. Introducing the magnitudes

$$\begin{aligned} &\ddot{\boldsymbol{g}} = \begin{pmatrix} \ddot{\boldsymbol{g}}_N \\ \ddot{\boldsymbol{g}}_T \end{pmatrix} \ ; \quad \boldsymbol{\lambda} = \begin{pmatrix} \boldsymbol{\lambda}_N \\ \boldsymbol{\lambda}_T \end{pmatrix} \ ; \quad \overline{\boldsymbol{w}} = \begin{pmatrix} \overline{\boldsymbol{w}}_N \\ \overline{\boldsymbol{w}}_T \end{pmatrix} \ ; \\ &\boldsymbol{W} = (\boldsymbol{W}_N\, ,\ \boldsymbol{W}_T) \ ; \quad \boldsymbol{N}_G = (\boldsymbol{H}_R\, ,\ 0) \end{aligned} \tag{71}$$

we can write eq. (70)

$$\boldsymbol{M}\ddot{\boldsymbol{q}} - \boldsymbol{h} - (\boldsymbol{W} + \boldsymbol{N}_G)\boldsymbol{\lambda} = 0 \ , \qquad \ddot{\boldsymbol{g}} = \boldsymbol{W}^T \ddot{\boldsymbol{q}} + \overline{\boldsymbol{w}} \ . \tag{72}$$

Outside the transition events and thus for a not changing contact configuration the relative accelerations $\ddot{\boldsymbol{g}}$ are zero. In this case the equations (72) have a solution for $\ddot{g}$ and $\boldsymbol{\lambda}$, which writes

$$\begin{aligned} \boldsymbol{\lambda} &= -\left[\boldsymbol{W}^T\boldsymbol{M}^{-1}(\boldsymbol{W}+\boldsymbol{N}_G)\right]^{-1}(\boldsymbol{W}^T\boldsymbol{M}^{-1}\boldsymbol{h}+\overline{\boldsymbol{w}}) , \\ \ddot{q} &= \boldsymbol{M}^{-1}\left[\boldsymbol{h}+(\boldsymbol{W}+\boldsymbol{N}_G)\boldsymbol{\lambda}\right] . \end{aligned} \tag{73}$$

The first equation of (73) may also be expressed in the form $\boldsymbol{A\lambda}+\boldsymbol{b}=0$ with $\boldsymbol{A}$ and $\boldsymbol{b}$ following from (73).

We combine now the equations of motion in the form (70) with the contact equations (55) and (62) which results in the following set containing all unilateral processes in the contacts under consideration:

$$\begin{aligned} &\boldsymbol{M}\ddot{\boldsymbol{q}} - [\boldsymbol{W}_N+\boldsymbol{H}_R,\ \boldsymbol{W}_T]\begin{pmatrix}\boldsymbol{\lambda}_N\\ \boldsymbol{\lambda}_T\end{pmatrix} - \boldsymbol{h}(\boldsymbol{q},\dot{\boldsymbol{q}},t)=0 \\ &\begin{pmatrix}\ddot{\boldsymbol{g}}_N\\ \ddot{\boldsymbol{g}}_T\end{pmatrix} = \begin{pmatrix}\boldsymbol{W}_N^T\\ \boldsymbol{W}_T^T\end{pmatrix}\ddot{\boldsymbol{q}} + \begin{pmatrix}\overline{\boldsymbol{w}}_N\\ \overline{\boldsymbol{w}}_T\end{pmatrix} \\ &-\ddot{\boldsymbol{g}}_N^T(\boldsymbol{\lambda}_N^*-\boldsymbol{\lambda}_N)\le 0\ ;\ \boldsymbol{\lambda}_N\in C_N\ ;\ \forall\boldsymbol{\lambda}_N^*\in C_N\ ;\ I\in I_N \\ &-\ddot{\boldsymbol{g}}_{Ti}^T(\boldsymbol{\lambda}_{Ti}^*-\boldsymbol{\lambda}_{Ti})\le 0\ ;\ \boldsymbol{\lambda}_{Ti}\in C_{Ti}(\lambda_{Ni})\ ;\ \forall\boldsymbol{\lambda}_{Ti}^*\in C_{Ti}(\lambda_{Ni})\ ;\ i\in I_T \end{aligned} \tag{74}$$

The system (74) is not solvable in this form. Therefore the variational inequalities are transformed into equalities. From this we get a nonlinear system of equations which represent linear or nonlinear complementarity problems depending on the type of contact, i.e. plane or spatial contacts. Linear complementarity problem can be solved by algorithms related to linear programming methods, for example Lemke's algorithm, nonlinear complementarity problems require iterative algorithms [11], [12], [24].

2.4 IMPACTS WITH FRICTION

Impacts with and without friction play an important role in many machines and mechanisms. They are a research topic at the author's institute since many years starting with impacts without friction [29] and finally proceeding to a theory which describes impacts with friction in a rather general way [8]. This new theory has convincingly been verified by a large variety of experiments performed by [2]. In the following we shall give a short survey.

We assume as usual that an impact takes place in an infinitesimal short time and without any change of position, orientation and all non-impulsive forces. Nevertheless and virtually we zoom the impact time, establish the equations for a compression and for an expansion phase and then apply these equations again for an infinitesimal short time interval. The evaluation has to be performed on a velocity level which we realize by formal integration of the equations of motion, the constraints and the contact laws (eqs. 74). Denoting the beginning of an impact, the end of compression and the end of expansion by the indices A, C, E, respectively, we get for $\Delta t = t_E - t_A$

$$
\begin{aligned}
\boldsymbol{M}(\dot{\boldsymbol{q}}_C - \dot{\boldsymbol{q}}_A) - (\boldsymbol{W}_N \boldsymbol{W}_T)\begin{pmatrix} \boldsymbol{\Lambda}_{NC} \\ \boldsymbol{\Lambda}_{TC} \end{pmatrix} &= 0 \\
\boldsymbol{M}(\dot{\boldsymbol{q}}_E - \dot{\boldsymbol{q}}_C) - (\boldsymbol{W}_N \boldsymbol{W}_T)\begin{pmatrix} \boldsymbol{\Lambda}_{NE} \\ \boldsymbol{\Lambda}_{TE} \end{pmatrix} &= 0 \\
\text{with } \boldsymbol{\Lambda}_i = \lim_{t_E \to t_A} \int_{t_A}^{t_E} \boldsymbol{\lambda}_i \mathrm{dt}
\end{aligned} \tag{75}
$$

Here $\boldsymbol{\Lambda}_{NC}, \boldsymbol{\Lambda}_{TC}$ are the impulses in the normal and tangential direction which are transferred during compression, and $\boldsymbol{\Lambda}_{NE}, \boldsymbol{\Lambda}_{TE}$ those of expansion. Defining $\dot{\boldsymbol{q}}_A = \dot{\boldsymbol{q}}(t_A)$; $\dot{\boldsymbol{q}}_C = \dot{\boldsymbol{q}}(t_C)$; $\dot{\boldsymbol{q}}_E = \dot{\boldsymbol{q}}(t_E)$ we express the relative velocities as

$$
\begin{aligned}
\begin{pmatrix} \dot{\boldsymbol{g}}_{NA} \\ \dot{\boldsymbol{g}}_{TA} \end{pmatrix} &= \begin{pmatrix} \boldsymbol{W}_N^T \\ \boldsymbol{W}_T^T \end{pmatrix} \dot{\boldsymbol{q}}_A + \begin{pmatrix} \tilde{\boldsymbol{w}}_N \\ \tilde{\boldsymbol{w}}_T \end{pmatrix}, \\
\begin{pmatrix} \dot{\boldsymbol{g}}_{NE} \\ \dot{\boldsymbol{g}}_{TE} \end{pmatrix} &= \begin{pmatrix} \boldsymbol{W}_N^T \\ \boldsymbol{W}_T^T \end{pmatrix} \dot{\boldsymbol{q}}_E + \begin{pmatrix} \tilde{\boldsymbol{w}}_N \\ \tilde{\boldsymbol{w}}_T \end{pmatrix}, \\
\begin{pmatrix} \dot{\boldsymbol{g}}_{NC} \\ \dot{\boldsymbol{g}}_{TC} \end{pmatrix} &= \begin{pmatrix} \boldsymbol{W}_N^T \\ \boldsymbol{W}_T^T \end{pmatrix} \dot{\boldsymbol{q}}_C + \begin{pmatrix} \tilde{\boldsymbol{w}}_N \\ \tilde{\boldsymbol{w}}_T \end{pmatrix}.
\end{aligned} \tag{76}
$$

Considering in a first step the **Compression Phase** and combining the equations (75) and (76) we come out with

$$
\begin{pmatrix} \dot{\boldsymbol{g}}_{NC} \\ \dot{\boldsymbol{g}}_{TC} \end{pmatrix} = \underbrace{\begin{pmatrix} \boldsymbol{W}_N^T \\ \boldsymbol{W}_T^T \end{pmatrix} \boldsymbol{M}^{-1} \begin{pmatrix} \boldsymbol{W}_N \\ \boldsymbol{W}_T \end{pmatrix}^T}_{\boldsymbol{G}} \cdot \begin{pmatrix} \boldsymbol{\Lambda}_{NC} \\ \boldsymbol{\Lambda}_{TC} \end{pmatrix} + \begin{pmatrix} \dot{\boldsymbol{g}}_{NA} \\ \dot{\boldsymbol{g}}_{TA} \end{pmatrix}, \tag{77}
$$

where $\boldsymbol{G}$ is called the matrix of projected inertia. It consists of four blocks $\boldsymbol{G}_{NN} \ldots \boldsymbol{G}_{TT}$. Equation (77) allows to calculate the relative velocities $\dot{\boldsymbol{g}}_{NC}$ and $\dot{\boldsymbol{g}}_{TC}$ at the end of the compression phase, depending from the velocities at the beginning of the impact $\dot{\boldsymbol{g}}_{NA}$ and $\dot{\boldsymbol{g}}_{TA}$ under the influence of the contact impulsions $\boldsymbol{\Lambda}_{NC}$ and $\boldsymbol{\Lambda}_{TC}$. To calculate these impulsions two impact laws in normal and tangential direction are necessary. As already indicated magnitudes of relative kinematics and constraint forces (here impulses) are complementary quantities. In normal direction these are $\dot{\boldsymbol{g}}_{NC}$ and $\boldsymbol{\Lambda}_{NC}$. In tangential direction we have the relative tangential velocity vector $\dot{\boldsymbol{g}}_{TC}$ and the friction saturation $(\boldsymbol{\Lambda}_{TC} - (\text{diag}\mu_i)\boldsymbol{\Lambda}_{NC})$. Decomposing the tangential behavior we obtain [2]:

$$
\begin{aligned}
\Lambda_{TCV,i} &= \Lambda_{TC,i} + \mu_i \Lambda_{TN,i} \\
\dot{g}_{TC,i} &= \dot{g}_{TC,i}^{+} - \dot{g}_{TC,i}^{-} \\
\Lambda_{TCV,i}^{(+)} &= \Lambda_{TCV,i} \\
\Lambda_{TCV,i}^{(-)} &= -\Lambda_{TCV,i} + 2\mu_i \boldsymbol{\Lambda}_{NC,i}
\end{aligned} \tag{78}
$$

Together with equation (77) this results in a Linear Complementary Problem (LCP) in standard form $\boldsymbol{y} = \boldsymbol{A}\boldsymbol{x} + \boldsymbol{b}$ with $\boldsymbol{x} \geq 0, \boldsymbol{y} \geq 0$ and $\boldsymbol{x}^T\boldsymbol{y} = 0$:

$$\underbrace{\begin{pmatrix} \dot{\boldsymbol{g}}_{NC} \\ \dot{\boldsymbol{g}}_{TC}^{+} \\ \boldsymbol{\Lambda}_{TCV}^{(-)} \end{pmatrix}}_{\boldsymbol{y}} = \underbrace{\begin{pmatrix} \boldsymbol{G}_{NN} - \boldsymbol{G}_{NT}\boldsymbol{\mu} & \boldsymbol{G}_{NT} & 0 \\ \boldsymbol{G}_{TN} - \boldsymbol{G}_{TT}\boldsymbol{\mu} & \boldsymbol{G}_{TT} & \boldsymbol{E} \\ 2\boldsymbol{\mu} & -\boldsymbol{E} & 0 \end{pmatrix}}_{\boldsymbol{A}} \underbrace{\begin{pmatrix} \boldsymbol{\Lambda}_{NC} \\ \boldsymbol{\Lambda}_{TCV}^{(+)} \\ \dot{\boldsymbol{g}}_{TC}^{-} \end{pmatrix}}_{\boldsymbol{x}} + \underbrace{\begin{pmatrix} \dot{\boldsymbol{g}}_{NA} \\ \dot{\boldsymbol{g}}_{TA} \\ 0 \end{pmatrix}}_{\boldsymbol{b}} \tag{79}$$

$\boldsymbol{\mu}$ is a diagonal matrix, containing the friction coefficients of the contacts. The problem can be solved numerically. The velocities $\dot{\boldsymbol{g}}_{NC}, \dot{\boldsymbol{g}}_{TC}$ and the impulsions $\boldsymbol{\Lambda}_{NC}, \boldsymbol{\Lambda}_{TC}$ are either part of the result or can be obtained by transformation (78) and by $\boldsymbol{\Lambda}_{TC} = \boldsymbol{\Lambda}_{TCV}^{(+)} - \boldsymbol{\mu}\boldsymbol{\Lambda}_{NC}$.

In the compression phase kinetic energy of the colliding bodies is stored as potential energy. During the **expansion phase** parts of this energy are transformed back. This regaining process is governed by two coefficients of restitution in normal and tangential directions ε_N and ε_T respectively.

In the case of multiple impacts Poisson's hypothesis does not guarantee impenetrability of the bodies [8]. So the law is enhanced by an additional condition. The normal impulsion during phase of restitution Λ_{NE} is in the minimum $\varepsilon_E\Lambda_{NC}$ in each contact, but can be arbitrary high to avoid penetration. In this case the bodies remain in contact after the impact. This impact law with a complementary character is drawn in Fig. 8.

In tangential direction the impact law is ruled by three effects: At first a minimum impulsion $\varepsilon_T\Lambda_{TC}$ must be transferred. On the other hand, the impulsion must not exceed the friction limit. Between these limits a tangential relative velocity $\dot{g}_{TE0}$ is prescribed. This velocity allows a restitution of stored energy, if during the phase of restitution sticking occurs [2]. The existence of $\dot{g}_{TE0}$ was not introduced in Glocker's dissertation [8] and also not in Moreau's impact model [10], but from Beitelschmidt's measurements [2] it could be derived as a necessary correction. With

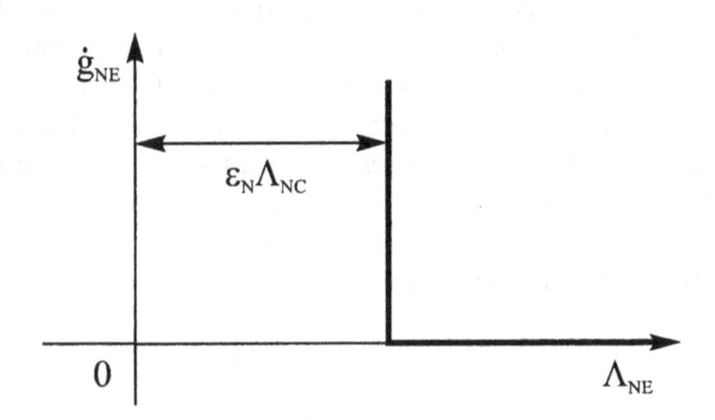

Figure 8: Impact law for the phase of expansion in normal direction

$$\dot{\boldsymbol{g}}_{TE0} = \boldsymbol{G}_{TN}\boldsymbol{\varepsilon}_N\boldsymbol{\Lambda}_{NC} + \boldsymbol{G}_{TT}\boldsymbol{\varepsilon}_N\boldsymbol{\Lambda}_{TC} \tag{80}$$

one can calculate $\dot{g}_{TE0}$ for all contacts and the tangential impact law looks like drawn in Fig. 9. $\boldsymbol{\varepsilon}_N$ and $\boldsymbol{\varepsilon}_T$ are diagonal matrices containing the different coefficients for all contacts.

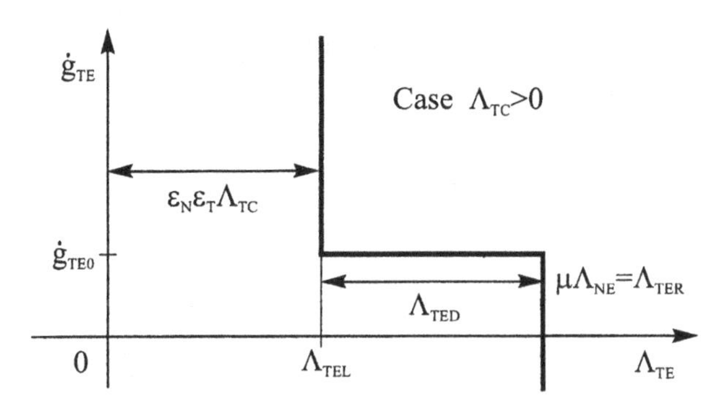

Figure 9: Impact law for the phase of restitution in tangential direction

To formulate the equation for the phase of restitution as a LCP similar to the compression phase the two matrices

$$\begin{aligned} \boldsymbol{S}^+ &= \operatorname{diag}\left(\tfrac{1}{2}\left(1+\operatorname{sign}(\boldsymbol{\Lambda}_{TC,i})\right)\right) \quad , \\ \boldsymbol{S}^- &= \operatorname{diag}\left(\tfrac{1}{2}\left(1-\operatorname{sign}(\boldsymbol{\Lambda}_{TC,i})\right)\right) \end{aligned} \tag{81}$$

are introduced. After some transformations similar to those of the compression phase the LCP writes

$$\begin{pmatrix} \dot{\boldsymbol{g}}_{NE} \\ \dot{\boldsymbol{g}}^+_{TEV} \\ \boldsymbol{\Lambda}^{(-)}_{TEV} \end{pmatrix} = \begin{pmatrix} \boldsymbol{G}_{NN} - \boldsymbol{G}_{NT}\boldsymbol{S}^-\boldsymbol{\mu} & \boldsymbol{G}_{NT} & 0 \\ \boldsymbol{G}_{TN} - \boldsymbol{G}_{TT}\boldsymbol{S}^-\boldsymbol{\mu} & \boldsymbol{G}_{TT} & \boldsymbol{E} \\ \boldsymbol{\mu} & -\boldsymbol{E} & 0 \end{pmatrix} \begin{pmatrix} \boldsymbol{\Lambda}_{NP} \\ \boldsymbol{\Lambda}^{(+)}_{TEV} \\ \dot{\boldsymbol{g}}^-_{TEV} \end{pmatrix} + \boldsymbol{b} \tag{82}$$

with

$$\boldsymbol{b} = \begin{pmatrix} \boldsymbol{G}_{NN}\varepsilon_N\boldsymbol{\Lambda}_{NC} + \boldsymbol{G}_{NT}\boldsymbol{S}^+\varepsilon_N\varepsilon_T\boldsymbol{\Lambda}_{TC} - \boldsymbol{G}_{NT}\boldsymbol{S}^-\boldsymbol{\mu}\varepsilon_N\boldsymbol{\Lambda}_{NC} + \dot{\boldsymbol{g}}_{NC} \\ \boldsymbol{G}_{TT}(\boldsymbol{S}^+ - \boldsymbol{E})\varepsilon_N\varepsilon_T\boldsymbol{\Lambda}_{TC} - \boldsymbol{G}_{TT}\boldsymbol{S}^-\boldsymbol{\mu}\varepsilon_N\boldsymbol{\Lambda}_{NC} + \dot{\boldsymbol{g}}_{TC} \\ \boldsymbol{\mu}\varepsilon_N\boldsymbol{\Lambda}_{NC} - \varepsilon_N\varepsilon_T|\boldsymbol{\Lambda}_{TC}| \end{pmatrix} . \tag{83}$$

After solution the velocities $\dot{\boldsymbol{g}}_{NC}, \dot{\boldsymbol{g}}_{TC}$ and the impulsions $\boldsymbol{\Lambda}_{NC}, \boldsymbol{\Lambda}_{TC}$ are either part of the result or can be obtained by the transformations:

$$\dot{\boldsymbol{g}}_{TE} = \dot{\boldsymbol{g}}^+_{TEV} - \dot{\boldsymbol{g}}^-_{TEV} + \dot{\boldsymbol{g}}_{TE0} \tag{84}$$

$$\begin{aligned} \boldsymbol{\Lambda}_{NE} &= \boldsymbol{\Lambda}_{NP} + \varepsilon_N\boldsymbol{\Lambda}_{NC} \\ \boldsymbol{\Lambda}_{TE} &= \boldsymbol{\Lambda}^{(+)}_{TEV} + \boldsymbol{\Lambda}_{TEL} = \boldsymbol{\Lambda}^{(+)}_{TEV} + \boldsymbol{S}^+\varepsilon_N\varepsilon_T\boldsymbol{\Lambda}_{TC} - \boldsymbol{S}^-\boldsymbol{\mu}\boldsymbol{\Lambda}_{NE} \end{aligned} \tag{85}$$

If the impulsions during the two phases of the impact are known, one can calculate the motion $\dot{\boldsymbol{q}}_E$ of the multibody system at the end of the impact.

$$\dot{\boldsymbol{q}}_E = \dot{\boldsymbol{q}}_A \boldsymbol{M}^{-1} \left(\boldsymbol{W}_N(\boldsymbol{\Lambda}_{NC} + \boldsymbol{\Lambda}_{NE}) + \boldsymbol{W}_T(\boldsymbol{\Lambda}_{TC} + \boldsymbol{\Lambda}_{TE}) \right) \tag{86}$$

The above impact theory has been verified in [2]. More than 600 throwing measurements have been performed. Figure 10 depicts a typical result for a material pair PVC/PVC. The comparison with measurements is excellent, as in many other cases.

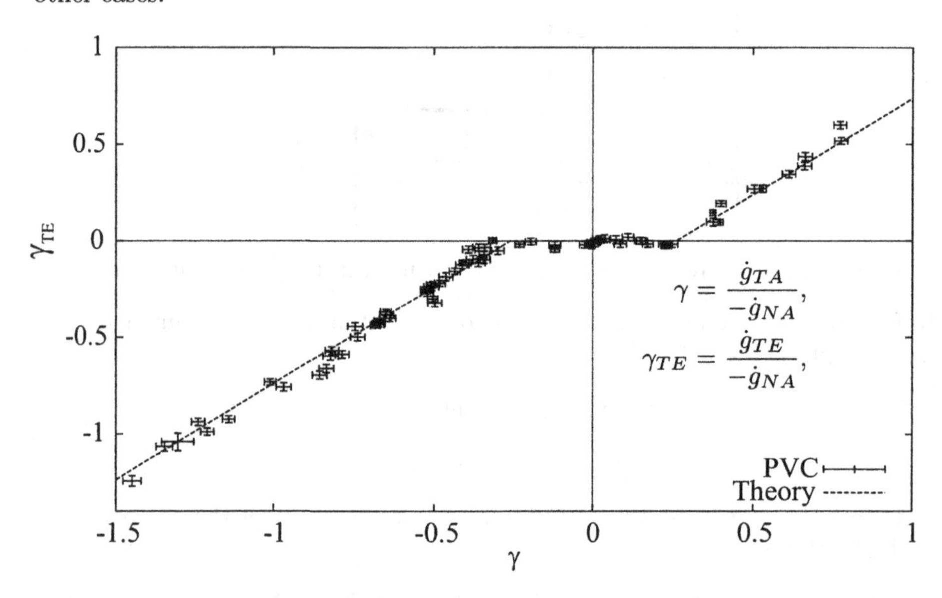

Figure 10: Dimensionless tangential relative velocity after versus before the impact, PVC-body

3. Applications

From a large variety of industrial applications we shall present in the following three typical examples, namely chains, vibration conveyors and chimney dampers. We do not go into details, because all applications have been published elsewhere. Therefore we shall give only a short summary for each case. In all cases the theoretical methods are based on those of chapter 2.

3.1 CVT-CHAINS [40]

The dynamics of chains has been investigated in [6], [7], [39], [40]. The dissertation [7] considers roller chains where the main problem consists in many degrees of freedom, usually some hundred, on one side and from this in a large number of contacts with the guides and the sprocket wheels on the other side. An additional problem arises by the dynamical behavior of the tension devices which require detailed and complicated models.

Another type of chains are the CVT-chains applied in continuous variable transmissions, where they transmit power by friction. At the time being there exist three different chain configurations, the rocker pin chains by P.I.V. in Germany, the metal-pushed V-belt-chain by van Doorne in the Netherlands and the Borg-Warner chain from the US. We shall focus here on the PIV-chain, which was investigated in all details in the doctoral thesis [40]. Figure 11 represents a typical CVT-configuration. A complete CVT-gear consists of two discs with hydraulically shifted sheaves as indicated in Figure 11. One can change the gear-ratio continuously by opening the discs in one pulley and at the same time closing it in the second pulley. Therefore the continuous variable properties are achieved by axially moving sheaves that are shifted hydraulically. This device also applies the necessary forces to the chain.

The mechanical model includes for the plane motion of the chain altogether 63 links with 189 degrees of freedom. Each pulley possesses two translational and one rotational degrees of freedom with respect to rigid body motion. Pitching effects of the axially movable pulley are regarded by additional degrees of freedom. Both pulleys are additionally modeled as elastic bodies applying a Ritz-approach within the context of elastic multibody theory [3], [4], [40]. Special care has been taken to solve the problem of unilateral contacts in connection with elastically deformed surfaces.

The results were computed for stationary operation with constant driving speed and output torque on geared level. Computing time for this case amounts about 8 hours on a SUN workstation. Figure 12 shows the contact forces acting on a pair of rocker pins during one revolution and the tensile force in the related chain link.

As long as the chain link is part of a strand, no contact forces work on its rocker pins. When it comes into contact with one of the pulleys, the pins are pressed between the two sheaves, and hence the normal force increases. Its amplitude depends on the geometry of the sheaves and the transmitted power. The frictional force is a function of this normal force and the relative velocity between the pulley and the pins. It is split into one radial and one circumferential component. The radial contact force coincides with a radial movement of the chain link that equals a power dissipation. In contrast, the circumferential contact force causes the changes of the tensile force in the corresponding chain link, which leads to different tensile force levels in the two strands that agree with the transmitted torque.

Due to the mechanical model, the simulation provides an integrated value of the tensile force in the plates of a chain link, whereas measurement was performed for the tensile force in a clasp plate (left-hand side of Figure 13). (Measurements were performed by G. Sauer and K.T. Renius, Lehrstuhl für Landmaschinen, Technische Universität München.) Therefore, it is necessary to determine the distribution of the tensile force on the plates of the chain links. Using the results of the dynamic simulation shown in Figure 12 and modeling the pair of rocker pins as bending beams and the plates as linear springs, we get the graph of the clasp plate in the right-hand side of Figure 13. The comparison of simulation and measurement

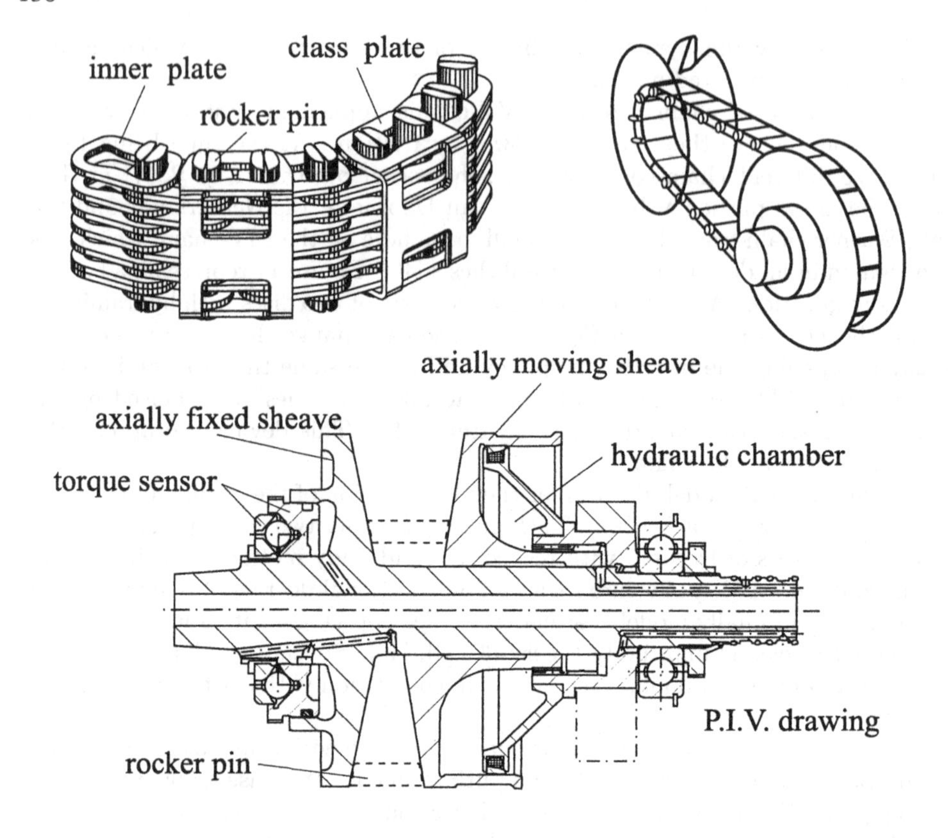

Figure 11: Typical rocker pin chain configuration

confirms the mechanical model.

3.2 VIBRATION CONVEYOR [44]

Vibratory feeders are used in automatic assembly to feed small parts. They are capable to store, transport, orient and isolate the parts. An oscillating track with frequencies up to 100 Hz excites the transportation process, which is mainly based on impact and friction phenomena between the parts and the track. Vibratory feeders are applied for a wide variety of parts and for lots of different tasks. In the majority of cases, the parts are available as a sort of bulk material that is stored in a container. The transportation process, starting in this reservoir, is often combined with orienting devices that orient parts, or select only these parts having already

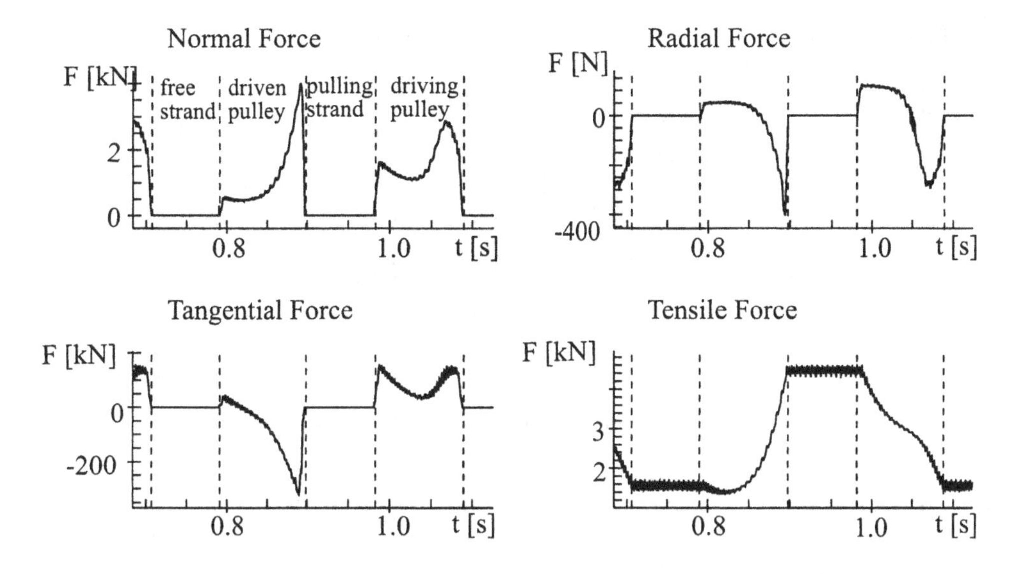

Figure 12: Forces acting on the rocker pins and in a chain link

the right orientation (Figure 14 shows an example of a vibratory bowl feeder with an orienting device). Each kind of parts, with its special geometry and mechanical properties, requires an individual adaption of the feeder. This individual tuning comprises the development of suitable track and orienting device geometries and the adjustment of the excitation parameters frequency and amplitude. Due to the complex mechanics of the feeding process this design is usually done by trial and error without any theoretical background. A complete dynamical model of the transportation process allows a theoretical investigation and consequently an improvement in the properties of the feeder.

Friction and impact phenomena between the parts and the track are the most important mechanical properties of transportation processes. Consequently, the required dynamical model has to deal with unilateral constraints, dry friction and multiple impacts. The mechanical model of the vibratory feeder can be split in two parts: the transportation process and the base device. The dissertation [44] focuses on modeling and simulation of the transportation process. The modeling of the base device can be done with well known standard technics for multibody systems. Friction and impact effects have a fundamental importance for the transportation of parts. Changing contact configuration between the parts and the track and also between the parts itself are characteristic for the feeding process. The contacts appear either continuously for a certain time interval, or for a discrete time (impact). A structure varying multibody system with unilateral constraints

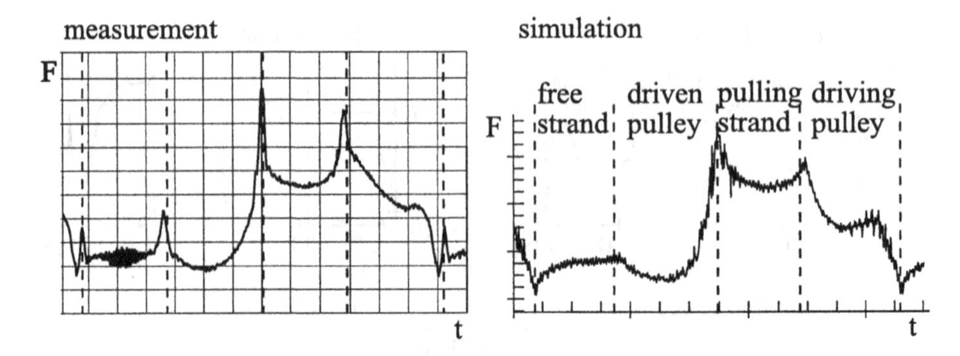

Figure 13: Tensile force in the clasp plate

with friction is an ideal technique for modeling the feeding process. Its formulation results in a set of differential equations with inequality constraints requiring special mathematical and numerical methods (see chapter 2).

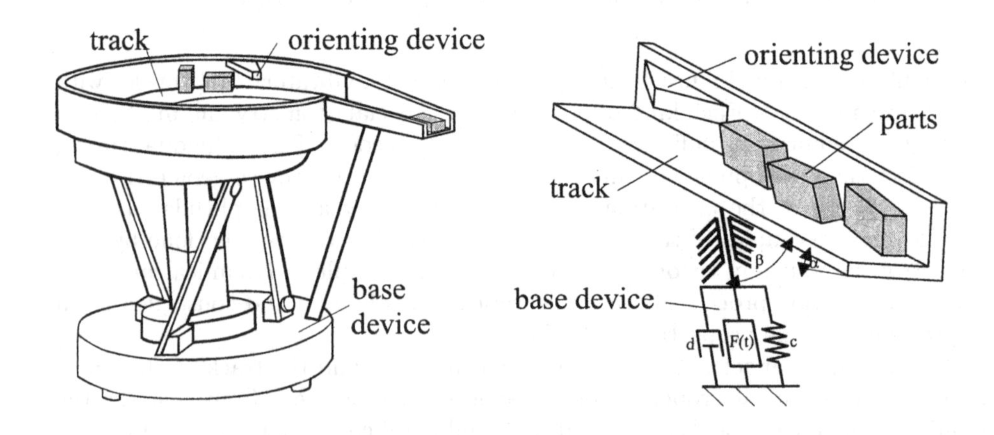

Figure 14: Vibratory bowl feeder and mechanical model

For the verification of the developed model of the transportation process an experimental vibratory feeder was built, allowing different measurements concerning the impact model and the average transportation rates. Figure 15 shows the principle of the used device. The track, fixed on leaf springs is excited with an electro magnetic shaker with a frequency about 50 Hz. The eigenfrequency of the system is at 52 Hz. The resulting vibration amplitude reaches a maximum value of about 2 mm. The track has an inclination angle $\alpha = 3^o$, the angle between the track and the direction of the vibration is $\beta = 15^o$. For the accurate contactless measurement of the motion of the transported part six laser distance sensors were applied. For the vibration measurement of the track an eddy current sensor is used.

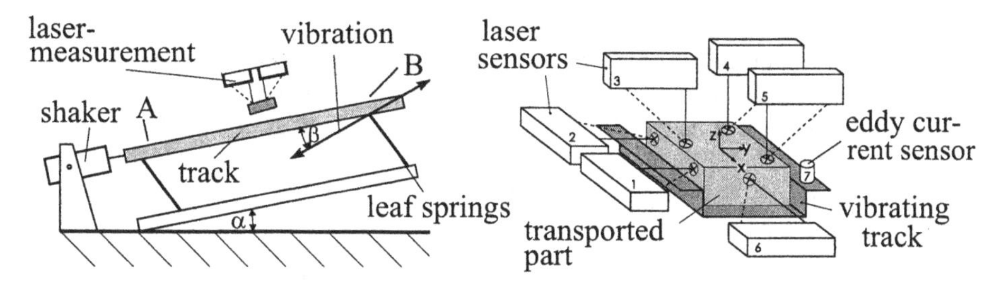

Figure 15: Test setup and part measurement

For a comparison of the theory and the measurements the averaged transportation rate was used. Figure 16 gives a result, which before the background of the complexity of the problem looks good. An interesting finding is the fact that the averaged transportation velocity does not depend very much on the number of parts and also not on the type of modeling, plane or spatial [44]. Therefore the design of vibration conveyors can be carried through considering one part only. For the layout of orienting devices we need of course a spatial theory.

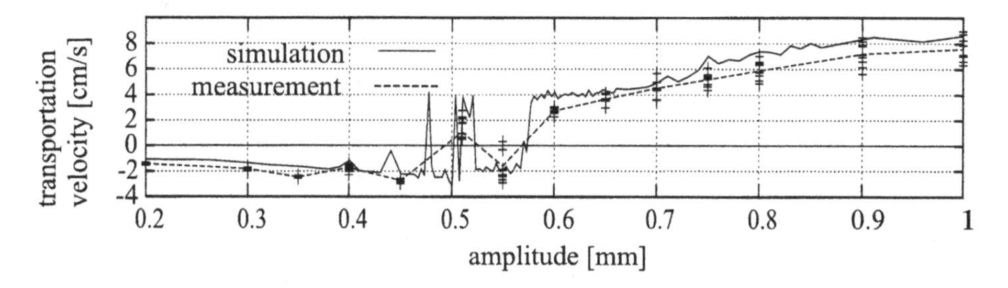

Figure 16: Simulation and measurement of the average transportation ratewith $\varepsilon = 0.84, \mu = 0.23$

3.3 CHIMNEY DAMPERS [34]

Towerlike structures like steel chimneys may be excited to vibrations by vortex streets. Such a mechanism becomes dangerous if the first eigenfrequency of the structure is small enough to be exited by not too large wind velocities which appear frequently. Oscillations of that type are damped significantly by a simple idea. A pendulum is attached to the chimney top. Its mass and length is tuned to the chimneys first eigenfrequency by classical analysis, and additionally optimal viscous damping is also evaluated by classical formulas, given for example by Den Hartog's method. In a second step the thus determined ideal damping behavior is not realized by viscous means but by a package of circular plates which are moved by the end of the pendulum rod through internal circular holes of the plates. The approximate realization of viscous damping by dry friction is accompanied by impacts of the pendulum rod in the plate holes and by stick-slip transitions between the plates. To achieve a best fit to the ideal viscous case an optimization of the complete pendulum-plate-system is carried through applying multibody theory with unilateral contacts.

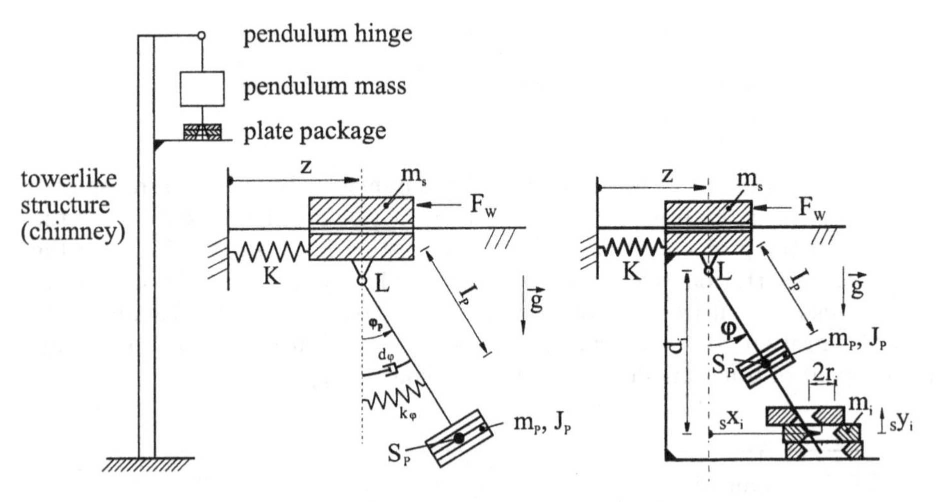

Figure 17: Principal configuration, viscous and plate model [34]

Figure 17 illustrates the basic principle. At the top of the chimney a pendulum is arranged which damps the oscillations. To achieve best damping efficiency the damper is optimized in two steps. In a first step a classical damper is assumed working with viscous damping. For the resonance frequency area such a damper can be optimized with respect to length, mass and viscous damping of the pendulum. As damping will be realized by dry friction within a plate package, the plates must be selected in such a way that they perform a damping effect coming as nearest as possible to the optimal viscous damping. Therefore, in a second step the plate package will be optimized with respect to damping efficiency.

To verify the theory two types of experiments have been performed. In a first test a pendulum-plate-configuration has been arranged in a car-like frame with wheels, which could be excited periodically (Fig. 18). Figure 18 depicts also the results which compare well with theory.

A second test has been performed with a real steel chimney, which was bended by a steel rope. By releasing suddenly this rope the chimney starts to vibrate. This process has been measured and compared with the corresponding theory based on the equations of chapter 2. Figure 19 illustrates a comparison between experiment and simulation, which also confirms the methods presented.

If we approximate the optimal viscous system of Fig. 17 (middle picture) by

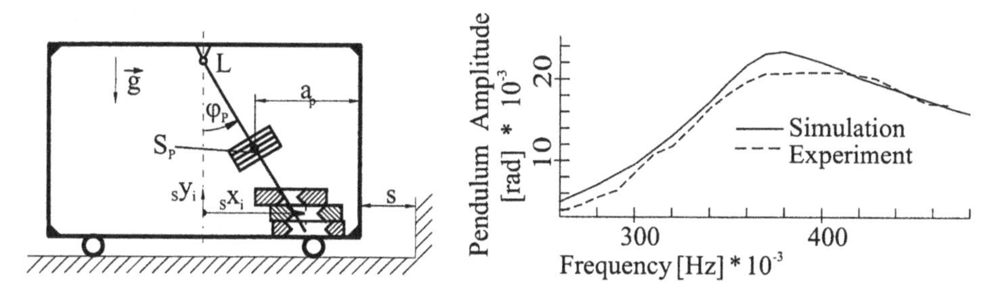

Figure 18: Vibrating test-setup and a comparison theory/ measurements

a package of plates, we can do that in a best way by optimizing the number of plates, their radii and thicknesses. We find that the number of plates affect the damping efficiency only partly. On the left side of Figure 20 the total mass of the plates is put constant, also the radius of the top at the ground damper plate. The number of plates is varied from one to eight. Doing so the distribution of the plate holes is kept linear. In the case of one plate the diameter of the plate hole is half as big as the largest one in all other cases.

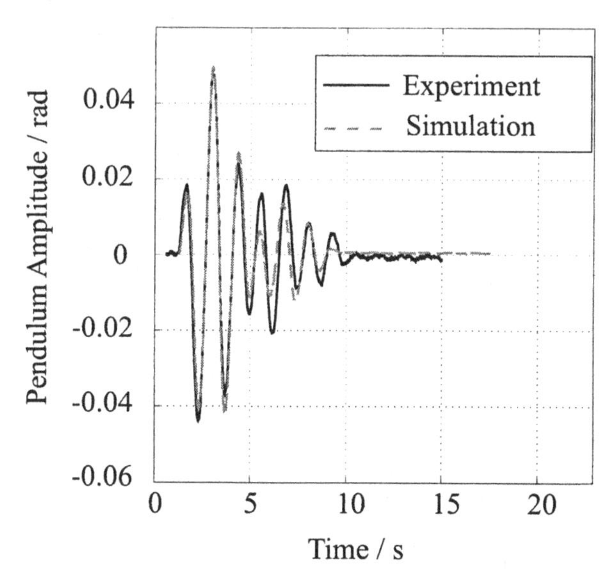

Figure 19: Pendulum vibrations in a real chimney (height 80 m, first bending eigenfrequency 0.76 Hz)

Figure 20 shows that the damping mechanism works best with one plate and worst with two plates. The reason lies in the fact, that in the case of two plates only the upper one is moved effectively. The ground plate is ineffective and hence its damping is too small. As an important result it should be noticed, that the application of a large number of plates does not make sense. Obviously the damping characteristic cannot be improved by using more than five or six plates.

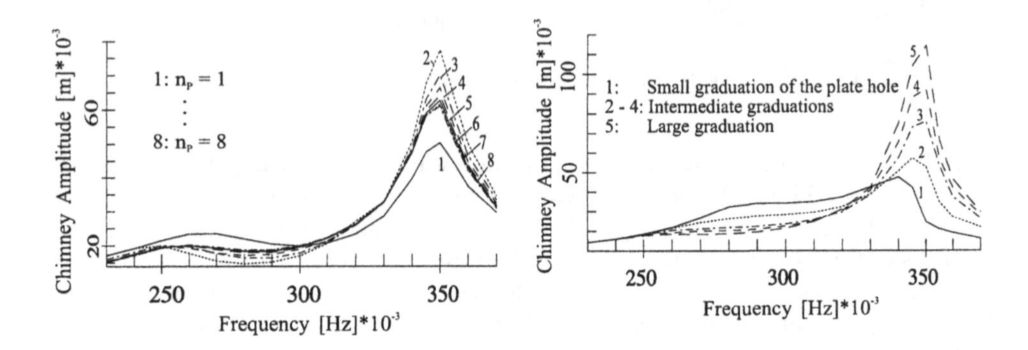

Figure 20: Variation of plate number (left) and of graduation (right) of the plate hole diameters

A further sensible parameter is the distribution of hole diameters over all plates, called graduation. The right graph of Figure 20 shows with number 1 the behavior of a system with small and with graph number 5 with a big graduation. The other curves belong to some intermediate graduations. Keeping the total mass of the plates constant an increasing graduation (from 1 to 5) leads to a decrease of damping and for the considered configuration to rising chimney displacements.

4. Summary

Classical theory of rigid or elastic multibody systems dynamics may be characterized by d'Alembert's principle in the form of Lagrange and extended by Jourdain allowing to project the equations of motion of all interconnected bodies into their free directions according to their individual constraints. These constraints are bilateral in classical theory and do not include contact phenomena.

If we want to consider dynamical problems with unilateral contacts, especially multibody systems with multiple contacts, and if we further want to model such contacts by unilateral constraints, we must enlarge multibody theory by certain rules describing unilateral features. One fundamental property, sometimes addressed to as Signorini's law, consists in the fact, that for each contact either quantities of relative kinematics are zero and the corresponding constraint forces are not zero, or vice versa. This establishes a linear or nonlinear complementarity problem for each of the contacts, which has to be added to the classical multibody formalism. In a more mathematical sense the nonsmooth character of contact dynamics may be formulated in terms of variational or hemi-variational inequalities depending on the convex or non-convex features of the related energies. Representations of that kind allow elegant mathematical developments, which however have again to be decomposed for practical applications.

Paper gives a survey of multibody dynamics including unilateral contacts. It starts with a short representation of the projection method in multibody dynamics, proceeds with the relative kinematics of contacts, discusses then the problems of multiple contacts and establishes the appropriate equations of motion. The important case of impacts with and without friction is presented in a final theoretical chapter. Theory has been applied to many practical problems. Paper gives three examples from industry, which confirm the relevancy and the usefulness of the presented methods.

References

1. **Baraff, D.:** Issues in Computing Contact Forces for Non-Penetrating Rigid Bodies. Algorithmica, 10, 292-352, 1993

2. **Beitelschmidt, M.:** Reibstöße in Mehrkörpersystemen. Dissertation TU-München, Lehrstuhl B für Mechanik, 1998

3. **Bremer, H.:** Dynamik und Regelung mechanischer Systeme. Teubner Studienbücher, Mechanik, B.G. Teubner, Stuttgart, 1988

4. **Bremer, H.; Pfeiffer, F.:** Elastische Mehrkörpersysteme. Teubner Studienbücher, Mechanik, B.G. Teubner, Stuttgart, 1992

5. **Brogliatto, B.:** Nonsmooth Impact Dynamics. Springer, London, 1996

6. **Fritz, P.; Pfeiffer, F.:** Dynamics of High Speed Roller Chain Drives. Proceeding of the 1995 Design Engineering Technical Conference, Boston September 17-21, Vol. 3, Part A, DE-Vol. 84-1, 573-584, 1995

7. **Fritz, P.:** Dynamik schnellaufender Kettentriebe. Fortschrittberichte VDI, Reihe 11, Nr. 253, VDI-Verlag, Düsseldorf 1998

8. **Glocker, Ch.:** Dynamik von Starrkörpersystemen mit Reibung und Stößen. Fortschritt-Berichte VDI, Reihe 18: Mechanik/Bruchmechanik, Nr. 182, VDI-Verlag, Düsseldorf, 1995

9. **Glocker, Ch.; Pfeiffer, F.:** Multiple Impacts with Friction in Multibody Systems. Nonlinear Dynamics 7, 471-497, 1995

10. **Jean, M.; Moreau, J.J.:** Unilaterality and Dry Friction in the Dynamics of Rigid Body Collections. Proceedings Contact Mechanics International Symposium, October 7-9, EPFL, Lausanne, Switzerland, A. Curnier, Ed., PPUR, 31-48, 1992

11. **Klarbring, A.:** Mathematical Programming and Augmented Lagrangian Methods for Frictional Contact Problems. Proceedings Contact Mechanics International Symposium, October 7-9, EPFL, Lausanne, Switzerland, A. Curnier, Ed., PPUR, 409-422, 1992

12. **Klarbring, A.:** Mathematical Programming in Contact Problems. Computational Methods for Contact Problems, M.H. Aliabadi, Ed., Elsevier, 1994

13. **Kwak, B.M.:** Complementarity Problem Formulation of Three-Dimensional Frictional Contact. ASME Journal of Applied Mechanics 58, 134-140, 1991

14. **Kwak, B.M.; Park, J.K.:** Three-Dimensional Frictional Analysis Using the Homotopy Method. ASME Journal of Applied Mechanics 61, 703-709, 1994

15. **Lötstedt, P.:** Coulomb Friction in Two-Dimensional Rigid Body Systems. Zeitschrift für Angewandte Mathematik und Mechanik 61, 605-615, 1981

16. **Lötstedt, P.:** Mechanical Systems of Rigid Bodies Subject to Unilateral Constraints. SIAM J. Appl. Math., 42, 281-296, 1982

17. **Mangasarian, O.L.:** Equivalence of the Complementarity Problem to a System of Nonlinear Equations. SIAM Journal of Applied Mathematics 31(1), 89-92, 1976

18. **Martins, J.A.C.; Oden, J.T.; Simoes, F.M.F.:** A Study of Static and Kinetic Friction. Int. J. Engng. Sci., Vol. 28, No. 1, 29-92, 1990

19. **Meitinger, Th.:** Modellierung und Simulation von Montageprozessen. Fortschritt-Berichte VDI, Reihe 2, Nr. 476, VDI-Verlag Düsseldorf 1998

20. **Monteiro Marques, M.D.P.:** Differential Inclusions in Nonsmooth Mechanical Problems. Birkhäuser, Basel, 1993

21. **Moreau, J.J.:** Application of Convex Analysis to Some Problems of Dry Friction. Trends in Applications of Pure Mathematics to Mechanics, Vol. 2, London, 1979

22. **Moreau, J.J.:** Une formulation du contact a frottement sec; application au calcul numerique. Technical Report 13, C.R. Acad. Sci. Paris, Serie II, 1986

23. **Moreau, J.J.:** Unilateral Contact and Dry Friction in Finite Freedom Dynamics. Non-Smooth Mechanics and Applications, CISM Courses and Lectures, Vol. 302, Springer Verlag, Wien, 1988

24. **Murty, K.G.:** Linear Complementarity, Linear and Nonlinear Programming. Sigma Series in Applied Mathematics (ed. White, D.J.), Heldermann Verlag, Berlin, 1988

25. **Oden, J.T.; Martins, J.A.C.:** Models and Computational Methods for Dynamic Friction Phenomena. Computer Methods in Applied Mechanics and Engineering, 52, 527-634, 1985

26. **Newton, I.:** Principia, Corol. VI, 1687

27. **Panagiotopoulos, P.D.:** Inequality Problems in Mechanics and Applications, Birkhäuser, Boston, Basel, Stuttgart, 1985

28. **Panagiotopoulos, P.D.:** Hemivariational Inequalities. Springer Verlag, Berlin, Heidelberg, 1993

29. **Pfeiffer, F.:** Mechanische Systeme mit unstetigen Übergängen. Ing. Arch. 54, 232-240, 1984

30. **Pfeiffer, F.:** Multibody Dynamics with Multiple Unilateral Contacts. Proceedings of the XIXth International Congress of Theoretical and Applied Mechanics, Kyoto, Japan, August 25-31, 1996

31. **Pfeiffer, F.; Glocker, Ch.:** Multibody Dynamics with Unilateral Contacts. Wiley Series in Nonlinear Science, John Wiley & Sons, Inc., New York, 1996

32. **Pfeiffer, F.:** Assembly processes with robotic systems. Robotics and Autonomous Systems 19, 151-166, 1996

33. **Pfeiffer, F.; Fritz, P.; Srnik, J.:** Nonlinear Vibrations of Chains. J. of Vibration and Control, 3: 397-410, 1997

34. **Pfeiffer, F.; Stiegelmeyr, A.:** Damping Towerlike Structures by Dry Friction. Proc. of DETC '97, ASME Design Eng. Techn. Conf., 1997

35. **Poisson, S.D.:** Traité de méchanique. Bachelier, Paris, 1833

36. **Rockafellar, R.T.:** Convex Analysis. Princeton University Press, 1972

37. **Rossmann, Th.; Pfeiffer, F.:** Efficient Algorithms for Nonsmooth Dynamics. Symposium on Nonlinear Dynamics and Stochastic Mechanics 1997, International Mechanical Engineering Congress and Exposition, Dallas, Texas, November 16-21, 1997

38. **Rosmann, Th.:** Eine Laufmaschine für Rohre. Fortschrittberichte VDI, Reihe 8, Nr. 732, VDI-Verlag Düsseldorf, 1998

39. **Srnik, J.; Pfeiffer, F.:** Dynamics of CVT Chain Drives: Mechanical Model and Verification. Proc. of DETC '97, ASME Design Eng. Techn. Conf., 1997

40. **Srnik, J.:** Dynamik von CVT-Keilkettengetrieben. Fortschritt-Berichte VDI, Reihe 12, Nr. 372, VDI-Verlag Düsseldorf 1999

41. **Wösle, M.; Pfeiffer, F.:** Dynamics of Multibody Systems Containing Dependent Unilateral Constraints with Friction. Journal of Vibration and Control 2(2), 161-192, 1996

42. **Wösle, M.:** Dynamik von räumlichen strukturvarianten Starrkörpersystemen. Fortschrittbereichte VDI, Reihe 19, Nr. 213, VDI-Verlag, Düsseldorf, 1997

43. **Wolfsteiner, P.; Pfeiffer, F.:** The Parts Transportation in a Vibratory Feeder. IUTAM Symposium on Unilateral Multibody Dynamics, München, 1998

44. **Wolfsteiner, P.:** Dynamik von Vibrationsförderern. Dissertation TU-München, Lehrstuhl B ür Mechanik, 1999

FRICTIONAL CONTACT OF SPATIAL MULTIBODY SYSTEMS WITH KINEMATIC LOOPS

J. PFISTER and P. EBERHARD
Institute B of Mechanics, University of Stuttgart,
Pfaffenwaldring 9, 70550 Stuttgart, Germany,
[jp,eberhard]@mechb.uni-stuttgart.de

Abstract. The presented investigation deals with the mathematical formulation of spatial holonomic multibody systems with kinematic loops subject to frictional contact. The equations of motion are given as a minimal set of differential equations, and the contact formulations are incorporated using an Augmented Lagrange method. The contact calculation is divided into continual contact and impact. While the normal impact is modelled by a modified Poisson hypothesis, the tangential contributions of both the impact and the continual contact are based on Coulomb's friction law. Stick-slip effects are taken into account. For the sake of verification several examples are investigated and some results are presented.

1. Introduction

The Multibody System (MBS) method provides means for an accurate and efficient investigation of engineering issues encountered in the fields of vehicle dynamics, aeronautics, robotics, biomechanics and mechanical engineering. It is the basic idea of MBS to transform thc actual system into an adequate model consisting of interconnected rigid bodies. The non-deformability of the discrete bodies resembles the fundamental idealization underlying this approach. Therefore, the MBS is most suitable to systems which show only small internal body distortions but undergo large translational and rotational motions.
Depending on the body interconnections, which in general resemble bilateral constraints, three different topological system configurations are distinguished. There are chain-, tree- and closed-loop structures. The governing equations of motion of MBS with closed-loop structures can either be given as a set of differential-algebraic equations called descriptor notation or as a minimal set of differential equations. In [1] and [2] different symbolical-numerical methods are presented for the solution of MBS con-

W. Schiehlen and M. Valášek (eds.), Virtual Nonlinear Multibody Systems, 141–154.

taining kinematic loops using only a minimal set of governing equations. These works also deal with the proper selection of the independent coordinates which should be used in order to avoid singularities during the solution process.
Unilateral constraints resulting from impacts and continual contacts occur in almost every technical application and are subject of many investigations, see, for example, [13], [10]. In [3], [4] and [5] an Augmented Lagrange formulation applied to linear-elastic, quasistatic contact problems is formulated. In [6] an extension of this approach to holonomic MBS subject to continual contact with dry friction can be found. If two bodies approach each other with a non-vanishing relative velocity and collide, impacts occur in both normal and tangential direction of the contact interface. Based on an impact theory found in [7] an Augmented Lagrangian formulation for three-dimensional frictional impact problems is presented in [8]. Within the scope of this paper the frictional contact and impact formulations of [6] and [8] are employed to holonomic MBS with kinematic loops. First a minimal set of differential equations characterizing the motion of holonomic MBS with kinematic loops is given. Then, the continual contact and impact formulations are incorporated into these equations. After presenting some numerical results the paper ends with a short summary and a survey of future work.

2. Mathematical Formulation of Kinematic Loops

In this paper holonomic, rheonomic or scleronomic multibody systems with kinematic loops are investigated. After properly releasing the kinematic loops, the corresponding non-minimal equations of motion and the involved loop-closing conditions are determined. This approach results in subsystems with tree structure, see Fig. 1. In this context a set of f generalized coordinates $\mathbf{y}$ is introduced for the dynamic description of the open loop structure. This set is divided into f^I independent generalized coordinates $\mathbf{y}_I$ characterizing the closed-loop structure and f^D dependent coordinates $\mathbf{y}_D$, which are functions of $\mathbf{y}_I$

$$\mathbf{y} = \begin{bmatrix} \mathbf{y}_I \\ \mathbf{y}_D(\mathbf{y}_I) \end{bmatrix}. \tag{1}$$

Referring to Fig. 1, the corresponding independent and dependent coordinates are $\mathbf{y}_I = [y_1, y_2]^T$ and $\mathbf{y}_D = [y_3(y_2), y_4(y_2)]^T$, respectively. The index-3 descriptor notation characterizing the system behavior, consists of the non-minimal equations of motion

$$\mathbf{M}(\mathbf{y},t) \cdot \ddot{\mathbf{y}} + \mathbf{k}(\mathbf{y},\dot{\mathbf{y}},t) - \mathbf{q}(\mathbf{y},\dot{\mathbf{y}},t) - \mathbf{C}^T(\mathbf{y},t) \cdot \boldsymbol{\lambda}_c = \mathbf{0} \tag{2}$$

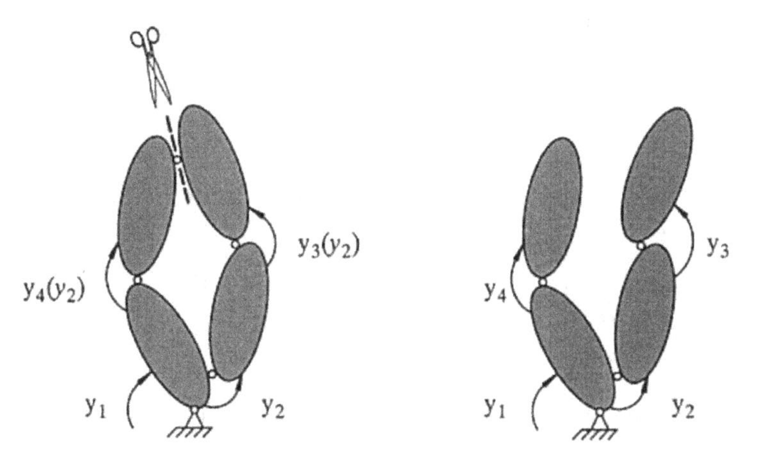

Figure 1. Transforming a closed-loop structure into an open-loop structure

and the implicit algebraic equations

$$\mathbf{c}\left(\mathbf{y}_I, \mathbf{y}_D, t\right) = \mathbf{0} \in I\!R^{f^D} \tag{3}$$

which ensure a closed-loop structure, as shown in [2]. The introduced abbreviations are the symmetric, positive definite mass matrix $\mathbf{M} \in I\!R^{f \times f}$, the generalized Coriolis forces $\mathbf{k} \in I\!R^f$, and the generalized applied forces $\mathbf{q} \in I\!R^f$. The vector $\boldsymbol{\lambda}_c \in I\!R^{f^D}$ embodies the generalized reaction forces due to closing the kinematic loops and the distribution matrix $\mathbf{C}\left(\mathbf{y}, t\right) \in I\!R^{f^D \times f}$ is defined by

$$\mathbf{C}\left(\mathbf{y}, t\right) = \left[\begin{array}{cc} \mathbf{C}_I & \mathbf{C}_D \end{array}\right] = \left[\begin{array}{cc} \dfrac{\partial \mathbf{c}}{\partial \mathbf{y}_I} & \dfrac{\partial \mathbf{c}}{\partial \mathbf{y}_D} \end{array}\right] = \frac{\partial \mathbf{c}\left(\mathbf{y}, t\right)}{\partial \mathbf{y}}. \tag{4}$$

The first time derivative of the loop-closing equations is found to be

$$\dot{\mathbf{c}} = \mathbf{C}_I \cdot \dot{\mathbf{y}}_I + \mathbf{C}_D \cdot \dot{\mathbf{y}}_D + \frac{\partial \mathbf{c}}{\partial t} = \mathbf{0} \tag{5}$$

and the corresponding second time derivative reads

$$\ddot{\mathbf{c}} = \mathbf{C}_I \cdot \ddot{\mathbf{y}}_I + \mathbf{C}_D \cdot \ddot{\mathbf{y}}_D + \dot{\mathbf{C}} \cdot \dot{\mathbf{y}} + \frac{d}{dt}\frac{\partial \mathbf{c}}{\partial t} = \mathbf{0}. \tag{6}$$

Equations (5) and (6), respectively, can be solved for the generalized velocities and accelerations, respectively,

$$\dot{\mathbf{y}} = \mathbf{J}_c \cdot \dot{\mathbf{y}}_I + \bar{\boldsymbol{\xi}}_c, \tag{7}$$

$$\ddot{\mathbf{y}} = \mathbf{J}_c \cdot \ddot{\mathbf{y}}_I + \boldsymbol{\xi}_c. \tag{8}$$

The emerging Jacobian matrix $\mathbf{J}_c \in I\!R^{f \times f^I}$ takes the form

$$\mathbf{J}_c = \begin{bmatrix} \mathbf{E} \\ -\mathbf{C}_D^{-1} \cdot \mathbf{C}_I \end{bmatrix} \tag{9}$$

where $\mathbf{E} \in I\!R^{f^I \times f^I}$ is the identity matrix. The vectors $\bar{\boldsymbol{\xi}}_c \in I\!R^f$ and $\boldsymbol{\xi}_c \in I\!R^f$ are found to be

$$\bar{\boldsymbol{\xi}}_c = \begin{bmatrix} \mathbf{0} \\ -\mathbf{C}_D^{-1} \cdot \dfrac{\partial \mathbf{c}}{\partial t} \end{bmatrix} \quad \text{and} \quad \boldsymbol{\xi}_c = \begin{bmatrix} \mathbf{0} \\ -\mathbf{C}_D^{-1} \cdot \left(\dot{\mathbf{C}}_I \cdot \dot{\mathbf{y}} + \dfrac{d}{dt}\dfrac{\partial \mathbf{c}}{\partial t} \right) \end{bmatrix}, \tag{10}$$

with the zero vector $\mathbf{0} \in I\!R^{f^I}$. The non-minimal equations of motions (2) and Eqn. (6) are referred to as the index-1 descriptor notation of the mechanical problem. Substituting Eqn. (8) into Eqn. (2) and applying the Principle of d'Alembert by utilizing the orthogonality of $\mathbf{J}_c$ and $\mathbf{C}$, compare with [12],

$$\mathbf{J}_c^T \cdot \mathbf{C}^T = \mathbf{0}, \tag{11}$$

yields the minimal set of governing equations

$$\mathbf{M}_d(\mathbf{y}_I, t) \cdot \ddot{\mathbf{y}}_I + \mathbf{k}_d(\mathbf{y}_I, \dot{\mathbf{y}}_I, t) - \mathbf{q}_d(\mathbf{y}_I, \dot{\mathbf{y}}_I, t) = \mathbf{0}. \tag{12}$$

The abbreviations denote

$$\mathbf{M}_d = \mathbf{J}_c^T \cdot \mathbf{M} \cdot \mathbf{J}_c \qquad \in I\!R^{f^I \times f^I}, \tag{13}$$

$$\mathbf{k}_d = \mathbf{J}_c^T \cdot (\mathbf{k} + \mathbf{M} \cdot \boldsymbol{\xi}_c) \qquad \in I\!R^{f^I}, \tag{14}$$

$$\mathbf{q}_d = \mathbf{J}_c^T \cdot \mathbf{q} \qquad \in I\!R^{f^I}. \tag{15}$$

Investigations dealing with both the proper selection of the independent coordinates with respect to singular configurations and the determination of the corresponding implicit loop-closing equations are found, e.g., in [1] and [2].

3. Contact of Multibody Systems with Kinematic Loops

Taking into account unilateral constraints, additional algebraic equations are required. They serve for the description of the contact kinematics given by the normal distances, the relative velocities and the relative accelerations of the contact points. Figure 2 shows two spatial bodies in the state of separation together with the right-handed coordinate system $\{P_1, \mathbf{n}, \mathbf{t}_1, \mathbf{t}_2\}$ composed of unit vectors. The normal vector $\mathbf{n}$ is perpendicular to the plane touching the surface of body 1 at point P_1, while the tangent vectors $\mathbf{t}_1$ and

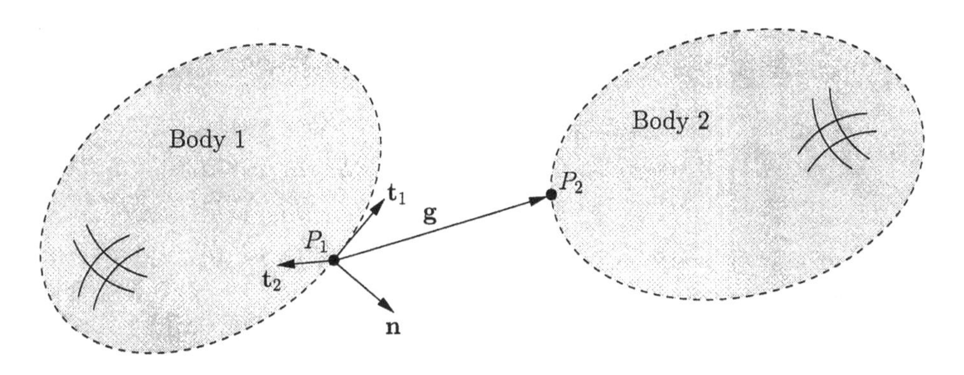

Figure 2. Two bodies in the state of separation

$\mathbf{t}_2$ are within this plane. The vector $\mathbf{g}$ is directed from the material point P_1 to point P_2 of body 2, which are assumed to have the smallest distance of all surface points. In this case the normal distance between these two bodies is given by

$$g_N = \mathbf{n} \cdot \mathbf{g}. \tag{16}$$

The relative velocities in normal and tangential direction, $\dot{g}_N \in \mathbb{R}$ and $\dot{\mathbf{g}}_T \in \mathbb{R}^2$, are obtained by projecting the difference between the absolute velocities of P_1 and P_2 onto the normal and tangent vectors. Taking into account Eqn. (7) one finally gets

$$\dot{g}_N = \mathbf{w}_N \left(\mathbf{n}, \mathbf{J}_{TP_1}, \mathbf{J}_{TP_2}\right) \cdot \left(\mathbf{J}_c \cdot \dot{\mathbf{y}}_I + \bar{\boldsymbol{\xi}}_c\right) + \hat{w}_N \left(\mathbf{n}, \bar{\mathbf{v}}_{P_1}, \bar{\mathbf{v}}_{P_2}\right), \tag{17}$$

$$\dot{\mathbf{g}}_T = \mathbf{w}_T^T \left(\mathbf{t}_i, \mathbf{J}_{TP_1}, \mathbf{J}_{TP_2}\right) \cdot \left(\mathbf{J}_c \cdot \dot{\mathbf{y}}_I + \bar{\boldsymbol{\xi}}_c\right) + \hat{\mathbf{w}}_T \left(\mathbf{t}_i, \bar{\mathbf{v}}_{P_1}, \bar{\mathbf{v}}_{P_2}\right). \tag{18}$$

The constraint matrices $\mathbf{w}_N \in \mathbb{R}^f$ and $\mathbf{w}_T \in \mathbb{R}^{f \times 2}$ depend on the Jacobian matrices of translation $\mathbf{J}_{TP_1}$ and $\mathbf{J}_{TP_2}$, whereas $\hat{w}_N \in \mathbb{R}$ and $\hat{\mathbf{w}}_T \in \mathbb{R}^2$ are functions of the local velocities $\bar{\mathbf{v}}_{P_1}$ and $\bar{\mathbf{v}}_{P_2}$ evaluated at the corresponding material points. They take the form

$$\mathbf{w}_N = \mathbf{n} \cdot \left(\mathbf{J}_{TP_2} - \mathbf{J}_{TP_1}\right), \qquad \hat{w}_N = \mathbf{n} \cdot \left(\bar{\mathbf{v}}_{P_2} - \bar{\mathbf{v}}_{P_1}\right), \tag{19}$$

$$\mathbf{w}_T^T = \begin{bmatrix} \mathbf{t}_1 \cdot \left(\mathbf{J}_{TP_2} - \mathbf{J}_{TP_1}\right) \\ \mathbf{t}_2 \cdot \left(\mathbf{J}_{TP_2} - \mathbf{J}_{TP_1}\right) \end{bmatrix}, \qquad \hat{\mathbf{w}}_T = \begin{bmatrix} \mathbf{t}_1 \cdot \left(\bar{\mathbf{v}}_{P_2} - \bar{\mathbf{v}}_{P_1}\right) \\ \mathbf{t}_2 \cdot \left(\bar{\mathbf{v}}_{P_2} - \bar{\mathbf{v}}_{P_1}\right) \end{bmatrix}. \tag{20}$$

The relative accelerations $\ddot{g}_N \in \mathbb{R}$ and $\ddot{\mathbf{g}}_T \in \mathbb{R}^2$ are given by differentiating Eqns. (17) and (18) with respect to time, compare with [9],

$$\ddot{g}_N = \mathbf{w}_N \cdot \left(\mathbf{J}_c \cdot \ddot{\mathbf{y}}_I + \boldsymbol{\xi}_c\right) + \bar{w}_N, \tag{21}$$

$$\ddot{\mathbf{g}}_T = \mathbf{w}_T^T \cdot \left(\mathbf{J}_c \cdot \ddot{\mathbf{y}}_I + \boldsymbol{\xi}_c\right) + \bar{\mathbf{w}}_T. \tag{22}$$

While in [13] the planar case with contacting polygons is considered, we now treat the spatial case assuming that the system bodies are polyhedra bounded entirely by planes. Then, the fundamental surface theorems, see also [9], provide for $\bar{w}_N$ and $\bar{\mathbf{w}}_T$ the following formulations

$$\bar{w}_N = \mathbf{n} \cdot (\bar{\mathbf{a}}_{P_2} - \bar{\mathbf{a}}_{P_1} - \tilde{\boldsymbol{\omega}}_1 \cdot (\mathbf{v}_{P_2} - \mathbf{v}_{P_1} + \dot{u}\mathbf{t}_1 + \dot{v}\mathbf{t}_2)), \tag{23}$$

$$\bar{\mathbf{w}}_T = \begin{bmatrix} \mathbf{t}_1 \cdot (\bar{\mathbf{a}}_{P_2} - \bar{\mathbf{a}}_{P_1} - \tilde{\boldsymbol{\omega}}_1 \cdot (\mathbf{v}_{P_2} - \mathbf{v}_{P_1} + \dot{u}\mathbf{t}_1 + \dot{v}\mathbf{t}_2)) \\ \mathbf{t}_2 \cdot (\bar{\mathbf{a}}_{P_2} - \bar{\mathbf{a}}_{P_1} - \tilde{\boldsymbol{\omega}}_1 \cdot (\mathbf{v}_{P_2} - \mathbf{v}_{P_1} + \dot{u}\mathbf{t}_1 + \dot{v}\mathbf{t}_2)) \end{bmatrix}. \tag{24}$$

The appearing translational velocities of P_1 and P_2 are denoted by $\mathbf{v}_{P_1}$ and $\mathbf{v}_{P_2}$, while the corresponding local accelerations are $\bar{\mathbf{a}}_{P_1}$ and $\bar{\mathbf{a}}_{P_2}$. The tilde operator applied to the rotational velocity vector of body 1, $\boldsymbol{\omega}_1 = [\omega_1, \omega_2, \omega_3]^T$, yields the skew-symmetric expression

$$\tilde{\boldsymbol{\omega}}_1 = \begin{bmatrix} 0 & -\omega_3 & \omega_2 \\ \omega_3 & 0 & -\omega_1 \\ -\omega_2 & \omega_1 & 0 \end{bmatrix}, \tag{25}$$

whereas $\dot{u}$ and $\dot{v}$ are defined as

$$\dot{u} = \mathbf{t}_1 \cdot (\mathbf{v}_{P_2} - \mathbf{v}_{P_1} - \tilde{\boldsymbol{\omega}}_1 \cdot \mathbf{g}), \quad \dot{v} = \mathbf{t}_2 \cdot (\mathbf{v}_{P_2} - \mathbf{v}_{P_1} - \tilde{\boldsymbol{\omega}}_1 \cdot \mathbf{g}). \tag{26}$$

To model systems of dynamically changing contact configurations, the following index sets are introduced, see [11], [7] and [6],

$$\begin{array}{lll} I_P = \{1, 2, \ldots, n_P\} & \text{with } n_P \text{ elements} & \rightarrow \text{possible contacts,} \\ I_I = \{i \in I_P \mid g_{Ni} = 0\} & \text{with } n_I \text{ elements} & \rightarrow \text{impact,} \\ I_C = \{i \in I_I \mid \dot{g}_{Ni} = 0\} & \text{with } n_C \text{ elements} & \rightarrow \text{continual contacts,} \\ I_S = \{i \in I_C \mid \dot{g}_{Ti} = 0\} & \text{with } n_S \text{ elements} & \rightarrow \text{sticking.} \end{array} \tag{27}$$

The index set I_P contains all n_P potential contact points. Contacts, where after the collision detection an impact calculation is required, are gathered in I_I. Members belonging to the set I_C are subject to continual contact and I_S is composed of all possibly sticking contact points. Due to changes in the kinematical state of the unilateral constraints the composition of the index sets I_I, I_C and I_S varies with time, see Fig. 3. Using the Lagrange multiplier method the non-minimal equations of motion of the continual contact problem for systems including kinematic loops read

$$\mathbf{M} \cdot \ddot{\mathbf{y}} + \mathbf{k} - \mathbf{q} - [\mathbf{W}_N + \mathbf{H}_R] \cdot \boldsymbol{\lambda}_N - \mathbf{W}_T \cdot \boldsymbol{\lambda}_T - \mathbf{C}^T \cdot \boldsymbol{\lambda}_c = \mathbf{0} \tag{28}$$

with the constraint matrices $\mathbf{W}_N = \{\mathbf{w}_{Ni}\} \in I\!R^{f^b \times n_I}, i \in I_C$, and $\mathbf{W}_T = \{\mathbf{w}_{Ti}\} \in I\!R^{f^b \times 2n_S}, i \in I_S$. The constraint forces in normal direction are given by $\boldsymbol{\lambda}_N = \{\lambda_{Ni}\} \in I\!R^{n_C}$, $i \in I_C$, while the vector $\boldsymbol{\lambda}_T = \{\boldsymbol{\lambda}_{Ti}\} \in I\!R^{n_S}$, $i \in I_S$, embodies all potentially sticking tangent constraint forces. The $n_C - n_S$ sliding contacts are incorporated into Eqn. (28) by means of

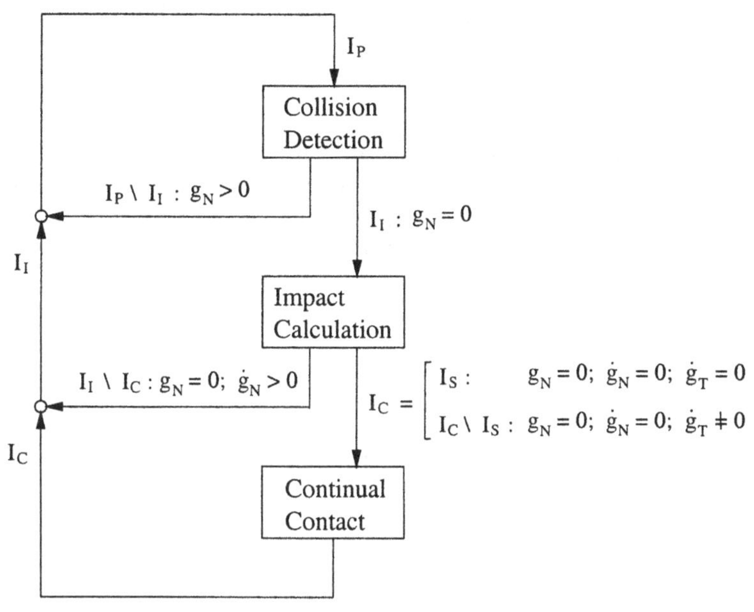

Figure 3. State transitions of rigid body contacts

$\mathbf{H}_R \in I\!R^{f^b \times n_I}$. Applying Coulomb's friction law and taking into account that the sliding forces act in opposite direction to the relative tangential motion, the columns of $\mathbf{H}_R$ are suitably composed of zeros and

$$\frac{\mu_i}{|\dot{\mathbf{g}}_{Ti}|}\mathbf{w}_{Ti} \cdot \dot{\mathbf{g}}_{Ti}, \quad i \in I_C \setminus I_S, \tag{29}$$

with the coefficient of sliding friction μ_i. Equations (21), (22) and (28) provide $f + f^D + n_c + 2n_s$ equations for the solution of the frictional contact problem with kinematic loops. However, the number of unknowns $\ddot{\mathbf{y}}$, $\ddot{\mathbf{g}}_N$, $\ddot{\mathbf{g}}_T$, $\boldsymbol{\lambda}_N$, $\boldsymbol{\lambda}_T$ and $\boldsymbol{\lambda}_c$ is $f + f^D + 2\,(n_c + 2n_s)$. Therefore, $n_c + 2n_s$ contact laws are required. The n_c contact laws in normal direction can be stated as a complementary problem

$$\ddot{\mathbf{g}}_N \geq \mathbf{0}, \quad \boldsymbol{\lambda}_N \geq \mathbf{0}, \quad \ddot{\mathbf{g}}_N \cdot \boldsymbol{\lambda}_N = 0 \tag{30}$$

with vanishing normal contact forces ($\lambda_{Ni} = 0$) in the state of separation ($\ddot{g}_{Ni} \geq 0$) and purely compressive forces ($\lambda_{Ni} \geq 0$) in the case of continual contact ($\ddot{g}_{Ni} = 0$) for all $i \in I_C$, see [7], [11] and [6]. Using Coulomb's law of friction, the $2n_S$ tangential contact laws of potentially sticking contact points can be expressed on acceleration level, compare with [6]. For $i \in I_S$ they read

$$\begin{aligned} |\boldsymbol{\lambda}_{Ti}| &< \mu_{0i}\lambda_{Ni} \qquad \text{if } |\ddot{\mathbf{g}}_{Ti}| = 0 \quad \rightarrow \text{sticking,} \\ |\boldsymbol{\lambda}_{Ti}| &= \mu_{0i}\lambda_{Ni} \qquad \text{if } |\ddot{\mathbf{g}}_{Ti}| > 0 \quad \rightarrow \text{transition to sliding,} \end{aligned} \tag{31}$$

where μ_{0i} is the coefficient of static friction obeying the following condition, see [6],

$$\lim_{|\dot{\mathbf{g}}_{Ti}|\to 0} \mu_i\left(|\dot{\mathbf{g}}_{Ti}|\right) = \mu_{0i}. \tag{32}$$

Using an Augmented Lagrangian formulation the contact laws given by the inequalities (30) and (31) can be transformed into equations which represent projections of the contact forces onto corresponding convex sets. In this case the contact law in normal direction can be expressed by means of the functional $\mathbf{\Pi}_N(\boldsymbol{\lambda}_N, \ddot{\mathbf{y}}_I) = \{\Pi_{Ni}(\lambda_{Ni}, \ddot{\mathbf{y}}_I)\} \in I\!R^{nc}$, $i \in I_C$, and Eqn. (21), see [6],

$$\lambda_{Ni} = \Pi_{Ni}(\lambda_{Ni}, \ddot{\mathbf{y}}_I) = \begin{cases} 0 & \text{if } \tau_{Ni} < 0, \\ \tau_{Ni} & \text{if } \tau_{Ni} \geq 0, \end{cases} \tag{33}$$

where

$$\tau_{Ni}(\lambda_{Ni}, \ddot{\mathbf{y}}_I) = \lambda_{Ni} - r\ddot{g}_{Ni} = \lambda_{Ni} - r\left(\mathbf{w}_{Ni}\cdot\mathbf{J}_c\cdot\ddot{\mathbf{y}}_I + \mathbf{w}_{Ni}\cdot\boldsymbol{\xi}_c + \bar{w}_{Ni}\right). \tag{34}$$

The arbitrary factor $r \in I\!R$ must soley obey the condition $r > 0$. The contact law in tangential direction can be stated by introducing the functional $\mathbf{\Pi}_T(\boldsymbol{\lambda}_N, \boldsymbol{\lambda}_T, \ddot{\mathbf{y}}_I) = \{\mathbf{\Pi}_{Ti}(\lambda_{Ni}, \boldsymbol{\lambda}_{Ti}, \ddot{\mathbf{y}}_I)\} \in I\!R^{2ns}$, $i \in I_S$, and Eqn. (22)

$$\boldsymbol{\lambda}_{Ti} = \mathbf{\Pi}_{Ti}(\lambda_{Ni}, \boldsymbol{\lambda}_{Ti}, \ddot{\mathbf{y}}_I) = \begin{cases} \boldsymbol{\tau}_{Ti} & \text{if } |\boldsymbol{\tau}_{Ti}| < \mu_{0i}\lambda_{Ni}, \\ \mu_i\lambda_{Ni}\dfrac{\boldsymbol{\tau}_{Ti}}{|\boldsymbol{\tau}_{Ti}|} & \text{if } |\boldsymbol{\tau}_{Ti}| \geq \mu_{0i}\lambda_{Ni} \end{cases} \tag{35}$$

with the abbreviation

$$\boldsymbol{\tau}_{Ti}(\lambda_{Ni}, \boldsymbol{\lambda}_{Ti}, \ddot{\mathbf{y}}_I) = \boldsymbol{\lambda}_{Ti} - r\ddot{\mathbf{g}}_{Ti} = \boldsymbol{\lambda}_{Ti} - r\left(\mathbf{w}_{Ti}^T\cdot\mathbf{J}_c\cdot\ddot{\mathbf{y}}_I + \mathbf{w}_{Ti}^T\cdot\boldsymbol{\xi}_c + \bar{\mathbf{w}}_{Ti}\right). \tag{36}$$

After substituting Eqns. (8), (33) and (35) into Eqn. (28), the principle of d'Alembert given by Eqn. (11) yields a minimal set of equations, which can be used for the description of continual contact problems with kinematic loops. By extending Eqn. (12) it follows

$$\mathbf{M}_d\cdot\ddot{\mathbf{y}}_I + \mathbf{k}_d - \mathbf{q}_d - \mathbf{J}_c^T\cdot[\mathbf{W}_N + \mathbf{H}_R]\cdot\mathbf{\Pi}_N - \mathbf{J}_c^T\cdot\mathbf{W}_T\cdot\mathbf{\Pi}_T = \mathbf{0}. \tag{37}$$

The remaining unknowns $\ddot{\mathbf{y}}_I$, $\boldsymbol{\lambda}_N$ and $\boldsymbol{\lambda}_T$ are found by a two-stage iterative calculation process using Newton's method. In a first step $\ddot{\mathbf{y}}_I$ is obtained from Eqn. (37) while the contact forces remain unchanged. They are updated in a succeeding second step according to Eqns. (33) and (35). These two steps are repeated for convergence. Then, an integration scheme is applied resulting in updated velocities and positions $\dot{\mathbf{y}}_I$ and $\mathbf{y}_I$. Evaluation

of the loop closing equations (3) and (5) finally yields $\mathbf{y}_D$ and $\dot{\mathbf{y}}_D$. This solution method is suitable for problems with independent and dependent constraints involving non-unique contact forces.
The impact calculation process is similar to the continual contact problem. Therefore, it is not described in all its details. The interested reader is referred to [8]. The impact is separated into a compression and an extension phase. The governing equations of motion of the compression phase are obtained by time integration of Eqn. (28) taking into account all elements of the index set I_I. In the course of the time integration over the very short time interval of contact, $\mathbf{k}$ and $\mathbf{q}$, respectively, yield infinitesimally small terms which are neglected and the contact forces are transformed into normal and tangential impacts. In [7] a tangential impact law based on Coulomb's friction law is presented for the planar case and in [8] applied to spatial problems. The impact laws in both normal and tangential direction can be expressed by functionals on velocity level similar to Eqns. (33) and (35), see [8]. Making in addition use of the contact kinematics given by Eqns. (17) and (18), the loop-closing conditions (5) and the orthogonality condition (11) the compression phase is completely described mathematically. The unknown generalized velocities and the normal and tangential impacts at the end of the compression phase are evaluated by a two-stage iterative calculation process using Newton's method. For the expansion phase a modification of Poisson's hypothesis is applied, see [7], [11] and [8], which allows multiple impacts. Similar to the compression phase the governing equations are expressed on velocity level and the impact laws in normal and tangential direction are established by appropriate projections of the contact impacts onto convex sets, see [8]. In conjunction with the contact kinematics, the loop-closing conditions and the data of the compression phase a sufficient number of equations is available for the determination of the generalized velocities and impacts at the end of the expansion phase.

4. Example

The presented mathematical formulations of MBS with kinematic loops subject to frictional contacts have been implemented into a research software. To prove the reliability and accuracy of the algorithm several examples have been studied by the authors. In the following a pendulum of three bodies and a gear fixed to the ground are considered, see Fig. 4. The topology of the pendulum is characterized by a tree-structure whereas the gear shows one kinematical loop. By arbitrarily disconnecting a bilateral gear joint, an open-loop structure of the system is obtained. Its kinematics is completely described by a set of $f = 6$ relative angles with $f^I = 4$

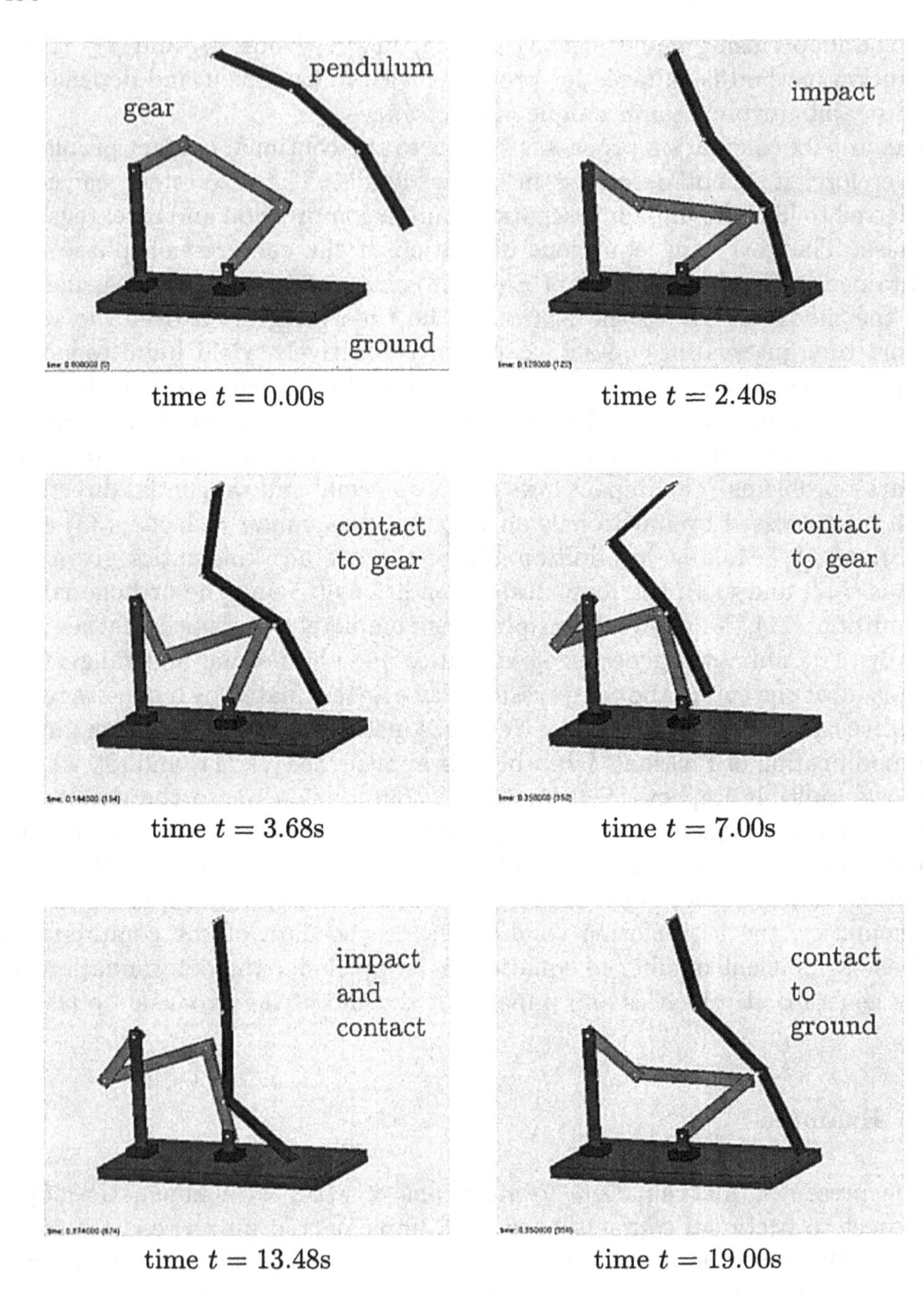

Figure 4. Frictional contact of a MBS with one kinematic loop

independent and $f^D = 2$ dependent coordinates. Therefore, 2 algebraic equations are required for imposing a closed-loop structure on the system. Gravitational forces are applied to all bodies which move with an initial velocity. The initial value problem is solved by the explicit Euler integration scheme with constant step size. The chronology of the system motion is illustrated in Fig. 4. It can be seen, that the loop-closing conditions are satisfied throughout the integration process. Frequently changing partly-elastic impacts and continual contacts occur between the different members of the system[1].
Figures 5 and 6 show the frequency of occurence of impacts and continual contacts with respect to time. The number of impacts n_I and continual contacts n_C differ at some instants in the course of the time integration. This means, that contact points separated after the impact calculation. To show more distinctly that the number of continual contacts varies during the simulation process the period of time from $t = 9$s to $t = 12$s of Fig. 6 is depicted on a larger scale in Fig. 7. Within this period the number of continual contacts ranges between 1 and 6 entailing dynamical changes in the contact formulation. In consequence of these changes the kinematic and kinetic equations which are used for the mathematical description of multibody systems with contact show a time dependent structure.

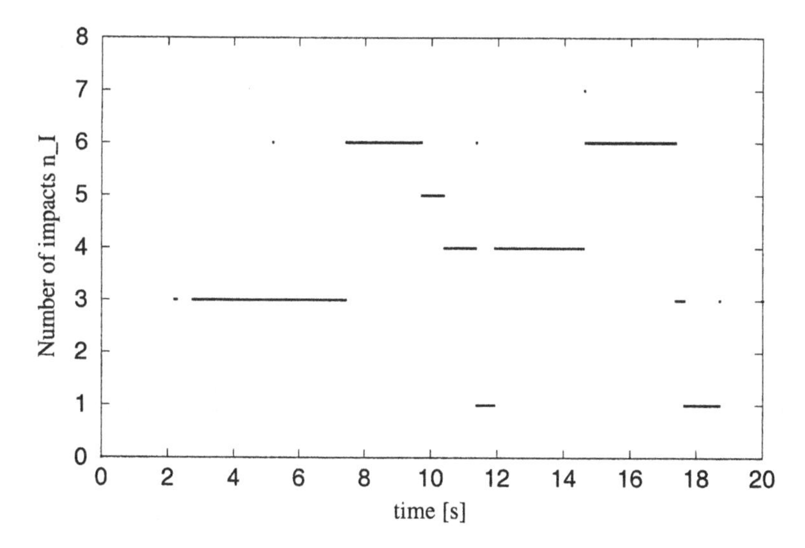

Figure 5. Number of impacts with respect to time

[1] For a video of the motion see `http://www.mechb.uni-stuttgart.de/staff/Eberhard/video/mechanism.mpg`

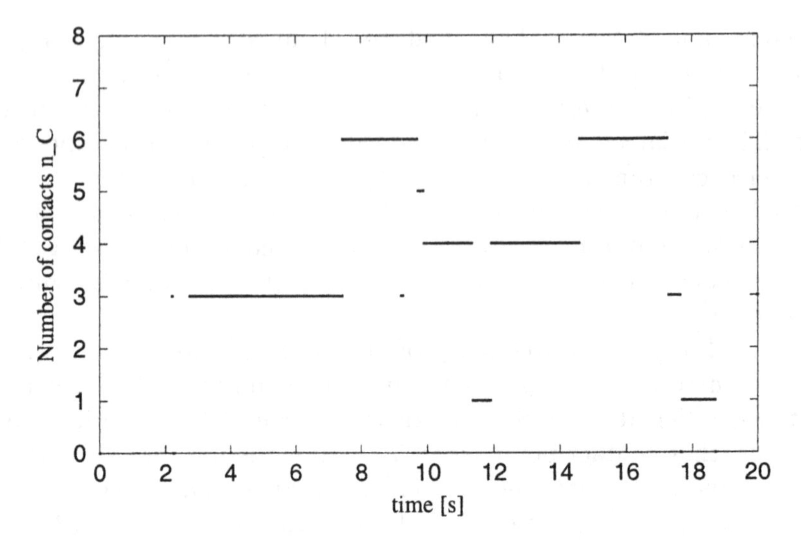

Figure 6. Number of continual contacts with respect to time

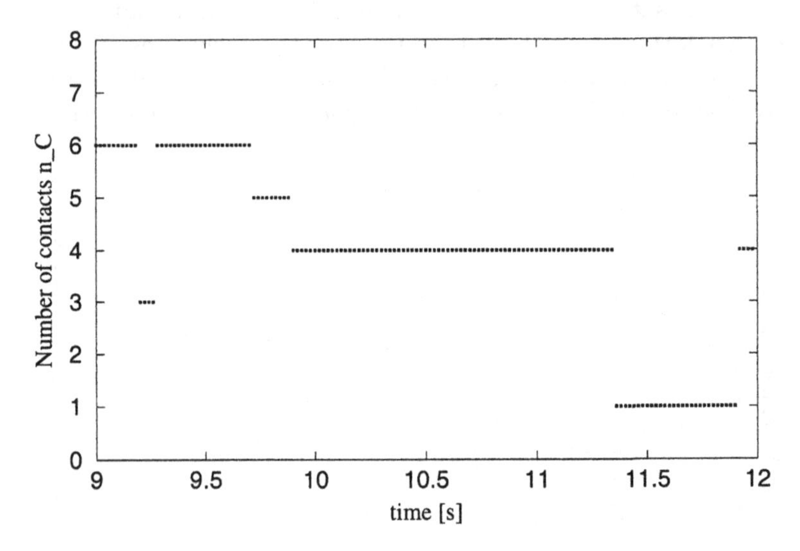

Figure 7. Number of continual contacts from Fig. 6 on a larger scale

5. Conclusions

In this paper frictional contact problems of spatial MBS with kinematic loops are investigated. The equations of motion of bilaterally constrained MBS with kinematic loops are described by a minimal set of differential equations. Unilateral constraints are taken into account using the contact formulations introduced in [6] and [8], which are adapted to systems with kinematic loops. The contact problem is divided into impacts and continual contacts. The impact calculation is performed on velocity level while continual contacts are evaluated on acceleration level. The shown example involving bilateral constraints with one kinematic loop and frequently changing impacts/contacts with sliding and sticking friction serves as a verification of the introduced approach and its successful implementation. Future investigations will include frictional contact problems of flexible MBS. As proposed in [14] a floating frame of reference formulation will be used to obtain the kinematics and kinetics of flexible systems subject to both large rigid body motions and large deformations. The rigid body motion is described by a set of minimal coordinates, see [15]. Flexible bodies are discretized by isoparametric finite elements. The nodal coordinates of the elements define the arising deformations. The contact formulation will be based on a penaltization/homogenization of the contact forces using Coulomb's friction law.

References

1. Leister, G. (1992) *Beschreibung und Simulation von Mehrkörpersystemen mit geschlossenen kinematischen Schleifen*, VDI Fortschritt-Berichte, Reihe 11, No. 167, VDI-Verlag, Düsseldorf.
2. Schirm, W. (1993) *Symbolisch-numerische Behandlung von kinematischen Schleifen in Mehrkörpersystemen*, VDI Fortschritt-Berichte, Reihe 11, No. 198, VDI-Verlag, Düsseldorf.
3. Alart, P. and Curnier, A. (1991) A Mixed Formulation for Frictional Contact Problems Prone to Newton Like Solution Methods, *Computer Methods in Applied Mechanics and Engineering*, Vol. **92**, 353–375.
4. Simo, J.C. and Laursen, T.A. (1992) An Augmented Lagrangian Treatment of Contact Problems Involving Friction, *Computers & Structures*, Vol. **42**, No. 1, 97–116.
5. Klarbring, A. (1992) Mathematical Programming and Augmented Lagrangian Methods for Frictional Contact Problems, in A. Curnier (Ed.) *Proceedings Contact Mechanics International Symposium, EPFL, Lausanne, Switzerland, October 7-9, 1992*, 409–422.
6. Wösle, M. (1997) *Dynamik von räumlichen strukturvarianten Starrkörpersystemen*, VDI Fortschritt-Berichte, Reihe 18, No. 213, VDI-Verlag, Düsseldorf.
7. Pfeiffer, F. and Glocker, C. *Multibody Dynamics with Unilateral Contacts*, Wiley, New York.
8. Wolfsteiner, P. and Pfeiffer, F. (1997) Dynamics of a Vibratory Feeder, *Proceedings ASME Design Engineering Technical Conferences, September 14-17, 1997*, Sacramento, California.
9. Meitinger, T. (1998) *Dynamik automatisierter Montageprozesse*, VDI Fortschritt-Berichte, Reihe 2, No. 476, VDI-Verlag, Düsseldorf.
10. Eberhard, P. (2000) *Kontaktuntersuchungen durch hybride Mehrkörpersystem / Finite Elemente Simulationen*, Shaker, Aachen.
11. Glocker, C. (1995) *Dynamik von Starrkörpersystemen mit Reibung und Stößen*, VDI Fortschritt-Berichte, Reihe 18, No. 182, VDI-Verlag, Düsseldorf.

12. Schiehlen, W. (1986) *Technische Dynamik*, Teubner, Stuttgart.
13. Pfister, J. and Eberhard, P. (2002) Frictional Contact of Flexible and Rigid Bodies, *Granular Matter*, Vol. **4**, No. 1, 25–36.
14. Pfister, J. (2001) Einige Grundlagen der elastischen Mehrkörpersysteme, *Report ZB–129*, University of Stuttgart, Institute B of Mechanics.
15. Melzer, F. (1994) *Symbolisch-numerische Modellierung elastischer Mehrkörpersysteme mit Anwendung auf rechnerische Lebensdauervorhersagen*, VDI Fortschritt-Berichte, Reihe 20, No. 139, VDI-Verlag, Düsseldorf.

VIRTUAL ASSEMBLY IN MULTIBODY DYNAMICS

W. SCHIEHLEN and C. SCHOLZ
Institute B of Mechanics, University of Stuttgart,
Pfaffenwaldring 9, D-70550 Stuttgart, Germany
E-mail: [wos,cs]@mechb.uni-stuttgart.de

1. Introduction

The dynamic analysis of complex engineering systems requires the modelling of components from different engineering fields. The modular decomposition of the global system is based on engineering intuition of corresponding engineering disciplines. Once the interfaces of the subsystems have been defined these subsystems can be modeled independently from each other. The exchange and modification of a subsystem is also independent from any other component as well as different and independent software tools can be used for every engineering discipline.

The overall simulation or virtual assembly, respectively, is achieved by coupling of all subsystems in order to obtain the global system behaviour. For mechatronic systems the coupling of subsystems may be realized on three different levels of model description as illustrated in Figure 1.

In the physical model description a system is represented by a physical model, e.g. in case of multibody systems there are rigid bodies, massless joints and coupling force elements. The mathematical model description is a representation of a system by mathematical equations which can be derived from the physical model description, e.g. the equations of motion of a multibody system. The simulation results of the mathematical model description are considered as the behavioral model description, e.g. the trajectories of position and velocity of the bodies.

Coupling of models in the behavioral model description is referred to as simulator coupling, modular simulation or virtual assembly, respectively. The simulation of the global system is realized by a time discrete linker and scheduler which combines the inputs and outputs of the corresponding subsystems and establishes communication between the subsystems to discrete time instants. Therefore, it is possible to use different software packages for each subsystem and then to link the solvers together.

W. Schiehlen and M. Valášek (eds.), Virtual Nonlinear Multibody Systems, 155–172.

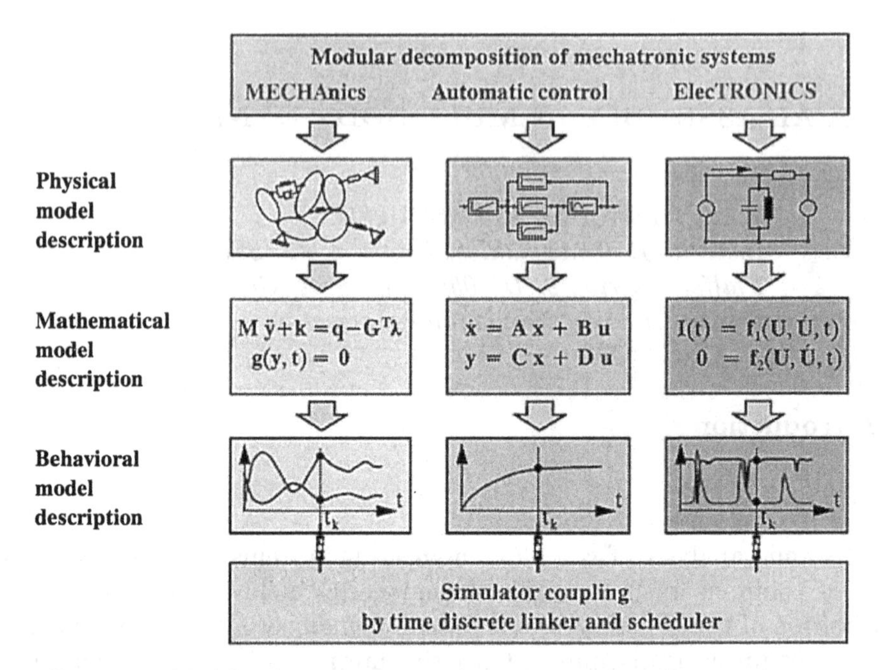

Figure 1. Modular decomposition and simulation of mechatronic systems

2. Modular modelling of dynamical systems

The global system may be decomposed into subsystems which are described in the block representation introduced in the following. On this basis the feed–through property of subsystems is analyzed. The global system is formed by interconnections of the corresponding blocks, and conditions for algebraic loops in the global system structure are given. The modeling procedure of subsystems is described and will be shown to be independent of the global system structure, see also Kübler and Schiehlen [5].

2.1. BLOCK REPRESENTATION OF SUBSYSTEMS

The block representation of dynamic systems is widely used in system dynamics and control theory, see for example Kailath [3] and Ogata [9], where systems are characterized by a block with input vector $\boldsymbol{u}$ and output vector $\boldsymbol{y}$, Figure 2. It is assumed that the outputs $\boldsymbol{y}$ have no feedback to the inputs $\boldsymbol{u}$ and are uniquely determined by

$$\boldsymbol{y} = S(\boldsymbol{u}) \tag{1}$$

where S is some operator or vector function completely describing the system behavior.

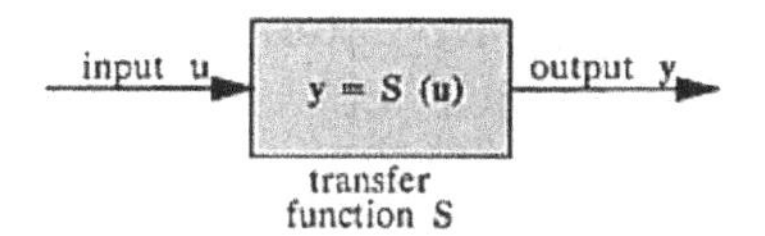

Figure 2. Block representation of a subsystem

The input–output description (1) can be used for a dynamical analysis without knowledge of the internal structure of the system which leads to the well known "black–box" representation allowing access to the system only by means of the input and the output terminals. On this basis, systems are classified due to their feed-through property. If input changes result in output changes without any delay by internal dynamics, the system is called a feed–through system. Otherwise the system has no feed–through.

2.2. GLOBAL SYSTEM STRUCTURE

A global system consists of subsystems with interconnections between the corresponding blocks, Figure 3. The global system structure is determined by all interconnections and can be represented by the coupling equations

$$\boldsymbol{u} = \boldsymbol{L}\boldsymbol{y}, \quad L_{ij} \in \{0; 1\} \tag{2}$$

with the global input and the global output vector

$$\boldsymbol{u} = \begin{bmatrix} \boldsymbol{u}^I \\ \vdots \\ \boldsymbol{u}^N \end{bmatrix}, \quad \boldsymbol{y} = \begin{bmatrix} \boldsymbol{y}^I \\ \vdots \\ \boldsymbol{y}^N \end{bmatrix}. \tag{3}$$

The elements of the incidence matrix $\boldsymbol{L}$ are zero or one representing not existing or existing interconnections of corresponding inputs and outputs.

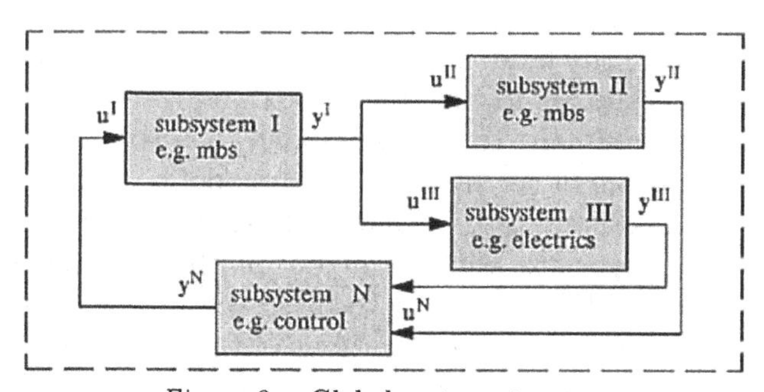

Figure 3. Global system structure

If any interconnections in the system result in a closed loop of subsystems all of which are feed–through systems, then an algebraic loop does exist. In this case the explicit determination of all inputs with the coupling equations (2) turns into the implicit problem

$$\boldsymbol{u} = \boldsymbol{L}\boldsymbol{y}(\boldsymbol{u}) \leadsto \boldsymbol{f}(\boldsymbol{u}) = \boldsymbol{u} - \boldsymbol{L}\boldsymbol{y}(\boldsymbol{u}) = \boldsymbol{0}. \tag{4}$$

Consequently, algebraic loops do not allow explicit determination of all inputs only with knowledge of the global system structure given by the coupling equations (2).

The introduced modular description of systems also includes hierarchical system structures. On the one hand each subsystem can be considered as a system which is again formed by subsystems, on the other hand each system may represent a subsystem of a hierarchically higher system.

Since the internal structure of a subsystem is independent of the global system structure, this approach supports in particular interchangeability and reusability of system blocks.

2.3. MODELING OF SUBSYSTEMS

The presented modular description of systems allows for independent and parallel modeling of the internal dynamics of each subsystem. In general, the modeling procedure of a subsystem comprises again three levels of model description, Figure 4. In the physical description the engineering system is represented by a physical model, such as rigid bodies, joints and coupling elements in the case of multibody systems. The inputs and outputs of the physical model are also physical quantities, such as forces or motion of bodies. The mathematical description is an abstract representation of the physical model by equations, e.g. the equations of motion of multibody systems, with continuous input and output quantities. In the behavioral description the dynamic response of the subsystem is obtained as a result of given input signals and the hidden internal behavior. Therefore, the behavioral description corresponds to the "black–box" representation mentioned in Section 2.1. In general, the dynamic response has to be determined by means of time simulation of the mathematical model leading to discretized input and output quantities in the behavioral description.

To choose an appropriate physical model of the considered technical system represents a problem which has to be solved by the engineer. Then, in many engineering disciplines the equations of the mathematical description and the dynamic response of the system by means of time simulation can be generated in a systematic and consequently computer–aided way with knowledge of the physical model. However, a clear input–output representation is often missing.

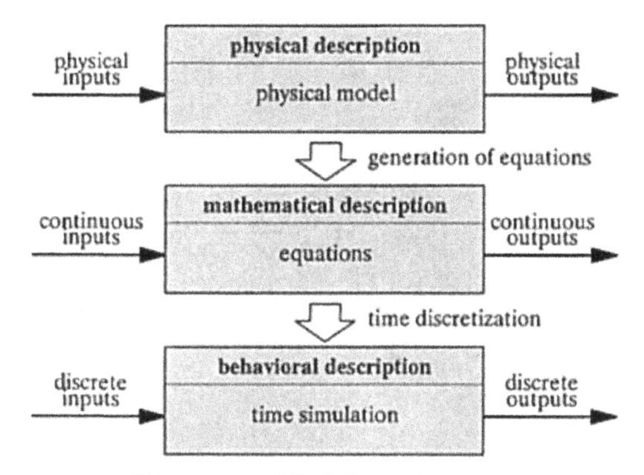

Figure 4. Modeling of a subsystem

3. Modeling of multibody systems

The general block representation of dynamical systems with input and output quantities is now applied to a multibody system. The modeling procedure of a multibody system block is described in accordance to the model description introduced in Section 2.3.

3.1. PHYSICAL MODEL DESCRIPTION

A multibody system consists of rigid bodies, joints and coupling elements. A body may degenerate to a particle or to a body without inertia. The ideal joints are massless and include rigid joints or scleronomic constraints, respectively, as well as joints with completely given motion, also referred to as rheonomic constraints. Joints result in constraint forces and torques acting on bodies. Massless coupling elements such as springs and dampers depend on kinematical variables and result in applied forces and torques. Purely time varying forces and torques are also applied quantities. The topology of the multibody system is arbitrary, i.e. chains, trees and closed loops may occur.

The block representation introduced in Section 2.1 with corresponding inputs and outputs is applied to a multibody system in Figure 5. Physical inputs include purely time varying applied forces and torques $\boldsymbol{u}_f(t)$, prescribed motion of bodies consisting of positions $\boldsymbol{u}_b(t)$, velocities $\dot{\boldsymbol{u}}_b(t)$ and accelerations $\ddot{\boldsymbol{u}}_b(t)$, and prescribed motion of coupling elements $\boldsymbol{u}_c(t)$, $\dot{\boldsymbol{u}}_c(t)$, $\ddot{\boldsymbol{u}}_c(t)$.

Physical outputs are the absolute motion of bodies consisting of positions $\boldsymbol{y}_f(t)$, velocities $\dot{\boldsymbol{y}}_f(t)$ and accelerations $\ddot{\boldsymbol{y}}_f(t)$ resulting from the applied forces and torques $\boldsymbol{u}_f(t)$. Additional outputs are constraint forces and torques $\boldsymbol{y}_b(t)$ from bodies and applied forces and torques $\boldsymbol{y}_c(t)$ from

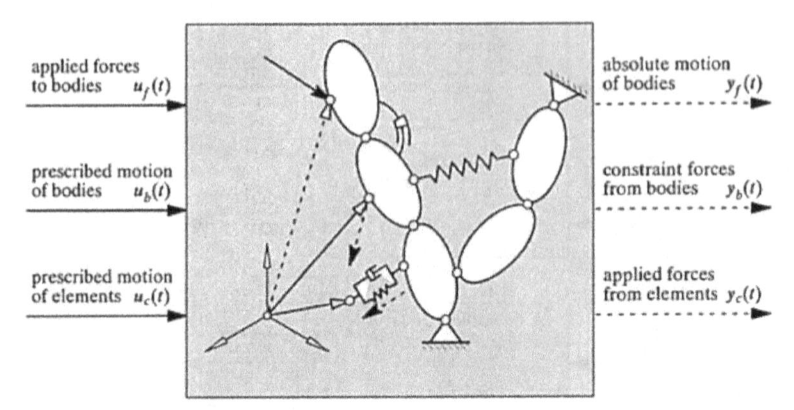

Figure 5. Physical model description of a multibody system block

coupling elements resulting from the prescribed motion of bodies $\boldsymbol{u}_b(t)$, $\dot{\boldsymbol{u}}_b(t)$, $\ddot{\boldsymbol{u}}_b(t)$ and coupling elements $\boldsymbol{u}_c(t)$, $\dot{\boldsymbol{u}}_c(t)$, $\ddot{\boldsymbol{u}}_c(t)$, respectively.

3.2. MATHEMATICAL MODEL DESCRIPTION

The equations of motion and the equations of reaction are derived for a multibody system block considered as an ordinary multibody system, including holonomic constraints and proportional or proportional–differential forces, [12, 13]. The mathematical representation of the multibody system block is also transformed to the more general representation of dynamical systems in the state space. Determination of output quantities of a multibody system block is discussed and the feed–through property is investigated on this basis.

3.2.1. *Kinematics*

A multibody system block with p rigid bodies and q holonomic, rheonomic constraints, resulting in $f = 6p - q$ degrees of freedom, is considered. The absolute position of the system is given relative to the inertial frame by the 3×1–translation vector of the center of mass C_i and the 3×3–rotation tensor for each body $i = 1(1)p$ by

$$\boldsymbol{r}_i = \boldsymbol{r}_i(\boldsymbol{z}, \boldsymbol{u}_m, t), \qquad (5a)$$

$$\boldsymbol{S}_i = \boldsymbol{S}_i(\boldsymbol{z}, \boldsymbol{u}_m, t) \qquad (5b)$$

where $\boldsymbol{z}$ is the $f \times 1$–position vector summarizing the f generalized coordinates of the system, $\boldsymbol{u}_m = \begin{bmatrix} \boldsymbol{u}_b^T & \boldsymbol{u}_c^T \end{bmatrix}^T$ represents the input vector of prescribed motions of bodies and coupling elements and t denotes time. The absolute velocities are found by time differentiation with respect to the inertial frame:

$$\begin{aligned} \boldsymbol{v}_i &= \boldsymbol{J}_{Ti}(\boldsymbol{z}, \boldsymbol{u}_m, t)\dot{\boldsymbol{z}} + \bar{\boldsymbol{v}}_i(\boldsymbol{z}, \boldsymbol{u}_m, \dot{\boldsymbol{u}}_m, t), & \text{(6a)} \\ \boldsymbol{\omega}_i &= \boldsymbol{J}_{Ri}(\boldsymbol{z}, \boldsymbol{u}_m, t)\dot{\boldsymbol{z}} + \bar{\boldsymbol{\omega}}_i(\boldsymbol{z}, \boldsymbol{u}_m, \dot{\boldsymbol{u}}_m, t). & \text{(6b)} \end{aligned}$$

The $3 \times f$–matrices $\boldsymbol{J}_{Ti}$ and $\boldsymbol{J}_{Ri}$ denote the Jacobian matrices and characterize the virtual translation and rotation displacement of the system. The absolute accelerations are obtained by a second time differentiation:

$$\begin{aligned} \boldsymbol{a}_i &= \boldsymbol{J}_{Ti}(\boldsymbol{z}, \boldsymbol{u}_m, t)\ddot{\boldsymbol{z}} + \bar{\boldsymbol{a}}_i(\boldsymbol{z}, \dot{\boldsymbol{z}}, \boldsymbol{u}_m, \dot{\boldsymbol{u}}_m, \ddot{\boldsymbol{u}}_m, t), & \text{(7a)} \\ \boldsymbol{\alpha}_i &= \boldsymbol{J}_{Ri}(\boldsymbol{z}, \boldsymbol{u}_m, t)\ddot{\boldsymbol{z}} + \bar{\boldsymbol{\alpha}}_i(\boldsymbol{z}, \dot{\boldsymbol{z}}, \boldsymbol{u}_m, \dot{\boldsymbol{u}}_m, \ddot{\boldsymbol{u}}_m, t). & \text{(7b)} \end{aligned}$$

3.2.2. *Newton–Euler equations*

For the application of Newton's and Euler's equations to multibody systems the free body diagram has to be used. Kinematical joints have to be replaced by adequate constraint forces and torques. The ideal applied forces and torques depend only on the kinematical variables of the system. They are independent of the constraint forces and torques. Then, the Newton–Euler equations of the complete system can be represented in the inertial frame as

$$\begin{aligned} \bar{\bar{\boldsymbol{M}}}(\boldsymbol{z}, \boldsymbol{u}_m, t)\bar{\boldsymbol{J}}(\boldsymbol{z}, \boldsymbol{u}_m, t)\ddot{\boldsymbol{z}} &+ \bar{\boldsymbol{q}}^c(\boldsymbol{z}, \dot{\boldsymbol{z}}, \boldsymbol{u}_m, \dot{\boldsymbol{u}}_m, \ddot{\boldsymbol{u}}_m, t) \\ &= \bar{\boldsymbol{q}}^e(\boldsymbol{z}, \dot{\boldsymbol{z}}, \boldsymbol{u}_f, \boldsymbol{u}_m, \dot{\boldsymbol{u}}_m, t) + \bar{\boldsymbol{q}}^r(\boldsymbol{z}, \boldsymbol{u}_m, t) \quad (8) \end{aligned}$$

where the $6p \times 6p$–block diagonal matrix

$$\bar{\bar{\boldsymbol{M}}} = \text{diag}\,\{\, m_1\boldsymbol{E} \;\ldots\; m_p\boldsymbol{E} \;\; \boldsymbol{I}_1 \;\ldots\; \boldsymbol{I}_p \,\} \quad (9)$$

summarizes the masses m_i with the 3×3–identity matrix $\boldsymbol{E}$ and the 3×3–inertia tensors $\boldsymbol{I}_i$, and where the global $6p \times f$–Jacobian matrix

$$\bar{\boldsymbol{J}} = \left[\, \boldsymbol{J}_{T1}^T \;\ldots\; \boldsymbol{J}_{Tp}^T \;\; \boldsymbol{J}_{R1}^T \;\ldots\; \boldsymbol{J}_{Rp}^T \,\right]^T \quad (10)$$

contains the Jacobian matrices $\boldsymbol{J}_{Ti}^T$ and $\boldsymbol{J}_{Ri}^T$ of translation and rotation, respectively. The $6p \times 1$–force vectors $\bar{\boldsymbol{q}}^c$ and $\bar{\boldsymbol{q}}^e$ represent the Coriolis forces and the ideal applied forces. The $6p \times 1$–vector of constraint forces and torques can be expressed as

$$\bar{\boldsymbol{q}}^r(t) = \bar{\boldsymbol{Q}}(\boldsymbol{z}, \boldsymbol{u}_m, t)\boldsymbol{\lambda} \quad (11)$$

with the $q \times 1$–vector of generalized constraints $\boldsymbol{\lambda}$ and the global $6p \times q$–distribution matrix $\bar{\boldsymbol{Q}}$. The Newton–Euler equations (8) represent $f + q = 6p$ differential algebraic equations for f generalized coordinates $\boldsymbol{z}$ and q generalized constraints $\boldsymbol{\lambda}$.

3.2.3. *Equations of motion*

The equations of motion of holonomic systems are found according to d'Alembert's principle by premultiplication of the Newton–Euler equations (8) by the transposed Jacobian $\bar{\boldsymbol{J}}^T$ as

$$\boldsymbol{M}(\boldsymbol{z}, \boldsymbol{u}_m, t)\ddot{\boldsymbol{z}} + \boldsymbol{k}(\boldsymbol{z}, \dot{\boldsymbol{z}}, \boldsymbol{u}_m, \dot{\boldsymbol{u}}_m, \ddot{\boldsymbol{u}}_m, t) = \boldsymbol{q}(\boldsymbol{z}, \dot{\boldsymbol{z}}, \boldsymbol{u}_f, \boldsymbol{u}_m, \dot{\boldsymbol{u}}_m, t), \quad (12)$$

where $\boldsymbol{M} = \bar{\boldsymbol{J}}^T \bar{\bar{\boldsymbol{M}}} \bar{\boldsymbol{J}}$ is the symmetric and positive definite $f \times f$–inertia matrix and the $f \times 1$–vectors $\boldsymbol{k}$ and $\boldsymbol{q}$ describe the generalized Coriolis forces and the generalized applied forces, respectively. The constraint forces and torques are eliminated due to the orthogonality $\bar{\boldsymbol{J}}^T \bar{\boldsymbol{Q}} = \boldsymbol{0}$ and therefore the number of equations is reduced from $6p$ to f for the free motion of the system. Consequently, the equations of motion (12) represent f ordinary differential equations.

3.2.4. *Equations of reaction*

According to d'Alembert's principle premultiplication of the Newton–Euler equations (8) by $\bar{\boldsymbol{Q}}^T \bar{\bar{\boldsymbol{M}}}^{-1}$ results in q purely algebraic equations of reaction

$$\boldsymbol{N}(\boldsymbol{z}, \boldsymbol{u}_m, t)\boldsymbol{\lambda} + \hat{\boldsymbol{q}}(\boldsymbol{z}, \dot{\boldsymbol{z}}, \boldsymbol{u}_f, \boldsymbol{u}_m, \dot{\boldsymbol{u}}_m, t) = \hat{\boldsymbol{k}}(\boldsymbol{z}, \dot{\boldsymbol{z}}, \boldsymbol{u}_m, \dot{\boldsymbol{u}}_m, \ddot{\boldsymbol{u}}_m, t) \quad (13)$$

for q generalized constraints $\boldsymbol{\lambda}$, where $\boldsymbol{N} = \bar{\boldsymbol{Q}}^T \bar{\bar{\boldsymbol{M}}}^{-1} \bar{\boldsymbol{Q}}$ is the symmetric and positive definite $q \times q$–reaction matrix and $\hat{\boldsymbol{q}}$ and $\hat{\boldsymbol{k}}$ are $q \times 1$–vectors.

3.2.5. *State–space formulation*

The state–space formulation is the most common mathematical description of dynamic systems, see Kailath [3] and Ogata [9], which consists of the state equations to describe the internal dynamics and the output equations to determine output quantities of the system, Figure 6.

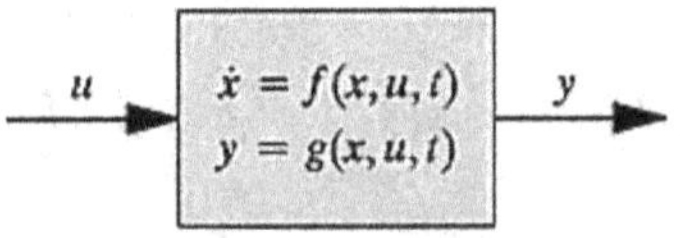

Figure 6. Mathematical model description in state–space formulation

The input vector of a multibody system block consists of applied forces $\boldsymbol{u}_f(t)$, prescribed motion of bodies $\boldsymbol{u}_b(t)$, $\dot{\boldsymbol{u}}_b(t)$, $\ddot{\boldsymbol{u}}_b(t)$ and prescribed motion of coupling elements $\boldsymbol{u}_c(t)$, $\dot{\boldsymbol{u}}_c(t)$, $\ddot{\boldsymbol{u}}_c(t)$:

$$\boldsymbol{u} = \begin{bmatrix} \boldsymbol{u}_f^T & \boldsymbol{u}_b^T & \dot{\boldsymbol{u}}_b^T & \ddot{\boldsymbol{u}}_b^T & \boldsymbol{u}_c^T & \dot{\boldsymbol{u}}_c^T & \ddot{\boldsymbol{u}}_c^T \end{bmatrix}^T. \quad (14)$$

The state vector

$$\boldsymbol{x} = \begin{bmatrix} \boldsymbol{z}^T & \dot{\boldsymbol{z}}^T \end{bmatrix}^T \quad (15)$$

is composed of generalized coordinates and velocities.

Due to the symmetric and positive definite inertia matrix, the equations of motion (12) can be transformed to the nonlinear state equations

$$\dot{\boldsymbol{x}} = \boldsymbol{f}(\boldsymbol{x}, \boldsymbol{u}, t), \qquad \boldsymbol{x}(t_0) = \boldsymbol{x}_0. \tag{16}$$

The output vector is formed with the absolute motion of bodies $\boldsymbol{y}_f(t)$, $\dot{\boldsymbol{y}}_f(t)$ and $\ddot{\boldsymbol{y}}_f(t)$, constraint forces and torques $\boldsymbol{y}_b(t)$ from bodies and applied forces and torques $\boldsymbol{y}_c(t)$ from coupling elements:

$$\boldsymbol{y} = \left[\boldsymbol{y}_f^T \ \dot{\boldsymbol{y}}_f^T \ \ddot{\boldsymbol{y}}_f^T \ \boldsymbol{y}_b^T \ \boldsymbol{y}_c^T \right]^T. \tag{17}$$

The nonlinear output equations can be written as

$$\boldsymbol{y} = \boldsymbol{g}(\boldsymbol{x}, \boldsymbol{u}, t). \tag{18}$$

The components of the output vector $\boldsymbol{y}$ are available as the kinematical quantities (5), (6) and (7) as functions of the generalized accelerations $\ddot{\boldsymbol{z}}$ determined from the equations of motion (12) and as the constraint forces (11) with the generalized constraints calculated by the equations of reaction (13). Applied forces are available from the vector $\bar{\boldsymbol{q}}^e$ of the Newton–Euler equations (8).

3.2.6. *Feed–through property*

In Section 2.1 the feed–through property of a system block was defined. In the mathematical description the feed–through property is characterized by an explicit dependency of the outputs on the inputs in the output equations (18).

The state–space formulation of multibody system blocks with different input quantities is illustrated in Figure 7. Applied forces and torques as inputs $\boldsymbol{u}_f$ have no feed–through on absolute position $\boldsymbol{y}_f$ and velocity $\dot{\boldsymbol{y}}_f$ of bodies. Likewise, accelerations of prescribed motion of coupling elements $\ddot{\boldsymbol{u}}_c$ have no feed–through on applied forces from coupling elements. All other output quantities of multibody systems are explicitly dependent on the corresponding inputs, Figure 7.

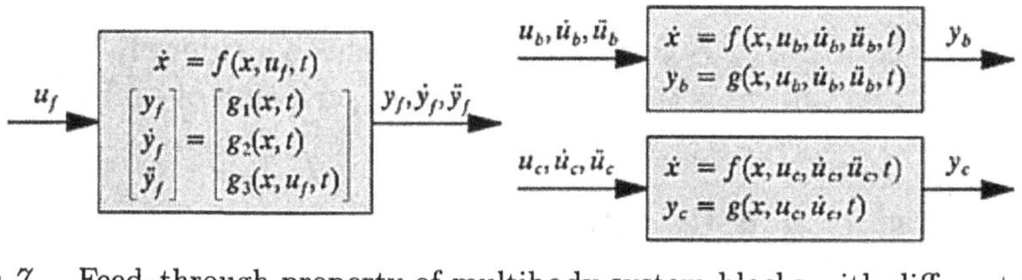

Figure 7. Feed–through property of multibody system blocks with different inputs

3.3. BEHAVIORAL MODEL DESCRIPTION

In behavioral description the dynamic response of the subsystem is obtained as a result of given input signals and the hidden internal behavior. Therefore, the behavioral description corresponds to the "black–box" representation introduced in Section 2.1. In general, the dynamic response has to be determined by means of time simulation of the mathematical model. This involves numerical integration of the state equations and evaluation of the output equations at discrete time instants, see Kübler and Schiehlen [5].

4. Simulator Coupling

In general there are two options for the simulation of systems given in the modular description introduced in Section 2. On one hand simulation tools which are also based on this description, often referred to as block simulators, can be used, such as ACSL [8] or SIMULINK [7]. On the other hand coupling of different simulation tools is an option.

The advantage of block simulators is the use of standard numerical integration methods with additional solution of algebraic loops. With sufficiently exact solution of algebraic loops, the properties of numerical integration methods remain unchanged. The disadvantages are the requirements to supply all subsystems in one simulation tool and to use the same numerical method for all subsystems.

Coupling of simulation tools allows subsystems to be simulated in different programs avoiding the mentioned disadvantages of block simulators. Although standard solvers can be used for each subsystem additional numerical problems arise from the coupling of the solvers.

It has been shown by Kübler [4] that zero–stability for non–iterative simulator coupling is only guaranteed, if algebraic loops do not exist between the subsystems. Otherwise, instability of the modular simulation may occur. To obtain zero–stability in the general case, if simulators are coupled for modular simulation, the algebraic loops may be treated in each global time step by iterative methods.

4.1. NON–ITERATIVE SIMULATOR COUPLING

The parallel integration of N coupled subsystems is considered. The basis of the following investigation is the time discrete description of each subsystem i:

$$\boldsymbol{x}^i_{k+1} = \boldsymbol{\Phi}^i(\boldsymbol{\varphi}^i, m^i, \tilde{\boldsymbol{u}}^i), \tag{19a}$$

$$\boldsymbol{y}^i_{k+1} = \boldsymbol{g}^i(\boldsymbol{x}^i_{k+1}, \tilde{\boldsymbol{u}}^i_{k+1}, t_{k+1}), \qquad i = 1(1)N. \tag{19b}$$

The inputs of each subsystem are calculated by the algebraic coupling equations (2) which can be evaluated without approximation error at each global time t_{k+1}:

$$\boldsymbol{u}^i_{k+1} = \boldsymbol{L}^i \boldsymbol{y}_{k+1}. \tag{19c}$$

The coupled integration (19) for $N = 2$ subsystems is illustrated in Figure 8.

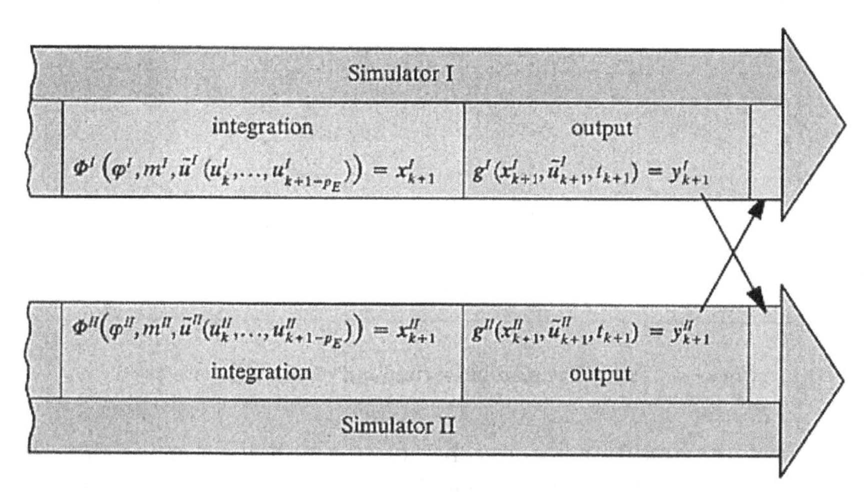

Figure 8. Non–iterative simulator coupling of two subsystems

However, the coupled integration (19) is zero–stable only, if algebraic loops do not exist between the subsystems.

4.2. ITERATIVE SIMULATOR COUPLING

Integration of each subsystem is carried out according to Section 4.1 with extrapolation of the inputs, followed by the coupled solution of the output equations of all subsystems.

For an arbitrary number of simulators, each subsystem i now can be written as

$$\boldsymbol{x}^i_{k+1} = \boldsymbol{\Phi}^i(\varphi^i, m^i, \tilde{\boldsymbol{u}}^i), \tag{20a}$$

$$\boldsymbol{y}^i_{k+1} = \boldsymbol{g}^i(\boldsymbol{x}^i_{k+1}, \boldsymbol{u}^i_{k+1}, t_{k+1}), \qquad i = 1(1)N. \tag{20b}$$

In contrast to equations (19b) no extrapolation is needed for the input vector $\boldsymbol{u}^i_{k+1}$ which is now found by iteration.

The inputs of the output equations can be eliminated using the coupling equations $\boldsymbol{u}^i_{k+1} = \boldsymbol{L}^i \boldsymbol{y}_{k+1}$, leading to the system of nonlinear algebraic equations for the outputs of the global system

$$\boldsymbol{y}_{k+1} = \boldsymbol{\Psi}(\boldsymbol{y}_{k+1}) \tag{21}$$

which has to be solved iteratively for each global time step.

In Figure 9 iterative simulator coupling is illustrated for $N = 2$ subsystems.

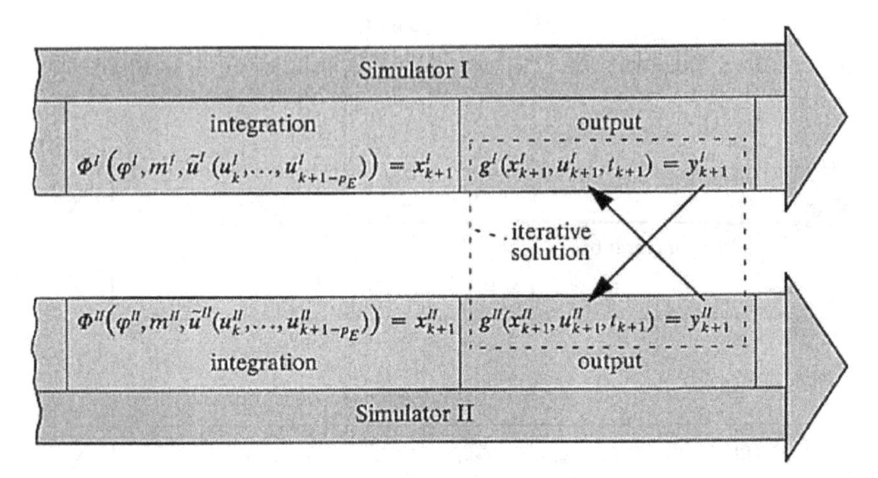

Figure 9. Iterative simulator coupling of two subsystems

Now the only condition for stability of the global system is the use of zero–stable integrators, leading to the statement that iterative simulator is zero–stable, if zero–stable integrators are used. This statement also holds in the general case for one–step and multi–step methods as well as for output equations are nonlinearly dependent on the inputs.

The system of nonlinear algebraic equations (21) has to be solved for each global time step by an iterative method, in general, see Dennis and Schnabel [2], Ortega and Rheinboldt [10]. By the Gauß–Jacobi and the Gauß–Seidel method local convergence is not guaranteed. Newton's method is locally convergent, however, the Jacobian is not available in modular simulation, in general. Consequently, a quasi–Newton method which uses an approximation to the Jacobian has to be used. One of the most reliable quasi–Newton methods is Broyden's method, see Broyden [1], Dennis and Schnabel [2], which uses a secant approximation to the Jacobian and is also locally convergent.

5. Communication step size control

For an efficient computation of systems changing strongly their dynamic behaviour during the numerical solution there are integration methods with automatic step size control available. Therefore, nearly all simulation tools consist of such integration methods. Executing a simulator coupling it is possible to use the integration methods of each coupled tool but there is also a need to control the global time step for the communication between the subsystems.

For a communication step size control two different error estimation methods based on the classical step size control are regarded. The Richardson-extrapolation estimates an error by the difference of two solutions computed by the same integration method but with different step sizes. The error estimation based on the embedded formula uses the difference of two solutions computed by two integration methods with different integration order but the same step size.

In order to generate a communication step size control based on the embedded formula each subsystem i has to execute a global time step from t_k to t_{k+H} by two integration methods $\mathbf{\Phi}$ and $\mathbf{\Omega}$ of different order, see Figure 10.

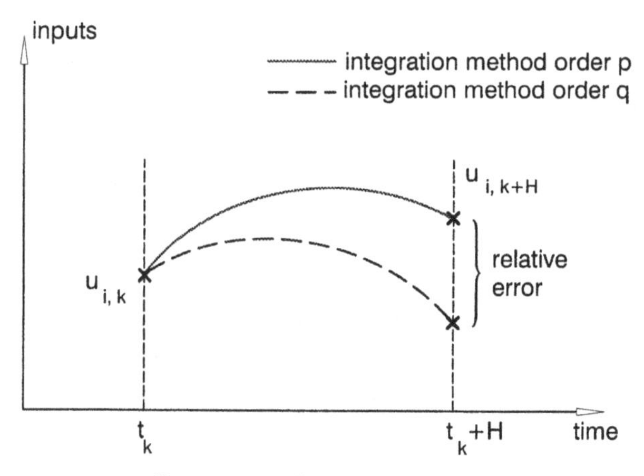

Figure 10. Embedded Formula

At the end of this time step the difference of the new input values serves for the error estimation written as

$$err_{mr} = \| \hat{\boldsymbol{u}}(t_k + H, \mathbf{\Phi}) - \bar{\boldsymbol{u}}(t_k + H, \mathbf{\Omega}) \| . \qquad (22)$$

For the communication step size control based on the embedded formula the subsystems need additional information, an access to the system only by means of the input and output terminals can not be guaranteed. Accordingly, the defined requirements for a simulator coupling can not be met.

For a communication step size control based on the Richardson-extrapolation each subsystem i has to execute a global time step from t_k to t_{k+H} by one integration method with two different step sizes, see Figure 11.

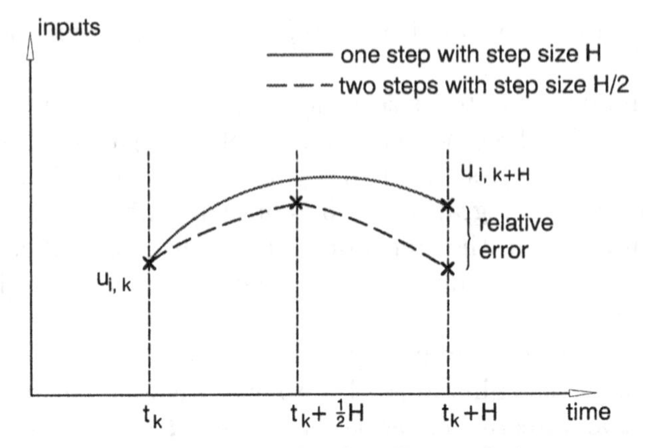

Figure 11. Richardson-Extrapolation

At the end of this time step the difference of the new input values serves for the error estimation leading to

$$err_{mr} = \| \boldsymbol{u}_H(t_k + H) - \boldsymbol{u}_{2\text{xH}/2}(t_k + H) \| \,. \tag{23}$$

In contrast to the embedded formula, the requirement not to interfere with the coupled tools is met. In Figure 12 the iterative simulator coupling with added communication step size control based on the Richardson-extrapolation is illustrated for $N = 2$ subsystems.

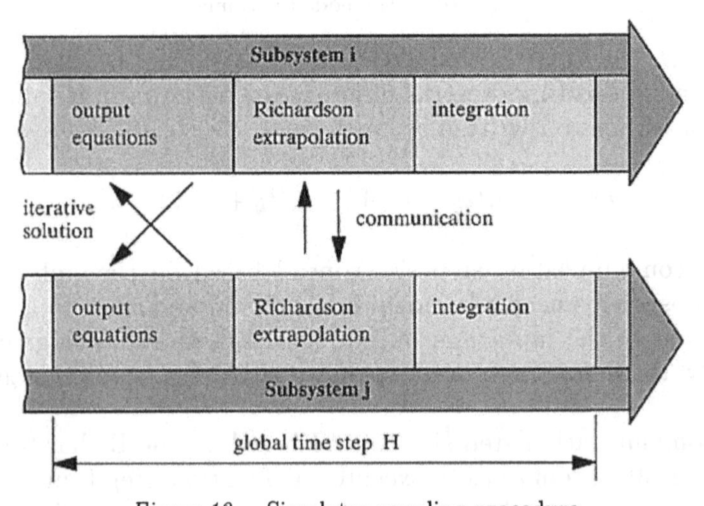

Figure 12. Simulator coupling procedure

6. Test example

The considered model of a vehicle suspension on a flexible track is illustrated in Figure 13. The model consists of a wheel mass which is connected by a spring-damper-combination to the vehicle body and by a second spring to the track mass connected by a third spring to the ground. This global system is decomposed into two subsystems. The first subsystem consists of the spring-damper-combination c_1,d_1, the wheel mass m_1 and the second spring c_2. The second subsystem consists of the track mass m_2 and the third spring c_s. Both subsystems are connected via a node point. The node motions as well as the force acting on the node are exchanged by the corresponding inputs and outputs of each subsystem, see Figure 13.

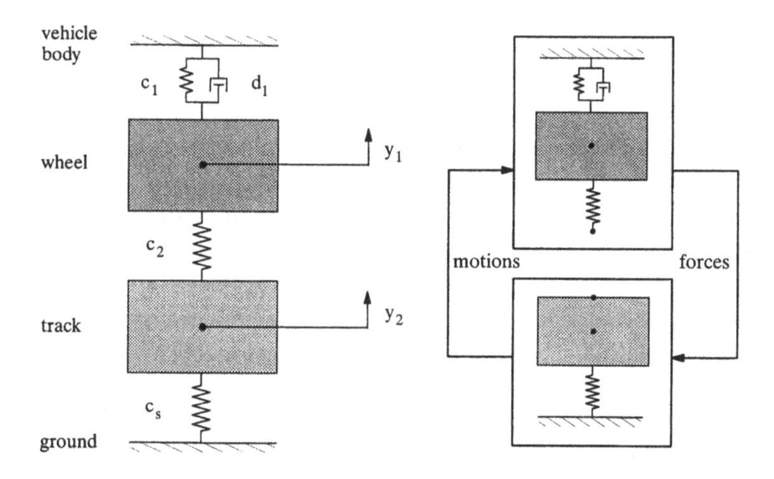

Figure 13. Model setup and decomposition

The simulation experiment setup with the given parameters illustrates Figure 14. With the given parameters the two eigenfrequencies of the system follow as $f_{eig,1} \approx 20Hz$ and $f_{eig,2} \approx 1.6kHz$. With this large difference in the two frequency values the coupled modules are characterized by large differences in their eigendynamic properties. Both subsystems are modelled with the tool NEWEUL [6] which generates the equations of motion. For the simulation the tool NEWMOS [11] was applied. Both subsystems use the Runge-Kutta integration method of 4. order with a local step size $h = 10^{-5}$ and the inputs are extrapolated by a method of 3. order. The error barriers for the control of the global time step are chosen as 10^{-3} for the relative and absolute error.

Comparing computations with and without communication step size control the difference of the computation time can be determined. The

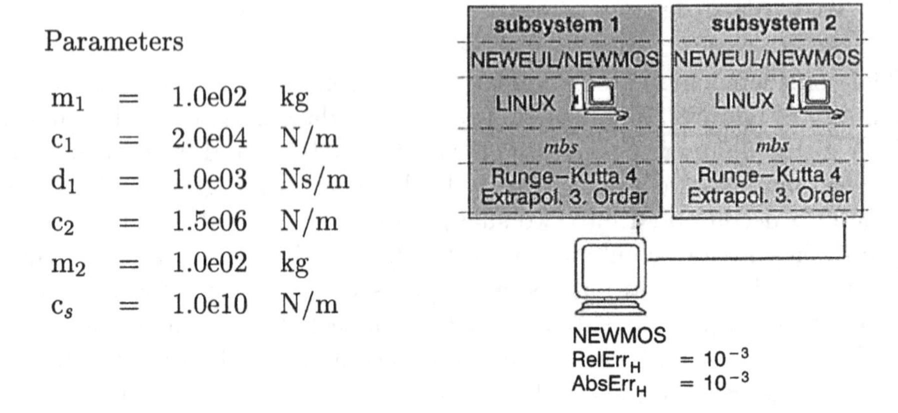

Figure 14. Simulation parameters and setup

simulation time with variable global step size is about $t_{sim} \approx 8s$ while the computation time with fixed global step size $H = h$ requires about $t_{sim} \approx 29s$. Changing the parameters resulting in a model with similar eigenfrequencies a reduction of computation time has not been observed. Accordingly an increase of efficiency can be expected in particular for subsystems described by large differences in their eigendynamics.

Regarding the properties of the communication step size control the global time step depends on the inputs of each module. For example, the size of the global time step depends mainly on the exchanged force acting on the node and less on the motion of the node. Figure 15 illustrates the behaviour of the spring force and the corresponding global time step.

The global time step adapts to large values in case of small force value changes. Vice versa having large changes of the force value small global time steps have to be used. Knowing the same behaviour from integration step size control the functionality of the communication step size control is approved by plausibility.

7. Summary

A block representation of multibody systems is presented in order to decompose a global system into subsystems. Then, the subsystems are coupled by interconnecting their inputs and outputs. Based on that a modular simulation can be executed.

The dynamical analysis of the global system is realized by a time discrete linker and scheduler which combines the inputs and outputs of the corre-

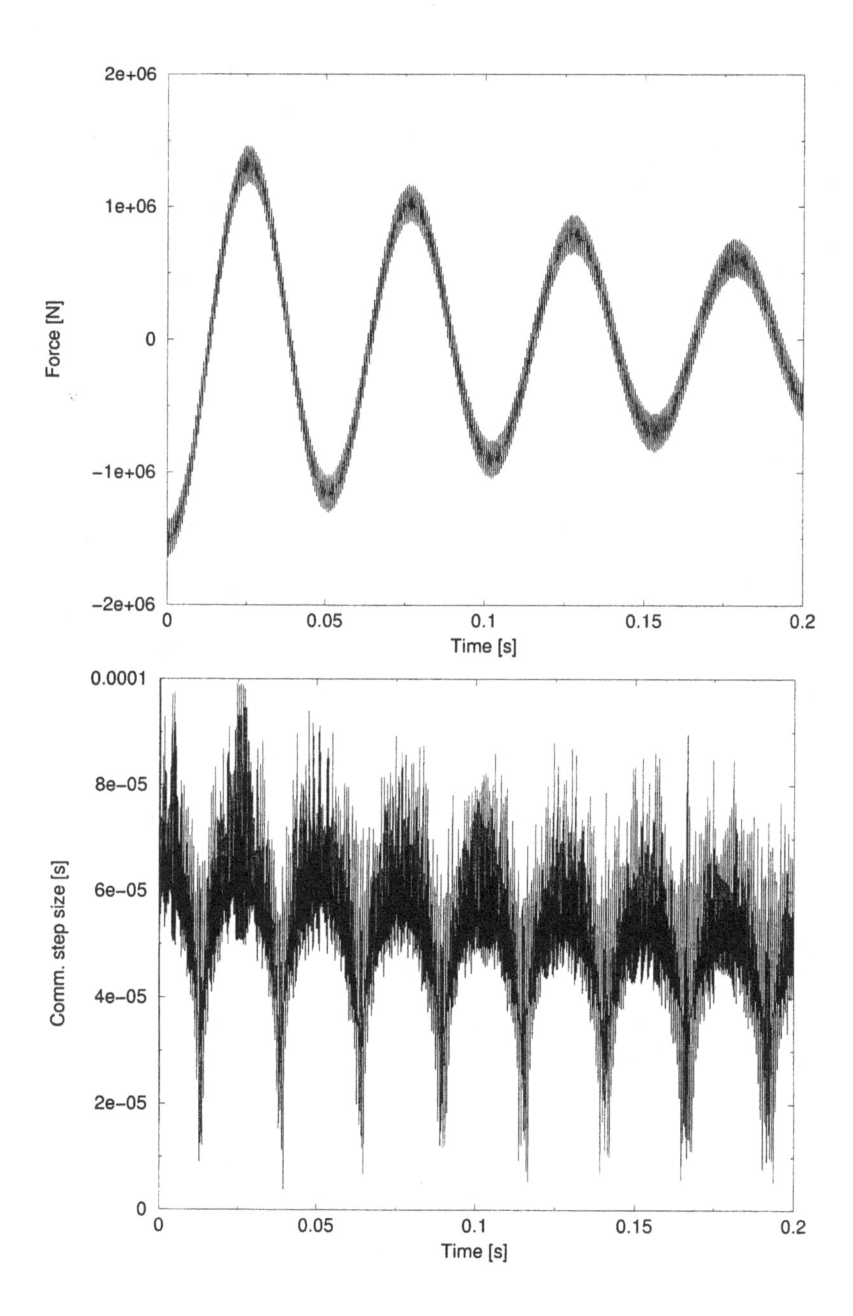

Figure 15. Simulation results

sponding subsystems and establishes communication between them. This modular simulation is stabilized by the iterative simulator coupling method. An increase of the numerical efficiency by an automatic communication step size control can be achieved. Therefore, two methods of step size control are discussed, and one of them is recommended for engineering applications.

The method proposed for modular modeling and simulation is applied for a vehicle supension on a flexible track. The advantages of the communication step size control are clearly described by this example. Thus, a communication step size control is recommended for virtual assembly in multibody dynamics.

References

1. Broyden, C.G. (1965) Solving Nonlinear Simultaneous Equations, Mathematics of Computation **19**, 577–593.
2. Dennis, J.E.; Schnabel, R.B. (1983) Numerical Methods for Unconstrained Optimization and Nonlinear Equations, Prentice Hall, Englewood Cliffs, New Jersey.
3. Kailath, Th. (1980) Linear systems, Prentice Hall, Englewood Cliffs, New Jersey.
4. Kübler, R. (2000) Modulare Modellierung und Simulation mechatronischer Systeme, Fortschritt-Berichte VDI, Reihe 20, Nr. 327, Düsseldorf.
5. Kübler, R.; Schiehlen, W. (2000) Modular Simulation in Multibody System Dynamics, Multibody System Dynamics **4**, 107-127.
6. Leister, G. (1993) Programmsystem NEWEUL'92 Universität Stuttgart, Institut B für Mechanik, Anleitung AN-32, Stuttgart
7. The Mathworks Inc. (1996) SIMULINK User's Manual.
8. Mitchell & Gauthier Assoc. (1985) ACSL Newsletter, Concord.
9. Ogata, K. (1967) State space analysis of control systems, Prentice Hall, Englewood Cliffs, New Jersey.
10. Ortega, J.M.; Rheinboldt, W.C. (1970) Iterative Solution of Nonlinear Equations in Several Variables, Academic Press, New York.
11. Rükgauer, A. (1997) Modulare Simulation mechatronischer Systeme mit Anwendung in der Fahrzeugdynamik, Fortschritt-Berichte VDI, Reihe 20, Nr. 248, Düsseldorf.
12. Schiehlen, W. (1986) Technische Dynamik, Teubner, Stuttgart.
13. Schiehlen, W. (1997) Multibody System Dynamics: Roots and Perspectives, Multibody System Dynamics **1**, 149–188.

NUMERICAL SIMULATION OF FLEXIBLE MULTIBODY SYSTEMS BY USING A VIRTUAL RIGID BODY MODEL.

M. PASCAL, T. GAGARINA, *Laboratoire CEMIF SC, Université d'Evry Val d'Essonne, 91020, EVRY,France*

1. Introduction

In the past decade, interest in multibody system dynamics has grown, leading to several computer codes [1] available for the generation of the motion equations. If in the beginning, only multibody systems with rigid components [2] were considered, weight minimization and large accelerations in many technical problems lead to an increasing tendancy for elastic vibrations. In recent years, many works [3] deal with the dynamical simulation of these systems in which each flexible component undergoes large rigid body motion but small elastic deformations. It results that the motion equations of the whole system involve ordinary differential equations coupled with partial differential equations. Approximated methods are used in which the displacement field of each flexible component is discretized by a Rayleigh-Ritz method leading to a new system with a finite number of degrees of freedom.
The aims of this work is to show that it is possible to use any existing rigid body codes for the dynamical simulation of elastic multibody systems provided that these codes are based on variational methods like Lagrange's equations or direct applications of d'Alembert principle. A first attempt for this possibility was made in [4] for planar interconnected flexible beams. In the present work, three dimensional interconnected flexible bodies are considered and the method used to identify the approximated model of the flexible multibody system with a fictitious rigid multibody system is rather different from the idea used in [4]. An example is shown, using the symbolic dynamical code AUTOLEV [5] devoted to rigid body simulation.

2. Problem Formulation

Let us consider a multibody system composed of several rigid or flexible bodies connected by p hinges. Each flexible component undergoes large rigid body motion coupled with small structural deformations.
Let us consider one flexible component (S) of the system and its fixed reference configuration (S_0) at $t = 0$ .The assumption of large displacement and small strains lead to the standard method in which an intermediate configuration $(\tilde{S})$ of the body is

W. Schiehlen and M. Valášek (eds.), *Virtual Nonlinear Multibody Systems*, 173–186.

introduced : $(\tilde{S})$ is deduced from the reference configuration (S_0) by means of a rigid transformation. For each material point M, we define the displacement field $\vec{u} = \overrightarrow{M'M}$ where M' is rigidly connected to $(\tilde{S})$. The position of M with respect to an inertial reference frame $(O_0, \vec{x}_0, \vec{y}_0, \vec{z}_0)$ is given by :

$$\vec{r} = \vec{c} + \vec{x} + \vec{u} \qquad \vec{c}(t) = \overrightarrow{O_0 O}\ ; \quad \vec{x} = \overrightarrow{OM'}\ ; \quad \vec{u} = \vec{u}(\vec{x}, t).$$

O is the origin of a « floating reference frame » [6] $(O; \vec{x}, \vec{y}, \vec{z})$ rigidly connected to $(\tilde{S})$, $\vec{\omega}$ is the angular velocity vector of the floating reference frame.
The kinetic energy of the flexible body (S) can be written as :

$$T = 1/2 \Big[m\, \dot{\vec{c}}^2 + \vec{\omega}.\, \vec{J}_0\, \vec{\omega} + \int_S \dot{\vec{u}}^2\ dm + 2\, \dot{\vec{c}}. \int_S \dot{\vec{u}}\ dm + 2\, \vec{\omega}. \int_S (\vec{x} + \vec{u}) \wedge \dot{\vec{u}}\ dm + 2\, \dot{\vec{c}}.\, \vec{\omega} \wedge \int_S (\vec{x} + \vec{u})\ dm \Big] \tag{1}$$

where $\vec{J}_0 = \int_S \left[(\vec{x} + \vec{u})^2\ \vec{E} - (\vec{x} + \vec{u}) \otimes (\vec{x} + \vec{u}) \right] dm$ is the inertia tensor in point O of the body (S) in its deformed configuration. The displacement field $\vec{u}$ is approximated by a finite number of mode shapes :

$$\begin{cases} \vec{u} = \underline{\vec{N}}\ (\vec{x})\ \underline{q}(t) \\ \underline{\vec{N}} = \left[\vec{N}_1, \ldots, \vec{N}_n \right] \quad \underline{q} = {}^t[q_1, \ldots, q_n] \end{cases} \tag{2}$$

It results for the kinetic energy the following approximated expression :

$$T = \frac{1}{2} \Big[m\, \dot{\vec{c}}^2 + \vec{\omega}.\vec{J}_0\, \vec{\omega} + {}^t\underline{\dot{q}}\ \underline{\mathrm{M}}\ \underline{\dot{q}} + 2\, \dot{\vec{c}}.\underline{\vec{\alpha}}\ \underline{\dot{q}} + 2\, \vec{\omega}.\, \underline{\vec{G}}\ \underline{\dot{q}} + 2\, \dot{\vec{c}}.(\vec{\omega} \wedge m\, \vec{x}_0) + 2\, \dot{\vec{c}}.\, \vec{\omega} \wedge \underline{\vec{\alpha}}\ \underline{q} + 2\, \vec{\omega}.\ {}^t\!\left(\underline{\vec{H}}\ \underline{q} \right) \underline{q} \Big] \tag{3}$$

with the following notations :

$$\begin{cases} \vec{J}_0 = \vec{I}_0 + {}^t\underline{q}\,\vec{\underline{\beta}} + {}^t\vec{\underline{\beta}}\,\underline{q} + {}^t\underline{q}\,\vec{\underline{\gamma}}\,\underline{q} \\ \vec{\underline{\beta}} = \int_S \left[\left({}^t\vec{\underline{N}}.\vec{x}\right)\vec{E} - {}^t\vec{\underline{N}} \otimes \vec{x}\right] dm \\ \vec{\underline{\gamma}} = \int_S \left[\left({}^t\vec{\underline{N}}.\vec{\underline{N}}\right)\vec{E} - {}^t\vec{\underline{N}} \otimes \vec{\underline{N}}\right] dm \\ \vec{\underline{G}} = \int_S \vec{x} \wedge \vec{\underline{N}}\, dm \end{cases} \qquad \begin{cases} \vec{\underline{H}} = \int_S {}^t\vec{\underline{N}} \wedge \vec{\underline{N}}\, dm \\ \underline{M} = \int_S {}^t\vec{\underline{N}}.\vec{\underline{N}}\, dm \\ \vec{\underline{\alpha}} = \int_S \vec{\underline{N}}\, dm \\ \vec{I}_0 = \int_S \left(\vec{x}^2\, \vec{E} - \vec{x} \otimes \vec{x}\right) dm \\ m\, \vec{x}_0 = \int_S \vec{x}\, dm \end{cases} \tag{4}$$

Let us assume that the body is subjected to body forces $\vec{f}$ and surface forces $\vec{F}$ applied on the boundary ∂S of the body. For a virtual velocity field defined by $\vec{V}^*(M) = \dot{\vec{c}}^* + \vec{\omega}^* \wedge (\vec{x} + \vec{u}) + \dot{\vec{u}}^*$ we obtain the corresponding virtual power of the applied forces :

$$\Im_1^* = \int_S \vec{f}.\vec{V}^*(M) dv + \int_{\partial S} \vec{F}.\vec{V}^*(M)\, d\sigma$$

Using the approximated value (2) of the displacement field, the following expression of $\Im_1^*$ is obtained :

$$\Im_1^* = \vec{R}.\dot{\vec{c}}^* + \left(\vec{M}_0 + \vec{\underline{B}}\,\underline{q}\right).\vec{\omega}^* + \underline{D}\,\dot{\underline{q}}^* \tag{5}$$

where

$$\begin{cases} \vec{R} = \int_S \vec{f}\, dv + \int_{\partial S} \vec{F}\, d\sigma & \vec{M}_0 = \int_S \vec{x} \wedge \vec{f}\, dv + \int_{\partial S} \vec{x} \wedge \vec{F}\, d\sigma \\ \underline{D} = \int_S \vec{f}.\vec{\underline{N}}\, dv + \int_{\partial S} \vec{F}.\vec{\underline{N}}\, d\sigma & \vec{\underline{B}} = \int_S \vec{\underline{N}} \wedge \vec{f}\, dv + \int_{\partial S} \vec{\underline{N}} \wedge \vec{F}\, d\sigma \end{cases} \tag{6}$$

Assuming linear constitutive laws, the elastic potential energy is approximated by $V_e = 1/2\ {}^t\underline{q}\ \underline{K}\ \underline{q}$, where $\underline{K} = \left(K_{ij}\right)$ is the n x n constant stiffness matrix.

3. Rigid body model

The rigid body model (Σ_R) is composed of a rigid body (S_R) rigidly connected to the floating reference frame $(O;\vec{x},\vec{y},\vec{z})$ and a total number of 3n materials points M_i of mass m_i defined by $\overrightarrow{OM_i} = \overrightarrow{X_i} + \overrightarrow{\underline{A_i}}\ \underline{q}$ (i=1,2,...,3n). In this formula, $\vec{X}_i$ is a vector fixed with respect to the floating reference frame while $\vec{\underline{A}}_i$ is a 1 x n matrix of whose the columns are vectors fixed with respect to the same reference frame. The absolute velocity of the material point M_i is given by $\vec{V}(M_i) = \dot{\vec{c}} + \vec{\omega} \wedge \overrightarrow{OM_i} + \underline{\vec{A}_i}\ \underline{\dot{q}}$

It results for the system (Σ_R) the following expression of the kinetic energy

$$\begin{cases} 2\tilde{T} = \left(m_0 + \sum_{i=1}^{3n} m_i \right) \dot{\vec{c}}^2 + \vec{\omega} . \left(\vec{K}_0 + \sum_{i=1}^{3n} \vec{J}_i \right) \vec{\omega} + 2\dot{\vec{c}} . (\vec{\omega} \wedge m_0\, \vec{y}_0) + \\ \sum_{i=1}^{3n} \left(\underline{\vec{A}_i}\ \underline{\dot{q}} \right) . \left(\underline{\vec{A}_i}\ \underline{\dot{q}} \right) \\ + 2\dot{\vec{c}} . \vec{\omega} \wedge \sum_{i=1}^{3n} \left(\vec{X}_i + \underline{\vec{A}_i}\ \underline{q} \right) + 2\dot{\vec{c}} . \sum_{i=1}^{3n} \underline{\vec{A}_i}\ \underline{\dot{q}} + 2\, \vec{\omega} . \sum_{i=1}^{3n} \left[\left(\vec{X}_i + \underline{\vec{A}_i}\ \underline{q} \right) \wedge \underline{\vec{A}_i}\ \underline{\dot{q}} \right] \end{cases}$$

where m_0 and $\vec{K}_0$ are the mass and the inertia tensor in point O of the rigid body (S_R) and $m_0\, \vec{y}_0 = \int_{S_R} \vec{x}\, dm$ defines the position of the center of mass of (S_R) with respect to point O. At last

$$\vec{J}_i = m_i \left[\left(\vec{X}_i + \underline{\vec{A}_i q} \right)^2 \vec{E} - \left(\vec{X}_i + \underline{\vec{A}_i q} \right)^2 \otimes \left(\vec{X}_i + \underline{\vec{A}_i q} \right) \right] \quad (i = 1, 2, \dots 3n)$$

By identification of the kinetic energy of (Σ_R) with the expression (3) of the kinetic energy of the flexible component (S), we obtain the following equations :

$$\int_S \underline{\vec{N}}\, dm = \sum_{i=1}^{3n} m_i\ \underline{\vec{A}_i} \tag{7}$$

$$\int_S \underline{\vec{N}} \otimes \vec{x} \mathrm{dm} = \sum_{i=1}^{3n} m_i\ \underline{\vec{A}_i} \otimes \vec{X}_i \tag{8}$$

$$\int_S {}^t\underline{\vec{N}} \otimes \underline{\vec{N}}\, \mathrm{dm} = \sum_{i=1}^{3n} m_i\ {}^t\underline{\vec{A}_i} \otimes \underline{\vec{A}_i} \tag{9}$$

In order to make the identification for the applied and elastic forces of the real system, let us assume that for the rigid body model, the applied forces on the part (S_R) is composed of a resultant force $\vec{\phi}$ and a resultant torque $\vec{\Gamma}_O$ applied in point O while each point M_i is submitted to the force:

$\vec{F}_i = \lambda_i\, \vec{X}_i - \underline{\vec{A}_i}\ \underline{q}$, $(\lambda_1, \dots, \lambda_n)$ are constants

The virtual power of the applied forces for the rigid body model (Σ_R) is given by

$$\Im_2^* = \dot{\vec{c}}^* . \left(\vec{\phi} + \sum_{i=1}^{3n} \vec{F}_i \right) + \vec{\omega}^* . \left(\vec{\Gamma}_0 + \sum_{i=1}^{3n} \overrightarrow{OM}_i \wedge \vec{F}_i \right) +$$

$$\sum_{i=1}^{3n} \left(\lambda_i \vec{X}_i - \underline{\vec{A}}_i \underline{q} \right) . \underline{\vec{A}}_i \underline{\dot{q}}^*$$

By identification with the virtual power of the applied and elastic forces of the real system :

$$\Im_2^* = \Im_1^* - {}^t\underline{q} \; \underline{K} \; \underline{\dot{q}}^*$$

we obtain the following relations :

$$\vec{R} = \vec{\phi} + \sum_{i=1}^{3n} \vec{F}_i = \vec{\phi} + \sum_{i=1}^{3n} \left(\lambda_i \vec{X}_i - \underline{\vec{A}}_i \underline{q} \right) \tag{10}$$

$$\vec{M}_0 + \underline{\vec{B}} \, \underline{q} = \vec{\Gamma}_0 + \sum_{i=1}^{3n} \overrightarrow{OM}_i \wedge \vec{F}_i = \vec{\Gamma}_0 + \sum_{i=1}^{3n} \left(\underline{\vec{A}}_i \underline{q} \wedge (\lambda_i + 1) \vec{X}_i \right) \tag{11}$$

$$\underline{D} = \sum_{i=1}^{3n} \lambda_1 \vec{X}_i . \underline{\vec{A}}_i \tag{12}$$

$$\underline{K} = \sum_{i=1}^{3n} {}^t\underline{\vec{A}}_i . \underline{\vec{A}}_i \tag{13}$$

The two first equations (10) and (11) define the resultant force $\vec{\phi}$ and the resultant torque $\vec{\Gamma}_0$ applied on (S_R) when λ_i, $\vec{X}_i$, $\underline{\vec{A}}_i$ are known.

We conclude that to obtain a rigid body model of the flexible component
we have to solve with respect to the unknown quantities $\left(m_i, \; \underline{\vec{A}}_i, \; \vec{X}_i, \; \lambda_i \right)$ the equations (7), (8), (9), (12) and (13).
The total number of unknown variables is
$X = 3n + 9n^2 + 9n + 3n = 9n^2 + 15n$.
The total number of equations to be fullfilled is :
$E = 3n + 9n + 3n\dfrac{(3n+1)}{2} + n + n\dfrac{(n+1)}{2} = 5n^2 + 15n$
It results that the number of unknown variables being greater than the number of equations, founding a rigid body model is always possible and there are several solutions.

4. Numerical results

An application of this method is done using the symbolic dynamical code AUTOLEV, which originally was developped for systems of rigid bodies only. In this code, the motions equations are obtained by using Kane's equations [7] which are in fact equivalent to Lagrange's equations.

In [4], a first method to use AUTOLEV for the dynamical simulation of planar interconnected flexible beams is proposed. In this work, the additionnal terms occuring from the deformations of the beam (discretized by Rayleigh-Ritz method) are considered as forces or torques applied to the undeformed configuration of the beam and forces applied to massless points whose degrees of freedoms are the elastic variables. These assumptions lead to define forces and torques involving velocity and acceleration of the elastic variables. This event introduces some impossibility for using AUTOLEV for the solution of inverse problem in flexible mechanisms.

Let us consider a multibody system where each flexible component is discretized by Rayleigh-Ritz method.
The motion equations of the whole system can be written in the following form [3] :

$$\begin{bmatrix} \underline{M_{rr}} & \underline{M_{rl}} \\ {}^{t}\underline{M_{rl}} & \underline{M_{ll}} \end{bmatrix} \begin{bmatrix} \underline{\ddot{q}_r} \\ \underline{\ddot{q}_l} \end{bmatrix} + \begin{bmatrix} \underline{H_R} \\ \underline{H_l} \end{bmatrix} + \begin{bmatrix} 0 & 0 \\ 0 & \underline{K} \end{bmatrix} \begin{bmatrix} \underline{q_r} \\ \underline{q_l} \end{bmatrix} = \begin{bmatrix} \underline{F_r} \\ \underline{F_l} \end{bmatrix} \tag{14}$$

where $\left(\underline{q_r} \quad \underline{q_l}\right)$ are respectively the set of rigid and flexible coordinates of the system, $\underline{M_{rr}}, \underline{M_{rl}}, \underline{M_{ll}}$ are submatrices included in the system mass matrix, $\underline{K}$ is the stiffness matrix, $\underline{H_R}, \underline{H_l}$ are the matrices which contain the Coriolis, Centrifugal and gravity forces, $\underline{F_r}, \underline{F_l}$ are the matrices of generalized applied forces.

If the flexible multibody system considered is a robotic manipulator including flexible components, the applied forces are produced by the motors contained in the joints : in this case, it is possible to assume $\underline{F_l} = 0$.

The inverse dynamical problem consists to obtain the driving torques which produce a specified trajectory of the end effector. Let

$$\underline{f}\left(\underline{q_r}, \underline{q_l}, t\right) = 0 \tag{15}$$

be the set of the kinematical constraints equations that describes the specified motion trajectories. Let us assume that these equations can be solved with respect to the rigid coordinates $\underline{q_r}$:

$$\underline{q_r} = \underline{g}\left(\underline{q_l}, t\right) \tag{16}$$

By substitution in the last part of equations (14) (with $\underline{F_l} = 0$), we obtain a system of differential equations for the determination of the flexible variables $\underline{q_l}$. By substitution of the results in the first part of (14) we obtain the driving generalized forces $\underline{F_r}$.

A first validation of the introduction of a pseudo rigid model for the simulation of a flexible multibody system is made by using as in [4] AUTOLEV for the dynamical simulation of a planar manipulator composed of two flexible beams OA and AB of length l_1 and l_2, mass m_1 and m_2, acted by a torque Γ_O applied at the fixed pinned joint O and a torque Γ_A applied at the pinned joint A (figure 1). The two links are made of steel, the dimensions and material properties of the system are the following :

$$l_1 = 0.15 \text{ m}, \ l_2 = 0.35 \text{ m}$$
$$m_1 = 0.0331 \text{ kg}, \ m_2 = 0.0772 \text{ kg}$$

The two links have the same diameter d=0,006 m. One bending mode with simply supported end conditions is used in the modeling of the first beam and one bending mode with clamped free end conditions is used for the second beam. We assume that the two torques Γ_O and Γ_A are constant ($\Gamma_O = \Gamma_A = 1$ Nm).

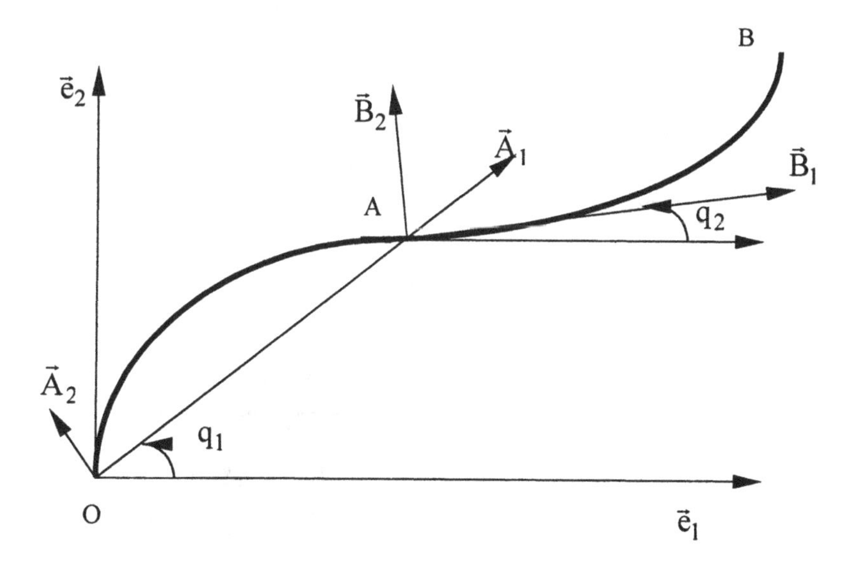

Figure1 : Flexible Manipulator

With respect to the fixed reference frame ($O ; \vec{e}_1 , \vec{e}_2$) the rigid variables $\underline{q_r}$ are the angle $q_1 = (\vec{e}_1, O\vec{A})$ and the angle $q_2 = (\vec{e}_1, A\vec{x}_1)$ ($A\vec{x}_1$ tangent in point A to the beam AB), while the elastic variables $\underline{q_l} = {}^t(q_3, q_4)$ are associated with the discretization of the deformations of the beams.
The corresponding values of the two rigid variables (q_1, q_2) and the two flexibles variables q_3, q_4 are plotted on figures 2,3,4,5.
We obtain the same curves by using the method of [4] and the method defined in this work.
Another application of this method is used for solving an inverse problem considered in [8]. We consider a planar slider crank mechanism (figure 6) composed of a rigid crankshaft OA, a flexible connecting rod AB and a slider block C. The dimensions and material properties of the two links OA and AB are the same as in the previous example, the mass M of the slider block is 0.1 kg.
The system is moving in the vertical plane $(O\vec{e}_1, O\vec{e}_2)$ ($O\vec{e}_2$ ascendant vertical), without friction in the pinned joints O and A and for the contact between the slider block and the ground. The system is acted by a torque $\Gamma\vec{e}_3$ ($\vec{e}_3 = \vec{e}_1 \wedge \vec{e}_2$) applied in point O in such a way that the motion of the slider block is a harmonic function of time defined by

$$x(B) = 0.35 - 0.8 l_2 \sin \omega t$$

where ω =100 rad/sec.

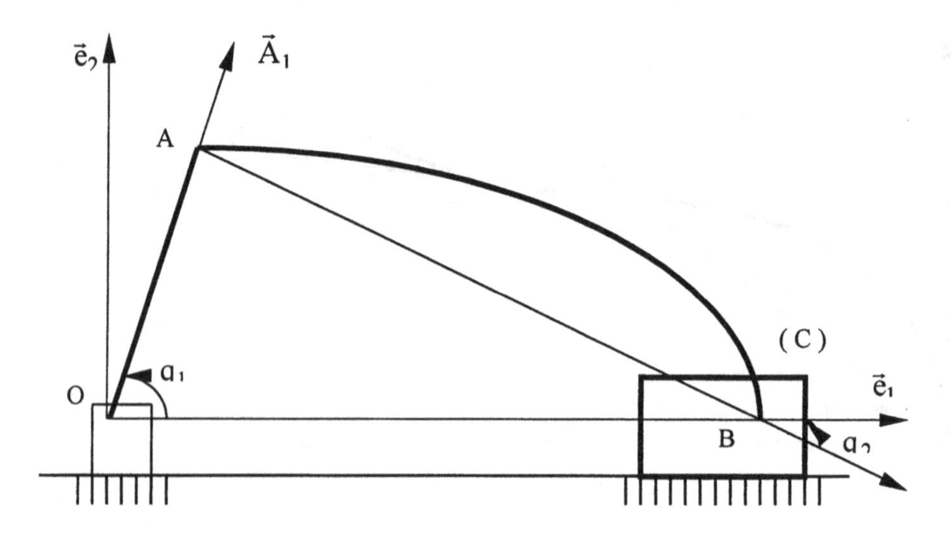

Figure 6: Slider Crank Mechanism

Assuming only bending vibrations, the connecting rod AB is discretized using two simply supported modes. In order to use the code AUTOLEV to obtain the driving torque Γ corresponding to the harmonic motion of the slider block, a rigid body model (Σ_R) of the system is built using the symbolic program MAPLE. The obtained rigid body model is solved by AUTOLEV to obtain the driving torque Γ corresponding to the harmonic motion of the slider block, a rigid body model (Σ_R) of the system is built using the symbolic program MAPLE. The obtained rigid body model is solved by AUTOLEV for the inverse problem. The two rigid variables q_1 and q_2 and the two flexible variables q_3 and q_4 are plotted on figures7-8-9-10. The torque Γ resulting from the harmonic motion of the slider block is plotted on figure 11 and compared to the corresponding torque obtained when the connecting rod is assumed to be rigid. The results obtained in figure 11 are identical to the results given in [8].

5. Conclusions

In this paper, we have shown that it is possible to use any dynamical codes devoted to rigid multibody systems in which the motion equations are obtained from d'Alembert principle for the simulation of flexible multibody systems in which the flexible components are discretized by a Rayleigh-Ritz procedure. Three dimensional elastic components are considered and not only elastic beams as in [4]. The method is based on the identification of the kinetic energy, potential energy and virtual work of the applied forces for the real system and a fictitious rigid multibody system. The main interest of this result is the possibility of using symbolic dynamical codes mainly available for rigid multibody systems for the dynamical simulation of flexible multibody systems Several simulations tests are done using the symbolic code AUTOLEV showing the efficiency of the method.

6. References

1.Schiehlen, W.(1990) *Multibody Systems Handbook*, Springer-Verlag, Berlin.

2.Wittenburg, J. (1977) *Dynamics of Systems of Rigid Bodies*, Teubner, Stuttgart.

3.Shabana, A.A.(1989) *Dynamics of Multibody Systems*, Wiley, Chichester.

4.Botz, M.and Hagedorn, P.(1990) Multiple Body Systems with Flexible Members, *Nonlinear Dynamics*. **1**, 433-447.

5.Schaechter, D.B.and Levinson, D.A., Kane,T.R.(1988) *Autolev User's Manual*. On Line Dynamics, Inc. Sunnyvale.

6.Pascal, M.(1995) Nonlinear Effects in Transient Dynamic Analysis of Flexible Multibody Systems, Proceedings of the 1995 ASME Design Eng. Techn. Conf., Boston, Massachussets, Sept. 17-20, **3**, part A, 75-86.

7. Kane, T.R.and Levinson, D.A.(1985) *Dynamics: Theory and Applications*, Mc Graw-Hill, New York.

8.Gofron M., Shabana A.A., (1993) Control Strusture Interaction in the Nonlinear Analysis of Flexible Mechanics Systems, *Nonlinear Dynamics*, **4**, 183-206.

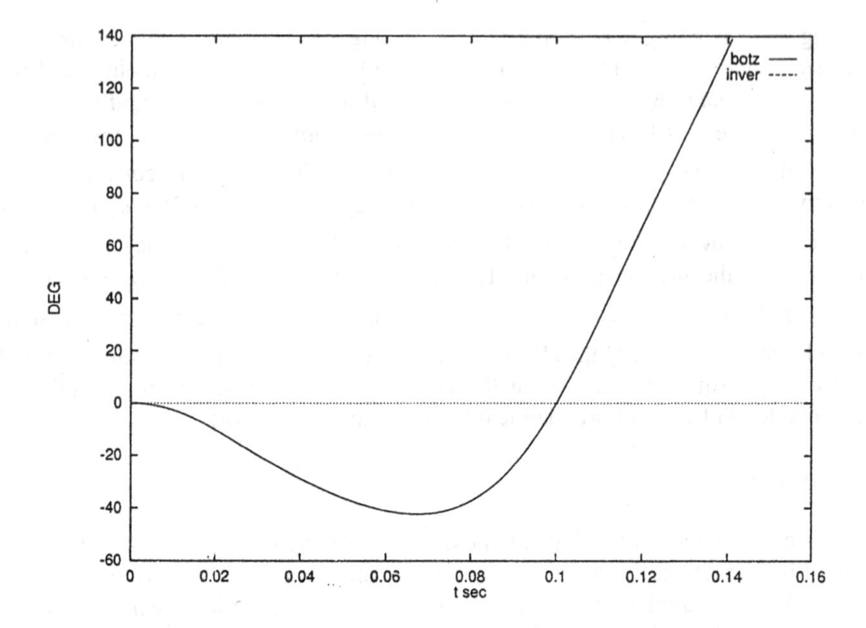

Figure 2: Comparison of the Method with [4]. Rigid Variable q_1

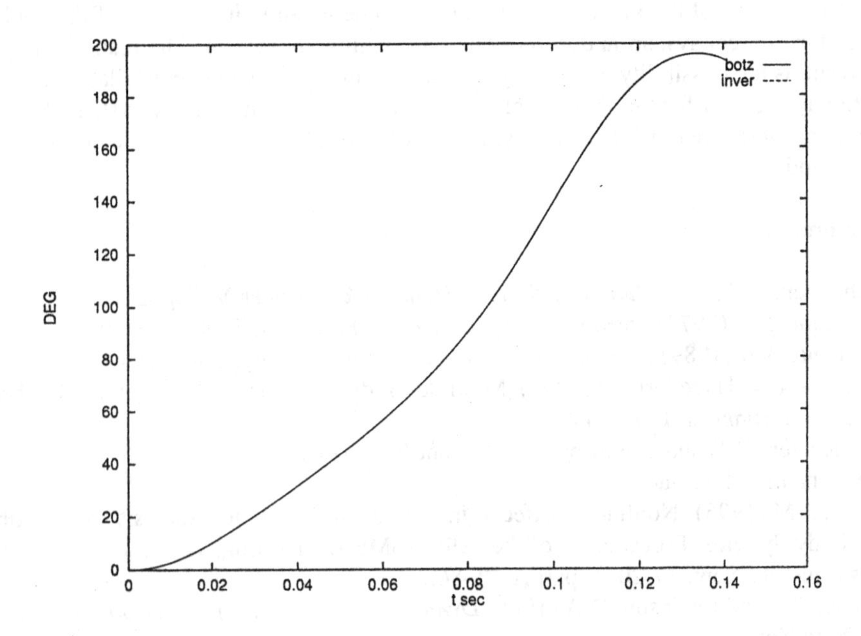

Figure 3: Comparison of the Method with [4]: Rigid Variable q_2

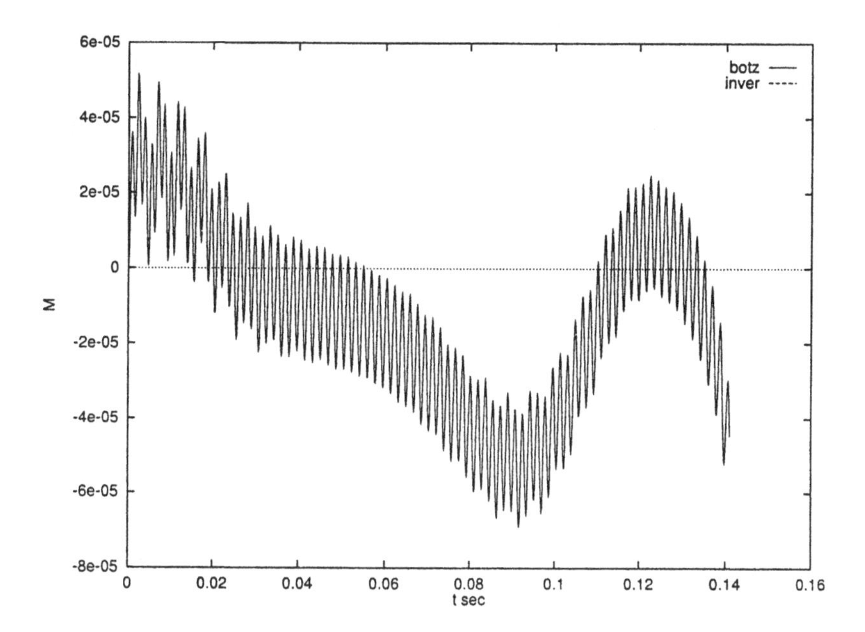

Figure 4: Comparison of the Method with [4]. Flexible Variable q_3

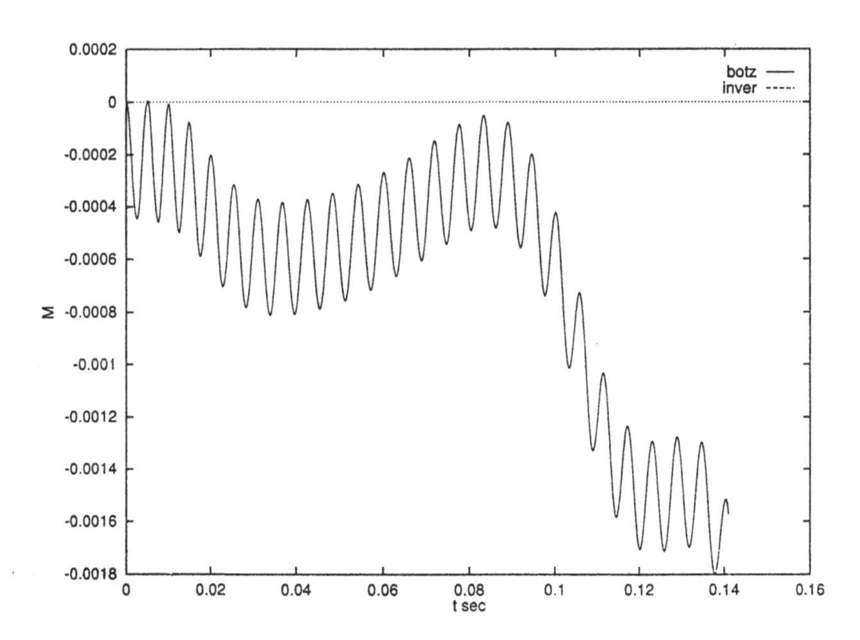

Figure 5: Comparison of the Method with [4]. Flexible Variable q_4

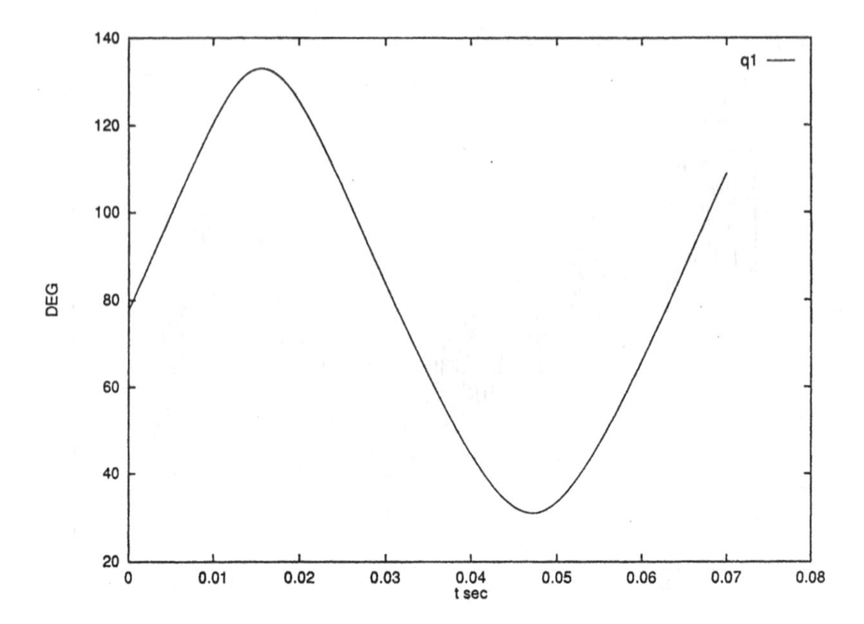

Figure 7: Slider Crank Mechanism. Rigid Variable q_1

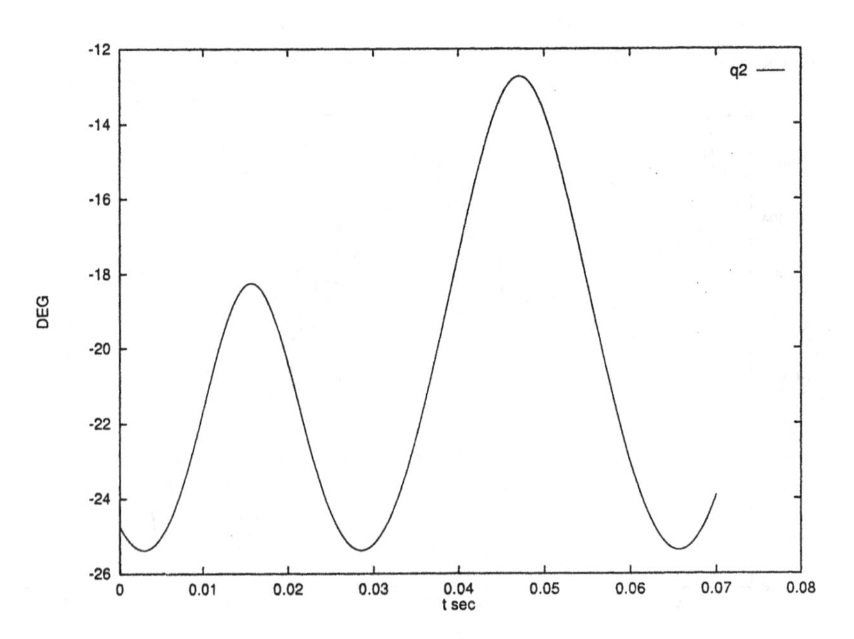

Figure 8: Slider Crank Mechanism. Rigid Variable q_2

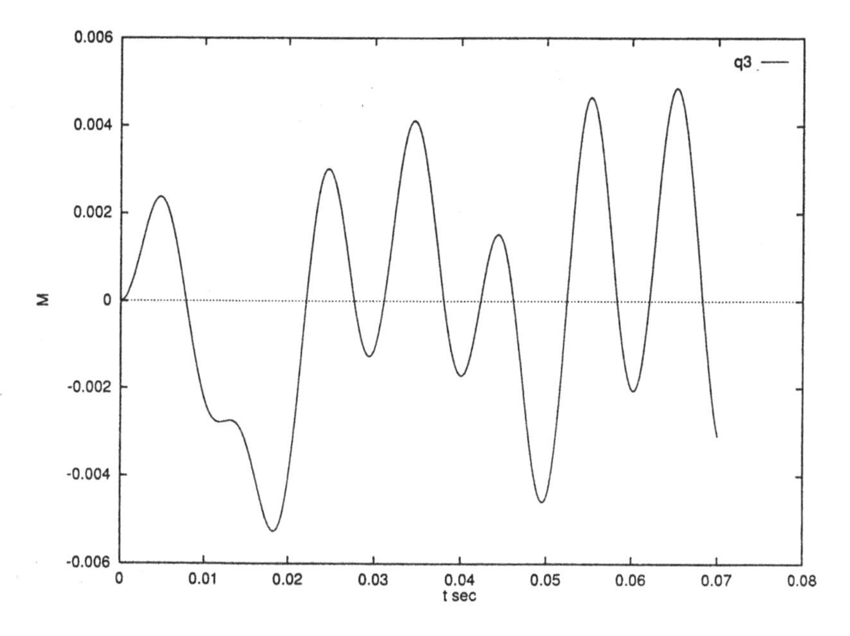

Figure 9: Slider Crank Mechanism. Flexible Variable q_3

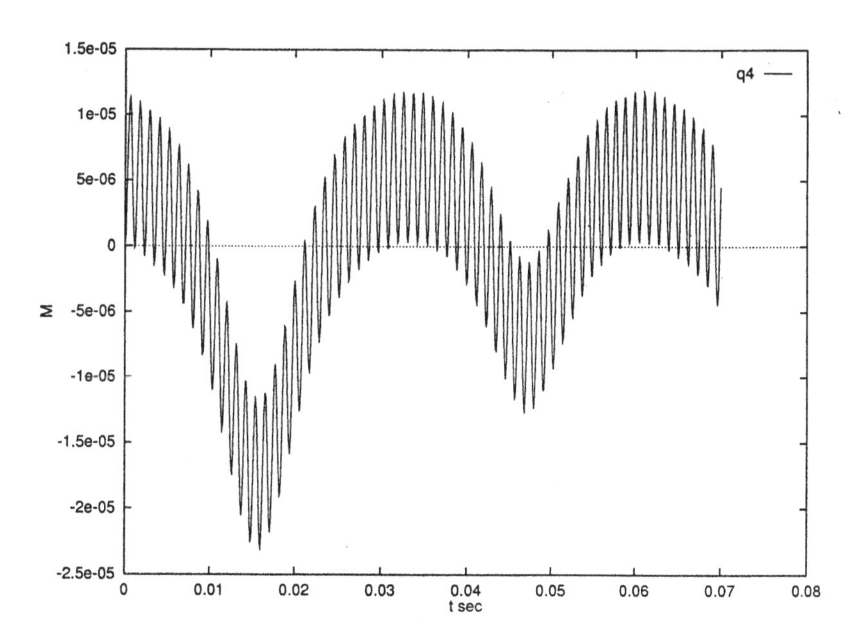

Figure 10: Slider Crank Mechanism. Flexible Variable q_4

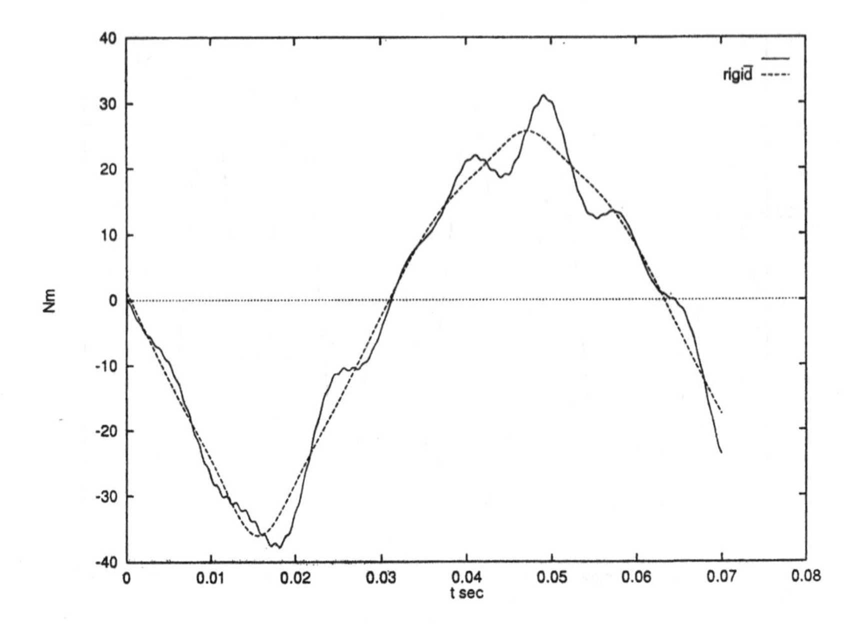

Figure 11: Slider Crank Mechanism. Driving Torque.

A VIRTUAL MULTIBODY AND FINITE ELEMENT ANALYSIS ENVIRONMENT IN THE FIELD OF AEROSPACE CRASHWORTHINESS

H. M. LANKARANI, G. OLIVARES and H. NAGARAJAN
hamid.lankarani@wichita.edu
National Institute for Aviation Research, Wichita State University, Wichita, KS 67260-0093, USA

Abstract – One important concept in crashworthiness analysis is how the contact/impact is treated or modeled. One method for predicting the impact response of multibody mechanical systems is to treat the local deformations as well as the contact forces as continuous. The application of this method requires the knowledge of the variation of the contact and frictional forces. Contact force models, including nonlinear visco-elastic, Hertzian-based, and visco-plastic models are presented. Computer-aided analysis tools typically used in the studies of the aircraft crashworthiness are then described. A methodology is then presented for the entire design cycle from airframe to the cabin, seat, restraint and egress system. The methodology incorporates a combination of multibody modeling and non-linear finite element analysis of the airframe, seat and the occupant as well as component testing in early design stages, and sled and/or full-scale testing in later stages of design evaluations. A Virtual Reality (VR) system is utilized for effective system visualization and to better understand the interaction between the various subsystems. Examples of the use of this methodology for some of the current crash safety issues are presented.

1. INTRODUCTION

A theory is often a general statement of principle abstracted from observation and a model is a representation of a theory that can be used for prediction. To be useful, a model must be realistic and yet simple to understand and easy to manipulate. These are conflicting requirements, for realistic models are seldom simple and simple models are seldom realistic. Often, the scope of a model is defined by what is considered relevant. Features or behavior that are pertinent must be included in the model and those that are not can be ignored. Modeling here refers to the process of analysis and synthesis to arrive at a suitable mathematical description that encompasses the relevant dynamic characteristics of the component, preferably in terms of parameters that can be easily determined in practice (component testing). The procedure for developing a model is often an iterative one. The cycle begins with identifying the purpose of the model and its constraints, as well as the kinds of simplifying assumptions or omissions that can be made, determining the means of obtaining parameters and insight into the discipline are essential to making appropriate simplifying assumptions. Whereas oversimplification

W. Schiehlen and M. Valášek (eds.), Virtual Nonlinear Multibody Systems, 187–212.

and omissions may lead to unacceptable loss of accuracy, a model that it's too detailed can be cumbersome to use.

Modeling and simulation are especially beneficial to solve aerospace crashworthiness applications, where the actual system does not exist or is too expensive, time consuming or hazardous to conduct, or when experimenting with an actual system can cause unacceptable disruptions. Changing the values of parameters, or exploring a new concept or operating strategy, can often be done more quickly in a simulation that by conducting a series of experimental studies on the actual system. The design of aircraft for improvement of its crashworthiness requires the knowledge and integration of several items. These items include the specifications and standards, human tolerance, injury criteria, energy absorption concepts in airframe design, seat legs, seat pan, seat cushion, restraint systems, surrounding structures, fire safety, economic and ergonomic considerations. One of the most important concept of crashworthiness analysis is how the contact/impact is treated or modeled.

Knowledge of contact/impact mechanics is a crucial step in the prediction or assessment of the crash response of a mechanical system or a structure. In an impact, non-linear contact forces of unknown nature are created, which act and disappear over a short period of time. The method of predicting the impact response of the mechanical systems can primarily be classified into two groups. In one, the impact forces are treated as a discontinuous event [1-3]. A simple form of this type of analysis is usually conducted for accident reconstruction purposes for which each vehicle is modeled as a single body or object. In this so-called, "discontinuous" or "piecewise" analysis methods, momentum-balance/impulse equations are formulated and solved for the departing velocities or momenta of the system after impact. One basic assumption underlying such analysis is that the impact occurs in such a short duration that the system configuration does not change during contact duration. This limitation constraints the applicability of the piecewise analysis method to crashworthiness problems. A more suitable method used for impact analysis of mechanical systems is treating the local deformations and the contact forces as continuous. The application of these methods requires the knowledge of the variations of the contact forces. Different models have been postulated to represent this variation.

The development of current aircraft crash dynamics standards dates back to the 1970's during which the product liability grew for small aircraft [4]. To address the crashworthiness characteristics of the transport category aircraft, small general aviation aircraft, and rotorcraft, Federal Aviation Administration (FAA) and National Aeronautics and Space Administration (NASA) initiated a wide range of research and development programs. These programs represented a concentrated effort to analyze the aircraft behavior and the occupant characteristics through interrelated studies of accident data, dynamic analyses of crash events, full-scale aircraft impact tests, and aircraft seat tests. A panel named GASP (General Aviation Safety Panel) was also formed in 1978 to make recommendations on crashworthiness requirements. They considered survivable accidents for which the floor remained intact. The results of these studies formed the basis for the development of crashworthiness design standards for civil aircraft [5]. These requirements are defined in the *Federal Aviation Regulations*

(FARs) Parts 23, 25, and 27 for general aviation aircraft, transport aircraft, and rotorcraft respectively [6-8]. These regulations were first proposed in 1982 and they became effective in 1988. In general, the FARs contains two distinct dynamic test conditions. Test-1 conditions, illustrated and described in Figure 1, require a seat inclination of 60 degrees in pitch and a mean velocity change of no less than 30 ft/sec (20 mph = 9 m/s), and is intended to evaluate the means provided to reduce the spinal loading and related injuries produced by the combined vertical/horizontal load environment typically generated by an aircraft crash event. Test-2 conditions, illustrated and described in Figure 1, require the inclination of the seat on the track by 10 degrees in yaw direction and a mean velocity change of no less than 42 ft/sec (29 mph = 13 m/s) in the longitudinal direction. To account for a reasonable floor warpage level that may occur during a crash, one of the seat tracks is misaligned by 10 degrees in pitch and the other one by 10 degrees in roll. The Test-2 conditions are intended to provide an assessment of the seat structural performance and the occupant restraint system. The deceleration pulses in both tests are triangular shaped.

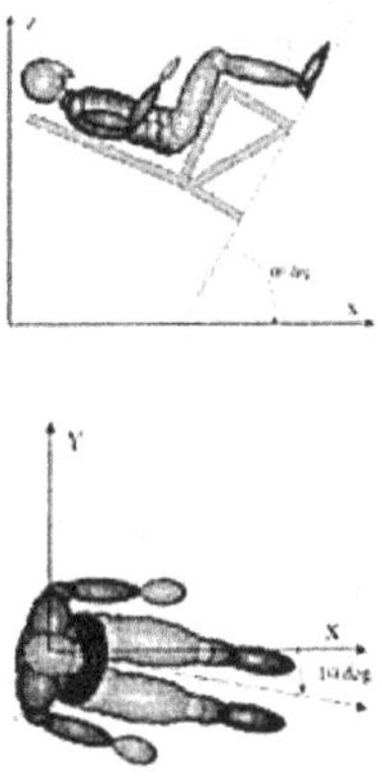

DYNAMIC TEST REQUIREMENTS	PART 23	PART 25	PART 27
TEST 1			
Test Velocity - ft/sec	31 (9.5 m/sec)	35 (10.7 m/sec)	30 (9.2 m/sec)
Seat Pitch Angle - Degrees	60	60	60
Seat Yaw Angle - Degrees	0	0	0
Peak Deceleration - G's	19/15	14	30
Time To Peak - sec	0.05/0.06	0.08	0.031
Floor Deformation - Degrees	None	None	10 Pitch/10 Roll
TEST 2			
Test Velocity - ft/sec	42 (12.8 m/sec)	44 (13.4 m/sec)	42 (12.8 m/sec)
Seat Pitch Angle - Degrees	0	0	0
Seat Yaw Angle - Degrees	±10	±10	±10
Peak Deceleration - G's	26/21	16	18.4
Time To Peak - sec	0.05/0.06	0.09	0.071
Floor Deformation - Degrees	10 Pitch/10 Roll	10 Pitch/10 Roll	10 Pitch/10 Roll
COMPLIANCE CRITERIA			
HIC	1000	1000	1000
Lumbar Load - lb	1500 (6675 N)	1500 (6675 N)	1500 (6675 N)
Strap Loads - lb	1750[1]/2000[2] (7787N[1]/8900N[2])	1750[1]/2000[2] (7787N[1]/8900N[2])	1750[1]/2000[2] (7787N[1]/8900N[2])
Femur Loads - lb	N/A	2250	N/A

[1] - passenger [2] - pilot

Figure 1 - Dynamic test requirements.

The pass/fail criteria are based on the data collected from accidents and the recommendation of the GASP on the frequency of major fatal injuries to specific body regions in aircraft accidents. The injury and pass/fail criteria in the FARs include the following:

a. Maximum compressive load measured between the pelvis and the lumbar spine (of the Part 572 Subpart B Hybrid II ATD, per FARs) must not exceed 1500 pounds m1p5e2F (6675 N).

b. Loads in the individual straps must not exceed 1750 pounds (7785 N) for pilot and 2000 pounds (8,900 N) for passengers.

c. The compressive load in the femur of Hybrid II ATD (Anthropomorphic Test Dummy) must not exceed 2250 pounds (10,000 N). This requirement is only for transport category (Part 25) aircraft.

d. The Head Injury Criteria (*HIC*), evaluation for the Hybrid II ATD must not exceed 1000 [9].

$$HIC = \left[(t_2 - t_1) \left\{ \frac{1}{(t_2 - t_1)} \int_{t_1}^{t_2} a(t) dt \right\}^{2.5} \right]_{max}$$

In contrast with the automotive industry, in the aerospace industry numerical simulation methods are primarily used at the very end of the product development process. Often they are applied to confirm the reliability of an already existing design, or sometimes for further design improvements by means of optimization methods. Numerical simulation methods are much more efficient when used at an early stage of the product development process. In this way, the number of hardware tests can be reduced since a profound knowledge of the model already exists. The intend on this paper is to give an overview of the different simulation techniques (multibody and finite element method approach) that could be implemented at the early design stages of the airframe, interior, and evacuation systems. Crashworthiness problems are best addressed in a systems approach utilizing combinations of CAE tools, component tests, sled and/or full-scale tests.

2. MODELING CONTACT FORCES

When two solids are in contact, deformation takes place in the local contact zone resulting in a contact force. This suggests that the contact force is directly related to the amount of local deformation or indentation of the two solids. The best-known force model for the contact between two spheres of isotropic material was developed by Hertz based on the theory of elasticity [10]. With radii R_i and R_j of the two spheres 'i' and 'j', and masses m_i and m_j, the contact force f follows the relation

$$f = K\,\delta^n, \tag{1}$$

where δ is amount of the relative penetration or indentation between the surfaces of the two spheres and n = 1.5. The generalized impact parameter "*K*" depends on the material properties and the "effective radius" *R* of the spheres:

$$K = \frac{4}{3\pi(h_i + h_j)}\sqrt{R}\,; \qquad R = \frac{R_i R_j}{R_i + R_j} \tag{2}$$

where the material parameters h_i and h_j are

$$h_k = \frac{1 - \nu_k^2}{\pi E_k} \quad k = i,j \tag{3}$$

Variables ν_k and E_k are, respectively, the Poisson's ratio and the Young's modulus associated with each sphere.

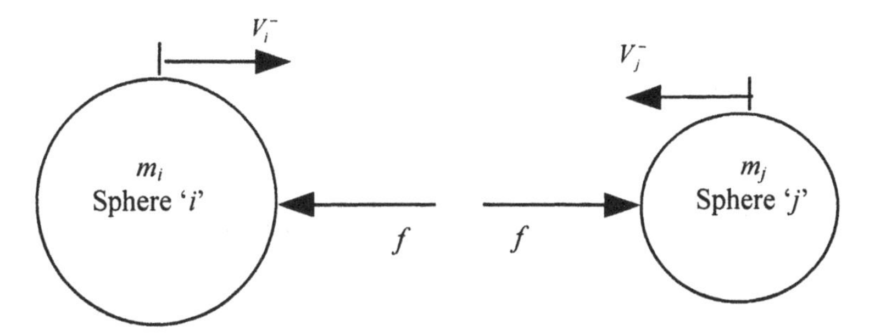

Figure 2. A direct-central impact of two spheres.

Consider now a situation for which the contact between the two spheres is caused by a direct-central collision. The two spheres have velocities V_i^- and V_j^- right before impact. We would like to determine the variations of the interaction force between the two spheres during the short period of contact. The normal direction **n** to the contact surfaces and a pair of forces f and $-f$ is are shown in Figure 2.

Generally, the two spheres will not rebound with the same initial velocities, because part of the initial kinetic energy is dissipated as permanent deformation, heat, etc. Apparently the contact force model of equation (1) cannot be used during both phases of contact, compression and restitution, since this would suggest that no energy is dissipated in the process of impact. One popular treatment is based on the idea that dissipation of energy occurs as internal damping of colliding solids. This assumption is valid for low impact velocities; i.e., those impact situations for which impact velocities are negligible compared to the propagation speed of deformation waves across the solids. The contact force model will then be in terms of a hysteresis damping as [11]

$$f = K\delta^n + \eta\,\delta^n\,\dot{\delta} \tag{4}$$

The damping factor can be explained in terms of restitution e, utilizing the energy and motion principles as,

$$\eta = \frac{3K(1-e^2)}{4\dot{\delta}^-} \tag{5}$$

The contact force in conjunction with the damping representation may be written in an alternative form as

$$f = K\delta^n\left[1 + \frac{3(1-e^2)}{4}\frac{\dot{\delta}}{\dot{\delta}^-}\right] \tag{6}$$

which shows the effect of impact speed and the coefficient of restitution on the variations of the contact force. The clear advantage of the Hertz contact force model, $f = K\delta^n$, with its damping representation in equation (6) over the Kelvin-Voigt viscoelastic model is its nonlinearity. The overall pattern of impact is far from linear,

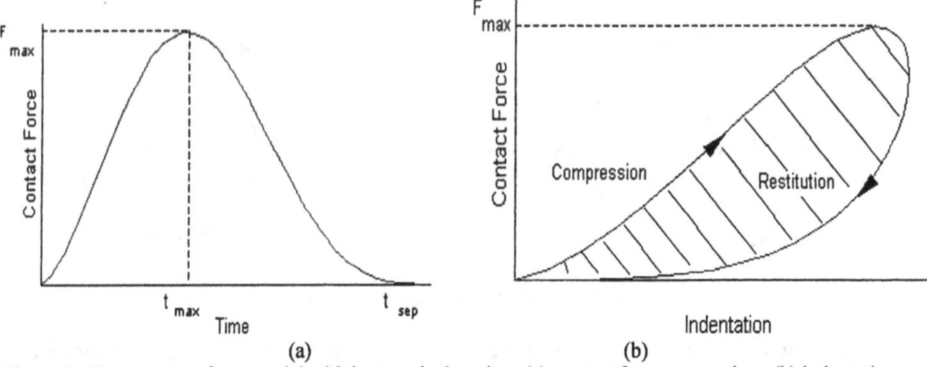

Figure 3. Hertz contact force model with hysteresis damping: (a) contact force versus time, (b) indentation versus indentation.

while the Kelvin-Voigt model and its damping representation are linear models. The solution for indentation corresponding to the linear models is a half-damped harmonic. This indicates that the bodies in impact must exert tension on each other right before separation. On the other hand, the Hertzian contact force model predicts no tension on the bodies before separation, as observed from the solution for its corresponding indentation of Figure 3.

At higher impact velocities, the dissipation of energy is mostly in the form of local plasticity of the surfaces in contact. This means that some permanent indentation is left behind on the surfaces of the two spheres after separation, and that accounts for the energy loss in impact. This is not an unreasonable assumption for impact problems in which two metallic bodies with the initial relative velocity larger than $10^{-5}(E/\rho)^{0.5}$ collide [12], where ρ is the mass density and the quantity $(E/\rho)^{0.5}$ is the larger of two propagation speeds of the elastic deformation waves in the colliding solids. With this condition, the contact force loads according to equation (1) during the compression period, and unloads according to

$$f = f_m \left[\frac{\delta - \delta_p}{\delta_m - \delta_p} \right]^n \quad \text{during restitution} \tag{7}$$

where variable δ_p is the permanent indentation of the two spheres after separation. The shape of the hysteresis loop corresponding to this contact force model and the solution corresponding to the variation of the indentation with time is shown in Figure 3. As discussed earlier, the area enclosed by the contact force - indentation curve in Figure 4 represents the *energy loss*.

The proposed contact force model can be used for impact between two spheres, if the parameters in the model are known. The generalized parameter K may be evaluated from the radii and the material properties of the two spheres using equation (2). The remaining parameters are δ_m, f_m, δ_p, which can be determined by integrating the relative indentation equations of motion twice, substitution in the contact force expression, and integrating the contact force around the hysteresis loop and equating it to the kinetic energy loss, as [13]

$$\delta_m = \left[\frac{n+1}{2K} m_{eff} \dot{\delta}^{-2}\right]^{\frac{1}{(n+1)}} \tag{8}$$

$$f_m = K\delta^n \tag{9}$$

$$\delta_p = \frac{(n+1) m_{eff} \dot{\delta}^{-2}}{2 f_m}(1-e^2) \tag{10}$$

Hence maximum indentation on the two spheres and maximum contact force depend on the material properties, masses, radii, and velocities of the two spheres right before impact. The permanent indentation is evaluated from the initial approach velocities of the spheres and a known coefficient of restitution between the spheres.

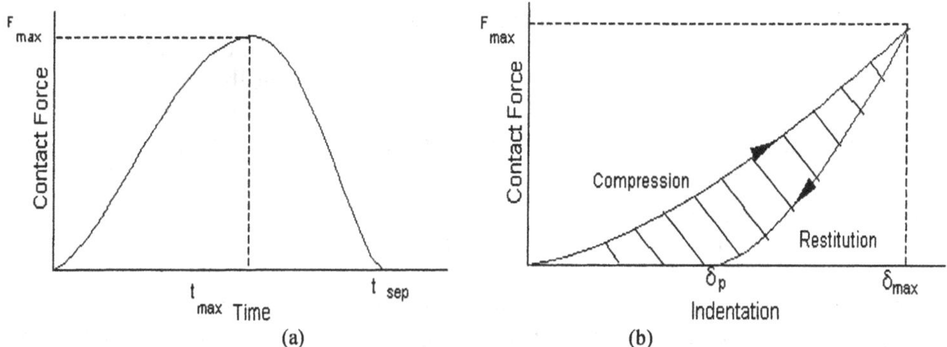

Figure 4. Hertzian contact force model with permanent indentation: (a)contact force versus time (b) contact force versus deformation.

A continuous analysis may now be performed by numerically integrating the equations of motion of the two spheres forward in time in conjunction with the developed contact force model. A solution is thus obtained in the form of positions, velocities, and accelerations of the spheres at any instant of time during the contact period. As a by-product of the preceding parameter identification process, one can approximate the duration of the contact period between the two spheres as [14]

$$\Delta t \cong 2\frac{\delta_m}{\dot{\delta}^-}\int_0^1 \frac{dz}{\sqrt{1-z^{n+1}}} = 2.94\frac{\delta_m}{\dot{\delta}^-} \tag{11}$$

which illustrates that the duration of the contact can be predicted from the initial speed of the impact and the geometry and material properties of the surfaces in contact.

The preceding discussion on the evaluation of the contact forces can easily be generalized to impact situations within a multibody system, as illustrated by the impact of two bodies '*i*' and '*j*' in the system of Figure 5. The points of contact on the two bodies are P_i and P_j, n is a unit vector in the normal direction, and t is a unit vector in the tangential direction to the contact surfaces of the two bodies. No matter which type of coordinates is used to assemble the equations of motion for the multibody system, the coordinates and the velocities of the bodies can be calculated, at any instant of time, from the solution of the equations of motion. For a known system configuration at the initial time of contact, the location of the contact points r_i^{P-} and r_j^{P-} and the components of the algebraic unit vectors n and t, with respect to a non-moving XYZ coordinate system, may be calculated. From the known body velocities, velocities of the contact points in the xyz coordinate system, $\dot{r}_i^{P-}$ and $\dot{r}_j^{P-}$, may also be calculated at that time. Hence, the indentation and the indentation velocity at the initial time of contact are

$$\delta = 0 \tag{12}$$

$$V_n^- = \dot{\delta}^- = n^T (\dot{r}_i^{P-} - \dot{r}_j^{P-}) \tag{13}$$

in which the symbol ^{T} performs the transpose operation and velocities of the contact points are projected in the normal direction to the contact surface. Similarly the relative tangential velocity V_t^- at the time of impact can be defined as $V_t^- = t^T (\dot{r}_i^{P-} - \dot{r}_j^{P-})$. The expression in equation (2) for the parameter K can be used for the contact between any two bodies if the local surfaces of contact are both spherically shaped. Similar expressions have been obtained by Hertz [10] and others [15, 16] for other shapes of the local contact surfaces such as sphere on plane, parallel cylinders, and plane on plane. Once the generalized parameter K is calculated, with a given coefficient of restitution 'e' and the initial approach velocity $\dot{\delta}^-$, all other parameters in the contact force model can be determined. The friction force is also evaluated from the normal contact forces and the law of limiting friction. With known variations of the contact and friction forces during the contact period, a continuous analysis of the system can be performed simply by adding these forces to the multibody system equations of motion. This analysis method provides accurate results, since all of the equations of motion are integrated over the period of contact. It thus accounts for the changes in the configuration and the velocities of the system during that period. To avoid computational inaccuracies and inefficiencies associated with the integration of the system equations of motion over the period of contact, an effective mass concept for a two-particle model of the system during the contact period as well as scaling of the time axis are sometimes performed [13].

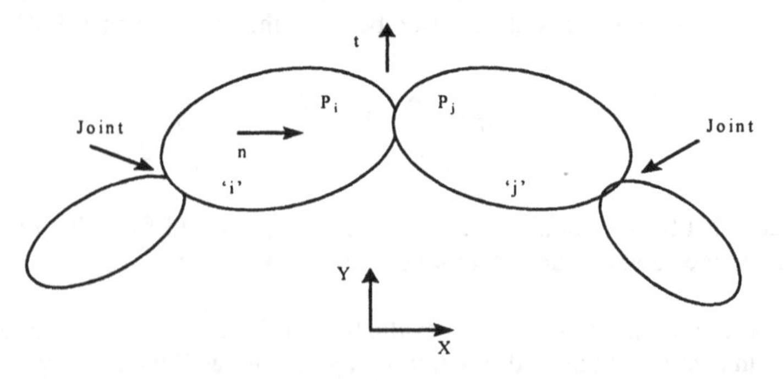

Figure 5. Schematic configuration of a multibody impact.

3. COMPUTER AIDED ENGINEERING TOOLS FOR SEAT, OCCUPANT AND AIRFRAME MODELING

To improve aircraft crash safety, conditions critical to an occupant's survival during a crash must be known. A large number of possible aircraft crash environments exist, and the impact sled testing may neither be possible nor feasible for some configurations. Cost and time are also other burdens of testing procedures. Furthermore, prior to testing, multiple simulations (analysis) must be conducted to better define the experimental testing program. Rigorous analytical techniques are necessary for design of crashworthy aircraft airframes, seats, occupant surroundings, and restraint systems. Validated analytical models also reduce the necessity of fabrication of design modifications. Some of the non-linear finite element analysis codes used for modeling and analysis of aircraft seats and interiors (structures) are LS-DYNA, MSC/DYTRAN, MARC and PAM-CRASH. Analysis codes that model the dynamic response of a human or ATD during a crash are known as gross motion simulators. These models are comprised basically of kinematically connected body segments with joint stiffness and contact forces between penetrating segments or segments in contact with the surrounding. Some of the existing body gross motion simulators include: ROS, CAL2D, HSRI, MVMA2D, SIMULA, PROMETHEUS, CAL3D, UCIN, SOM-LA/TA, ATB, and MADYMO [17]. Currently, the codes mostly used for reconstruction of aircraft crash scenarios are SOM-LA/TA, ATB, and MADYMO. More advanced ATD and human finite element models are also available these days for detailed assessment of the injury potential to different body regions. A systems approach, shown in Figure 6, is used for seat design problems utilizing nonlinear finite element tools with experimental component tests to validate the models, and iterative analysis for the final seat design. Each element in the load path, namely, the seat legs, pan, and cushion, is analyzed individually as well as its function in the entire seat model.

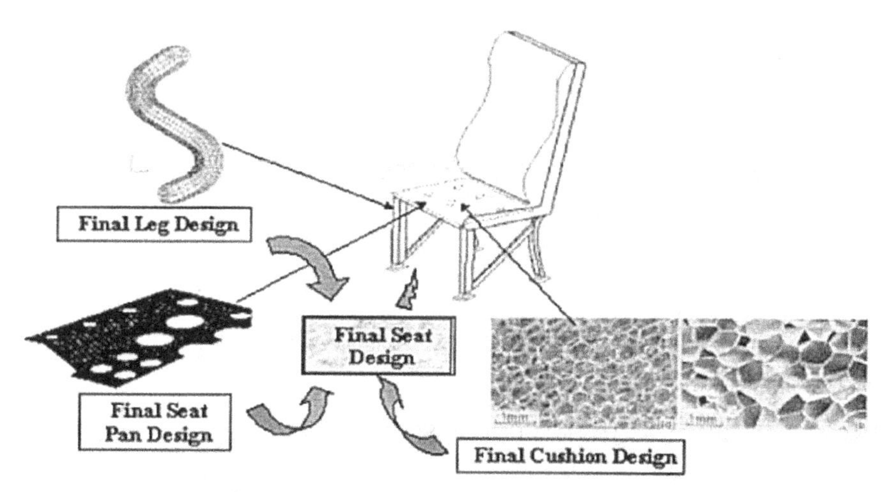

Figure 6. Systems approach in designing an aircraft seat.

The survivability of the occupant in the event of an aircraft accident primarily depends on a number of key features like maintaining a survivable volume around the occupant, which is the passenger compartment, having effective restraints features to properly restraint the occupant within the survivable volume, limiting the occupant loads by having energy absorbing structures and seats to have effective evacuation systems in place for the mitigating of post-crash hazards. A methodology, shown in Figure 7, is used for the development of safety systems in an aircraft utilizing finite element tools, multibody modeling techniques and virtual reality tools for the safe egress of the occupant from an aircraft in the event of a crash. The methodology uses the responses from the aircraft response with the impact surface, obtained from finite element modeling of the impact, being applied to the occupants and designing effective systems for the safety of the occupant during the event using multibody modeling tools and the modeling the egress of the occupant from the aircraft using virtual reality tools.

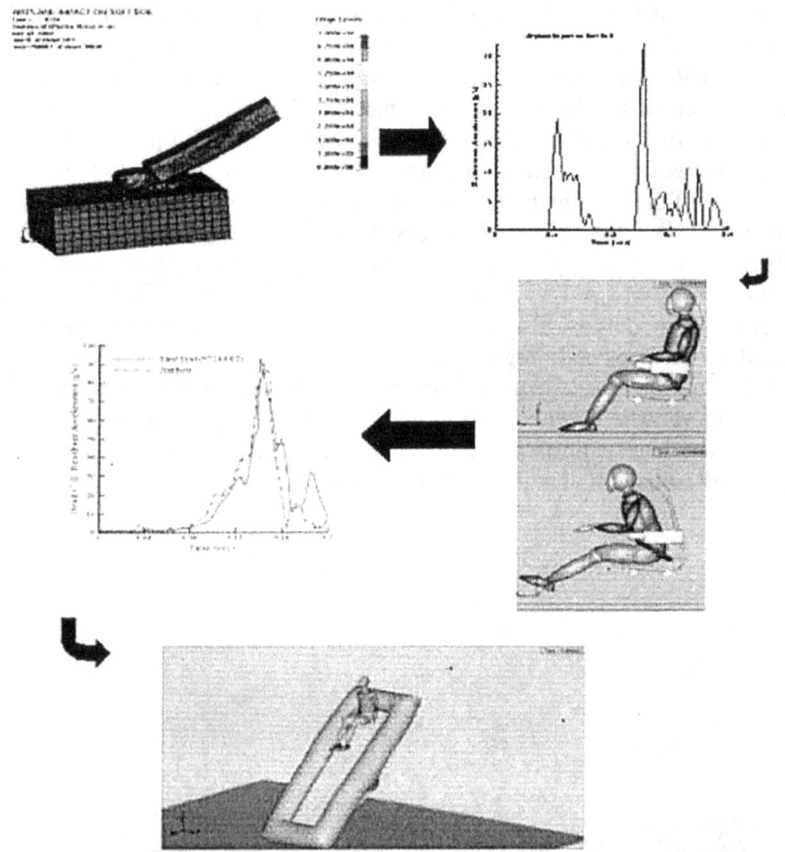

Figure 7. Virtual development cycle for aerospace safety systems.

4. AIRFRAME CRASHWORTHINESS DESIGN

Crash-resistant features and design methodologies, such as Aircraft Crash Survival Design Guide [18] are based on hard surface impacts. Accident data, however, indicates that only 18% of potentially survivable civilian crashes occur on hard, prepared surfaces [19]. The statistics are even lower for the army at 7% and the Navy at zero percent [20].

In contrast, 51% of civilian and 83% of Navy crashes occur on water and soft soil. The crash-resistant subsystems designed for rigid surface impacts, such as landing gears or sub floors, are not as effective in soft soil and water since the structure undergoes different loading conditions. Hard surface impacts introduce distributed loads into the stiffest structural members, such as keel beams and frames.

Nowadays, there is no reliable procedure for predicting the interaction processes that will occur between soil or water and an aircraft. While the in-flight behavior of the aircraft can be predicted with an optimum degree of precision, the same cannot be said for the phenomenological behavior that takes place at impact. This is due primarily to the lack of knowledge and to the complexity of the soil behavior to allow a reliable prediction of reactive forces against various geometries of the impact when subjected to a dynamic loading situation. The functional relationships and behavioral characteristics of soil and water are much more complex than for other common engineering materials, thus making generalized behavior conditions extremely difficult to determine with reliable accuracy.

4.1 ANALYSIS OF AND AIRCRAFT DROP TEST ON SOFT SOIL

The soil model is assumed to be of uniform structure and it does not have different layers. A material model capable of showing an eroding effect was used to represent these properties. This material model is of type 5 in LS-DYNA [21]. It provides a simple model for foam and soils whose material properties are not well characterized.

The soil is idealized as a rectangular block. The rectangular block is meshed using 8 node hexahedron solid elements. The element formulation, based on Lagrangian formulation, is used for the soil impact simulation. A material model capable of showing an eroding effect was used to represent these properties. It provides a simple model for soils and foams whose material properties are not well characterized. It simulates the crushing through volumetric deformations. If the yield stress is too low this soil model gives nearly fluid like behavior. It is governed by a pressure dependent flow rule,

$$\Phi = J_1 - [a_0 + a_1 p + a_2 p^2] \tag{14}$$

where Φ is Plastic yield function, J_1 is Yield stress, a_0, a_1, a_2 are empirical constants and p is pressure.

The constants a_1 and a_2 are kept zero to eliminate the pressure dependence of the yield strength. The pressure cutoff for tensile fracture and bulk modulus of the soil are specified since the loading and unloading functions of soil are not known. Figure 8 shows the volumetric strain versus the pressure curve for a solid model. The properties for the soil model used is listed in table 1.

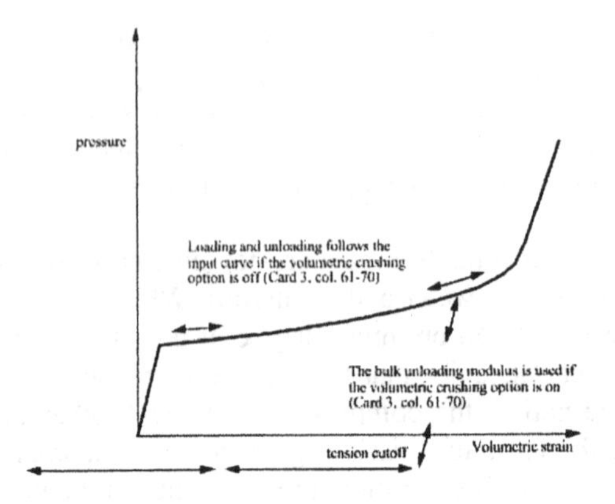

Figure 8. Volumetric Strain versus pressure curve for soil model.

Table 1. Soil properties.

Mass Density	1.5142 E-04 lbm/in^3
Shear Modulus	980 lb/in^2
Bulk Modulus	24400 lb/in^2
Yield function constant	1.45
Pressure cutoff for tensile fracture	-0.08265

4.2 ANALYSIS OF AND AIRCRAFT DROP TEST ON SOFT SOIL.

Four identical four-place, high wings, single-engine airplane specimens with nominal masses of 1043 kg were crashed at the Langley Impact Dynamics Research Facility [22] under controlled free-flight test conditions(see figure 9). These tests were conducted with nominal velocities of 25 m/sec along the flight path at various flight-path angles, ground-contact pitch angles, and roll angles. Three of the airplane specimens were crashed on a concrete surface; one was crashed on soil. Each airplane had a gross mass of 1043 kg and the pilot, copilot were represented by anthropomorphic dummies. The airplane specimen, suspended by two swing cables attached to the top of the gantry, is drawn back and above the impact surface by a pullback cable to a height of about 49m as shown in figure 8.

The airplane contacted the soil impact surface on the nose gear with the velocity of 25.3 m/sec along a flight path angle of –32 degrees and at a roll angle of –1.5 degrees. Prior to the tests the engines were completely removed from the fuselage. The accelerations measured on the floor of the aircraft were nominally 20 to 25g's. The aircraft was modeled using the finite element method. The model is meshed using 4 node quadrilateral shell elements. The model had the same mass of 1043 kg as in the actual test. Figure 10 illustrates the crash simulation of the airplane during a negative pitch (node down) crash starting at the initial ground contact. The airplane contacted the soil impact surface with a velocity of 25.3 m/sec along a flight path angle of –32 degrees.

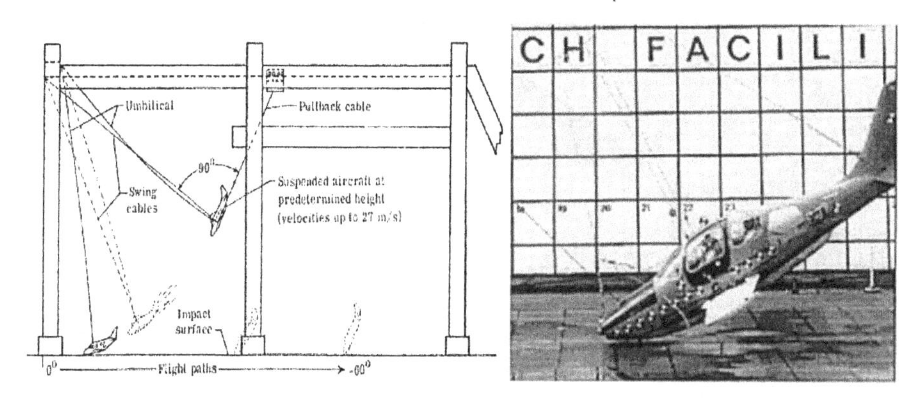

Figure 9. Full-scale crash impact test facility.

The nose began to buckle immediately after initial ground contact followed by the fuselage at 0.09 sec into the impact. The aft section of the fuselage failed at 0.16 sec. The airplane model began to plough into the soil creating a crater in the soil model throwing out the deleted nodes. The acceleration profile in the Z-direction had two spikes, the first one is when the nose just touches the soil, and the second one is due to the fuselage impact on the soft soil. The first one reached a maximum of 18g while the second one had reached 30g. The acceleration data was obtained at the center of mass of the aircraft model unlike in the drop test where it was measured at different locations on the cabin floor. Overall, the simulation had a reasonable representation of the actual airplane crash test.

The acceleration profiles from this test are used as input for the occupant simulation using multibody tools, to obtain the occupant responses during the impact. Aircraft systems like seats and restraints are suitably designed to reduce the injuries to the occupant.

4.3 HELICOPTER IMPACT ON WATER

Water is modeled using MAT_ELASTIC_ *, which is an isotropic elastic material and is available for beam, shell, and solid elements in LSDYNA. A specialization of this material allows the modeling of fluids. The FLUID option is valid for solid elements only [23]. The standard input deck requires: Material Identification, Mass Density, Young's Modulus, Poisson's Ratio, Bulk Modulus, Tensor viscosity coefficient and Cavitation Pressure. For the Fluid option the Bulk Modulus (K) has to be defined as Young's Modulus and Poisson's Ratio are ignored. With the fluid option fluid-like behavior is obtained where the Bulk Modulus, K, and pressure rate, p, are given by:

$$\mathrm{K} = \frac{E}{3(1-2\nu)} \tag{15}$$

$$\mathrm{p} = -\mathrm{K}\,\varepsilon_{\mathrm{ii}} \tag{16}$$

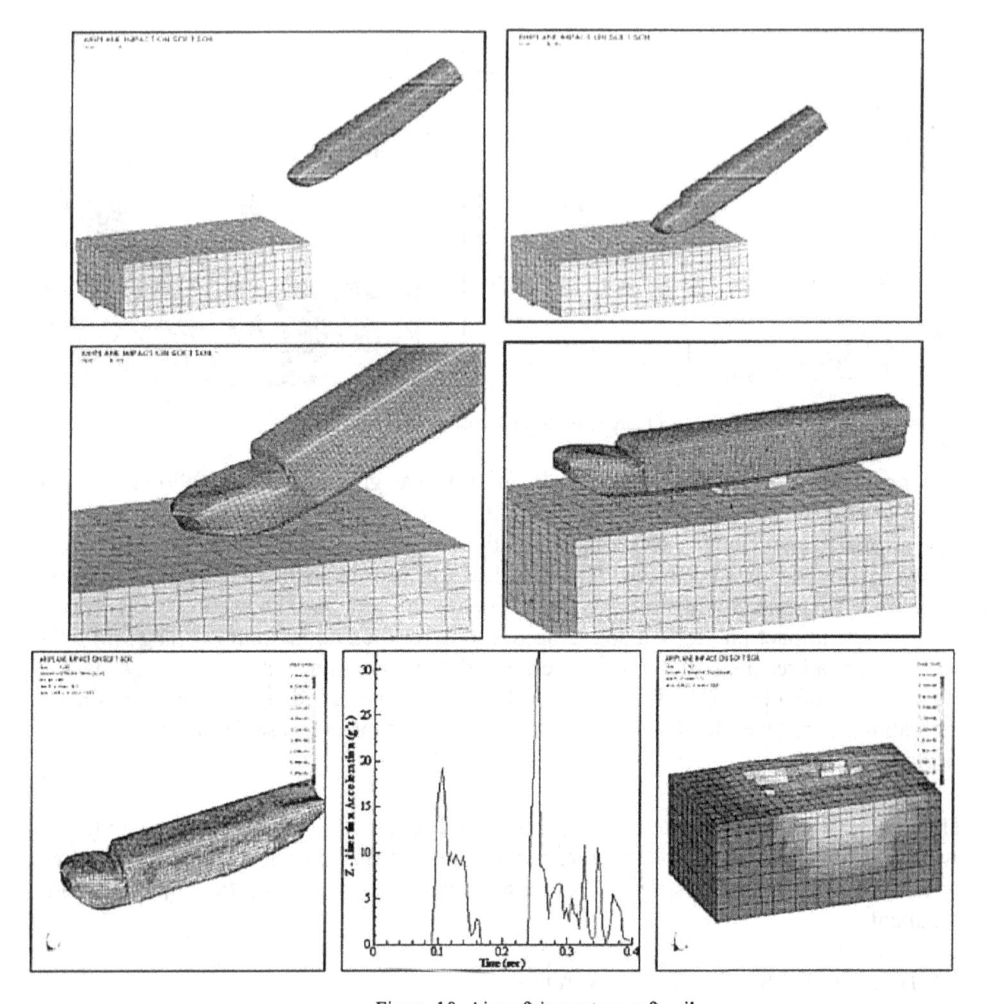

Figure 10. Aircraft impact on soft soil.

and the Shear Modulus is set to zero. A tensor viscosity is used which acts only for the deviatoric stresses, S_{ij}^{n+1}, given in terms of the damping coefficient as:

$$S_{ij}^{n+1} = V_C \, \Delta L \, a \, \rho \, \varepsilon_{ij} \tag{17}$$

Where ΔL, is a characteristic element length, 'a' is the fluid bulk sound speed, ρ is the fluid density, and ε_{ij} is the deviatoric strain rate. In this elastic material, co-rotational rate of the deviatoric Cauchy stress tensor is computed as:

$$s_{ij}^{\nabla^{n+1/2}} = 2G\dot{\varepsilon}_{ij}'^{\,n+1/2} \tag{18}$$

and pressure

$$p^{n+1} = -K \ln V^{n+1} \tag{19}$$

where G and K are the elastic shear and bulk moduli, respectively, and V is the relative volume, i.e., the ratio of the current volume to the initial volume. The axial and bending damping factors are used to damp down numerical noise. The update of the force resultants, F_i, and moment resultants, M_i, includes the damping factors:

$$F_i = F_i^n + \frac{(1 + DA)\Delta F_i^{n+1/2}}{\Delta t} \tag{20}$$

$$M_i = M_i^n + \frac{(1 + DA)\Delta M_i^{n+1/2}}{\Delta t} \tag{21}$$

The input properties for this model are:

Density: Density of Water is 0.001 kgs/cm^3 @ 3.98 °C [24]

Bulk Modulus: The Bulk modulus for a material refers to the ratio of pressure induced to the decrease in volume. This is the inverse of compressibility. For most practical purposes water may be considered as incompressible, but actually it is about 100 times as compressible as steel. Bulk Modulus = K for Water = 292,000 P(force)si @ temperature 32 °F and pressure 15 p(force)si . Converting to standard units of Kg/cm/sec; K = 2.06e+07 Kg/cm.sec^2

Viscosity Coefficient: The viscosity coefficient is a function of characteristic element length ΔL, 'a' the Fluid bulk sound speed, ρ is the fluid density, and ε_{ij} the deviatoric strain rate and deviatoric stresses.

$$S_{ij}^{n+1} = V_C \, \Delta L \, a \, \rho \, \varepsilon_{ij} \tag{22}$$

$V_C \, \Delta L \, a\rho$ = Absolute Viscosity (Dynamic Viscosity)

Absolute Viscosity @ 32 °F = 1.792 cp = 0.01792e-03 kg/(cm.s)

'a' (Fluid bulk sound speed) = $(K/\rho)^{1/2}$ = (2.06e+07/0.001)$^{1/2}$ = 143527 cm/s

Hence, $V.C = 1.248e\text{-}07 / \Delta L$ (ΔL depends upon the model in consideration)

Cavitation Pressure: Cavitation is defined as the process of formation of the vapor phase of a liquid when it is subjected to reduced pressures at constant ambient temperature. A liquid is said to cavitate when vapor bubbles are observed to form and grow as a consequence of pressure reduction. When the phase transition is a result of pressure change by hydrodynamic means, a two-phase flow composed of a liquid and its vapor is called a cavitating flow. From a purely physical-chemical point of view, of course, no distinction need be made between boiling and cavitation. Hence, Saturation pressure can be taken as the cavitation pressure. Saturation Pressure of Water @ 32 °F = 6.564 mill bars = 0.00669 Kg/cm^2

The water model developed in LSDYNA is validated against experimental results conducted by dropping a ball into a beaker of water. Figure 11 shows the process of validation of the water model using the experimental results and finite element model.

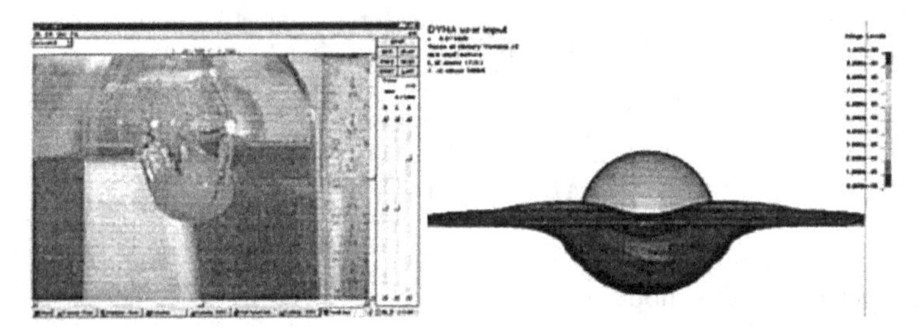

Figure 11. Validation of water model.

Simula Technologies at U.S Army Yuma Proving Ground utilizing a surplus Bell Helicopter UH-1H "Huey" airframe conducted a vertical dynamic test [25]. The test helicopter had been striped of nearly all components such as engine/transmission, tail boom, landing gear, etc., leaving the bare hull. The test weight was 2260 lbs. The test was purely a vertical drop of approximately 9 ft. measured from the lowest point of the helicopter belly to the water surface. This provided a calculated impact velocity of approximately 24 ft/sec (7.31 m/s). Fresh water was utilized with, no surface waves. The water depth at the impact point was approximately 90 inches. The peak pressure reading at the various sensors ranged from a low of 2.3 psi (15.8 KPa) to a maximum of 18.4 psi(126.9 Kpa). The peak accelerometer readings ranged from a minimum of 27.9g to a maximum of 69g.

The dimensions of the air and water models used in the eulerian simulation of ball impact on water were scaled up to accommodate the helicopter model. The element formulation of the model is the same. The mesh density and the total number of the solid elements remain the same. Helicopter is lagrangian solid here modeled with Belytschko-Tsay shell elements and air/water stays as the eulerian fluid. The boundary conditions also are similar to the eulerian ball impact simulation. The Helicopter's dimensions are approximately 500 cms (16') length, 150 cms (5') width including the wings, and 110 (4') cms height. The helicopter model was auto meshed with triangular elements. The quad elements were giving excessive warpage, aspect and skew around the edges of the base. The total number of triangular elements in the helicopter is 2592 (Belytschko-Tsay elements). The thickness of the shell elements is 0.5 cms. 1 point is chosen for through the shell integration. The helicopter was scaled to fit into the air-water model. The helicopter was translated in x, y and z directions for proper orientation also. The helicopter is a not a rigid body in this case. It has been given the properties of Aluminum 7075-T6 with a Plastic Kinematic Material Model.

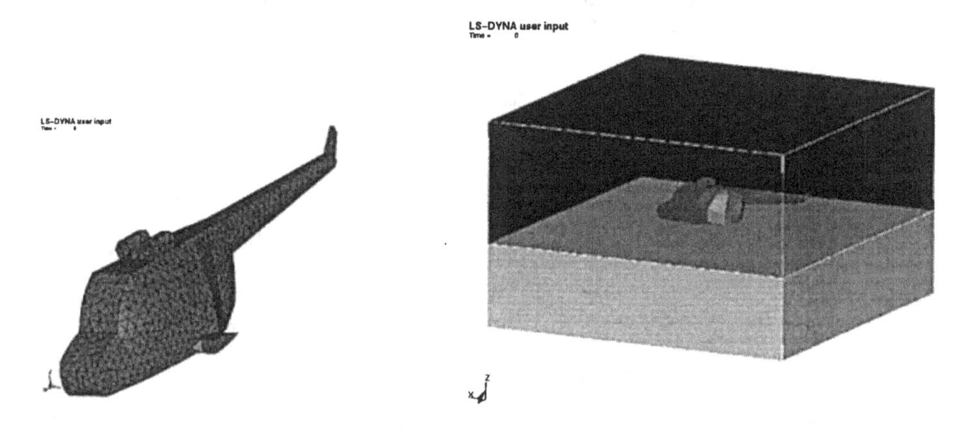

Figure 12. Helicopter model.

Helicopter is coupled with the air and water both (constrained penalty coupling). The helicopter has been impacted with a vertical Z-velocity of 24 ft/sec (731.52 cm/sec). The weight of the helicopter is 2260 lbs. Figure 12. shows the helicopter model and figure 13 shows the simulation of the helicopter model impacting the water surface.

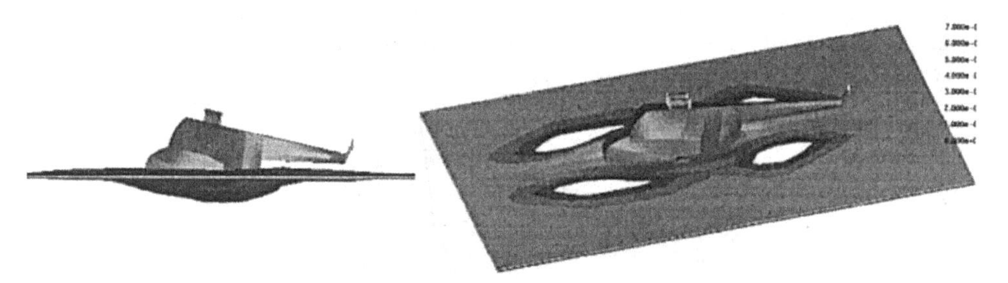

Figure 13. Simulation frame.

The water depth at the impact point is 240 cms, which is close to 90 inches. The impact attitude is pure flat (no pitch or roll). The termination time for the analysis is 0.5 seconds and 770 d3plots were obtained to gather more data points during impact. The peak value of acceleration is around 77'g' and a second peak of 23'g'comes a little later (Figure 14a.). Then slowly the acceleration drops down to zero when the helicopter is sinking down with a constant velocity. There's another acceleration peak of about 3-4'g' when the wings impact the water surface. Since, the velocity has already reduced to a minimal value after the initial impact, the value of the deceleration because of the wing impact is meager.

If we compare the actual test conditions, every aspect was kept same. The mass of the helicopter, the depth of the water at the point of impact, the shape of the helicopter, velocity of impact etc. were chosen as exact as possible. The final results are really comparable. Referring Figure 14b, the maximum acceleration (peak) is 70g (for GAC) and in our simulation (Fig 14a), it is around 77g. The difference can be attributed to the atmospheric pressure conditions, shape of the helicopter and the structural difference.

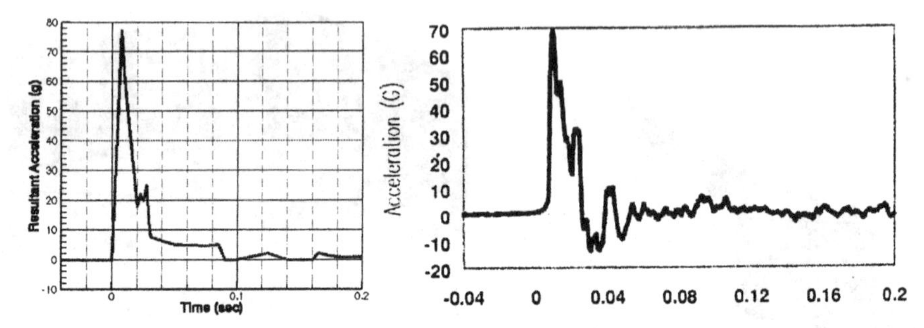

Figure 14a. Acceleration time history. Figure 14b. Acceleration time history – GAC.

The helicopter sustained major structural damage from the water impact. Figure 15. shows stressed state of helicopter at initial touchdown. The stresses increase till the time the whole body comes in initial contact with the water surface. Subsequently, the stresses drop as the helicopter sinks down with a constant velocity. The maximum effective (von-mises) stress induced on to the helicopter base is 1.030e+06 kg/(cm.sec^2), which is the yield stress for Aluminum 7075-T6. So the zones represented in red are yielding (failure) while impacting. It is observed that the stresses are high only while impacting. The stresses in the wing also exceed the yield stress (1.03e+06), when impacting. Again, after the impact permanent deformation takes place in the wings, but the stresses reduce as the whole body sinks.

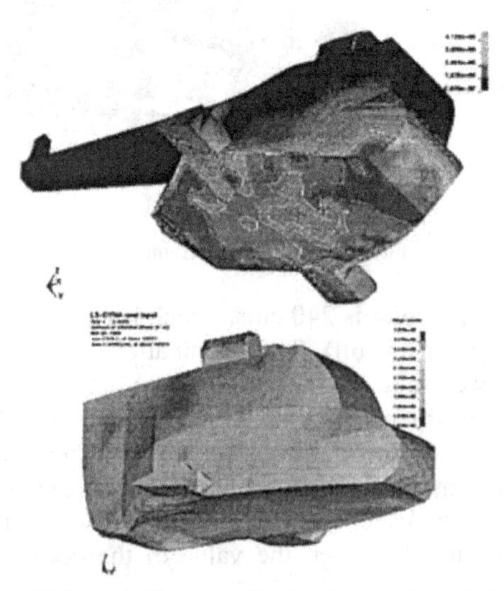

Figure 15. Stressed state of Helicopter (Contours of V- M stress) and the deformation in the Helicopter.

5. CAE AND VR TOOLS FOR CABIN INTERIOR MODELING AND VISUALIZATION

Both the multibody and the finite element method offer their specific advantages and disadvantages for cabin interior crashworthiness design applications. The multibody approach is particularly attractive due to its capability of simulating in very efficient way complex kinematical connections, while the finite element method offers the capability of describing (local) structural deformations and stress distribution. The following example application will give an overview on how the coupling of these two methods in conjunction with component testing is applied throughout the system development cycle, as shown in Figure 16.

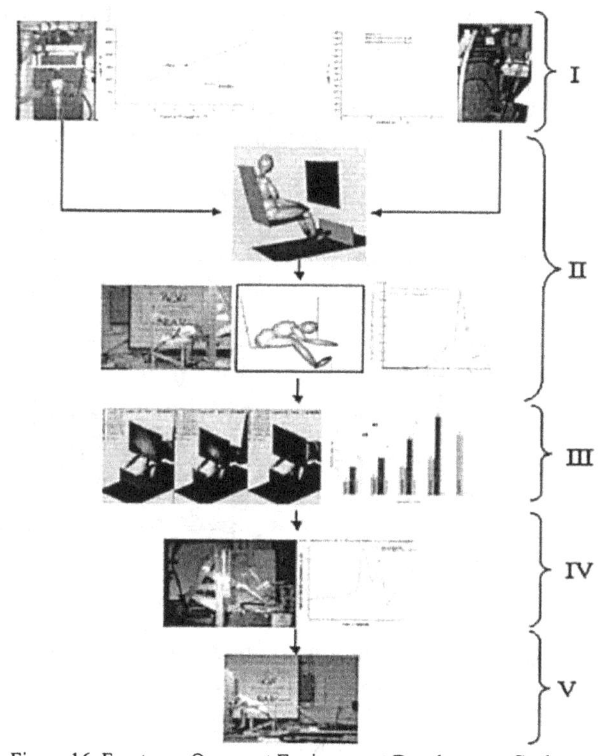

Figure 16. Front-row Occupant Environment Development Cycle.

Airline seat manufacturers have experienced some difficulty in satisfying the Head Injury Criteria (HIC) contained in the FAR 25.562 [7] 16g dynamic seat test requirements for so-called "front-row seats" that are located behind a bulkhead or cabin class divider. Traditionally sled testing has been used to develop aircraft interiors. Due to the high cost, test set up time, and complexity of developing the system by using sled testing, a Multibody/FEA approach combined with component and sled testing is presented. Initially three different lap belt (nylon, polyester 8%, and polyester 19%), three bulkheads (Aluminum, standard Nomex honeycomb, and modified Nomex honeycomb) materials were available; and six possible seat setback configurations (28, 30, 33, 35, 38, and 40 inches) to choose from. If we were to develop this system the traditional way (sled testing), we would have to test 54 different configurations, and since usually two sled tests are conducted per configuration, we would have a total of 108 sled tests. A program like this could take up to three months of sled testing, and hundreds of thousands of dollars in components. In order to avoid the high cost and the long development cycle, CAE and VR tools in conjunction with component and sled testing can reduce the development cycle to a month of simulation work, including 6 sled tests for model validation, 4 component tests, and 2 final system certification tests.

5.1. *PHASE I AND II:* MATERIAL COMPONENT TEST, MODEL DEFINITION AND VALIDATION

A Multibody/FEA model of the seat/ATD/bulkhead test configuration is shown in Figure 17. The rigid seat was represented as two planes that are fixed in space. One plane represents the seat pan while the other represents the seat back. The contact forces between these planes and the appropriate anthropomorphic test dummy (ATD) body segments were represented by the Hertzian-type model of contact (6) or defined in terms of the appropriate loading and unloading curves obtained through component testing.

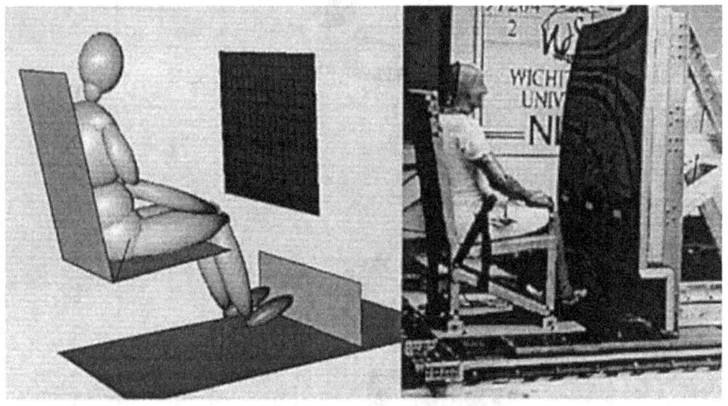

Figure 17. Multibody/fea model and sled test setup.

The 50th percentile Hybrid II ATD multibody model was placed in the seat. A two-point restraint system was modeled using belt properties that were representative of the system used in the sled tests. The anchor points of the belt were located at the intersection of the seat pan and seat back as shown in Figure 8. The floor was also modeled by means of a rigid plane. An additional rigid plane was placed in front of the legs just below the bulkhead. The bulkhead was modeled using 180 quadrilateral shell elements representing a 30 x 24 in. (76.2 x 60.9 cm) bulkhead surface. An elasto-plastic material model was used to define the behavior of the bulkhead materials. The stress-strain curve used in the model were acquired from a series of tensile tests of using an MTS servo-hydraulic test stand and linear impactor load/deflection tests since this data was not available in literature. Prior to conducting the parameter studies the models were validated with sled test data for the three different bulkhead configurations. As shown on figure 18 there is good correlation between the models and the sled tests.

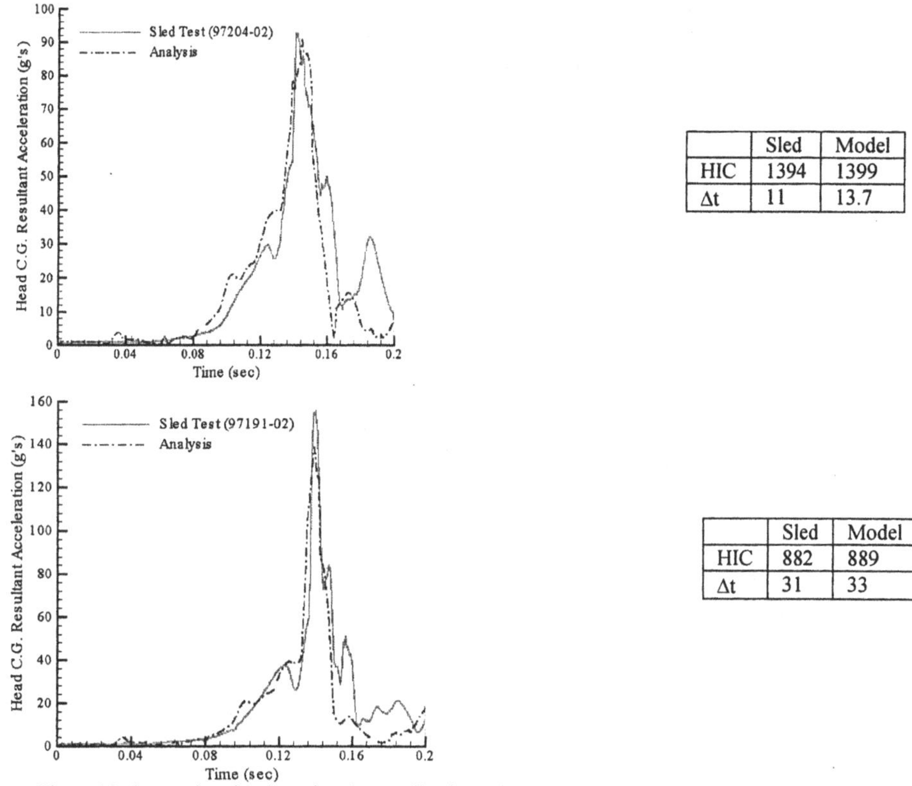

	Sled	Model
HIC	1394	1399
Δt	11	13.7

	Sled	Model
HIC	882	889
Δt	31	33

Figure 18. Comparison head acceleration profiles from sled and MADYMO for a production bulkhead.

5.2 *PHASE III:* PARAMETER STUDY

The purpose of this parameter study was to observe the injury criteria and kinematics of the ATD under various seat setback conditions (28, 30, 33, 35, 38, and 40 inches), various lap belt (nylon, polyester 8%, and polyester 19%), and bulkhead material configurations (aluminum panel, standard Nomex honeycomb, and modified Nomex honeycomb). These three factors were identified as the most important parameters influencing the magnitude of the HIC values. This is due to the fact that these are the parameters that dictate the head impact angle, head impact velocity, head impact acceleration, and the translation of the lower torso. From the parameter study we concluded that the lower the head impact angle, the higher the peak acceleration and HIC values become. The stiffness of the lap belt material and the seat setback distance are the parameters that dictate the magnitude of the Head Impact Angle. Belts with higher stiffness allow less movement of the lower torso. When we have the same setback distance and different lap belt materials we can observe that the stiffer lap belt material the larger the Head Impact Angle becomes, this is due to the fact that a stiffer belt allows less forward lower torso displacement prior to ATD/Bulkhead impact hence allowing more rotation of the upper torso prior to impact. From the sled test analysis and the computer models we can observe that the longer the seat setback distance the more critical the rearward translation of the torso after initial head impact becomes. This movement of the lower torso during impact becomes more critical for longer seat setback distances as well as when softer materials are used to the bulkhead structure construction. A sample of the results from the parametric study is shown in Figures 19 and 20.

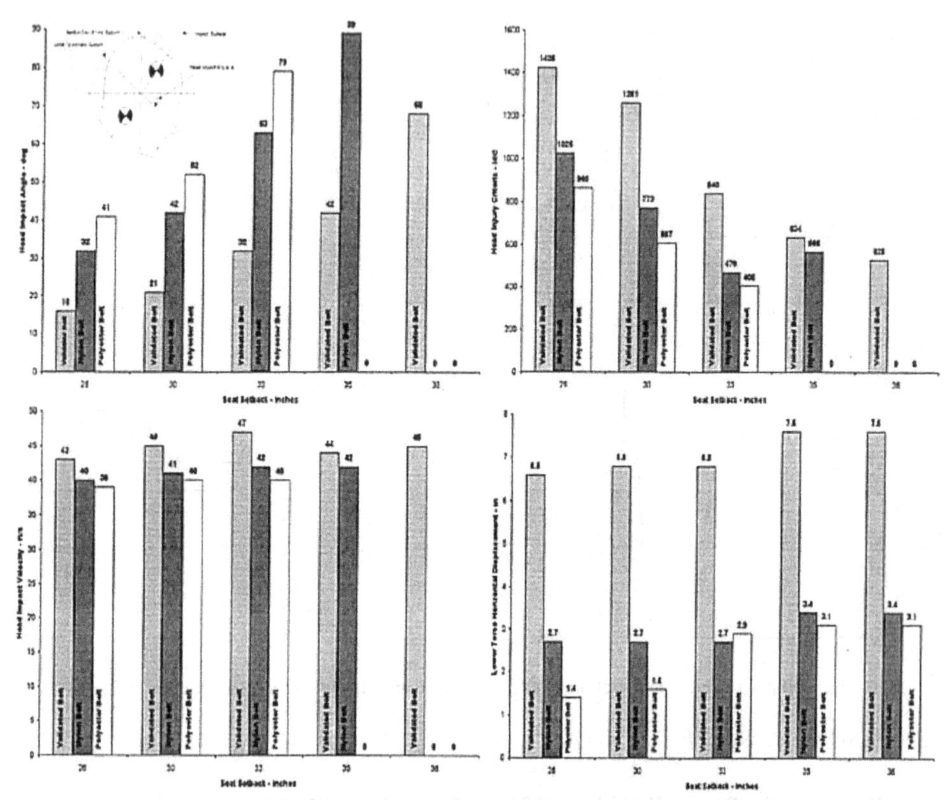

Figure 19. Head Impact Velocity for the Aluminum Panel Parameter Study.

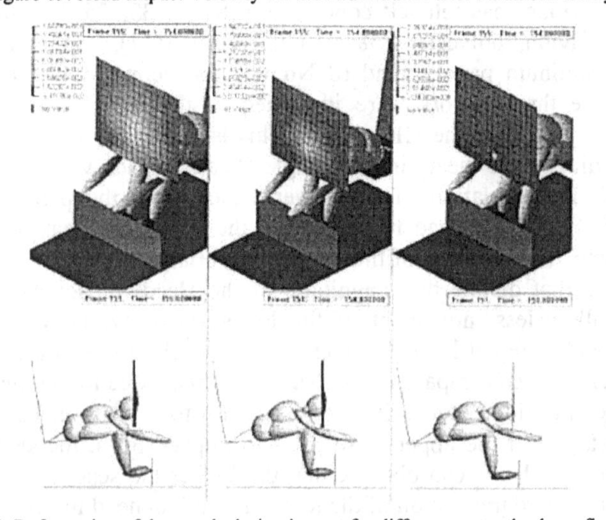

Figure 20. Deformation of the panels during impact for different seat setback configurations.

From the parameter study it was observed that the configuration that meets all the design requirements (biomechanical performance, cost, legroom, and component availability) as well as FAR 25 requirements was the aluminum bulkhead with a seat setback distance of 35 (0.89meters) inches as shown in the table bellow.

5.3 *PHASE IV AND V:* COMPONENT TEST AND CERTIFICATION SLED TEST OF THE FINAL DESIGN

Due to the high cost, test set up time, and complexity of a sled test, NIAR's Enhanced Head Injury Criteria Component Tester was used to validate the final design prior to conducting the final certification tests. This component tester is an inverted pendulum type impactor. It consists of an accelerator, pendulum arm, support arm, ATD head/neck assembly, signal processing electronics, and a computerized control and data acquisition system. As shown on table 2 the component test confirms the results obtained from the simulation design, consequently we concluded that the system was ready to proceed with phase V, the sled test certification process.

Table 2. Final System Configuration.

Parameter	**Sled**	**Component**	**Simulation**
Bulkhead Material	AL 2024 -O	AL 2024-O	AL 2024 -O
Seat Setback Distance – in	35	35	35
Head Impact Velocity –ft/s	45	44	44
Head Impact Angle – deg	38°	38°	42°
Head C.G. Peak Acceleration – g	143	143	140
HIC	694	685	634
HIC Window (Δt) – ms	23.7	21.4	22.8
Average Head Acceleration –g	61	63	60

The results from the sled test certification process confirm the results of both the component and the simulation model see table 2 and figure 21. By using this Multibody/FEA technique in conjunction with component testing and minimal sled testing we were able not only of selecting the proper system configuration to meet the system requirements, but also through simulation we gained a better understanding on the ATD (by using 3D Virtual Reality visualization of the system) responses to the different input parameters.

6. APPLICATION OF CAE AND VR TOOLS FOR VIRTUAL TESTING OF EVACUATION SYSTEMS

Throughout the more than 40 years since the birth of this industry, governmental requirements have continuously increased and placed stricter regulations on the design of evacuation systems. As aircraft became larger and carried more people, the systems became more sophisticated and complex. Demand for improvements such as increased safety, greater evacuee throughput, decreased inflation time and weight reductions are constantly forcing the industry forward [26]. This specific CAE and VR application

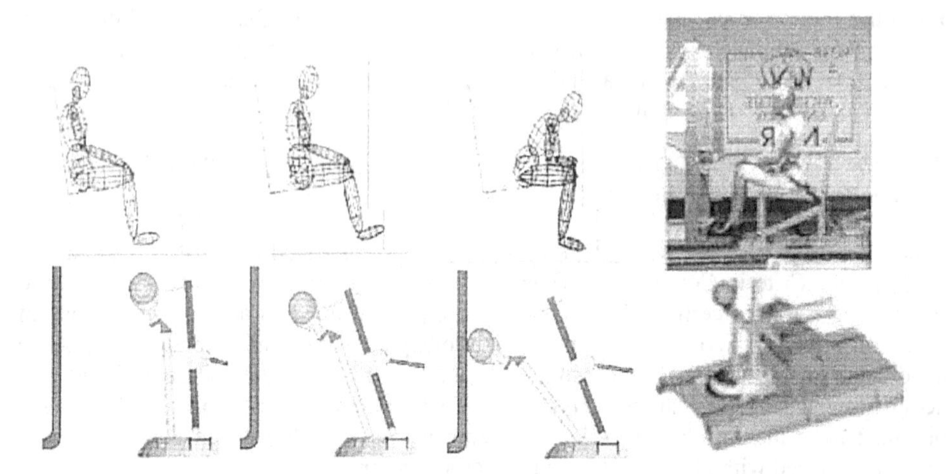

Figure 21. Kinematic sequence of a full-scale sled test and component test simulation

shows how we could analyze the performance of the evacuation system under different ambient conditions (0 C, 23 C, and 44 C).

Due to the size of an evacuation system a test like this would be very difficult and expensive to perform in a temperature chamber, but using FEA techniques to describe the evacuation system, and a multibody model of the ATD we can easily evaluate the performance of the system. A model of the evacuation system was created using MADYMO. The slide model consists of 9800 triangular elements. A typical evacuation system inflator mass flow rate was defined to inflate the slide. Appropriate contacts were defined between the multibody ATD and the evacuation system surfaces.

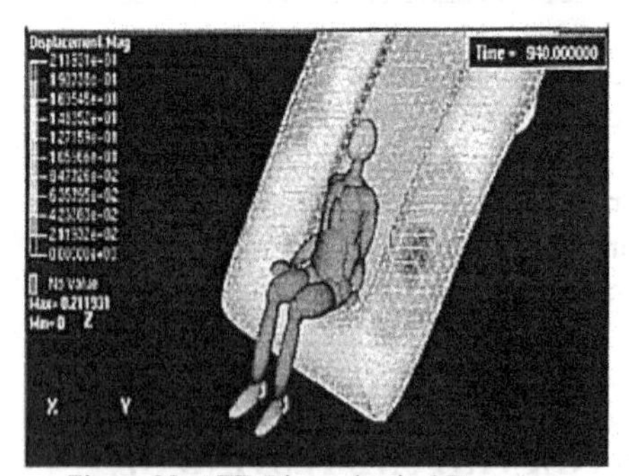

Figure 22. ATD Biomechanical Response.

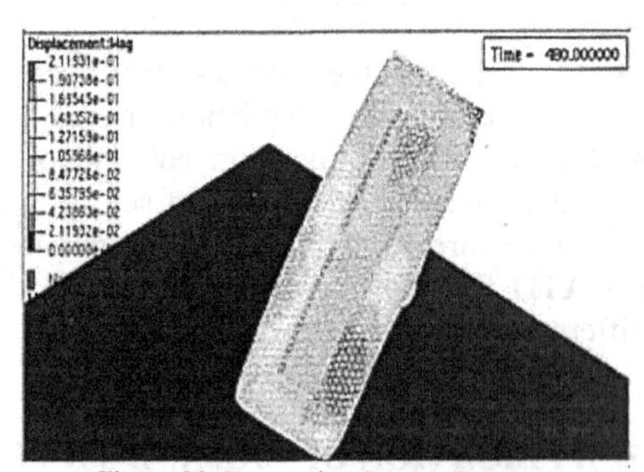

Figure 23. Evacuation System Response.

The biomechanical response of the ATD under these different ambient conditions is shown on Figure 22, as we can observe for both the cold and hot ambient conditions the head acceleration is greater than for nominal ambient conditions. For the cold ambient condition case this is due to the reduction in the internal pressure and total volume of the bag due to the effect of lower temperatures (see Figure 23), this makes the impact with the ground more severe. On the other hand for the hot load case, the increase in pressure and slide volume make the slide stiffer hence increasing the sliding velocity of the ATD, and increasing the rebound of the head at the end of the slide.

In order to gain a better understanding of the evacuation system performance the National Institute for Aviation Research's Virtual Reality facilities were used. This VR facility allows the user to visualize the results of the simulation in a screen 7 ft high and 15 ft wide (see figure 24). Virtual Reality offers a way for engineers to visualize, manipulate and interact with computers and extremely complex data. VR improves visualization of the system by allowing the user to co-exist in the same space as the system model therefore gaining a better appreciation of the system geometry and performance. VR also improves interaction with design in terms of more intuitive model manipulation and functional experimentation, the designer can effectively interact with the product model directly rather than using the conventional 2-D mouse and cursor. Another advantage of particular importance is the sense of scale, which can only be conveyed by immersing the designer in the "design". The simulation technique showed on this section could be easily implemented to conduct 'virtually' all the required product testing to ensure it's optimum functionality prior to building a physical model. The data generated by the multibody simulation of the various crash event provides a myriad amount of data which needs to be effectively understood. The virtual reality environment provides a suitable platform for the effective viewing of these data and also in understanding their significance.

Figure 24. Virtual reality control room and visualization of the evacuation simulation model at the NIAR.

7. CONCLUSIONS

This paper presents examples of the use of a systems approach methodology to some of the current aircraft crashworthiness problems. It makes use of appropriate injury criteria, component performance tests, full-scale sled tests, and the presented some of the latest CAE and VR tools. The modeling of the aircraft impact/accident is accomplished by using finite element analysis and multibody dynamic simulation tools as well as projections onto a virtual reality environment. The importance of CAE tools in aircraft crashworthiness problems is demonstrated via examples. This provides much insight into the nature of the seat, occupant, and airframe responses individually and collectively. The main purpose of utilizing this methodology is the anticipation that due to myriad amount of data that is generated by the finite element and multibody simulation CAE and VR tools need to be effectively utilized for analysis during the design phase.

8. REFERENCES

1. Greenwood, D.T. (1965) *Principles of Dynamics*, Prentice-Hall, Englewood Cliffs, New Jersey.
2. Meirovitch, L. (1970) *Methods of Analytical Dynamics*, McGraw-Hill, New York.
3. Kane, T.R. (1968) *Dynamics*, Holt, Rinehart & Winston, New York
4. J W Olcott, The development of dynamic performance standards for federal aviation aircraft seats, *Business and Commercial Aviation*, 1995.
5. R F Chandler, Crash injury protection in civil aviation, in A M Nahum and J W Melvin (eds.), *Accidental Injury Biomechanics and Prevention*, Springer-Verlag, New York, 151-185, 1993.
6. *Title 14 U.S. code of Federal Regulations, Part 23*, Amendment 23-39, Section 23.562, published in the Federal Register of August 14, 1988, effective date of September 14, 1988.
7. *Title 14 U.S. code of Federal Regulations, Part 25*, Amendment 25-64, Section 25.562, published in the Federal Register of May 17, 1988, effective date of June 16, 1988.
8. *Title 14 U.S. code of Federal Regulations, Part 27*, Amendment 27-25, Section 27.562, published in the Federal Register of November 13, 1989, effective date of December 13, 1989.
9. SAE, AS8049, *Performance standards for seats in civil rotorcraft and transport airplanes*, Aerospace Standards, 1990.
10. Hertz, H. (1895) *Gesammelte Werke* **1**, Leipzig, Germany.
11. Hunt, K.H. and Grossley, F.R.E. (1975) Coefficient of restitution interpreted as damping in vibroimpact, *ASME J. of Applied Mechanics* **7**, 440-445.
12. Love, A.E.H. (1944) *A Treatise on the Mathematical Theory of Elasticity*, 4th ed., Dover Publications, New York.
13. Lankarani, H.M. and Nikravesh, P.E. (1993) Continuous contact force models for impact analysis in multibody systems, *International J. of Nonlinear Dynamics* **5**, 193-207, Kluwer Academic Publishers, Dordrecht.
14. Lankarani, H.M. and Nikravesh, P.E. (1992) A Hertzian contact force model with permanent indentation in impact analysis of solids, ASME Advances in Design Automation, Design Technical Conferences, Scottsdale, AZ.
15. Goldsmith, W. (1960) *Impact, the Theory and Physical Behavior of Colliding Solids*, E. Arnold Ltd., London.
16. Lankarani, H.M. (1988) Canonical equations of motion and estimation of parameters in the analysis of impact problems, Ph.D. Dissertation, University of Arizona, Tucson, AZ.
17. TNO. *MADYMO Theoretical manual, Version 5.3*, TNO, Delft, Netherlands, 1998.
18. Ft. Ecstis, Aircraft Crash Survival Design guide, December 1989,Simula Inc., U.S. Army AVSCOM TR 89-D-22 (Volume A- E), US Army Aviation, Research and Technology Activity, VA
19. Coltman, J.W., et al., " Analysis of Rotorcraft Crash Dynamics for Development of Improved Crashworthiness Design Criteria", DOT/FAA/CT-85/11, U.S. Department of Transportation, Federal Aviation Administration
20. Sareen, Ashish K. (Bell Helicopter Textron, Inc) Smith, Michael R. | Hashish, Emam "Crash analysis of an energy-absorbing sub floor during ground and water Impacts" Annual Forum Proceedings - American Helicopter Society v 2 May 25-May 27 1999 American Helicopter Soc p 1603-1612 0733-4249
21. LS-DYNA, Ver.960, User Manual, April 1994
22. Victor L. Vaughan, Jr., and Robert J. Hayduk, " Crash Test of Four Identical High Wing Single Engine Airplanes", Report No. NASA TP-1699, August 1990, NASA Langley Research center, Hampton, VA 23665
23. John O. Hallquist, " Material Models", LS-DYNA Theoretical Manual, April 1994
24. www.matweb.com/SpecufucMaterial.asp?bassnum=DWATR0&group=General
25. Marvin K. Richards, E. Allen Kelly, "Development of a Water Impact Dynamic Test Eacility and Crash Testing a UH-iH Aircraft onto a Water Surface", U.S. Army Yuma Proving Ground, Yuma, Arizona.
26. M M Sadeghi, 'Evolution and design of vehicle structures for crash protection - A system approach', Crashworthiness of transportation systems: structural impact and occupant protection, *Proceeding of the NATO advanced study institute*, Volume II, July 1996.

VIRTUAL PROVING GROUND SIMULATION FOR HIGHWAY SAFETY RESEARCH AND VEHICLE DESIGN

EDWARD J. HAUG, L. D. CHEN, YIANNIS PAPELIS and DARIO SOLIS
Center for Virtual Proving Ground Simulation, The University of Iowa
Iowa City, Iowa, USA

Abstract

Fundamentally new ground vehicle virtual proving ground (VPG) capabilities, made possible by the National Advanced Driving Simulator, whose development has recently been completed by the US Department of Transportation and The University of Iowa, are presented. Highway safety research and vehicle and equipment design applications using revolutionary new virtual proving grounds are reviewed and capabilities of the simulator are summarized. Technological developments enabling these new capabilities are presented, including high fidelity real-time dynamic simulation techniques, computer image generation enhancements, precision motion control capabilities, and virtual environment modeling tools.

I. INTRODUCTION

Aircraft flight simulators have evolved to a high level of maturity during the six decades since E.A. Link invented his initial flight training simulator [1]. While flight simulators are most commonly viewed as training tools, significant use is made of advanced flight simulators in the process of aircraft development.

1.1. NASA VERTICAL MOTION BASE FLIGHT SIMULATOR

The most advanced flight simulator shown in Fig. 1 is the Vertical Motion Simulator (VMS) operated by NASA, at its Ames facility in California. This simulator has an ideal motion envelope for flight; namely travels of 60 feet vertical, 40 feet lateral, and 6 feet longitudinal. The frequency response of this flight simulator, however, is far lower than is required for high fidelity ground vehicle simulation.

W. Schiehlen and M. Valášek (eds.), Virtual Nonlinear Multibody Systems, 213–232.

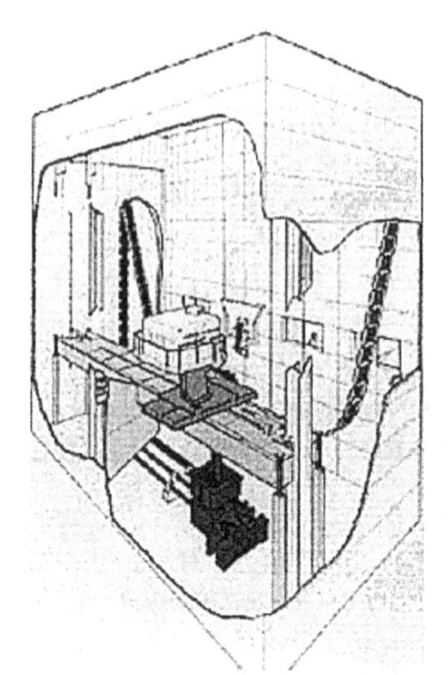

Figure 1. NASA Vertical Motion Simulator

1.2. CONCEPTUAL NATIONAL ADVANCED DRIVING SIMULATOR

The advanced state-of-the-art in aircraft simulation stood in stark contrast with that of ground vehicle simulation until the 1990s. A number of developments occurred during that decade in high fidelity ground vehicle driving simulation, leading to development of the National Advanced Driving Simulator (NADS) shown conceptually in Fig. 2.

Figure 2. Conceptual Rendering of the National Advanced Driving Simulator

This simulator represents vehicle behavior, in response to driver and roadway input, at an engineering level of fidelity that makes it capable of advanced ground vehicle safety research and vehicle system and subsystem design. Its capabilities include (1) a motion envelope of 64 by 64 feet horizontally that creates motion cues to within the threshold of human perception, (2) a wrap-around, high resolution graphics capability, (3) a three-dimensional audio system that provides vehicle and road noises, (4) a vehicle dynamic simulation system that predicts motion of the vehicle at an engineering level of fidelity, and (5) cabs located in the simulator dome, inside which the driver functions provide driver-vehicle interfaces found in typical cars and trucks. All simulator subsystems operate in real time, in response to actions taken by the driver, who is fully immersed in a driving experience that is virtually indistinguishable from performance in an actual vehicle on an actual roadway.

The purpose of this paper is to outline (1) fundamentally new human-centered research that is enabled by the NADS and (2) technological developments that have enabled development and implementation of the NADS, the most advanced ground vehicle driving simulator ever developed. To set the stage for presentation of enabling technologies, experimental capabilities of the NADS are illustrated through description of typical experiments being carried out with the simulator.

1.3. DAIMLER-CHRYSLER DRIVING SIMULATOR

The previously most advanced ground vehicle simulator shown in Fig. 3 is operated by Daimler-Chrysler in Berlin, Germany [2,3]. The motion system consists of a six-degree-of-freedom hexapod, but with somewhat lower frequency response characteristics than are ideal for ground vehicle virtual prototyping. This hexapod is mounted on a single-degree-of-freedom lateral track that moves 20 feet, providing one-dimensional braking or lane change acceleration cues, but not both simultaneously.

Figure 3. Daimler-Chrysler Driving Simulator

1.4. ARMY MILITARY VEHICLE SIMULATORS

The US Army Tank-automotive and Armaments Command (TACOM) operates a family of military vehicle simulators in Warren Michigan, with state-of-the-art motion capabilities appropriate for harsh off-road military vehicle applications. Their most advanced simulator, the Turret Motion Base Simulator shown in Fig. 4(a), supports hardware-in-the-loop simulation with a payload of 25 tons. Both this simulator and the new Ride Motion Simulator shown in Fig. 4(b) have high frequency and acceleration capabilities that are beyond the capability of any other known ground vehicle simulator.

(a) Turret Motion Base Simulator

(b) Ride Motion Simulator

Figure 4. Army Military Vehicle Simulators

2. THE NATIONAL ADVANCED DRIVING SIMULATOR

The National Highway Traffic Safety Administration of the US Department of Transportation and The University of Iowa have created the NADS, to enable research that will enhance US highway safety and vehicle system design [4,5]. The Department of Transportation invested approximately $69 million in development of the NADS system, and The University of Iowa provided $11 million in software and a building to house the NADS. Development of the simulator system was completed in early 2002, and the NADS is now operational on The University of Iowa's Oakdale campus. It is being used for both highway safety research and vehicle system engineering.

The overall goal of the NADS project is to achieve fundamental improvements in highway safety, transportation efficiency, and enhance international competitiveness of US vehicle and equipment manufacturers. Specific objectives of the project include conduct of research using the revolutionary new NADS simulator that will

(1) lead to safer highways, significantly reducing the number of crashes on US highways that currently lead annually to approximately 42,000 lives lost and a cost of $230 billion

(2) enhance vehicle and equipment product development effectiveness, hence international competitiveness, for an industrial sector that accounts for 11 percent of US Gross Domestic Product.

In a related development, the National Science Foundation awarded a multi-university Industry/University Cooperative Research Center (I/UCRC) for Virtual Proving Ground Simulation to the Universities of Iowa and Texas-Austin in 1997. The goal of the Center during its first five years of operation was to create a virtual proving ground capability for the NADS in order to support advanced vehicle engineering. The initial proving ground capability now operating with the NADS was created during the period 1997-2001. In early 2002, the National Science Foundation awarded the Center a second five year grant, to build upon and advance the proving grounds, carrying out advanced vehicle system engineering research with industrial and government sponsors. The universities are cooperating with the National Science Foundation, the Department of Defense, and industry to develop advanced virtual proving grounds and associated simulation research programs using the NADS that will lead to fundamental advances in vehicle and equipment product development effectiveness. Strong support for this research has been obtained from the agricultural, construction, and military vehicle and equipment off-road manufacturing sectors.

The purpose of this section is to provide a concrete definition of the NADS system, including simulation environments that are available to researchers, and to place NADS capabilities in context with existing ground vehicle driving simulators. The NADS, as it currently exists on the University of Iowa campus, is shown in Fig. 5. It contains nine advanced technology modules that interact with the driver to create a highly realistic simulated driving environment, as follows.

Figure 5. National Advanced Driving Simulator

2.1. VISUAL SYSTEM

A revolutionary new Evans & Sutherland Harmony visual system provides the driver with a highly realistic 360-degree field of view, similar to the environment shown in Fig. 6, including rear view mirror images. The visual system database includes a full range of current and new highway traffic control devices (signs and signals), three-dimensional objects that vehicles encounter (animals, potholes, concrete joints, pillars), high-density multiple-lane traffic interacting with the driver's vehicle, common intersection types, and roadway weather environments.

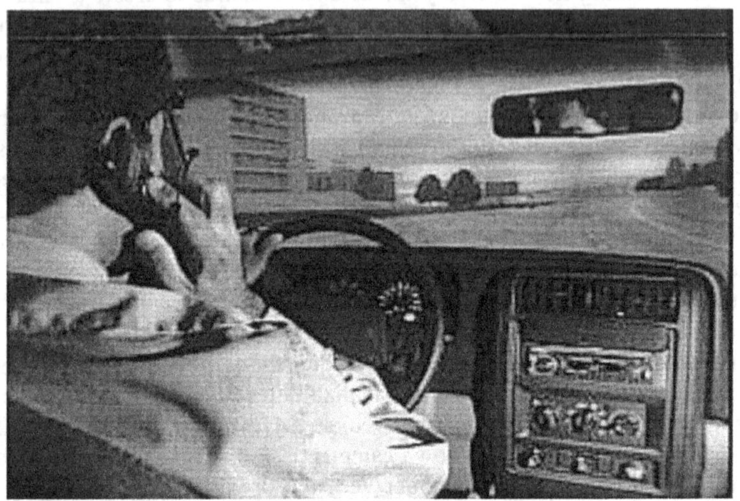

Figure 6. Driver in Simulator

The Harmony image generator (IG) displays imagery on a 7.3-m (24-ft) dome, inside which the driver views the scene from within a cab, as shown in Fig. 6. The IG consists of seven advanced graphics engines (AGEs), each containing four graphics processors (GPs). One of the AGEs contains four render processors (RPs), and each of the remaining AGEs contains three RPs. The RPs feed fifteen output channels of imagery, each driving one view port. Each view port corresponds to a single Barco LCD projector. Three view ports are used for a high-resolution (1.1 arc minutes per optical line) inset in the forward field of view, and the remaining 12 view ports each cover 30 degrees of horizontal field of view. Projectors for the forward field of view display images at 1024 x 768 resolution, whereas projectors for the rear field of view display images at 640 x 480 resolution. The visual system has a sustained polygon throughput of 21,000 polygons per frame, at a 60Hz update rate.

The NADS visual system provides the driver with realistic fields of view, including rearview images that are viewed through the actual vehicle mirrors. Advanced anit-aliasing and texture projection modes, combined with Phong shading for pixel illumination, provide unique capabilities. The actual scenery can vary greatly, depending on project requirements. On-road databases include highways with interchanges and appropriate surrounding scenery, rural roads with curves, urban-arterial roads, residential areas, and industrial parks. All roads are built to highway design standards and include operational traffic lights, signs, and railroad crossings. Off-road environments include geo-specific proving grounds and imaginary locations appropriate for agricultural work. In addition to static scenery, the visual system can, under the control of the scenario system, display numerous moving objects, including vehicles, trucks, motorcycles, pedestrians, and miscellaneous other objects.

2.2. MOTION SYSTEM

The motion system shown in Fig. 5 provides the largest translational motion envelope ever developed for a driving simulator (64 feet square). It also provides 360-degree horizontal turning, pitch and roll, and high-frequency cues that duplicate vehicle motion cues to the driver, over the full range of driving maneuvers. The motion system consists of several assemblies. The X-axis assembly is responsible for moving the X-axis beam along the 64-foot lateral space. Six steel belts, each driven by four 100HP A/C motors that are controlled by variable frequency drive (VFD) controllers, are responsible for generating the force necessary for X-axis motion. One end of each belt is attached to one side of the X-beam, and the other end wraps around a drum that is driven by two VFDs. The belt is tensioned across the simulator bay, where it wraps around the second pair of VFDs located on the other side of the bay, eventually attaching to the opposite side of the X-axis beam. The two pairs of VFDs work in a push-pull configuration to move the beam as commanded. All six belts work cooperatively to move the X-axis beam.

Located on top of the X-axis beam is a single steel belt with its own set of two pairs of VFDs. This belt is responsible for the Y-axis motion of the carriage, in a push-pull configuration similar to each of the six belts on the X-axis. Mounted on top of the carriage on the Y-axis is a six hydraulic actuator hexapod, on top of which is a ring that can yaw +/- 330 degrees about the center point. The dome, along with projectors, the vibration actuators and vehicle cab, are mounted on top of the yaw ring. The nine degree-of-freedom motion system can exert up to 0.6 g along the X and Y axes and 1.0 g along the Z axis. The turntable and hexapod can exert a maximum rotational acceleration of 120 degrees $/\text{sec}^2$ along the pitch, yaw, and roll axes. Vibration actuators provide a maximum of +/- 0.2 inch displacement, at 35 hertz, with a maximum velocity of +/- 8 inches/sec.

2.3. CONTROL FEEL SYSTEM

Control feel systems for steering, brakes, clutch, transmission, and throttle provide the driver with a realistic feel of the road and vehicle system response to driver inputs, over the full vehicle maneuvering and operating range. The control feel system is capable of representing automatic and manual control characteristics such as power steering, existing and experimental drivetrains, anti-lock braking systems, and headway control systems.

2.4. AUDIO SYSTEM

The audio system provides realistic three-dimensional sound sources that are coordinated with other sensory systems. The audio database includes sounds emanating from current and new design highway surfaces, contact with three-dimensional objects that vehicles encounter (potholes, concrete/tar joints, pillars, etc.), high-density multiple-lane traffic, vehicle operation (engine, brake, and wind noise), and the roadway due to changes in the weather conditions.

2.5. VEHICLE DYNAMICS

High fidelity vehicle dynamics software [6-9], provided by The University of Iowa, accurately represents vehicle motions and control feel conditions in response to driver control actions, road surface conditions, and aerodynamic disturbances. Vehicle dynamics models of the scope shown in Fig. 7 simulate light passenger cars and trucks, heavy trucks and buses, and off-road wheeled and tracked vehicles. The models encompass normal driving conditions and extreme maneuvers encountered during crash avoidance situations, including spinout and incipient rollover.

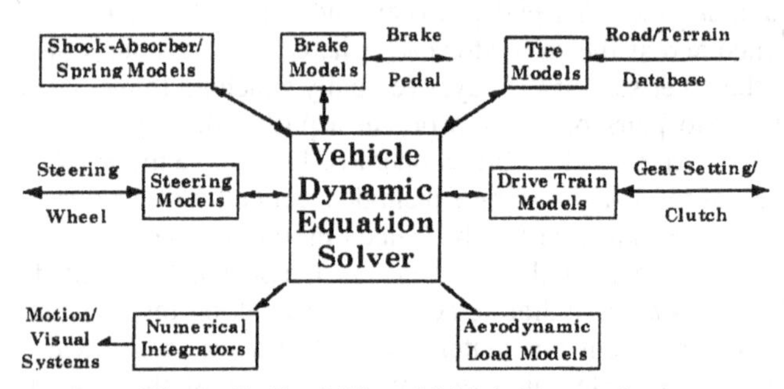

Figure 7. Real-Time Vehicle Dynamics

2.6. SCENARIO AUTHORING

The NADS scenario control and authoring module [10,11] developed by The University of Iowa, shown schematically in Fig 8, simulates traffic elements in real time such as passenger cars, motorcycles, trucks, buses, rail-based vehicles and pedestrians, each with autonomous, reactive behaviors.

Figure 8. Autonomous Vehicle Control

Specialized runtime agents simulate operation of traffic control devices, weather effects including wind and fog, and ambient traffic around the simulator driver. A suite of Windows NT-hosted graphical tools allows rapid development of scenes and scenarios, without explicit programming, for use by experimenters to develop and test scenarios before proceeding to the NADS facility.

Scenario development tools allow comprehensive control of the virtual environment, based on a rich set of conditions. Autonomous and scripted traffic can be utilized to accommodate a broad range of project requirements. Scene authoring is supported by a tile mosaic tool (TMT) that allows combination of predefined components of virtual environments. A library of approximately 70 tiles is currently available, and additional tiles are continually being added. The TMT and ISAT tools are closely linked and provide unique capabilities for the rapid development of virtual environment and associated scenarios.

2.7. VEHICLE CAB SYSTEM

Vehicle cabs in which the driver functions consist of actual vehicle cabs, configured to fit within the visual dome on the motion system. Four cabs were initially delivered with the NADS system, including cabs for a Chevrolet Malibu, Ford Taurus, Jeep Cherokee, and class-8 Freightliner truck. The cabs have a full range of standard, optional, and new design vehicle instrumentation interfaces. Cabs allow rapid interchangeability, in order to meet an unprecedented high experimental efficiency during NADS operation.

2.8. COMPUTER SYSTEM

A system of dozens of computers controls all aspects of NADS operations. Databases defining vehicle characteristics, the visual driving environment, audio characteristics of the environment, and roadway characteristics that influence performance of the vehicle are highly integrated, to enable the researcher to create and supervise controlled experiments.

2.9. SIMULATION DEVELOPMENT MODULE

A NADS-like simulator, called the simulation development module for off-line development of experiments and virtual environments, contains all the basic elements of the NADS, except motion. This very capable unit operates separate from the motion-based NADS simulator, thus enabling experimenters to cost-effectively preview scenarios and drive through scenes to assure themselves that the experiment is well structured before carrying out experiments on the full NADS system.

3. NADS-BASED VIRTUAL PROVING GROUND EXPERIMENTS

Since its initiation of operation in early 2002, three sets of experiments have been carried out to initiate highway safety and virtual proving ground engineering applications. The first set of experiments involved verification of simulator performance in highway safety crash avoidance limit maneuvers involving spin-out and related vehicle maneuvers in the NADS, shown in Fig. 9, that represent extreme limits of vehicle performance with interaction of the driver. Qualitatively excellent results have been obtained, correlating well with quantitative measures of performance of the vehicle. These results provide the foundation for a current set of highway safety experiments being carried out by The University of Iowa for the US National Highway Traffic Safety Administration.

Figure 9. Highway Safety Crash Avoidance Limit Maneuvers

The initial virtual proving ground experiment involved linking the Army Ride Motion Simulator shown in Fig. 10(a) with the NADS shown in Fig. 10(b), to demonstrate the feasibility of high fidelity engineering simulator interaction, coupled with computer-aided engineering design change analysis and verification with the driver in the loop. Army and NADS test drivers operated high fidelity simulated vehicles on a simulated Army proving ground, including the capability to see each other's vehicle and interact, as shown in Figs. 11(a) and 11(b). In addition to achieving excellent network-based interaction of engineering fidelity simulations, design characteristics of the vehicle functioning in the Ride Motion Simulator in Fig. 11(a) were modified in a CAE environment as the simulations occurred, to evaluate the driver's interaction with nominal and modified vehicle designs. Positive results indicate the feasibility of interactive simulation of vehicle designs with the driver in the loop, greatly reducing the time required for vehicle system development.

(a) Army Ride Motion Simulator (b) National Advanced Driving Simulator

Figure 10. Networking Advanced Driving Simulators

(a) Army Driver In Ride Motion Simulator (b) Driver In NADS Simulator

Figure 11. Interactive Simulation with Engineering Fidelity Simulators

A second virtual proving ground experiment that was carried out in early April of 2002 focused on validation of a NADS-based agricultural virtual proving ground. This project, carried out with Deere & Co., involved a John Deere tractor shown in Fig. 12(a) with the NADS simulator cab shown in Fig. 12(b).

The tractor and a proving ground were modeled at an engineering level of fidelity, and experiments were carried out on the actual proving ground shown in Fig. 13(a). The model of the tractor was exercised, with the operator in the loop, on the simulated proving ground in the NADS, as shown in Fig. 13(b). Initial qualitative and quantitative performance analysis indicates that an adequate level of correlation exists between the actual and virtual proving ground to form the basis for equipment design and evaluation using the virtual proving ground.

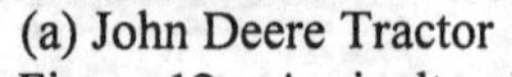

(a) John Deere Tractor (b) NADS Simulator Cab

Figure 12. Agricultural Tractor and NADS Simulator Cab

(a) Tractor Operator On Test Course (b) Tractor Operator In Simulator

Figure 13. Validation of NADS-Based Agricultural Proving Ground Simulation

4. TECHNOLOGIES ENABLING VEHICLE VIRTUAL PROVING GROUNDS

Technologies that enable design level of fidelity virtual proving grounds are summarized in this section. They range from computational methods for high fidelity dynamics and tire-road surface/terrain interaction to synthetic environment modeling for on- and off-road simulation to managing databases and synthetic environment modeling tools.

4.1. COMPUTATIONAL METHODS FOR VEHICLE DYNAMICS AND TIRE-ROAD SURFACE/TERRAIN INTERACTION

4.1.1. High Fidelity Dynamics

Even though the human can respond cognitively only to cues in a frequency range up to 2-5 Hz, transient events lead to the requirement that actuators and controllers, with which the driver interacts, must accurately predict transient dynamic effects at high frequency. For example, large transient forces that can arise in operation on rutted roads require high frequency compensation in the steering system, in order to avoid serious problems with the driver's ability to control the vehicle. Such control systems often respond with transients involving frequencies up to 100 Hz.

Similarly, anti-lock brake and active suspension systems experience response of the wheel assembly up to 20 g, leading to transient effects requiring controller compensation in excess of 80 Hz. Even though hydraulic pulse rates in anti-lock brakes are typically below 7 Hz, the reaction times of controllers and actuators are typically below 20 milliseconds. Since such components interact to first order with wheel spin, which contains stiff transients, one millisecond or less numerical integration step size is required in simulation with explicit numerical integrators to predict anti-lock brake subsystem performance that influences the driver's ability to control the vehicle. This translates to a frequency content above 80 Hz. Similarly, active suspension performance in the presence of potholes and short wavelength roadway undulations such as rumble strips requires that frequencies of 50 Hz or more be accounted for in designing the control system to maintain vehicle control by the driver.

Automated Highway System concepts of the future place an even greater demand on high simulation fidelity for support of device design to interact effectively with the driver. The automotive design industry will accept a driving simulator for use in subsystem design on a virtual proving ground only if extremely high fidelity, equivalent to a frequency response up to 100 Hz, is accounted for in subsystem simulations.

4.1.2. Real-Time Modeling And Simulation Methods

Innovative vehicle design technologies based on advanced control algorithms are commonly used in modern vehicles to achieve optimum power transfer from mechanical and electromechanical powertrains through the wheels to the road. The interaction between the tires and the road is governed by multiple sources of high frequency excitation coming from roughness of the road profile, tire flexible modes of vibration, and powertrain actuator response. In order to resolve this high frequency content, high fidelity dynamic algorithms, coupled with modern and efficient numerical integration techniques [12], are necessary to guarantee the stability of the numerical solution used to predict motion cues for the driver within the simulator environment. In emerging vehicle subsystems such as hybrid-electric powertrains, the vehicle system engineer will be required to conduct human factors and engineering-based studies on the interaction of such subsystems with the driver. These systems are very demanding in terms of the numerical approach to accurately predict their transient performance.

For the foregoing reasons, advanced dynamics formulations and implicit and multirate numerical integration techniques have been developed and tested for use in the NADS. Advanced numerical methods presented in a Special Session of the 1998 ASME Design Automation Conference [13,14] provide some of the capability required for simulating modern high frequency vehicle subsystems for design in a virtual proving ground.

In order to determine what vehicle modeling and simulation features are normally used by vehicle manufacturers during analysis and design, NADS and LMS International held a two-day workshop in October of 2001. During this workshop, it was found that the on-road vehicle developer community was calling for better representation and integration of flexible bodies into the vehicle models, and for higher fidelity bushings, shock absorbers, and tire component models. Some manufacturers do not even consider vehicle models without bushings for their dynamic evaluations. For this reason, a task to develop numerical techniques to efficiently deal with bushings in vehicle simulation was started. Implicit integrators with exact Jacobian matrices for bushing elements were formulated and are currently being implemented to address the real-time simulation of vehicle models containing bushings. Body flexibility, normally left for non-realtime simulation analysis, is being included into current simulation codes at NADS to allow for body compliance that affects vehicle performance and its interaction with the driver to first order.

4.1.3. Tire-Road/Soil Interface

To meet virtual proving ground goals of simulating both on- and off-road vehicles, at a design level of fidelity, a capability that includes not only driving vehicles over hard rigid soils but also over soft deformable soils must be provided. One of the key core technology developments required to achieve this capability is significantly enhanced tire-soil interaction modeling. This technology requires development in three main thrusts; (1) flexible tire-hard soil interaction, (2) rigid tire-deformable soil interaction, and (3) flexible tire-deformable soil interaction. A similar categorization of track-road/soil interface modeling will be required, but has not yet been formulated. Focusing the technology development in this fashion helps identify critical elements during vehicle simulation over highways and compact rural roads, mild off-road applications, and severe soft terrain off-road applications.

Within thrust (1), kinematic tire information calculated by the multibody dynamics module is coupled with terrain profile information to compute forces generated at the tire-road interface. Longitudinal and lateral forces at the tire-road interface are computed, based on the slip developed in the respective directions. Normal forces are computed using geometrical information that is given in terms of position and velocity of the wheel hub and shape of the terrain profile under each wheel, and combining it with stiffness and damping of the tire. The approach in this thrust is the most commonly used for both on- and off-road vehicle simulation, due to the large percentage of applications for which the rigid terrain assumption is applicable. Also, it involves the tire but not the soil, which is more difficult to represent. The complexity of the models under this thrust is a function of the amount of compliance and vibration modes in the tire. Its compliance is due to the amount of deformation, and the existence of vibration modes depends on the size and type of tire. Figure 14 shows three contact models for thrust (1). Single point contact queries the terrain at the point directly below the wheel attachment point. The intersection between the tire profile and the terrain determines the tire deformation length that is used to compute the normal force. This approach is widely used for on-road flat surfaces, but is not recommended for off-road applications. The distributed contact model queries the terrain in multiple points under the wheel attachment point to determine the shape of the contact. An averaging technique is then used to calculate the effective contact force between the tire and the terrain. The number of segments in the model is chosen based on the roughness and spatial frequency of the terrain. The full 2D model is based on a finite element representation, using beam elements for both radial and tangential members. This model allows for a distributed stiffness parameter representation of the tire that can be used to improve the single-parameter model used in the two previous models. It also includes tire inertial effects, which are critical for many applications.

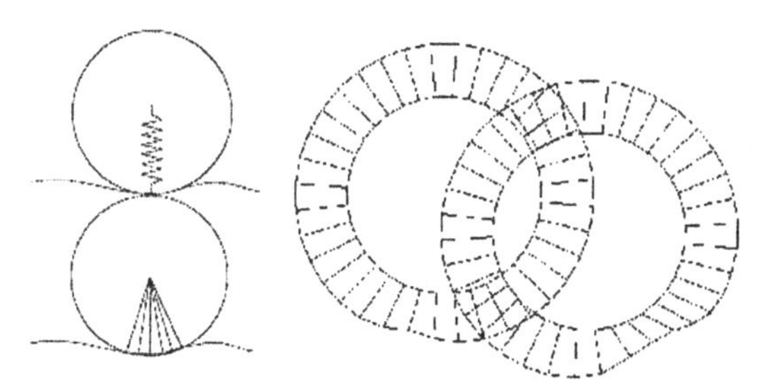

Figure 14. One Point, Distributed, And Full 2D Contact Models

Thrust (2) concentrates on applications for which tire deformation is small, compared to soil deformation. Analytical soil mechanics equations are used to compute tire forces due to deformation. This is made possible by the simple shape of the undeformed tire, which greatly simplifies the calculation. Figure 15 shows a schematic of a rigid wheel on a deformable terrain. Variations of this approach include the addition of a tire deformation section of known geometry. This known shape allows for integration of the normal pressure exerted by the soil on the tire, using analytical expressions, and provides a correction for tire deformation that extends the applicability of this approach to relatively harder soils or tires with lower pressures. This approach is also shown in Fig. 15.

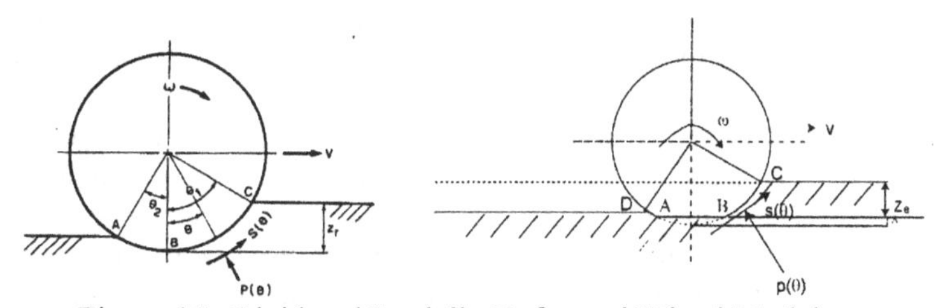

Figure 15. Rigid and Partially Deformed Wheel Models

Finally, thrust (3) addresses the more general problem of deformable tire and soil. A general nonlinear finite element approach is applied to solve for forces generated within the tire-soil interface and the resulting terrain profile deformation, as shown in Fig. 16.

On going activity to develop a better predicton capability for off-road vehicle simulation is targeting the soil formulations needed in the finite element representation for different soil types. This high-fidelity model is being used to improve the understanding of the traction phenomena and to develop better real-time models. This thrust also supports determining the trafficability of off-road vehicles over a given terrain.

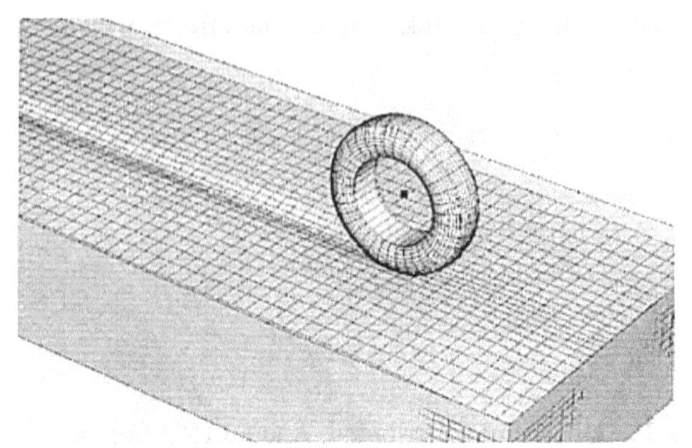

Figure 16. Three Dimensional Tire and Soil Finite Element Model

4.1.4. Vehicle System/Subsystem Simulation For Design

Due to the increasing complexity of vehicle and component design, as well as the nature of technologies that are being integrated into modern vehicles, the traditional design approach that depends on building and testing many physical prototypes is not cost effective for vehicle product development. It has become clear to developers throughout the automotive industry that the use of computer simulation as a tool to aid the designer is no longer an option, but a necessity. The simulation-based-design approach has varying degrees of implementation that range from modeling and simulating components, such as antilock brakes and headway control devices, to modeling, simulating, and integrating an entire vehicle system [15]. Depending on system complexity and size, this integrated environment requires a large amount of computational resources to produce a solution in a timely manner. Complete vehicle systems are now being used to reduce the number of physical tests in the design optimization process. In off-road applications, recent developments in tire-soil modeling and simulation, coupled with integrated vehicle simulation environments, have enabled a higher degree of dynamic predictability for complete vehicle systems. This new capability is being used with good results on rigid surfaces, but further development is necessary to incorporate soil compliance more effectively. Figure 17 shows a settling run of a High Mobility Multipurpose Wheeled Vehicle model simulated in DADS, integrated with four high-fidelity tires modeled using the ABAQUS software.

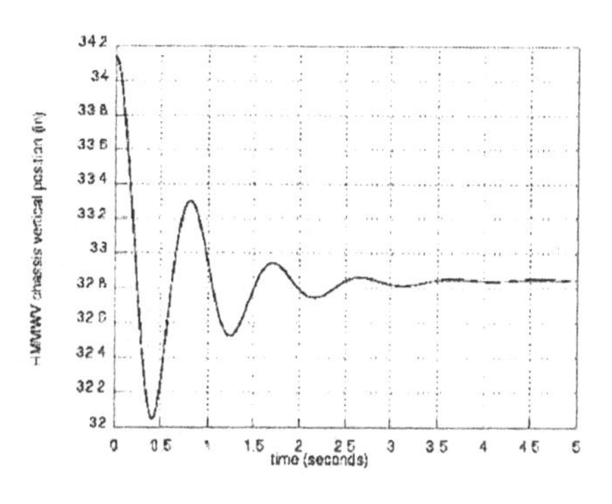

Figure 17. Integrated High-Fidelity Simulation Environment

A vehicle system and subsystem modeling environment for construction equipment was created for virtual proving ground applications [16]. More recently, a complete VPG was designed, implemented, and used to conduct experiments in the NADS for agricultural tractors.

4.2. SYNTHETIC ENVIRONMENTS

The creation of realistic synthetic driving environments poses many challenges [10]. In particular, driving simulation requires scene databases of large geographic regions with properly modeled roads, high-fidelity terrain, natural foliage, and appropriate cultural features. Because of the nearness of surroundings, synthetic driving environments must be modeled at a much higher resolution than flight terrain databases. Synthetic environments must also support the creation and execution of scenarios involving vehicular and pedestrian traffic that meet experimenter requirements [11]. The NADS synthetic driving environment comprises multiple correlated databases, each providing data about the same synthetic environment, but in views that are specialized to the interrogating subsystem. Specialized views are provided to support image generation, autonomous traffic elements, audio cue generation, vibration special effects, and tire-road interface.

On- and off-road applications have differing database requirements. For example, road surfaces are typically hard, allowing use of static database structures for roadway networks. In contrast, off-road terrain must be able to support real-time terrain updates corresponding to tire rut generation and other ground-vehicle interactions. Furthermore, road surfaces are typically much smoother and more well defined than off-road terrain. This results in the use of different representations for on- and off-road surfaces. On-road modeling is well developed. Accurate real-time modeling of dynamic off-road synthetic environments is less well developed. The following briefly summarizes features of NADS on- and off-road synthetic environments being created.

4.2.1. On-road synthetic environment

The most complex view of the NADS synthetic environment database is the Virtual Roadway Environment Database (VRED). VRED contains information about the logical and physical layout of roadways and can manage real-time interrogation of static and dynamic objects. Logical roadway information encompasses enough detail to support navigation and conforming to traffic rules. It includes the number, direction, and type of lanes, including standard, automated highway systems, and high occupancy vehicles; the nature of lane separators; shoulder information; road markers; and a complete intersection model that supports an arbitrary number of incident roads.

The physical roadway representation incorporates a hybrid model that uses three-dimensional splines with two-dimensional continuity for the roadway surfaces and a uniform grid of elevation and material posts used for intersections, special effects, and other non-standard roadway terrain. Any number of disjoint elevation grid maps can be included in a synthetic environment, and they can overlap the splinar representation. Road holes, seams, or other localized perturbations can be modeled using overlapping terrain. The splinar representation utilizes two independent splines, one for the longitudinal roadway axis and one for the lateral axis. The lateral spline can change at any point along the longitudinal axis, allowing modeling of arbitrary variations in road geometry. To support real-time interrogation of the terrain elevation by the tire module, VRED uses specialized data structures that utilize two-dimensional hashing algorithms to quickly access the spline or grid parameters.

4.2.2. Off-road synthetic environment

The terrain database component of the NADS synthetic environment provides a high-resolution representation of ground surface characteristics, including elevation, surface type, roughness, and friction, for both on-road and off-road applications. Off-road applications involving driving over extended regions of soft soil present severe modeling challenges. First, the driveable area that must be represented at high resolution generally becomes greater when vehicles are permitted to leave the confines of well-defined road paths. Second, off-road applications involving agricultural construction, and military vehicles often involve dynamic interaction of wheels and soft soil, thus making it necessary to support real-time update of the terrain database, as determined by the tire-soil interaction subsystem. Third, some applications include significant interactions between the ground and parts of the vehicle other than the tires, such as when rocks or bumps impact the vehicle undercarriage. Thus, off-road synthetic environments must support detection and characterization of geometric contact between vehicles and the ground and other objects.

These requirements make real-time modeling of off-road environments a considerable challenge. It is not feasible, for visual display purposes, to represent the entire database at the highest required resolution. Researchers are developing dynamic terrain modeling techniques based on variable resolution meshes that may be updated in real time as terrain updates are computed. The techniques being developed require tight integration of the terrain and visual databases, in order to achieve real-time display of terrain updates at appropriate fidelity.

4.2.3. Management of the synthetic environment

An arbitrary number of static and dynamic objects are managed in real time within the synthetic simulator environment (see Fig. 8). Static objects that are used to model traffic signs, poles, vegetation, structures, or any other necessary obstruction can be associated with a particular location on a road, or on off-road terrain. Dynamic objects represent participants that can move freely. VRED is responsible for storing the state of moving objects and providing, in real time, information about their location with respect to a specialized network coordinate system. For on-road applications, this coordinate system utilizes the road, lane, and parameterized distance along a road as the three coordinates for locating objects in space. Autonomous object behaviors use this coordinate system, because it is much simpler (and computationally cheaper) to compare the relative position of vehicles along a roadway by using the parameterized distance, as opposed to using vector-based methods. For off-road applications, the situation is more complex and remains to be investigated.

4.2.4. Authoring synthetic environments

The creation of synthetic on-road driving environments containing road networks that are consistent with civil engineering standards is a difficult process that requirs substantial effort by scene database specialists. Researchers are developing tools to support rapid construction, modification, refinement, and debugging of scene databases.

5. CONCLUSIONS

A high-fidelity real-time simulation capability and emerging ground vehicle driving simulators outlined in this paper provide the foundation for highway safety research and vehicle subsystem and system design, at an engineering level of fidelity. Together with the virtual proving ground simulation capability enabled very recently by the National Advanced Driving Simulator for real-time interaction of the vehicle and driver, a revolutionary new environment is now available for designing and testing vehicle subsystems and components. Realistic load histories obtained from vehicle system virtual proving ground simulation can now be used to design and test component hardware in the laboratory. This new capability serves two purposes; (1) it improves component design and testing by generating realistic duty cycle information early in the design cycle, and (2) it provides hardware measurements to validate on-line models to improve overall simulation accuracy and realism.

6. References

1. Link, E.A. (1930) Combination Training Device for Student Aviators and Student Entertainment Apparatus, *US Patent Specification 127,82.*
2. Drostol, J., and Panik, F. (1985), The Daimler-Benz Driving Simulator-a Tool for Vehicle Development, *SAE Paper No. 850334.*
3. Kading, W., and Hoffmeyer, F. (1995) The Advanced Daimler-Benz Driving Simulator, *SAE Paper No. 950175.*
4. Haug, E.J., et. Al (1990) Feasibility Study and Conceptual Design of a National Advanced Driving Simulator, *Report DOT-HS-807586*, National Highway Traffic Safety Administration, Washington, DC.
5. Haug, E. J., Cremer, J., Papelis, Y., Solis, D., and Ranganathan, R. (1998) Virtual Proving Ground Simulation for Vehicle Design, *ASME Design Automation Conference, Special Session on Virtual Proving Ground Simulation.*
6. Bae, D.S., and Haug, E.J. (1987) A Recursive Formulation for Constrained Mechanical System Dynamics: Part I, Open Loop Systems, *Mechanics of Structures and Machines*, **15**, 3, 354-382.
7. Bae, D.S., and Haug, E.J. (1987) A Recursive Formulation for Constrained Mechanical System Dynamics: Part II, Closed Loop Systems, *Mechanics of Structures and Machines*, **15**, 4, 481-506.
8. Bae, S.D., Kuhl, J.G., and Haug, E.J. (1988) A Recursive Formulation for Constrained Mechanical System Dynamics: Part III, Parallel Processor Implementation, *Mechanics of Structures and Machines*, **16**, 2, 249-270.
9. Tsai, F.F., and Haug, E.J. (1991) Real-Time Multibody System Dynamic Simulation; Part II: Parallel Algorithm and Numerical Results, Mechanics of Structures and Machines, **19**, 2, 129-162.
10. Cremer, J., Kearney, J., and Papelis, Y. (1996) Driving Simulation: Challenges for VR Technology, *IEEE Computer Graphics and Applications*, 16-20.
11. Cremer, J. Kearney, J., Papelis, Y., and Romano, R. (1994) The Software Architecture for Scenario Control in the Iowa Driving Simulator, *Proceedings of the 4th Computer Generated Forces and Behavioral Representation Conference*, 373-381.
12. Haug, E.J., and Deyo, R.C. (eds.) (1991) *Real-Time Integration Methods for Mechanical System Simulation*, Springer-Verlag, Heidelberg.
13. Serban, R., and Haug, E. J. (1998) Globally Independent Coordinates for Real-Time Vehicle System Simulation, *ASME Design Automation Conference, Special Session on Virtual Proving Ground Simulation.*
14. Solis, D., and Schwarz, C. (1998) Multi-Rate Integration in Hybrid-Electric Vehicle Virtual Proving Grounds, *ASME Design Automation Conference, Special Session on Virtual Proving Ground Simulation.*
15. Haug, E.J., Choi, K.K., Kuhl, J.K., and Wargo, J.D. (1995) Virtual Prototyping Simulation for Design of Mechanical Systems, *Journal of Mechanical Design*, **117**, 63-70.
16. Grant, P., Freeman, J. S., Vail, R., and Huck, F. (1998) Preparation of a Virtual Proving Ground for Construction Equipment Simulation, *ASME Design Automation Conference, Special Session on Virtual Proving Ground Simulation.*

DESIGN OF REDUNDANT PARALLEL ROBOTS BY MULTIDISCIPLINARY VIRTUAL MODELLING

Z. ŠIKA, M. VALÁŠEK, V. BAUMA, T. VAMPOLA
Department of Mechanics
Faculty of Mechanical Engineering
Czech Technical University in Prague
Karlovo nám. 13, 121 35 Praha 2, Czech Republic
E-Mail: sika@felber.fsik.cvut.cz, valasek@fsik.cvut.cz

1. Introduction

The paper deals with the description of design methodology for redundant parallel robots based on multidisciplinary virtual modelling. The redundant parallel robots means redundantly actuated parallel robots. The parallel robots have many advantages as low moving masses, higher stiffness of truss structure, all drives on the frame, but they suffer from many problems like appearance of singularities and thus smaller workspace, collisions of links. These drawbacks of parallel structures can be removed by the principle of redundant actuation [4, 1]. This means that the platform is supported and driven by more bars with drives than the necessary number of degrees of freedom (DOF). This principle not only deletes the singularities from workspace as more combinations of links in number of DOF are not simultaneously in singular positions, but it brings further advantages, especially increased and more uniform dynamic capabilities, stiffness, accuracy.

The design of redundant parallel robots is an example of particularly complex design problem. The mutual dependencies of all parameters and components are especially large. The successful design methodology is possible only using virtual models and design complexity decomposition.

The used virtual models cover both mechanical including control and geometric properties. During the design there are conflicts between geometrical dimensions of robots and corresponding mechanical properties. The conflict includes collisions of robot links, non-existence of geometrical solutions of kinematics and insufficiency of mechanical properties like stiffness, dynamics, dexterity, accuracy etc. The design process has been resolved into three hierarchical levels. Each of these levels is characterized by certain problem simplification and special design conflict which should be resolved within the level.

W. Schiehlen and M. Valášek (eds.), Virtual Nonlinear Multibody Systems, 233–241.

2. Design Methodology

The design methodology of redundant parallel robots [1] follows the general engineering design methodology described in [2]. The design process is a hierarchical process as the technical products consist of hierarchy of components. The design process repeats the same outline at each design level. It consists of three nested loops:

- Selecting the lower level components from which the solution will be built.
- Proposing the structural arrangement of the selected components.
- Calculating parameter values so that the solution is complete, i.e. all requirements and constraints are fulfilled.

These three nested loops of component choice, structural arrangement and parameter choice also correspond to the nested design iterations and nested design optimisation. The component choice in the case of parallel robots means the decision about the fully parallel or hybrid concept, about the redundant/non-redundant concept, about the kind of link actuators, about the planar/spatial version of joints, about the kind of actuators (electrical/hydraulic, moving screw/direct electrical drive etc.), about the way of measurement etc. The structural design means the decision about the considered shape of components, about the way of their interconnections etc. After the structural design all decisions are transformed into numerical values of parameters. Their values are evaluated in terms of requirements and constraints.

The solution uses parts of mechatronic design methodology [3]. The most important methodological approach is the search for **Ideal Final Solution** and **Conflict Resolution** instead of conflict compromise that solves the given problem despite different conflicting constraints. This approach is looking for solutions that have advantageous values in criteria previously conflicting instead of just looking for tolerable compromise. The concept of redundant actuation is an example of such solution that keeps all advantages of parallel structures, removes problems with singularities and even improves the variations of main mechanical properties [4 - 6]. Certainly such principle is not found for each design task but in many cases just the idea of looking for ideal solution helps to overcome the local compromises. All steps of design and optimisation of robot properties has to be driven by concrete technological target of future machine from the very beginning state of design process. On the other hand "space of considered possibilities" should be held as wide as possible.

Specifically in case of redundant parallel robots the design process has been resolved into three hierarchical levels. Each of these levels is characterized by certain problem simplification and **special design conflict** which should be resolved within the level. (Quasi)optimum variants obtained as the best results of foregoing design optimisation level serve as starting variants for optimisation within the consecutive level. Certainly in the robot design there is mutual dependence between all parameters and thus feedback between levels is necessary. However mentioned decomposition into three subsequent design conflicts enables reasonably to simplify the design process:

- **Level of Geometric Conflicts:** Important properties of robot being designed besides the geometric requirements of DOFs, workspace and dexterity are represented by simple geometric conditions. For example, requested limits of stiffness and modal properties are taken into account by some conditions for robot leg thickness, build-up spaces for real joints or robustness of machine frame.

Optimisation of robot structure and dimensions try to harmonize several geometric requirements that are on the first try contradictory:

1. Workspace without collisions and kinematic singularities should be maximized.
2. Ratio between total build-up space of machine and useful (technological) workspace should be minimized.
3. Dimensions or build-up spaces of important machine elements should be sufficient.
4. Dexterity should be optimised (maximization and uniformity in workspace).

- **Level of Structural Conflicts:** The structural conflict comprehends more precisely formulated conflict between structural (stiffness and modal) properties of the whole machine and accessible dynamics (velocity, acceleration, jerk) of robot end-effector. Mutual interrelations of these properties are very complex and in addition other important aims of machine designers (like accuracy for higher speeds of operations) are heavily influenced by them. Basic requirements are as follows:
  1. Accessible dynamics (velocity, acceleration, jerk) of robot end-effector should be maximal and uniform for representative trajectories within the workspace.
  2. The first eigenfrequencies of the robot should be as high as possible and uniform for all possible robot positions in the workspace.
  3. Cumulative stiffness measured on the end-effector should be maximal and uniform for all possible robot positions in the workspace.
- **Level of Actuation Conflicts:** Behaviour of the whole machine depends on dynamic interactions among mechanical parts, electrical or hydraulic actuators and feedback control loops of actuators. Simulation of complex mechatronic system must be performed in order to predict potential problems arising here. Thoroughgoing fulfilment of previous two design levels is crucial for efficiency of final complex tuning. Basic requirements are as follows:
  1. Control loops must be stable without troublesome vibrations.
  2. Control loops of actuators must be tuned in order to make drives as dynamic as possible. Technological times of production should be minimized.
  3. Energy consumption of drives necessary for production should be minimized.
  4. Accuracy for high speed operations should be maximized.

The applied design methodology is heavily based on the efficient computational tools for mapping robot design parameters into design criteria (requirements and constraints) and following **multiobjective optimization** of the robot parameters like dimensions, drive parameters, control parameters. For mechanical properties there have been developed computational tools based on global dynamics [7]. There are also very important visualization tools especially for multiobjective design. For the design of redundant parallel robots the following computational tools are used:

1. **Workspace, dexterity and collisions evaluation.** The crucial property of the robot is the geometric and kinematic synthesis. The size of workspace limited

by geometric and collision constraints are evaluated and mapped in each position. The efficient analysis of collisions of arbitrary bodies has been implemented. The basic entities for the collision detection are general cuboids. The complex bodies are replaced-approximated by the composite bodies composed from many cuboids. The problem of collisions of cuboids has been solved in two stages. The first fast step evaluates potential possibility of collision. The second stage is initialised whenever the collision cannot be excluded. The penetrations of edges of one body and surfaces of second body have been detected during the second detailed stage of analysis. Collision can be visualized in 3D or 2D in basic planes of coordinate system. Besides that the occurence of singularity positions and generally the manipulability of the robot are evaluated.

2. **Stiffness and eigenfrequency (modal) evaluation.** The accuracy is dependent on the robot stiffness. There are evaluated the maps of robot stiffness and eigenfrequencies.
3. **Dynamic capability evaluation.** The limitations of dynamic capabilities of drives are transformed into the areas of accessible accelerations and velocities at the points of selected trajectory using methods of global dynamics [7]. Choosing several trajectories like straight lines with different slopes across the workspace and the circles with different radii enables to map the overall dynamic capabilities of the robot.
4. **Force transmission evaluation.** The accessible accelerations and velocities from previous step are achieved through particular driving forces. Their determination due to the actuator redundancy is not straightforward and simple [8]. The driving forces and corresponding reaction forces in joints and structural elements are transmitted through the robot structure and this force transmission and distribution is important for dimensioning of robot structural elements.
5. **Kinematic and elastostatic accuracy evaluation.** The accuracy is essential robot property. It is influenced by the properties of encoders and by the robot stiffness in relation to the external applied forces.
6. **Control design.** The control design is done by the methodology design by simulation.
7. **Overall simulation.** The designed properties are verified within overall simulation where especially the multibody, elastic and control properties are investigated in deep interaction.
8. **Multiobjective optimization.** The above listed performance criteria as well as others are subjected to the multiobjective optimization using the design parameters of the robot. The Pareto sets of conflicting criteria are computed and visualized.

3. Design Case Study

The case study is devoted to the investigation of improvement of mechanical properties of Sliding Delta robot (*Figure 1*), also called Uran. The robot Octaslide (*Figure 3*), the more complex (6 DOF motion of end-effector) modification of original robot has been designed as well. The main potential of improvement is based on the application of principle of redundant actuation. It brings for Sliding Delta mainly improvement of stiffness and dynamics, for Octaslide especially the elimination of singularities. The design was performed within many iteration loops. It is difficult to reconstruct the content of all of them in details. However all three nested levels of design conflicts were investigated and solved as follows. There were used computational tools mentioned above.

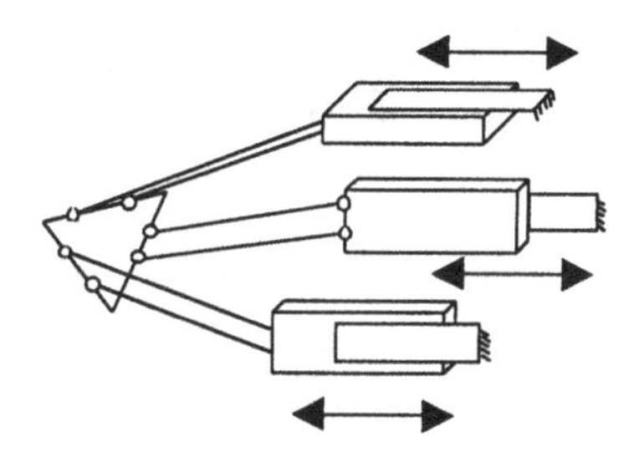

Figure 1. Original Sliding Delta (Uran) robot with sliding joints

4. Geometric Conflicts and their Solution

Initially the original structure from *Figure 1* had been extended into the redundant version on *Figure 2a.* Then the structural properties of designed robot were represented by simple geometric conditions. The critical value was the diameter of the legs in order to achieve reasonable stiffness. Simultaneously the lengths of legs had to be kept in values comparable with non-redundant version again due to comparable stiffness. The critical issue was the computation of accessible workspace due to the collisions and improvement of dexterity. The problems of finding parameters (dimensions of platform and legs) for simultaneously good workspace and dexterity could not been resolved on the level of parameter values and have led finally to the modification of the structure from initial version (*Figure 2a*) into the final one (*Figure 2b*).

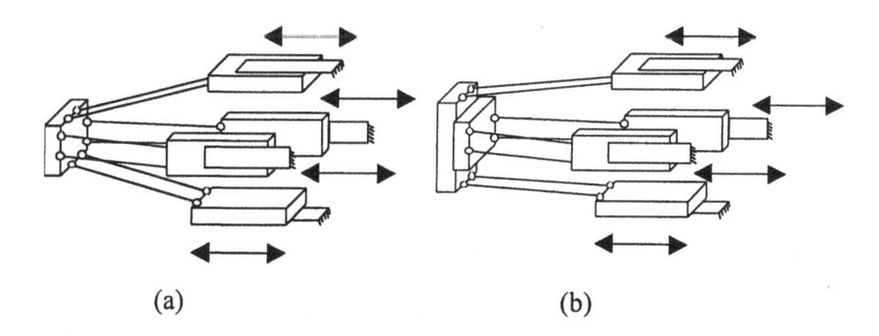

Figure 2. The redundant Sliding Delta: initial (a) and final (b) concepts

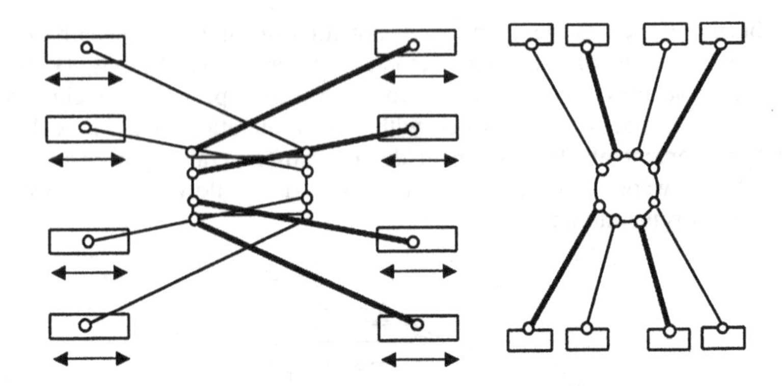

Figure 3. Scheme of robot Octaslide

5. Structural Conflicts and their Solution

The structural design is about the resolving of conflict between stiffness and dynamics. First, the stiffness of both non-redundant and redundant parallel structures are evaluated (*Figure 4*). This clearly demonstrates the significant improvement of stiffness by almost 50%.

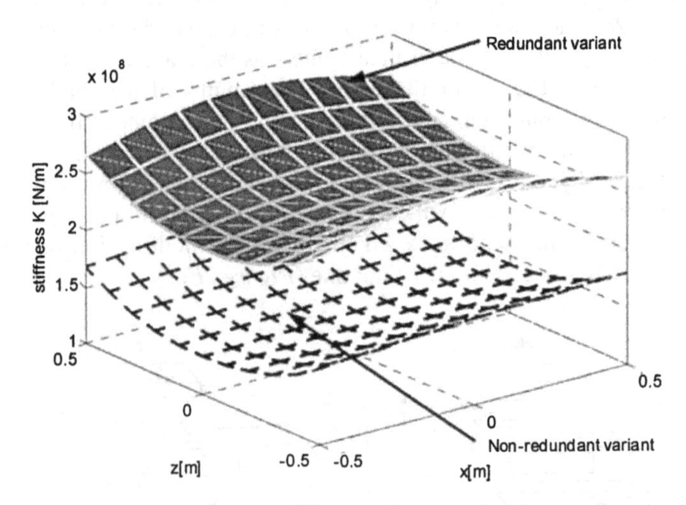

Figure 4. The comparison between stiffness of non-redundant and redundant parallel structure for one planar section in the workspace

Second, the dynamic capabilities are investigated. The limitations of dynamic capabilities of drives are transformed into the areas of accessible accelerations and velocities at the points of selected trajectory. Choosing circular trajectories with different radius the dynamic capabilities were evaluated (*Figure 5*). The accessible accelerations versus accessible velocities are plotted. Again the redundant actuation has proved significant improvement of dynamics by about 20%.

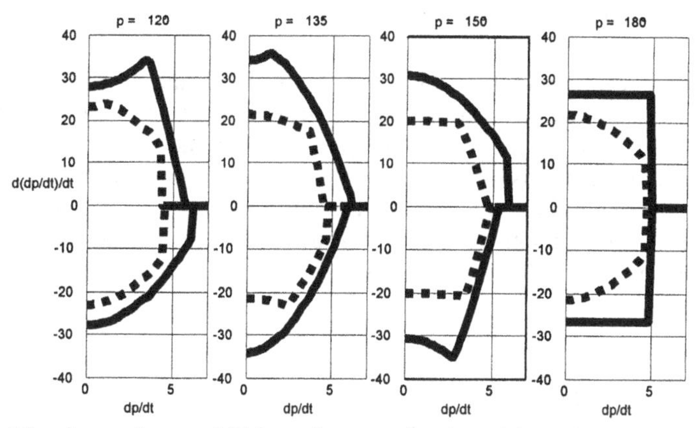

Figure 5. The dynamic capabilities of non-redundant (dash lines) and redundant (full line) parallel structures during the circular motion with radius 0.6 m (bounded areas cover accessible accelerations-vertical axis and accessible velocities-horizontal axis for 4 positions on the trajectory)

Then the conflict between stiffness (eigenfrequencies) and dynamics of end-effector (mass) for different variants of dimensions has been solved by multiobjective parameter optimisation. In short, on the border of space of possible solutions (Pareto set) increasing stiffness means increasing mass and decreasing acceleration capabilities. The genetic algorithms have been used for this task. Each point (*Figure 6*) represents one variant of setting of robot dimensions.

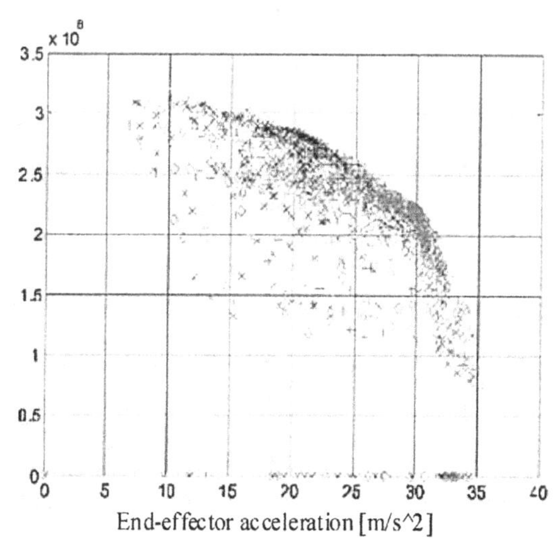

Figure 6. Results of multiobjective parameter optimisation of stiffness and dynamics (results of several optimisation processes displayed together)

6. Actuation Conflicts and their Solutions

The drive concept must be completed from the point of view of required dynamics, kinematics, dynamic accuracy and control strategy. The simplified scheme (*Figure 7*) describes two nested loops for slider position and motor angular velocity feedbacks. The end-effector position measurement can be considered for the upper-most feedback loop, nevertheless its practical realisation is not easy. The tuning of control gains completes the utilization of previously designed mechanical properties. The overall simulation is the final test of the whole redundant parallel robot conceptual design.

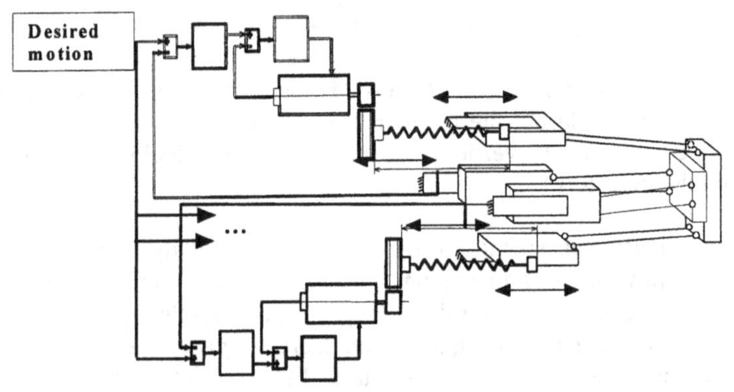

Figure 7. Simplified scheme of complex dynamic model including feedback control loops

7. Conclusions

The paper has briefly investigated the role of multidisciplinary virtual modelling for efficient design of complex mechatronic machines. It has been demonstrated on the design methodology for redundant parallel robots. The virtual models cover both mechanical and geometric properties. The basis for these virtual models is the decomposition of design process. The design process has been resolved into three hierarchical levels. Each of these levels is characterized by certain problem simplification and special design conflict which should be resolved within the level. Specific virtual models are necessary for each level. The computational tools related to these virtual models enable to parameterise the main design conflicts and solve them using multi-objective parameter optimisation. Proposed hierarchical methodology based on multidisciplinary virtual modelling proved to be useful and efficient for the design of complex mechatronic machines.

The proposed design methodology based on multidisciplinary virtual modelling has been demonstrated on the design case study of redundant Sliding Delta and Octaslide robots. They also demonstrate the application of the principle of redundant actuation that leads to the development of new robot parallel structures with promising properties.

Acknowledgment. The authors appreciate the kind support by the grant J04/98:212200003 "Development of algorithms of computational mechanics and their application in engineering".

8. References

1. Valášek, M., Šika, Z., Bauma, V., Vampola, T. (2001) Design Methodology for Redundant Parallel Robots, in *Proc. of AED 2001, 2nd Int. Conf. on Advanced Engineering Design*, Glasgow, 243-248.
2. Zdráhal, Z., Valášek, M. (1996) Modeling Tasks in Engineering Design, *Cybernetics and Systems* **27**, 105-118.
3. Valášek, M. (1998) Mechatronic System Design Methodology - Initial Principles Based on Case Studies, in J. Adolfsson and J. Karlsen (eds.), *Mechatronics 98*, Pergamon, Amsterdam, 501-506.
4. Šika, Z., Valášek, M., Miláček, S., Bastl, P. (1997) Synthesis and Analysis of Planar Redundant Parallel Robot, in J. Angeles and E. Zakhariev (eds.), *Proc. of NATO ASI Computational Methods in Mechanisms*, Varna, Vol. II, 353-362.
5. Kim J., Park F.C., Ryu S.J., Kim J., Hwang J.C., Park C., Iurascu C.C. (2001) Design and analysis of a redundantly actuated parallel mechanism for rapid machining, *IEEE Trans. On Robotics and Automation*, Vol. **17** (4), 423-434.
6. Kock, S., Schumacher, W. (1998) Regelungsstrategien für Parallelroboter mit redundanten Antrieben, in *VDI-Fachtagung "Neue Maschinenkonzepte mit parallelen Strukturen für Handhabung und Produktion"*, *VDI Berichte Nr. 1427*, Braunschweig, 155-164.
7. Valášek, M., Šika, Z. (2001) Evaluation of Dynamic Capabilities of Machines and Robots, *Multibody System Dynamics* **5**, 183-202.
8. Valášek, M., Šika, Z. (2001) Analysis of Required Driving Force Distribution in Redundant Parallel Robots, in *Proc. of AED 2001, 2nd Int. Conf. on Advanced Engineering Design*, Glasgow, 258-263.

KNOWLEDGE SUPPORT FOR VIRTUAL MODELLING AND SIMULATION

P. STEINBAUER, M. VALÁŠEK, Z. ZDRAHAL*, P. MULHOLLAND*, Z. ŠIKA

Department of Mechanics
Faculty of Mechanical Engineering
Czech Technical University in Prague
Karlovo nám. 13, 121 35 Praha 2, Czech Republic
E-Mail: valasek@fsik.cvut.cz

Knowledge Media Institute
The Open University
Milton Keynes, UK
E-Mail: z.zdrahal@open.ac.uk

1. Introduction

The paper describes an approach to providing knowledge support for virtual modelling and simulation (VMS). The design methodology, based on multidisciplinary virtual modelling and subsequent simulation, is essential for contemporary engineering design [1], but specifically for the design of complex mechatronic machines [3]. Additionally, current engineering design is multidisciplinary and therefore based on team work. The teams are often geographically distributed and would benefit from greater support for collaboration. Engineering design also draws heavily on previous experience and therefore it is essentially to build and maintain comprehensible and re-usable archives of previous cases. This is also true of simulation models.

Engineering design is in nature a knowledge intensive activity, but analysing the resulting documentation of such activities shows that there is a significant loss of knowledge, creating problems for team communication and future reuse. The simulation of dynamic systems, in particular the virtual nonlinear multibody system, is a prime example. The resulting simulation models do not include the majority of knowledge which was necessary and which was used for their development.

The simulation models include a large amount of knowledge, which is related to the development and usage of dynamic simulation models. However, there are no tools for storing, retrieving, reusing, sharing and communicating such knowledge. Therefore special tools are being developed for enriching traditional dynamic simulation models by the corresponding accompanying knowledge within EU funded project CLOCKWORK.

2. The Virtual Modelling and Simulation Methodology

The development of a virtual model and then its simulation typically goes through a number of tasks. These tasks are illustrated in our virtual modelling and simulation framework in Figure 1. The light arrows represent mappings between the different elements and the black arrows represent transformation processes between the tasks. The development of a simulation, as shown in Figure 1, involves five steps.

The first step is the analysis of the real world object (either actual or hypothesised) that

W. Schiehlen and M. Valášek (eds.), Virtual Nonlinear Multibody Systems, 243–252.

is motivating the simulation and modelling task. This real world object is investigated within a certain experimental frame concentrated on the aspects that need to investigated in order to answer some question. Answering this question is the objective of the investigation. The real world system is the model created in order to help answer the modelling question, such as how a real-world system would respond to particular stimuli. The real world object therefore has three elements:

- The real world system (either actual or hypothesised), of which we want to create a model to use to answer a modelling question, such as how a real-world system would respond to particular stimuli.
- The question (modelling question) is a question that is based within the real world and is to be answered using simulation.
- The solution is the interpretation of the interpreted output of the simulation task, this will be left blank until there is a set of results.

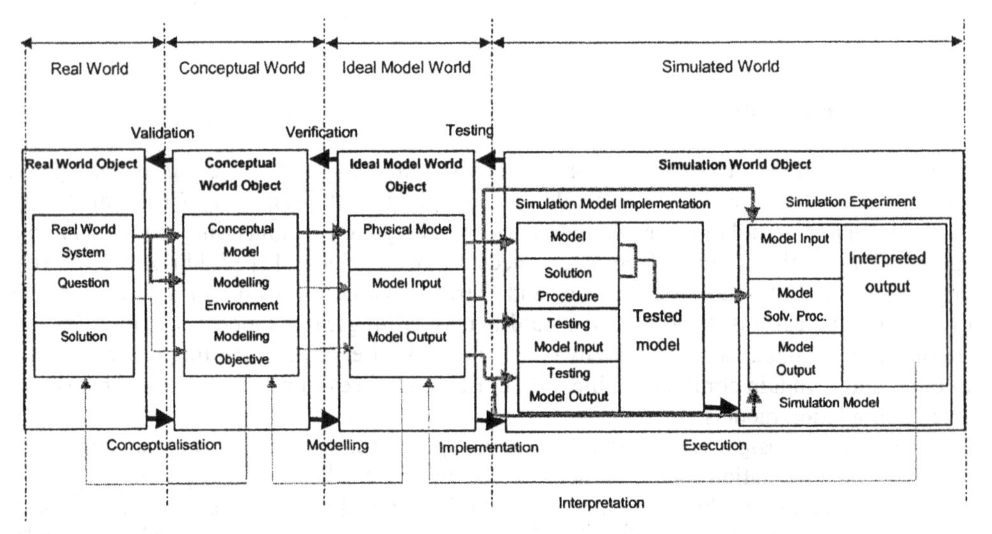

Figure 1. The Virtual Simulation and Modelling Framework.

The second step is the conceptual task, where the conceptual model is developed. The transformation from real world into conceptual world consists of creating a hierarchical break-down of the real world system being investigated (e.g. hierarchical description of real product components and its environment). This scheme should be accompanied by the description of function (physical interaction) as it is the basis of causal and functional explanation. As soon as this description is completed the conceptualisation is completed. Different assumptions are raised about the granularity of the model being developed. A crucial design decision is determination of what factors influence system behaviour and what system behaviours are to be incorporated into this task. The result of the conceptualisation is the conceptual model investigated within the modelling environment in order to answer the modelling objective:

- The conceptual model is an abstracted view of the real world system.
- The modelling objective is the conceptualisation of the question and form the objectives for the simulation task (i.e. what behaviour we are interested in? and how we will know that we have the correct answer from the results?)
- The modelling environment is the conceptualisation of the inputs of the real world

systems, providing a simulation environment to work within in the simulation task, typically the inputs are specified here.

Validation is the process of determining whether the conceptual model is an adequately accurate representation of the real world task.

The next task is physical modelling. The third step is the transformation from conceptual world into ideal model world. It consists of replacement of each element on the conceptual level by corresponding elements on the physical level. The elements of the physical model are ideal modelling elements (e.g. rigid body, ideally flexible body, ideal gas, ideal incompressible fluid, ideal capacitor etc.). The engineering sciences deal precisely with the ideal objects. Mathematically described exact prediction of the behaviour is available only for ideal objects. The art of engineering modelling and simulation is the capability to translate the behaviour of real world objects into the behaviour of a system consisting of ideal objects.

Again as soon as this replacement is completed the physical modelling is completed. In this stage, the majority of modelling assumptions are raised and many modelling modifications are being done. It is the way of "modelling it" and representing the real world system in a physical modelling manner. For example two typical modifications are provided: *Either* an element is decomposed into more elements (a small subsystem) (e.g. a body cannot be treated as rigid body and it is modelled as flexible body being modelled as subsystem) *or* several elements are replaced by one element as the detailed treatment of interactions is neglected (e.g. the influence of a car suspension mechanism can be neglected and the model consists only of chassis and wheels with flexibility of tyre and suspension spring). The modelling task results into the physical model, its input and the investigated output:

- The physical model is an accurate physical view of the conceptual model.
- The model input is the modelling conceptualisation of the inputs of the real world systems, providing a physical modelling of the real world stimuli to the investigated system, i.e. detailed physical description of modelling environment from conceptual task.
- The model output is the modelling conceptualisation of the objectives for the simulation task (i.e. how the interested behaviour is measured and evaluated ?)

The process of determining whether the physical model is an adequately accurate representation of the conceptual model is the verification.

The results of the modelling task are implemented in the form of a computer-executable set of instructions known as the simulation model. It is the transformation from physical world into simulation world. It consists of the replacement of each element on the physical modelling level by corresponding elements on the simulation level. Usually this replacement can be straightforward as simulation packages are trying to directly contain the ideal objects of physical models. However, always some modifications are necessary. The simulation task consists of two stages. The first stage deals with the implementation of the physical model using a particular simulation environment (simulation language and simulation software). The second stage deals with the simulation experiment, i.e. proper usage of the developed simulation model for the solution of the real world problem. Testing is the process of determining whether the implemented version of the model is an accurate representation of the physical model.

The linear design process is made up of numerous transitions across world objects of different levels of the framework, producing different iterations at each level. Each transition from one world to the next involves the adoption of specific assumptions pertaining to how the model should be re-represented in the next world. An example is shown in Figure 2. The designer develops the real world model, and then constructs an associated conceptual model which is transformed into an ideal model. A failure at this point leads the designer to construct an alternative model on the conceptual world model, meaning the problems identified in construction of the first ideal model can only be resolved by modifying the assumptions made when developing the first conceptual model. This second conceptual model is further developed into ideal and simulation models.

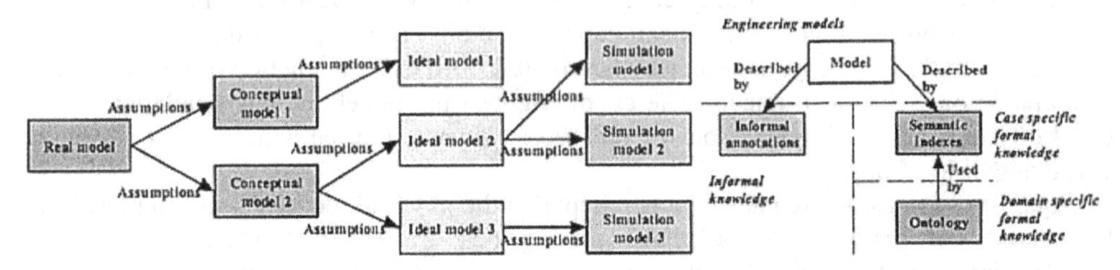

Figure 2. Iterative process of simulation model development

Figure 3. Formal and informal knowledge

3. Knowledge Support of Virtual Modelling and Simulation

As we have described above, the virtual modelling and simulation process proceeds through a number of steps, from real world to simulation world, and each of these worlds comprises a number of objects related to the model, its input and output. Additionally, the transitions between different steps creates different versions of modelling worlds, each with its own associated assumptions. Much of the knowledge used during the VMS process, and needed for its comprehension and reuse is lost. This knowledge is of two kinds – informal and formal. Informal knowledge takes the form of text and diagrams, and describes the model and its process of construction in the vocabulary and diagrammatic constructs of the VMS designer. We refer to this kind of knowledge as *informal annotations*. Formal knowledge describes the models and associated processes according to concepts represented in an ontology. An ontology is an explicit conceptualisation of a domain. The ontology therefore contains concepts applicable across some domain such as the simulation and modelling of mechatronic machinery. The associations between the ontology and specific simulation models are specific for each case. These associations are referred to *semantic indexes*. The relation between the model and formal and informal knowledge is shown in Figure 3.

A separate ontology is used to describe each world of the VMS framework. The real world is described in terms of a product ontology. This provides a description of the product whose development or investigation is the focus of the task. Currently, the product ontology takes the form of a simple taxonomy. The conceptual world is described according to a component ontology. The components described in the ontology are abstract component types, similar to design types in [2], that scope the components and/or behaviour under investigation. The ontology can be used to describe the structure of these component types and their functions. The objects of the ideal world and their interactions are described by an ideal object ontology. Finally, the simulation ontology can be used to describe the simulation world in terms of important simulation code objects used, their behaviour and interaction. These four ontologies, with examples, are outlined in Table 1. Further ontologies are used to describe mappings or transitions between worlds and the VMS framework itself.

World	Object ontology	Purpose	Examples
Real world	Product ontology	Classify the problem being solved and/or product being designed	Car product, heater product
Conceptual world	Component ontology	Classify the main components and/or behaviour of the required product	Frame, chassis, engine, geabox
Ideal model world	Ideal object ontology	Describe the idealised low level structure and behaviour of the product	Flexible body, revolute joint, beam
Simulated world	Simulation ontology	Describe the M&S code and its behaviour	Force element, continuous transfer function

TABLE 1. Ontologies of the four VMS worlds.

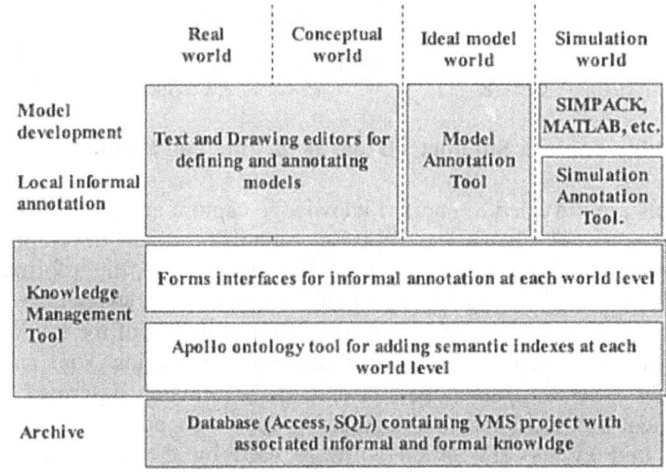

Figure 4. Toolkit for knowledge support of VMS

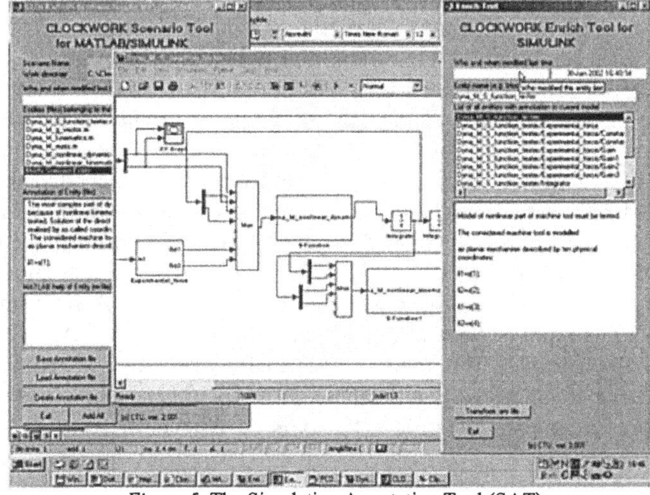

Figure 5. The Simulation Annotation Tool (SAT)

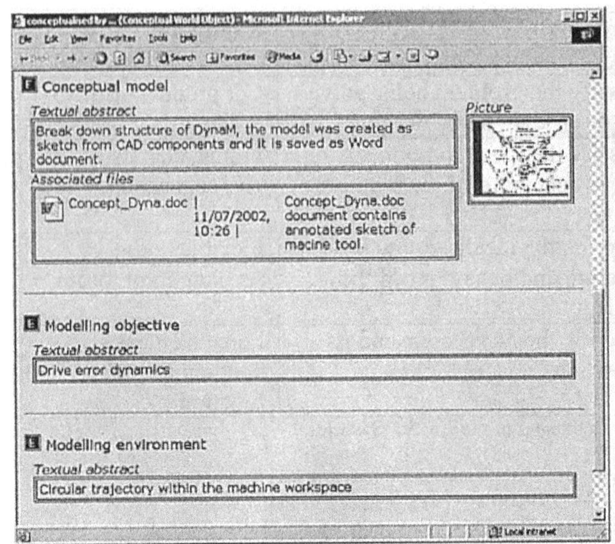

Figure 6. Knowledge management tool (KMT)

Figure 7. RWO: Picture of the machine tool DynaM

4. Knowledge Tools to Support Virtual Modelling and Simulation

A number of tools are provided to support knowledge capture and reuse at the four stages of the VMS framework [4]. The toolkit comprises support for model development, the local informal annotation of those models on the VMS designer's PC, the informal and formal description of those models using the web-based Knowledge Management Tool (KMT), and the archiving of the subsequent VMS projects. An overview of the toolkit is shown in Figure 4. The Modelling Annotation Tool, Simulation Annotation Tool and Knowledge Management Tool have been developed as part of the research described in this paper. Model development and local informal annotation is also supported by standard text and drawing editors and by the simulation software used by the VMS designer, such as MATLAB/SIMULINK or SIMPACK.

The Simulation Annotation Tool is used to provide informal annotation integrated with the designer's simulation software. Currently we have versions of the Simulation Annotation Tool integrated with MATLAB/SIMULINK and SIMPACK. A snapshot of the Simulation Annotation Tool integrated with MATLAB/SIMULINK is shown in Figure 5. The interface for adding informal annotations to single or groups of simulation object is shown to the top right side of the screen.

The Modelling Annotation Tool is used for describing the conceptual world and ideal world models and also associating additional annotations informal annotations with the models. Models and annotations developed locally by the VMS designer are then uploaded within the Knowledge Management Tool (KMT). Here further informal annotations can be supplied and semantic indexes can be associated with the models. A snapshot of the KMT interface is shown in Figure 6. Semantic indexes are added using the Apollo ontology tool (see http://apollo.open.ac.uk). The models, informal annotations, semantic indexes and associated ontologies are archived in a database currently either MS Access or SQL. In the next section we present an example of how the toolkit can be used.

5. Example: Machine Tool Analysis

The Clockwork methodology for simulation and modelling and the knowledge management support provided by the Clockwork tools is shown on the example of analysing properties of the parallel machine tool DynaM needed for the design of the tool controller.

5.1 REAL WORLD OBJECT (RWO)

Engineering simulation & modelling view
Following the Clockwork simulation and modelling methodology we define the Real World Object in terms of three descriptors: Real World System (RWS), Question and Solution. In the example, the Real World System is the description of the parallel machine tool DynaM with a detailed technical specification. The picture of the machine tool as shown in Figure 7 and the technical specification is included or referenced in the Knowledge Management Tool (KMT).

The focus of our investigation is defined as the Question associated with the RWO (see Figure 1). In this example the Question is to evaluate the *dynamic interaction between drives across the kinematic structure of the machine tool.* Question is also recorded in the KMT. Solution is a placeholder for conclusions of the simulation and modelling case.
Knowledge modelling view
For the RWO, the KMT provides a simple keyword indexing and classification system. In this example we associate with the RWS keywords: "machine tool" and "parallel kinematics".

5.2 CONCEPTUAL WORLD OBJECT (CWO)

Engineering simulation & modelling view
The conceptual model consists of a model structure with the specification of model components. In addition, CWO may include functional and causal models. The structural model used in the example is shown in Figure 8. At the CWO level, the Question of RWO has been converted into more detailed Modelling Environment and Modelling Objective of CWO (see Figure 1).

In the example, the Modelling Environment states that *we want to study circular trajectories* and the Modelling Objectives specifies that we assess the Question of RWO by *measuring drive error dynamics*. Conceptual models, Modelling Environment and Modelling Objective are recorded in the KMT.
Knowledge modelling view
In the example, the conceptual models have been elaborated under the following assumptions: the rotation and unbalance of the spindle is neglected, machine mounting is neglected and the detailed model of cutting processes is not considered. These assumptions set the applicability conditions for the conceptual model and therefore they are saved in KMT at the CWO level.

Objects of conceptual models can be used as semantic indexes. In general, we use components of knowledge models as semantic indexes when we want to represent modelling knowledge about the component potentially important for reuse. For example, if a new, sophisticated modelling technique of the left screw gear has been used, we may define a new semantic index *left-screw-gear* and associate it with the corresponding concept of the component ontology, say *screw_gear* – see Figure 8. Association of model components and classes in ontologies is saved in the KMT and used as a reference to the piece of important modelling knowledge for case retrieval and future reuse.

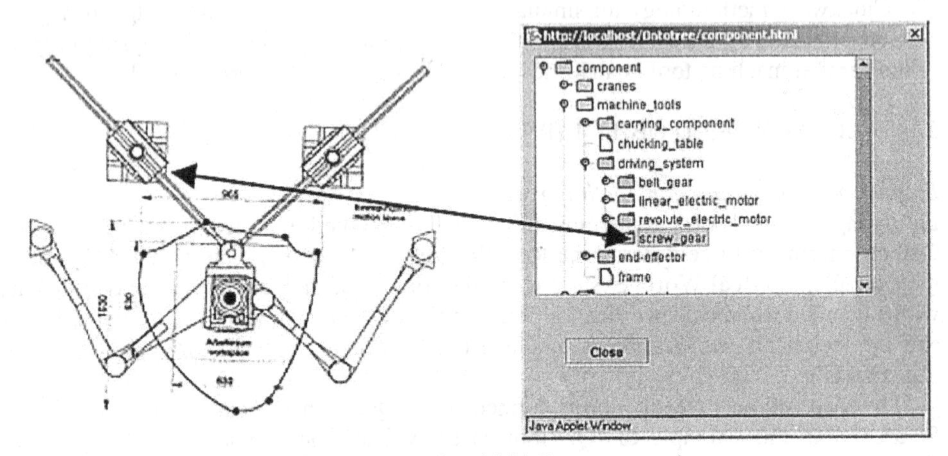

Figure 8. CWO: (a) Structural model (b) Component ontology

5.3 IDEAL MODEL WORLD OBJECT (MWO)

Engineering simulation & modelling view
In the Ideal Model World additional modelling decisions are made. Conceptual model is transformed into the Physical Model by representing its components using ideal modelling objects. Modelling Environment and Modelling Objective of CWO are further specified in terms of Model Input and Model Output. In our case, the Model Input is defined *as Cartesian co-ordinates of a circular trajectory* and the Model Output as the *difference between actual and requested drive position (control error)*. The Physical Model is shown in Figure 9.
Knowledge modelling view
The step from a conceptual model to a physical model requires additional modelling decisions, which play the role of assumptions associated with the MWO. For example, we assume that the whole mechanism will be treated as planar, we will consider joints as ideal and we do not consider the current feedback. These assumptions are saved as a part of MWO.

Similarly to CWO, important objects are selected as semantic indexes. They are associated with classes of the ideal object ontology. In our example, we consider the following semantic indexes: *rigid-body-1, ..., rigid-body-4, flexible-body-1, flexible-body-2, basic-revolute-joint, right-DC-motor* and *left-DC-motor*. An example of the assignment of a semantic index to a class is shown in Figure 9.

5.4 SIMULATION WORLD OBJECT (SWO)

Engineering simulation & modelling view
Simulation model implements the physical model of MWO in a selected simulation language. Usually, simulation languages are tightly connected to the problem domain and the class of physical models typically used for the domain. Our methodology encompasses a two-step development of simulation models: first, the simulation model is implemented

and its performance and correctness is assessed by running the model in various test conditions, then the model is applied to the modelling problem specified in higher levels (RWO, CWO and MWO). For example, the simulation model was implemented with SIMULINK ode23stiff problem solving procedure and tested with the input step function as a wanted position. The model behaviour was evaluated by measuring the time response of the position output. When the model correctness is satisfactorily verified, the model is run with the input and output corresponding to the original problem as specified in previous levels (RWO, CWO and MWO). Results are evaluated and propagated backwards to the higher-level models.

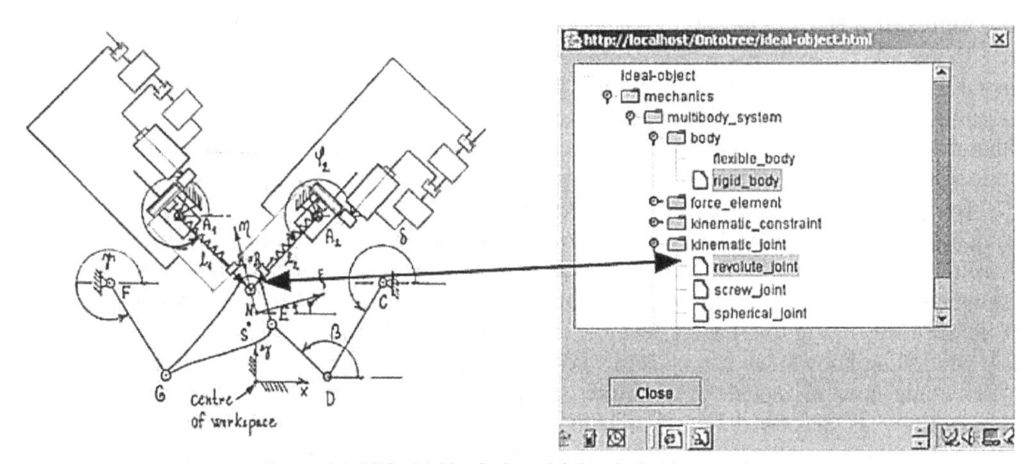

Figure 9. MWO: (a) Physical model (b) Ideal object ontology

Knowledge modelling view

Similarly as in the case of CWO and MWO, at the simulation model level the model has associated assumptions under which it has been designed and tested. Interesting modelling components are selected as semantic indexes and associated with the corresponding classes of the simulation ontology. This ontology consists of modelling primitives of the simulation environment, which implements the model. These primitives are usually provided by the developer of the simulation environment. In the described case we use the SIMULINK ontology derived from the SIMULINK documentation. In addition, knowledge annotations can be integrated with the Simulation Annotation Tool. This has been implemented for MATLAB/SIMULINK and SIMPACK. SIMULINK simulation model with annotations was shown in Figure 5.

5.5 INTER-WORLD RELATIONS

In addition assigning semantics to important modelling objects, as described above, the relationships between objects of different worlds are important for understanding the modelling process. The *tree structure* of models shown in Figures 8 and 9 is defined by the "*extends*" relation. This relation is defined automatically as the user extends the model to the next level.

In addition, important knowledge and modelling techniques may be represented as "is-be-modelled-as" and "is-implemented-as" relations. They state that a group of objects of a higher level is modelled or implemented by a group of objects at the lower level. Objects that enter relations are characterised by the semantic indexes. For example, an amplifier at the CWO level is-modelled-as an ideal amplifier and two non-linear functions (saturation for high output values and insensitivity for small input values) at the MWO level. Inter-model relations are important for retrieving relevant modelling knowledge.

5.6 QUERYING KNOWLEDGE MODELS

The knowledge models described in this section have been designed to facilitate collaboration, knowledge sharing and reuse. In the previous paragraphs we have described how to make modelling knowledge explicit and associate it with models at each level. Now we will outline how the relevant knowledge can be retrieved when needed.

In the KMT, the final query is composed of simple query components. The first and the most obvious query component is the keyword search. In KMT, all annotations and assumptions can be searched. However, keyword search is usually not enough focused and therefore does not provide good results. For this reason KMT introduces model annotations through ontologies, semantic indexes and inter-world relations described above.

Semantic indexes and ontologies allow the user to retrieve information by traversing ontologies. For example, the query “Show me an example of modelling kinematic joint” will retrieve the model of DynaM shown in Figure 9, because there is an association of the semantic index called basic-revolute-joint with the ideal modelling object revolute joint (see the arrow in Figure 9) and a revolute joint is a kind of kinematic joint.

Modelling knowledge can be retrieved through inter-world relations as follows: Query “Show me how to model an amplifier” returns the ideal amplifier with saturation and insensitivity as described before. Obviously, this inter-world relation can be used for different queries, e.g. “Show me all components which use for modelling saturation” etc. Various strategies can be defined for combining all available query components available in KMT. Usually, the major role is played by semantic query components and keyword search is used just to order retrieved cases.

6. Conclusions

The paper describes the concepts and tools for knowledge support of virtual modelling and simulation. These techniques are important for virtual modelling used in teams and for reuse typical for any engineering methodology. The developed tools have been tested on cooperative development of simulation models and are being currently tested on the reuse of simulation models. The accumulated experience suggests that the methodology and tools are applicable to other kinds of virtual modelling.

Acknowledgment. The authors appreciate the kind support by the EU IST 12566 Project CLOCKWORK.

7. References

1. Zdráhal, Z., Valášek, M. (1996) Modeling Tasks in Engineering Design, *Cybernetics and Systems* **27**, 105-118.
2. Schön, D.A. (1988) Designing: rules, types and worlds, *Design Studies* 9, 181-190.
3. Valášek, M. (1998) Mechatronic System Design Methodology - Initial Principles Based on Case Studies, In J. Adolfsson and J. Karlsen (eds.), *Mechatronics 98*, Pergamon, Amsterdam, pp. 501-506.
4. Valášek, M., Steinbauer, P., Zdráhal, Z. (2001) Knowledge Supported Conceptual Control Design Methodology Of Mechatronic Systems, In: *Proc. of 2nd Int. Conf. on Advanced Engineering Design, AED 01*, Univ. of Glasgow, Glasgow, pp. 237-242.

DAViD - A MULTYBODY VIRTUAL DUMMY FOR VIBRATIONAL COMFORT ANALYSIS OF CAR OCCUPANTS

P.P. VALENTINI
L. VITA
University of Rome "Tor Vergata" - Dept. of Mech. Eng.
Via del Politecnico, 1 - 00133 - Rome - Italy
email: valentini@ing.uniroma2.it; vita@ing.uniroma2.it

1. Introduction

The development of virtual simulators can avoid to set-up expensive test rigs, time-consuming tests, and is a winning strategy to be more competitive in road-vehicles market. Moreover vibrational comfort analysis is an important topic in vehicle design [1] and the possibility to perform virtual vibrational tests on the effects of changing some parameters is useful tool for the designer. A literature search reveals that most of the simulation models in the field are based on elementary linear models [12]. In some cases, finite elements are used, but this approach involves a large amount of parameters to be defined and managed. Thus the authors of this paper developed a virtual dummy model by means of multibody techniques. The formulation is the one described in Haug's text book [4]. The code, named DAViD (the acronym of Dynamic Automotive Virtual Dummy), can mimic the non linear behaviour of a 3D human body model and requires a very small set of body data. The model is completely parametric and can be automatically scaled to simulate a significant portion of population. The code can be also linked to experimental results of accelerometers time histories to perform multi-input analysis based on seat input (translational and rotational), steering wheel input and pedals input. Driver and occupants can be both simulated. It is possible to introduce non linear viscoelastic parameters to match the actual behaviour of cushion foams used in the manufacturing of seats. The model provides also an assessment of vibrational comfort computed in compliance with international standards. The results of the code DAViD have been compared with experimental ones acquired on a vibrational test rig.

W. Schiehlen and M. Valášek (eds.), Virtual Nonlinear Multibody Systems, 253–262.

2. Multibody Model

The developed model is based on a multibody dynamics approach [4]. In particular the whole model is made of 15 rigid elements, 12 of which define the dummy, and the remaining 3 describe the car environment. The dummy is composed of two feet, two legs, two thighs, the pelvis, two arms, two forearms an upper part that is formed by head, neck, shoulders and chest rigidly connected together. The other rigid bodies included in the model are seat, pedals and steering wheel. In order to represent the human body articulations, kinematics constraints and viscoelastic elements are used to connect each part of the dummy. There are two spherical joints between pelvis and thighs, two revolute joints with transverse axes between thighs and legs, two revolute joints with transverse axes between legs and feet, one prismatic joint with longitudinal axis between pelvis and upper part, two spherical joints between upper part and arms, two revolute joints with transverse axes between arms and forearms. The viscoelastic elements used in the dummy are one translational, between pelvis and the upper part to represent the stiffness of torso, and two rotational elements, between arm and forearm to reproduce the muscular elasticity of the elbow. All the data used for the dummy are taken from an internal biomechanical database (for more information contact the authors). The dummy interacts with the car environment by means of seat, pedal and steering wheel contact simulated by other viscoelastic elements (as it is explained in paragraph 3). The contact between hands and steering wheel and feet and platform car is simulated with four very stiff springs. There are no seat belts in the model because they do not affect the vibrational behavior in the field of small vibrations according to experimental results. The model can automatically scale geometric, mass properties and spring locations by means of changing few parameters (such as percentile). In fact the code is interlaced with an anthropometrical database. It is also possible to modify the backrest inclination and the hip-heel vertical position in order to change the configuration of the seat. The code can also manage several inputs at the same time. It can get input acceleration time histories acquired by experimental tests, as well as time histories on velocities and positions, filtering the signals in order to suppress noise. If necessary, forces and torques could be introduced as well as driving constraints.

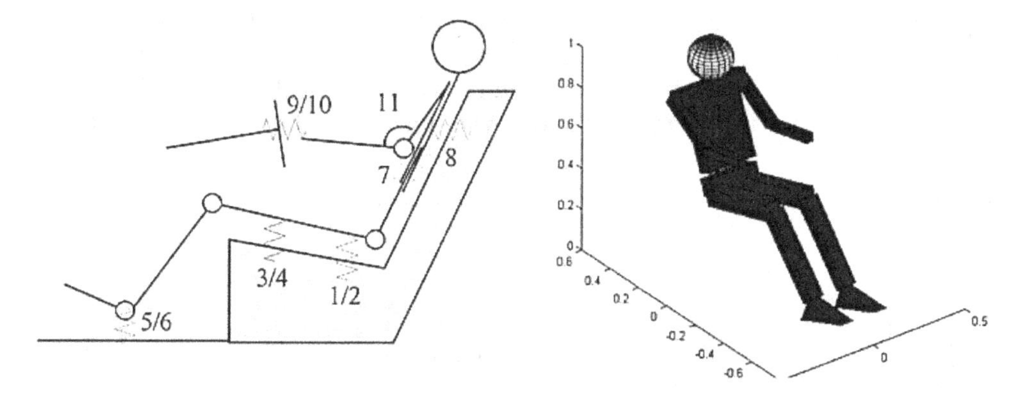

Figure 1. Viscoelastic elements (left) and complete 3-D dummy in Visualizer (right)

2.1 EQUATIONS OF THE MULTIBODY MODEL AND INTEGRATION

The equations of motion are deduced in the form of differential – algebraic system of index 3 that means the system is made up of the dynamic equations of each body of the model and the constraint equations on positions used in the model [4] [5]:

$$\begin{cases} [M]\{\ddot{q}\}+\left[\Psi_q\right]^T\{\lambda\}=\{F_e\} \\ \{\Psi\}=\{0\} \end{cases} \tag{1}$$

where $[M]$ is the global mass matrix; $\{\Psi\}$ is the vector of constraint equation; $\{\lambda\}$ is the vector of Lagrangian multipliers; $\{F_e\}$ is the vector of external generalized forces; $\{q\}$ is the vector of generalized coordinates. In our model there are 15 bodies, and 105 generalized coordinates. The spatial location of the i-th body is described with seven parameters (i.e. three for the position of the center of mass and the four Euler's parameters. The constraint equations used in the model can be divided into three groups:

- the first 15 equations (as many as the number of bodies in the model) are the normalization equations of the Euler's parameters;
- the second group of equations is made up of the scleronomic constraints.
- the last group we include a driving constraint at inclination of pedals w.r.t. the horizontal plane (first two of (3)); regarding pelvis we impose no translation along z axis and rotation around the same axis (last three of (3)).

The complete model has 24 d.o.f. The integration of the DAE system, as it is shown in equations (1), has been performed rearranging the system as a first order one in the unique unknown y which has not a physical meaning. Therefor the system to be integrated is in the following form (y' is the time derivative of y):

$$[K]\{y'\} = \{\phi(y)\} \tag{2}$$

where:

$$[K] = \begin{bmatrix} I & 0 & 0 & 0 \\ 0 & I & 0 & 0 \\ 0 & 0 & 0 & 0 \\ 0 & 0 & 0 & 0 \end{bmatrix} ; \{y\} = \begin{Bmatrix} q \\ \dot{q} \\ \ddot{q} \\ \lambda \end{Bmatrix} \text{ and } \{\phi(y)\} = \begin{Bmatrix} \{\dot{q}\} \\ \{\ddot{q}\} \\ \left\{[M]\{\ddot{q}\} + [\Psi_q]^T \{\lambda\} - \{F_e\}\right\} \\ \{\Psi\} \end{Bmatrix} \tag{3}$$

The system (3) is then solved by means of RADAU5 [9].

3. Experimental set-up tests

A key point of simulation is the contact between seat and occupant that influences the vibrational response of the dummy. For this purpose an experimental procedure has been performed to find the seat force - deflection curve. Special mats, equipped with pressure transducers, are put on several seats and a jury made by people belonging to different physical groups has sat on the instrumented seat. Pressure maps have been acquired. Then spring elements have been introduced in the model and anchored to the points of high pressure concentration. For the computation of stiffness, appropriate tests have been performed on cushions using standard dynamometer. These have shown a non linear behavior of polyurethane foams in their force/preload characteristic curves, an example these curves is shown in Fig.2. A second kind of tests were performed to check the correct dynamic response of dummy. Some car have been tested on standard tracks and accelerometers signals have been acquired at measurement point (Figure 4). The driving tests were of three different kinds: on a highway with constant speed of 100 km/h; on comfort and pavè track with respectively constant speed of 60 and 40 km/h.. Since the driving tests have shown important levels of variability according to track, driver and seat, the obtained signals have been replicated in a vibrational test rig, where the same seats have been mounted and the same driver has sat on. New signals have been acquired from SAE accelerometer pads placed on the cushion and on the backrestThe need for replicating these signals is due to obtain the repeatibility, and a standardization of the test procedures.

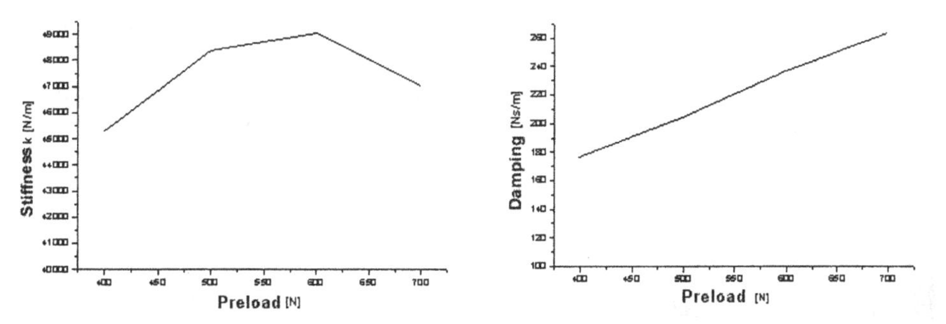

Figure 2. An example of non linear stiffness (on the left) and damping (on the right) of seat

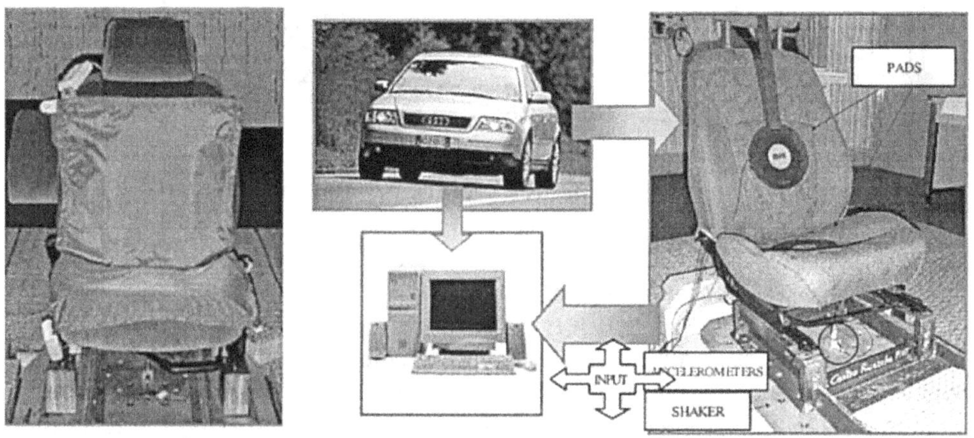

Figure 3. Experimental test rigs. Pressure mats (left) and vibrational shaker (right)

4. Experimental multi-input tests

The DAViD code can take directly accelerometers data files to simulate a multi-input configurations The user can impose input at the seat (6 d.o.f.), at steer wheel (4 d.o.f.) and at pedals (3 d.o.f.). The signals is pre-processed by a band pass filter. The cut – off frequencies are 0.5 Hz the lower and 50 Hz the higher, which is also the frequency range of interest in human vibration [1]. It is possible to run analysis directly acquiring data from a four axis shaker experimental test rig on which a car has been placed. The pneumatic actuators reproduce the track profiles and give vibrational inputs to the tires, and the response signals at six accelerometers has been collected. The pneumatic actuators were controlled by TEAM control system and their maximum vertical displacement was 50 mm and their frequency range is 0.5 – 100 Hz which is suitable for the vibrational analysis. The accelerometers, all with three axes of sensitivity have been placed as shown in Figure 4 three between the seat and the chassis, one on the steer column, one on the steer wheel, and one at the pedals. The accelerometer signals and the displacement of the actuators were acquired by DIFA 46 channels acquisition system and all the data were processed by LSM software.

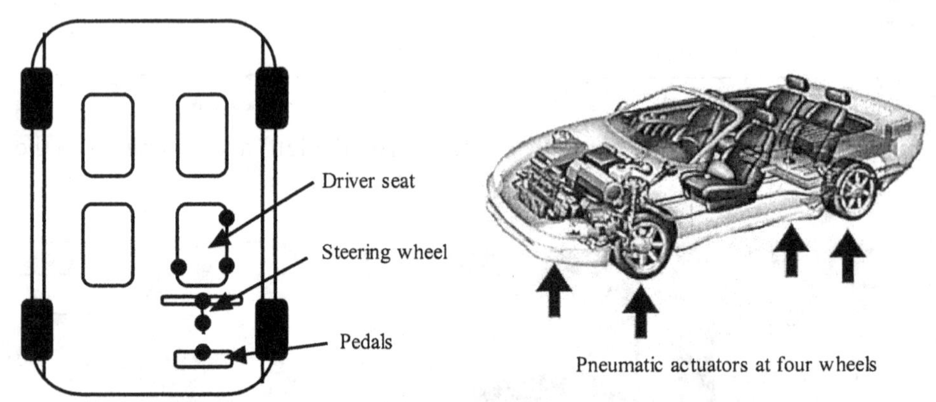

Figure 4. Layout of the experimental multi-input tests

5. Comfort assessment and virtual perceiving

The comfort assessment is important to predict the effect provoked by vibrations on the human body. Many car accidents happens because of tiredness, or disturbs to perception, that can be avoided decreasing the level of transmitted vibration. Three aspects of vibration are fundamentals: the *exposure time*, the *amplitude* and the *frequency* [1]. The consequences of vibration exposure are not simple: the perception of motion, the sensations it produces and the interference with health and activities are all complex phenomena. Various standards for assessing whole-body vibration have been promulgated. These standards attempt to define easy methods of quantifying complex vibration conditions, nevertheless no simple standard can offer evaluation procedure which can accurately predict all known effects of vibration on the body. However, to estimate the comfort of car occupants, the authors have followed the method prompted by BS 6841 norm. According to such norm the Vibration Dose Value *VDV* is defined in (6).

$$VDV = \left[\int_{t=0}^{t=T} a_w^{\,4}(t)\,dt \right]^{1/4} \tag{4}$$

where $a_w(t)$ is the frequency-weighted acceleration time-history and T is the period of time over which vibration is measured. The evaluation of (4) requires the weighting of acceleration time history, that can be approximated, as stated in the norm, with piecewise functions. To compute the overall VDV the vibrational signals have to be measured at three points: seat cushion-body interface, seat backrest-body interface and ground-feet interface.

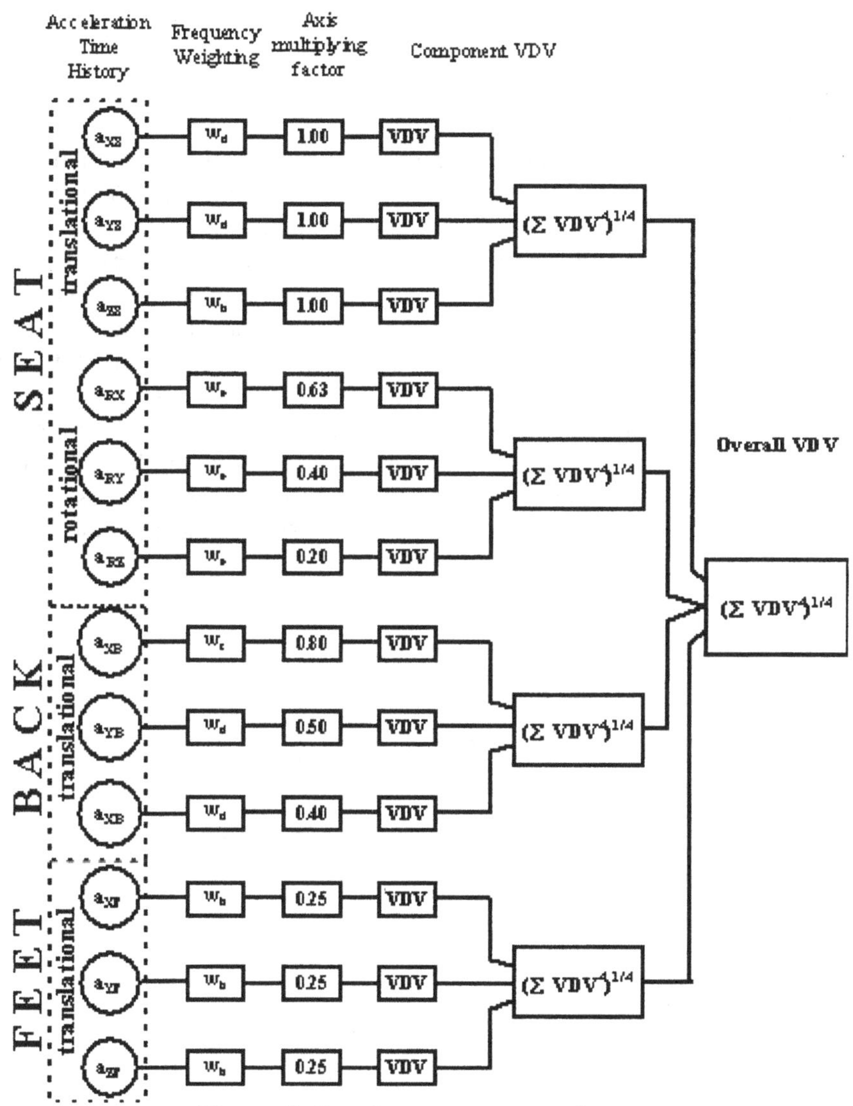

Figure 5. Comfort assessment scheme.

These time histories are then frequency weighted and scaled with different factors. For each weighted signals, VDV are computed and then an overall VDV is computed (Figure 5). The vibrational inputs refer to three points, seat cushion-body interface, seat backrest-body interface and ground-feet interface. For every point of measurement, more acceleration time histories are taken into account. For the seat, both of rotational and traslational acceleration are considered, for back and feet only translational components are considered.

6. Results

In this section some results obtained running DAViD simulation code are presented and compared with those experimentally acquired. The simulated test is a multi-input one. The simulation time is 10 seconds. In Figure 6 are compared the FFT of the experimental and computed time histories of pelvis (vertical acceleration) and upper part (horizontal acceleration). For the simulation we chose a driver belonging to 50 percentile, posed with the angle between legs and thighs of 105 deg, the head inclination of 18 deg, the neck inclination of 8 deg, the angle between the arms and the forearms of 42 deg, the pedals are at 20 deg. The little difference between the magnitudes can be justified taking into account that the experimental data are acquired on the skin of the occupants (i.e. where the SAE accelerometer pads were placed) that are not rigid and the numerical results are computed at the center of mass of rigid bodies. The seat VDV is 2.62; the back one is 3.86 and the feet one is 1.69 obtaining 4.08 as overall VDV.

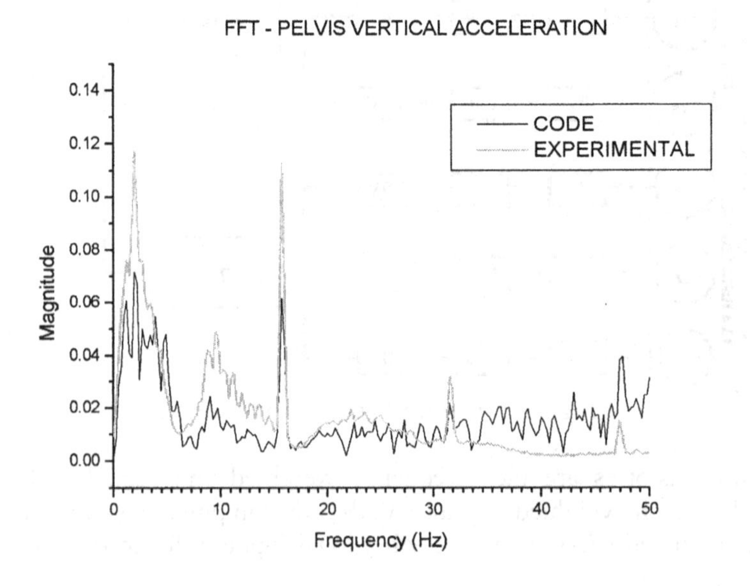

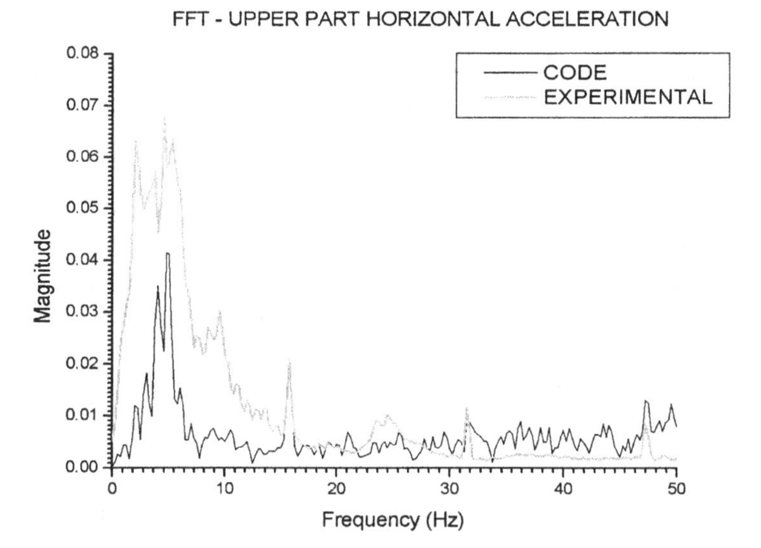

Figure 6. Comparison between computed and experimental FFTs

References

1. Griffin, M.J. (1990) *Handbook of Human Vibration*, Academic Press.
2. Silva, M.P.T., Ambrosio, J.A.C., Pereira, M.S. (1997) Biomechanical Model with Joint Resistance for Impact Simulation, *Multibody System Dynamics*, 65-84.
3. Kirchknopf, P. et al. Developement of a multibody calculation model for the passenger/seat system based on experimental results, *ATA Paper n.01A1085*.
4. Haug, E. J.(1989) *Computer-Aided Kinematics and Dinamics of Mechanical Systems*, Allyn and Bacon, vol. I, pp.48-104.
5. Pennestrì, E. (2002) *Dinamica Tecnica e Computazionale*, Casa Editrice Ambrosiana, Vol. 2.
6. Lin Wei, Griffin, M. J. (9-2000) Effect of subject weight on predictions of seat cushion transmissibility, *35th United Kingdom Group Meeting on Human Responses to Vibration, ISRV*, University of Southampton, England.
7. Lewis, C.H. (9-2000) Evaluating the vibration isolation of soft seats using an active anthropodynamic dummy *35th United Kingdom Group Meeting on Human Responses to Vibration, ISRV*, University of Southampton, England.

8. Lin Wei, Griffin, M.J. (9-2000) Effect of subject weight on predictions of seat cushion transmissibility, *35th United Kingdom Group Meeting on Human Responses to Vibration, ISRV*, University of Southampton, England.
9. Harrier, E., Wanner, G., (1991) *Solving Ordinary and Differential Equations II, Stiff and Differential-Algebraic Problems* Springler Verlag
10. Valentini, P.P. (A.Y. 1999-2000) Analisi delle modalità di implementazione di un manichino virtuale per la simulazione delle dinamiche vibrazionali del sistema uomo-sedile, Thesis, Univ. of Rome Tor Vergata.
11. Xuting Wu, Subhash Rakheia, Paul-Emile Boileau (1999) *Study of Human-Seat Interactions for Dynamic Seating Comfort Analysis*, Society of Automotive Engineers, Inc.
12. King, A.I., (5-1984) A Review of Biomechanical Models, *Journal of Biomechanical Engineering*, **vol.** 106.
13. Yi Gu (1999) A New Dummy for Vibration Transmissibility Measurement in Improving Riding Comfort, Society of Automotive Engineers, Inc.
14. King,A.I.(5-1984)A Review of Biomechanical Models, *Journal of Biomechanical Engineering*, vol.106
15. Amirouche, F.M.L. Xie, M. Patwardhan, A. (11-1994)Optimization of the contact damping and stiffness coefficients to minimize human body vibration, *Journal of Biomechanical Engineering*, vol. 116.
16. Yi Gu (1988) *A Comparison Test of Trasmissibility Response from Human Occupant and Anthropodynamic Dummy*, Society of Automotive Engineers, Inc.
17. Vita L. (A.Y. 2000-2001) Sviluppo ed implementazione di un codice di calcolo 3D per lo studio della dinamica per applicazione all'interfaccia uomo/veicolo, Thesis, University of Rome Tor Vergata.

DESIGN OF NONLINEAR CONTROL OF NONLINEAR MULTIBODY SYSTEMS

M. VALÁŠEK
Department of Mechanics
Faculty of Mechanical Engineering
Czech Technical University in Prague
Karlovo nám. 13, 121 35 Praha 2, Czech Republic
E-Mail: valasek@fsik.cvut.cz

1. Introduction

Multibody systems are the most usual models of mechatronic systems. Mechatronic systems use the synergy achieved by the combination of properties of mechanical (hardware) systems and control (software) systems applied by electronic devices. Therefore the control of multibody systems is very important topic for mechatronic design. Multibody systems have several crucial problems compared with general controlled systems. They make their control to be a challenge in many cases. Its general origin is the usual nonlinearity of multibody systems.

This paper deals with the description of these specific difficulties and with the description of main current approaches for their solutions.

2. Main Problems of Control of Multibody Systems

Multibody systems are the most often models of mechanical systems being controlled. They concentrate the mechanical functionality of machines to be enough precise on one hand and simultaneously to be enough concise for control on the other hand. The main difficulties of control of mechanical (multibody) systems are following:

- Mechanical (multibody) systems are usually nonlinear systems. The well established control methods are developed for linear systems, only recently there have been developed methods for control of nonlinear systems, however still not being so general and powerful as for linear systems.
- Nonlinearity is sometimes essential for the control of multibody systems. There exist multibody systems (e.g. nonholonomic systems) that being linearized about operation point become uncontrollable.
- Multibody systém undergo large displacements where linearization about operation point is impossible. The control problem is quite often not just stabilization around working (equilibrium) point, but the motion from one state into another one.
- Mechanical (multibody) systems in many cases have different number of inputs and outputs, or different number of inputs and degrees of freedom

W. Schiehlen and M. Valášek (eds.), Virtual Nonlinear Multibody Systems, 263–278.

(DOFs). The inputs are operated by actuators (drives). Therefore it can be spoken about under-actuated, equal-actuated and over-actuated systems. This is causing especially problems in combination with nonlinearity of mechanical systems.

- Multibody systems are quite often suitably described by redundant coördinates. Then the models are described by differential-algebraic equations (DAE). DAE cannot be generally and easily transformed into ordinary differential equations (ODE). Only ODE can be described by state space models. The absolute majority of current control methods suppose that the dynamic system being controlled is described by ODE and thus the traditional control methods are difficult to be applied.

Let us summarize the usual control problems to be solved for the control of multibody systems:

- The first fundamental problem is to stabilize the motion of multibody system around its equilibrium. An important subproblem is just to check the stability of some proposed control.
- The next problem is to move the multibody system from one position into another. This problem is transformed into other problems. The problem can be stated as the global stabilization of the system from the initial position around the final position as the stabilized equilibrium. By this way the problem is transformed into previous problem. The other problem statement is to stabilize the system around the given trajectory connecting both positions. It is well known trajectory tracking. For nonlinear multibody systems it can be a serious problem just to construct admissible trajectory connecting initial and final positions.
- The previous problem can be formulated as the determination of optimal trajectory with respect to some performance index.
- The problem could be just to find any admissible trajectory of the system connecting two positions of multibody system.

The stabilizing problems are addressed by several groups of methods:

- There were developed methods that exactly linearized the nonlinear system and thus enable after the linearizing transformation to apply standard control methods for linear systems.
- Another approach is to propose a control law based on some insight into the system and then to carry out the synthesis of control parameters for the nonlinear plant as control gain optimization by the technique of multiobjective parameter optimization (MOPO).
- Very promising method is to rewrite the nonlinear system into the form of linear system and then to apply the standard control methods for linear systems however for position (state) dependent system. It is another way for transforming the nonlinear system into linear-like one.

The problem of constructing suitable trajectory is formulated

- either directly as the optimal control problem that is generally not quite easy to be solved

- or in specifically difficult cases of under-actuated systems as time planning problem that requires first to construct the missing control actions in order to modify the system into equal-actuated one.

3. Simple Demonstrational Example

The above stated problems and further described solution methods can be demonstrated on the following simple problem. Despite its simplicity this example includes all serious problem of control of multibody systems.

The exemplary multibody system (*Fig. 1*) consists of two bodies connected by revolute joints in plane. It has two DOFs. The first joint is always actuated by controlled action M_1. The second joint can be or can be not actuated by the torque M_2. If it is actuated then the system is equal-actuated and methods for exact linearization can be applied. It is a case of rigid planar robot manipulator. If the second joint is not actuated then the system is under-actuated and there are problems of its control. If this not-actuated joint is equipped with spring and damper then it is a model of flexible robotic arm. If there is no spring in this not-actuated joint then it is a model of mechatronic toy with different control challenges.

The multibody system can be described by the independent coordinates (e.g. φ_1 and φ_2) or by the dependent coordinates (e.g. the cartesian coordinates of centre of mass and angles of links with respect to the frame or natural Cartesian coordinates of ends of links). According to that the control is formulated as traditional control of ODE or control of DAE.

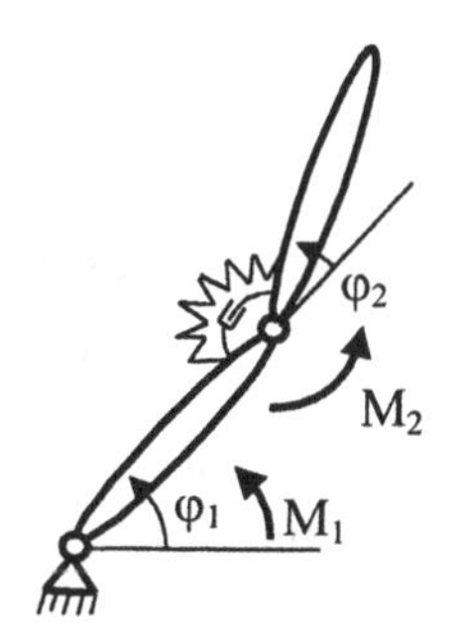

Figure 1. Example of simple multibody system with different control problems

4. Description of Nonlinear Multibody Systems

The (nonlinear) multibody systems including the control action are generally described by Lagrange equations of mixed type (e.g. [1])

$$\mathbf{M}(\mathbf{s})\frac{d^2\mathbf{s}}{dt^2} - \mathbf{\Phi}^T\boldsymbol{\lambda} = \mathbf{Q}(\mathbf{s},\frac{d\mathbf{s}}{dt}) + \mathbf{T}(\mathbf{s},\frac{d\mathbf{s}}{dt})\mathbf{u} \tag{1}$$

$$\mathbf{F}(\mathbf{s},\frac{d\mathbf{s}}{dt}) = \mathbf{0} \tag{2}$$

where **s** are the coordinates describing the multibody system, **u** the vector of input control variables, **M** is the mass matrix, **Q** are the generalized forces, **F** are the kinematic (holonomic or nonholonomic) constraints, **Φ** is the Jacobian corresponding to the kinematic constraints, **λ** are the Lagrange multipliers, **u** the control inputs (forces) acting through transmissions **T**. This fact that the general description of multibody systems uses differential-algebraic equations (DAE) is one of difficult problem of control synthesis for multibody systems.

These equations of motion for the purposes of control must be often transformed into the state space description. There are two possible ways how to do it. The first approach preserves the same variables **s** from original description and only eliminates the Lagrange multipliers **λ**. The kinematic constraints **F** in (2) are twice differentiated with respect to time

$$\mathbf{\Phi}\frac{d^2\mathbf{s}}{dt^2} = -(\mathbf{\Phi}\frac{d\mathbf{s}}{dt})_s\frac{d\mathbf{s}}{dt} \tag{3}$$

where subscript s means partial differentiation with respect to the coordinates **s**. The accelerations are expressed from the equations of motion (1) by multiplying it by the inverse of mass matrix M and they are substituted into (3). The equations for the Lagrange multipliers are thus derived. Solving for them and substituting back it is obtained the formulation of equations of motion of multibody system (1) in seemingly ordinary differential equations (ODE) only.

$$\begin{aligned}\frac{d^2\mathbf{s}}{dt^2} &= \mathbf{M}^{-1}(\mathbf{I} - \mathbf{\Phi}^T(\mathbf{\Phi}\mathbf{M}^{-1}\mathbf{\Phi}^T)^{-1}\mathbf{\Phi}\mathbf{M}^{-1})(\mathbf{Q}+\mathbf{Tu}) - \\ &\quad - \mathbf{M}^{-1}\mathbf{\Phi}^T(\mathbf{\Phi}\mathbf{M}^{-1}\mathbf{\Phi}^T)^{-1}(\mathbf{\Phi}\frac{d\mathbf{s}}{dt})_s\frac{d\mathbf{s}}{dt}\end{aligned} \tag{4}$$

Hence the state space description with state vector [**s**, d/dt **s**] in dependent coordinates **s** can be obtained. However, in such case the resulted equations are after linearization not controllable. It follows from the fact that in such case the dependent coordinates on the level of acceleration are related by the quadratic terms in velocities as follows from the second derivatives of kinematic constraints **F** in (3). Such equations are after linearization not controllable in usual positions with zero velocities. Nevertheless this simple formulation is sometimes useful, e.g. for optimal control in section 8.

The other approach is based on the description of multibody systems in independent coordinates **q**. The coordinates must be independent, e.g. the independent coordinates **q** are selected (constructed) from **s** generally

$$\mathbf{s} = \mathbf{r}(\mathbf{q})$$
$$\frac{d}{dt}\mathbf{s} = \mathbf{R}\frac{d}{dt}\mathbf{q} \tag{5}$$
$$\frac{d^2}{dt^2}\mathbf{s} = \mathbf{R}\frac{d^2}{dt^2}\mathbf{q} + \frac{d}{dt}\mathbf{R}\frac{d}{dt}\mathbf{q}$$

Substituting from (5) into the first time derivative of the kinematic constraint it is found out that this matrix **R** is the null space of the Jacobian **Φ**

$$\mathbf{\Phi R} = 0 \tag{6}$$

Thus multiplying the equations of motion (1) by **R**T eliminates again the Lagrange multipliers. Then using the state vector **x**=[**q**, d**q**/dt] the equations of motion can be transformed into the state space model

$$\frac{d}{dt}\mathbf{q} = \dot{\mathbf{q}}$$
$$\frac{d}{dt}\dot{\mathbf{q}} = (\mathbf{R}^T\mathbf{M}\mathbf{R})^{-1}(\mathbf{R}^T\mathbf{g} - \mathbf{R}^T\mathbf{M}\dot{\mathbf{R}}\dot{\mathbf{q}}) + (\mathbf{R}^T\mathbf{M}\mathbf{R})^{-1}\mathbf{R}^T\mathbf{T}\mathbf{u} \tag{7}$$

The theory of multibody systems and the solution of corresponding DAE describes several methods for deriving and obtaining the null space matrix **R** (e.g. [1]).

Now the dynamics of multibody systém can be described in the form of general nonlinear system in state space description

$$\frac{d\mathbf{x}}{dt} = \mathbf{f}(\mathbf{x}) + \mathbf{g}(\mathbf{x})\,\mathbf{u} \tag{8}$$

where **x**(nx1) is the state vector, **u**(mx1) is the control vector and generally **f(0)=0**. For the purpose of control it must be stated which variables **y** can be measured and used as the input to the control system. It is supposed their general dependence

$$\mathbf{y} = \mathbf{h}(\mathbf{x}) \tag{9}$$

Thus there is the output variable vector **y**(px1).

5. Exact Feedback Linearization of Nonlinear Systems

One of the fundamental approaches towards the control of nonlinear systems is the exact feedback linearization [2-5]. Its idea is simple. Considering the system (8) it is investigated the existence and the construction of a suitable (static) state feedback control

$$\mathbf{u} = \alpha(\mathbf{x}) + \beta(\mathbf{x})\,\mathbf{w} \tag{10}$$

and a transformation of variables

$$\mathbf{z} = \mathbf{T}(\mathbf{x}) \tag{11}$$

that together transform the original system (8) into a linear one

$$\frac{d\mathbf{z}}{dt} = \mathbf{A}\mathbf{z} + \mathbf{B}\mathbf{w} \tag{12}$$

The same problem can be investigated for the existence of a suitable dynamic state feedback control

$$\begin{aligned} \mathbf{u} &= \alpha(\mathbf{x},\mathbf{p}) + \beta(\mathbf{x},\mathbf{p})\mathbf{w} \\ \frac{d\mathbf{p}}{dt} &= \varphi(\mathbf{x}) + \eta(\mathbf{x})\,\mathbf{u} \end{aligned} \tag{13}$$

with a transformation of variables (11) that again transform the original system into a linear one (12). These problems are called input-state (exact feedback) linearization. And finally both these problems can be investigated for output feedback where instead of state variables $\mathbf{x}$ in feedbacks (10) or (13) there are used only output variables $\mathbf{y}$ (9). Then these problems are called input-output (exact feedback) linearization.

Then after transforming the original system into the equivalent linear one (12) there are applied the control techniques for linear systems like pole placement, LQR etc.

5.1 INPUT-OUTPUT EXACT FEEDBACK LINEARIZATION

The solution of the problem of input-output exact feedback linearization is simpler. The construction of possible linearizing feedback for the system (8)-(9) is the recursive application of the time differentiation rule

$$\begin{aligned} \mathbf{y}^{(1)} &= \frac{\partial \mathbf{h}}{\partial \mathbf{x}}(\mathbf{f}(\mathbf{x}) + \mathbf{g}(\mathbf{x})\,\mathbf{u}) = \mathbf{f}_1(\mathbf{x}) + \mathbf{g}_1(\mathbf{x})\,\mathbf{u} = L_{\mathbf{f}}\mathbf{h} + L_{\mathbf{g}}\mathbf{h}\,\mathbf{u} =_{def} \dot{\mathbf{z}}_1 =_{\text{def}} \mathbf{z}_2 \\ \mathbf{y}^{(k+1)} &= \frac{\partial \mathbf{z}_k}{\partial \mathbf{x}}(\mathbf{f}(\mathbf{x}) + \mathbf{g}(\mathbf{x})\,\mathbf{u}) = \mathbf{f}_{k+1}(\mathbf{x}) + \mathbf{g}_{k+1}(\mathbf{x})\,\mathbf{u} =_{def} \dot{\mathbf{z}}_{k+1} =_{\text{def}} \mathbf{z}_{k+2} \end{aligned} \tag{14}$$

There is supposed besides necessary smoothness that $\mathbf{g}_i=0$ for $i=1,\dots,r-1$ and $\mathbf{g}_r\neq0$ enabling to express

$$\mathbf{f}_r(\mathbf{x})+\mathbf{g}_r(\mathbf{x})\mathbf{u}=\mathbf{w}$$
$$\mathbf{u}=\frac{\mathbf{w}-\mathbf{f}_r(\mathbf{x})}{\mathbf{g}_r(\mathbf{x})} \tag{15}$$

and this leads to

$$\mathbf{y}^{(r)}=\mathbf{w} \tag{16}$$

This means that the original system is first transformed into the Brunovsky canonical form (14) and then into the Frobenius canonical form (16). These canonical forms prove the controllability of the system and enable to stabilize the system easily by pole placement.

The transformation (11) is

$$\mathbf{z}=\mathbf{T}(\mathbf{x})=[(\mathbf{h}(\mathbf{x}))^T,(L_\mathbf{f}\mathbf{h}(\mathbf{x}))^T,\ldots,(L_\mathbf{f}^{r-1}\mathbf{h}(\mathbf{x}))^T]^T \tag{17}$$

Here and in previous expressions there is used the formalism of Lie algebra

$$L_\mathbf{f}\mathbf{h}(\mathbf{x})=\frac{\partial\mathbf{h}}{\partial\mathbf{x}}\mathbf{f}(\mathbf{x}),\quad L_\mathbf{f}^{i+1}\mathbf{h}(\mathbf{x})=L_\mathbf{f}L_\mathbf{f}^{i}\mathbf{h}(\mathbf{x})$$
$$[\mathbf{f},\mathbf{g}](\mathbf{x})=\frac{\partial\mathbf{g}}{\partial\mathbf{x}}\mathbf{f}(\mathbf{x})-\frac{\partial\mathbf{f}}{\partial\mathbf{x}}\mathbf{g}(\mathbf{x}),\quad ad_\mathbf{f}^{0}\mathbf{g}(\mathbf{x})=\mathbf{g}(\mathbf{x}), \tag{18}$$
$$ad_\mathbf{f}^{1}\mathbf{g}(\mathbf{x})=[\mathbf{f},\mathbf{g}](\mathbf{x}),\quad ad_\mathbf{f}^{i+1}\mathbf{g}(\mathbf{x})=[\mathbf{f},ad_\mathbf{f}^{i}\mathbf{g}](\mathbf{x})$$

It can happen and it happens that $r<n$. In that case there is another dynamics in the system (8) besides (16). It is so called zero dynamics that is expressed from the original system stating the condition $\mathbf{y}=\mathbf{y}^{(1)}=\ldots=\mathbf{y}^{(r)}=\mathbf{0}$. The zero dynamics is not influenced by the transformation (17) and the control (15). The stability of this zero dynamics is decisive for the stabilizability of the whole system just by stabilizing the input-output dynamics (16) by the feedback (15) and pole placement.

5.2 INPUT-STATE EXACT FEEDBACK LINEARIZATION

This is the problem whether there exists such function $\mathbf{y}=\mathbf{h}(\mathbf{x})$ for which the input-output exact feedback linearization gives r=n. This means that the system is completely linearized and there is no zero dynamics with the danger of destabilizing the system. There have been proved the necessary and sufficient condition for its existence. This is the full rank (equal n) of the vector field $[\mathbf{g}, ad_\mathbf{f}\mathbf{g},\ldots,ad_\mathbf{f}^{n-1}\mathbf{g}]$ and the set $[\mathbf{g}, ad_\mathbf{f}\mathbf{g},\ldots,ad_\mathbf{f}^{n-2}\mathbf{g}]$ is involutive (this means that if two vectors $\mathbf{k}_1$ and $\mathbf{k}_2$ are from this set then $[\mathbf{k}_1,\mathbf{k}_2]$ is also from this set). However, the construction of such $\mathbf{h}(\mathbf{x})$ is not so easy as the computation of the input-output exact feedback linearization (14).

In all these derivations it is supposed that the dimension of **y** is equal to the dimension of **u** and **w**. This is certainly often violated that prevents to apply this method for the control of under-actuated systems.

5.3 DYNAMIC STATE EXACT FEEDBACK LINEARIZATION

There has been proved that the existence of dynamic exact feedback linearization is equivalent to the special property of the system (8)-(9) called flatness [6-8]. A dynamic system is flat if the system (8)-(9) can be transformed into the relation between states and outputs and inputs and outputs in the form

$$\mathbf{x} = \Phi(\mathbf{y}, \mathbf{y}^{(1)}, \ldots, \mathbf{y}^{(\alpha)}) \tag{19}$$

$$\mathbf{u} = \Psi(\mathbf{y}, \mathbf{y}^{(1)}, \ldots, \mathbf{y}^{(\alpha)}, \mathbf{y}^{(\alpha+1)}) \tag{20}$$

The equation (19) is inverted and substituted into (20) together with excessive derivatives of output variables **y** that are replaced by new subsystems of the form (16). This approach enable decide what kind of trajectories are admissible for the control of the investigated system.

Again in all these derivations it is supposed that the dimension of **y** is equal to the dimension of **u** and **w**. This is also violated for corresponding under-actuated systems (e.g. inverse pendulum). Thus only certain class of under-actuated systems can be controlled by this approach (e.g. crane with control of the cable length is flat [8-9], but crane without control of the cable length is not flat).

5.4 EXACT FEEDBACK LINEARIZATION OF DAE SYSTEMS

Interesting problem is the way of application of exact feedback linearization to DAE systems. The input-output exact feedback linearization can be applied immediately after elimination the Lagrange multipliers (4). One natural condition is that the output variables (9) must be selected as independent from the point of view of constraints (2). Therefore it is important to consider the choice of outputs based on the DOFs of the multibody system. However, the checking of condition for existence of input-state exact feedback linearization must be based on independent coordinates otherwise the transformation can loos full rank.

Another problem is what is the influence of exact feedback linearization constructed for independent coordinates on the remaining redundant ones [11]. The linearization of the input-output dynamics (16) means the establishment of superposition for nonlinear system. If two controls $\mathbf{w}_1$ and $\mathbf{w}_2$ cause two motions $\mathbf{y}_1$ and $\mathbf{y}_2$ of the multibody system described in independent variables then its linear combination $\alpha\mathbf{w}_1 + \beta\mathbf{w}_2$ leads to the motion $\alpha\mathbf{y}_1 + \beta\mathbf{y}_2$. This is not valid for the dependent output variables $\mathbf{y}_d$ because the constraint (2) is usually nonlinear

$$\begin{gathered}\mathbf{F}(\mathbf{y}, \mathbf{y}_d) = 0, \quad \mathbf{y}_d = \mathbf{F}^{-1}(\mathbf{y}), \\ \mathbf{F}^{-1}(\alpha\mathbf{y}_1 + \beta\mathbf{y}_2) \neq \alpha\mathbf{F}^{-1}(\mathbf{y}_1) + \beta\mathbf{F}^{-1}(\mathbf{y}_2)\end{gathered} \tag{21}$$

Nevertheless, if the linear combination moves to the desired position $\mathbf{y}_{des}$

$$\alpha \mathbf{y}_1 + \beta \mathbf{y}_2 = \mathbf{y}_{des} \tag{22}$$

then also the motion of dependent variables $\mathbf{y}_d$ ends in the corresponding desired position $\mathbf{y}_{d,\,des}$ as follows from (21)

$$\mathbf{y}_{d,des} = \mathbf{F}^{-1}(\mathbf{y}_{des}) \tag{23}$$

This fact is important for determining the suitable admissible (suboptimal) trajectories (e.g. extension of [10] for DAE in [11]).

6. Multiobjective Parameter Optimization

This method of control synthesis is the least restrictive one regarding the requirement on the description of nonlinear (multibody) system and the form of control law. However, the strength of claims about the properties of the synthetized control is relatively low. The synthesis procedure is following:

The Multi-Objective Parameter Optimization (MOPO) [12] is based on the design-by-simulaton. There is available the nonlinear simulation model of the controlled plant including the control law, the performance index usually in the form of integral performance index or maximum value and selected set of considered excitations. The control law is described in parametric form and its parameters are determined by the numerical optimization of the performance index evaluated by the simulation response of the plant to the considered excitations. Thus, by means of the MOPO approach the nonlinear models and models, which cannot be analytically treated, can be used. This approach enables not only to find parameters of nonlinear control of nonlinear plant, but it allows one also to find a satisfactory compromise among the performance criteria despite the fact that they conflict with each other. The MOPO approach is based on a search in the parameter space (Pareto optimality) by model simulation.

"Free system parameters", the tuning parameters, e.g. control coefficients, mass properties or installation positions, are varied within their given limits until an "optimal compromise" is found. In doing so the performance criteria (also called objective functions) $\mathbf{c}$ are weighted by user-defined weighting factors, the design parameters $\mathbf{d} \geq \mathbf{c}_{start}$. The optimization strategy is to minimize all weighted criteria $c_i(\mathbf{p})/d_i(\mathbf{p})$ in such a way that the currently *worst* criterion with the maximal value will be reduced: optimization tries to decrease the weighted criterion $c_i(\mathbf{p})/d_i(\mathbf{p}) \to 0$. The tuning parameters $\mathbf{p}$ are determined by solving a minmax-optimization problem with constraints and parameter restrictions

$$\alpha^* = \min_p(\max_i c_i / d_i) \tag{24}$$

$$c_i(\mathbf{p}) \leq d_i \ \text{ for } i \in \Psi \tag{25}$$

$$p_k^- \le p_k \le p_k^+ \tag{26}$$

The performance criteria (objective functions) may be free to be optimized or may be partially limited by a set Ψ of performance constraints, e.g. mechanical, hydraulic or electronic restrictions like power, pressure, current and voltage. The weighting factors or design parameters enable the user to adapt the criteria to adequate sizes and to determine the direction of optimization process by weighting some criteria more important $\left(c_l(\mathbf{p})/d_l(\mathbf{p})=1\right)$, others less important $\left(c_j(\mathbf{p})/d_j(\mathbf{p})<c_l(\mathbf{p})/d_l(\mathbf{p})\right)$. The parameter optimization is finished, when the maximum of all weighted criteria cannot be further decreased. The result is a point on the Pareto-optimal boundary (*Fig. 2*).

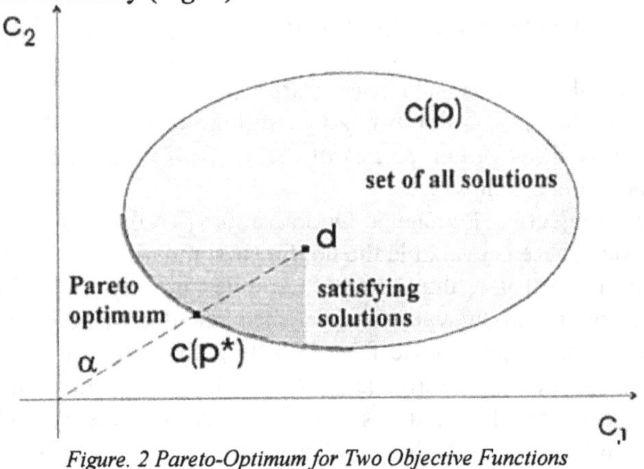

Figure. 2 Pareto-Optimum for Two Objective Functions

An example of application of MOPO synthesis for vehicle suspensions is in [13].

7. Nonlinear Linearization of Nonlinear Systems

This is another way of transforming a nonlinear system into a linear form where methods of linear control can be applied [14-19]. The dynamics of the nonlinear system is described by the equations (8). The multibody system must be described by independent coordinates, otherwise further described decomposed system is uncontrollable. The quadratic performance index of the control which is to be synthesized is the infinite horizon control

$$J=\int_0^\infty (\mathbf{x}^T\mathbf{Q}\,\mathbf{x}+\mathbf{u}^T\mathbf{R}\,\mathbf{u})\,dt \tag{27}$$

If there exists the decomposition of the system dynamics

$$\mathbf{f}(\mathbf{x})=\mathbf{A}(\mathbf{x})\mathbf{x} \tag{28}$$

and if the following assumptions [14] are valid

- The matrixes **Q** and **R** are positive definite.
- The matrixes **A** and **g** are analytic valued functions.
- There exists the control function **u**(t) and corresponding state trajectory **x**(t) for t∈<0,∞> which satisfy the system dynamics (8) and the performance index (27) is finite.
- The pair of marixes (**A(x)**,**g(x)**) is controllable and stabilizable for each **x** in the linear system sense, i.e.

$$rank[\mathbf{g(x)}, \mathbf{A(x)g(x)}, \ldots, \mathbf{A}^{n-1}\mathbf{(x)g(x)}] = n \tag{29}$$

- The state vector **x** is fully measured.

then there exists the control

$$\mathbf{u} = -\mathbf{K(x)x} \tag{30}$$

which minimizes the performance index (27). However, it has been shown that this control is only suboptimal.

The nonlinear gain matrix **K(x)** is determined as

$$\mathbf{K(x)} = \mathbf{R}^{-1}\mathbf{g}^T\mathbf{(x)P(x)} \tag{31}$$

where **P(x)** is the solution of the Riccatti equations

$$\mathbf{A}^T\mathbf{(x)P(x)} + \mathbf{P(x)A(x)} + \mathbf{Q} - \mathbf{P(x)g(x)R}^{-1}\mathbf{g}^T\mathbf{(x)P(x)} = \mathbf{0} \tag{32}$$

This control approach is called Nonlinear Quadratic Regulator (NQR) [17] or State Dependent Riccatti Equation (SDRE) control [15, 16].

The key problem is the computation of the decomposition **A(x)**. There have been derived some formulas for the computation of this decomposition, like [16]

$$\mathbf{A(x)} = \int_0^1 \left.\frac{\partial \mathbf{f}}{\partial \mathbf{x}}\right|_{\mathbf{x}=\lambda\,\mathbf{x}} d\lambda \tag{33}$$

But the computation of this formula for the multibody systems is very difficult because it requires the symbolic manipulation with the equations of motion which might be very large and complex and might include rational and trigonometric functions. Therefore it is not realistic to suppose to compute (33) for multibody systems.

Therefore there have been developed efficient algorithms [17] for computation of this decomposition that moreover enables to use the non-uniqueness of the decomposition (26) for control synthesis. The simplest procedure among them is following. Its idea is based on the following decomposition described for scalar function f of three variables x_1, x_2, x_3

$$
\begin{aligned}
f(x_1,x_2,x_3) &= \frac{f(x_1,x_2,x_3)-f(0,x_2,x_3)}{x_1}x_1 + \\
&+ \frac{f(0,x_2,x_3)-f(0,0,x_3)}{x_2}x_2 + \\
&+ \frac{f(0,0,x_3)-f(0,0,0)}{x_3}x_3
\end{aligned}
\tag{34}
$$

This decomposition must certainly solve the division by zero and it is dependent on the order of variables. The division by zero is equal according to l'Hospital rule in the limit to the corresponding derivative which for smooth functions exists and can be numerically computed by division for magnitudes of numerical accuracy of the computer.

The computation of this decomposition is not unique. It depends on the order of variables and besides that there are some free parameters for decomposition parametrization [17]. The influence of this ambiguity can be described and investigated by the recursive decomposition (34) applied to each addend of the previous decomposition step. There are 2^n such addends at the end for n variables which is very natural for polynomial functions f. However, the problem is that any addend f_i in any level of decomposition can be split into convex sum $f_i\lambda+f_i(1-\lambda)$ where λ can be any function. This makes the decomposition fully ambiguitious. The level of recursive decomposition and-or the free functions λ can be selected in order to improve controllability of decomposed system (8) with (28).

The main problem with NQR is that first it is not the optimum control of general nonlinear system as it has been shown in [18] and second there is still missing a proof of global stability of the control law (30) despite partial results [19]. There has not been found any case of unstable NQR control but general proof is missing. The practical experience is excellent and the stability can be checked for individual cases using the computed solution of the Riccati equation (30). The Lyapunov function V is proposed as

$$
V = \mathbf{x}^T \mathbf{P}(\mathbf{x})\mathbf{x} \tag{35}
$$

It is positive definite according to the properties of the solution of Riccati equation and the negativeness of its time derivative can be checked similarly in the same grid as the computed values of the gain $\mathbf{K}(\mathbf{x})$ deriving into

$$
\frac{dV}{dt} = -\mathbf{x}^T(\mathbf{Q}+\mathbf{P}(\mathbf{x})\mathbf{g}(\mathbf{x})\mathbf{R}^{-1}\mathbf{g}^T(\mathbf{x})\mathbf{P}(\mathbf{x}) - \frac{d\mathbf{P}(\mathbf{x})}{d\mathbf{x}}(\mathbf{A}(\mathbf{x})-\mathbf{g}(\mathbf{x})\mathbf{K}(\mathbf{x}))\mathbf{x})\,\mathbf{x} \le 0 \tag{36}
$$

The first two terms are proving stability and the third term is usually small.

An important practical issue of NQR is the implementation of the synthetized control law. One approach can be the real-time computation of Riccatti equation and the whole control. It is certainly very demanding of computational power but for some small systems reasonable.

The other approach is based generally on the look-up tables. It means that the solutions of NQR control are computed for some values of states off-line and for the on-line application the pre-computed gains are either retrieved or interpolated. Such approaches were already used for gain scheduling approaches. The problem is certainly the efficient choice of suitable states for off-line computation and the on-line interpolation between them. This approach is applicable even for large multibody systems.

Another important advantage of NQR approach is that the control can be extended into incorporating different limitations of admissible control sets (e.g. asymmetric interval of control variable for semi-active systems [20]).

This approach of NQR is applicable to all under-actuated (certainly equal-actuated) systems that satisfy the condition that the linearized system about the equilibrium $\mathbf{x}=\mathbf{0}$ is controllable. Even this condition is violated for important classes of under-actuated systems (e.g. some nonholonomic multibody systems). For this class of multibody systems the control is very difficult. The stabilizing control were developed for such systems having $\mathbf{f}(\mathbf{x})=\mathbf{0}$, but for general systems with $\mathbf{f}(\mathbf{x})\neq\mathbf{0}$ it is still an open problem.

8. Optimal Control of Multibody Systems

The optimal control in the formulation of Pontryagin principle of maximum is applied for nonlinear multibody systems especially for the purpose of trajectory synthesis. The traditional formulation of Pontryagin principle of maximum for ODE of the system (8) is well known. However, its formulation for DAE systems is quite new problem. There were several approaches to extend the maximum principle straightforward from ODE to DAE systems, however with incorrect results. Only recently [21] it has succeeded to derive the correct proven extension. It is based on the idea that the DAE system (1)-(2) is transformed into the ODE system (4) just for the formulation of optimality conditions. In other words it means that the optimality conditions raised on the Hamiltonian must not involve the Lagrange multipliers λ. The optimality conditions are formulated using the equivalent ODE problem, but the problem solution is done within DAE problem. Thus the efficient and/or convenient formulation of multibody model is maintained and only the optimality conditions are formulated based on ODE.

Let us determine the optimal control $\mathbf{u}$ such that it minimizes the performance index

$$J = l(\mathbf{x}(T)) + \int_0^T (h(t,\mathbf{x},\mathbf{y},\mathbf{u})\,dt \tag{37}$$

for the system generally described by the differential equations

$$\frac{\mathbf{dx}}{\mathrm{dt}} = \mathbf{f}(t, \mathbf{x}, \mathbf{y}, \mathbf{u}) \tag{38}$$

and algebraic equations

$$\mathbf{0} = \mathbf{g}(t, \mathbf{x}, \mathbf{y}) \tag{39}$$

during the motion in the time interval <0,T>. Here $\mathbf{x}$ describes so called slow variables (in case of multibody systems they are positions and velocities), $\mathbf{y}$ describes so called fast variables (in case of multibody systems they are Lagrange multipliers) and $\mathbf{u}$ describes the control variables.

The Hamiltonian is introduced as in traditional ODE problems

$$\mathrm{H} = -h(t, \mathbf{x}, \mathbf{y}, \mathbf{u}) + \boldsymbol{\psi}^T \mathbf{f}(t, \mathbf{x}, \mathbf{y}, \mathbf{u}) \tag{40}$$

with straightforward conjugate equations

$$\begin{aligned} \frac{d\boldsymbol{\psi}}{dt} &= -\frac{\partial H}{\partial \mathbf{x}} - \frac{\partial \mathbf{g}}{\partial \mathbf{x}}(t, \mathbf{x}, \mathbf{y})^T \boldsymbol{\mu} \\ \mathbf{0} &= -\frac{\partial H}{\partial \mathbf{y}} - \frac{\partial \mathbf{g}}{\partial \mathbf{y}}(t, \mathbf{x}, \mathbf{y})^T \boldsymbol{\mu} \end{aligned} \tag{41}$$

Then the optimality conditions of maximum principle are

$$H = \max_{(\mathbf{y}, \mathbf{u}) \in M(t, \mathbf{x})} H(t, \mathbf{x}, \mathbf{y}, \mathbf{u}) \tag{42}$$

This is the same as traditional maximum principle but the control together with the fast variable must be bound to the manifold M

$$M(t, \mathbf{x}) = \left\{ (\mathbf{y}, \mathbf{u}); \left[D^{i-1} \mathbf{g} \right] (t, \mathbf{x}, \mathbf{y}, \mathbf{u}) = \mathbf{0} \right\} \tag{43}$$

where operator D means the time derivative of the algebraic constraint $\mathbf{g}$ in (39) with the usage of differential equations (38) and i is the index of DAE.

9. Cyclic Control

The control, but just the synthesis of suitable admissible trajectory for under-actuated systems that are not flat, being linearized around equilibrium are loosing controllability or simply the system states go over the boundary of singularities, is a difficult problem. Then there is possible the approach of cyclic or invariant control [22].

Let solve the equation (15) for input variables **u** of the system (8) in the number equal to the number of output variables **y** in (9). The corresponding subsystems *i* and *j* will have the form (16) and (14)

$$
\begin{aligned}
\mathbf{y}_i^{(r_i)} &= \mathbf{w} \\
\mathbf{y}_j^{(r_j)} &= \mathbf{f}_{j,r_j}(\mathbf{x}) + \mathbf{g}_{j,r_j}(\mathbf{x})\,\mathbf{u} = \mathbf{f}_{j,r_j}(\mathbf{x}) + \mathbf{g}_{j,r_j}(\mathbf{x}) \frac{\mathbf{w} - \mathbf{f}_{i,r_i}(\mathbf{x})}{\mathbf{g}_{i,r_i}(\mathbf{x})}
\end{aligned} \tag{44}
$$

The new input variables **w** can be divided into two parts – arbitrary $\mathbf{w}_A$ and cyclic (invariant) $\mathbf{w}_I$ control

$$
\mathbf{w} = \mathbf{w}_A + \mathbf{w}_I \tag{45}
$$

The arbitrary control is used for the control of subsystem *i* and the cyclic (invariant) control is used for the control of those subsystems *j* that cannot be linearized as the subsystem *i*. The invariant control is constructed based on periodic functions $\{\mathbf{w}_{I,m}\}$ with period T in such way that it does not influence the state of subsystem *i* in times kT

$$
\begin{aligned}
& w_I = \sum_m k_m w_{I,m} \\
& \int_0^T w_{I,m} dt = 0,\ \iint w_{I,m} dt dt = 0, \ldots, \int \ldots \int_0^T w_{I,m} dt \ldots dt = 0
\end{aligned} \tag{46}
$$

By this way there are constructed new control variables $\mathbf{w}_I$ that influence only other subsystems than the subsystem *i*. Such control can be used for the path across the boundary of singularity etc. This control can be then combined with NQR control around the vicinity of equilibrium iff the system is controllable after linearization.

10. Conclusions

There have been developed many very powerful control approaches for general nonlinear systems. A special case are nonlinear multibody systems. Despite many new control methods there still remain open problems (e.g. stabilization of nonholonomic systems with offset). Further there are specific problems of application of control methods for multibody systems if they are described by redundant coordinates and DAE.

Acknowledment: The author appreciate the kind support by the grant MSMT J04/98: 212200003 "Algorithms of computational mechanics".

11. References

1. Stejskal, V., Valasek, M. (1996) *Kinematics and Dynamics of Machinery*, Marcel Dekker, New York
2. Isidori, A. (1989) *Nonlinear Control Systems*, Springer-Verlag, Berlin.
3. Slotine, J.J., Li, W. (1991) *Applied Nonlinear Control*, Prentice Hall, New Jersey
4. Fliess, M. (ed.) (1992) *Nonlinear Control System Design*, Pergamon Press.
5. Khalil, H.K. (1996) *Nonlinear Systems*, Prentice Hall, New Jersey.
6. Fliess, M. et al. (1995) Flatness and Dynamic Feedback Linearizability: Two Approaches, In: *Proc. Of 3rd European Control Conference*, Rome, pp. 649-654.
7. Delaleau, E., Rudolph, J. (1998) Control of Flat Systems by Quasi-static Feedback of Generalized States, *Int. J. of Control* 71 (1998), 5, pp. 745-765.
8. Rudolph, J., Delaleau, E. (1998) *Some Examples and Remarks on Quasi-Static Feedback of Generalized States*, Automatica 34 (1998), 8, pp. 993-999.
9. Woernle, Ch. (2002) *Control of Robotic Systems by Exact Linearization Methods*, In: P. Maisser, P. Tenberge (eds.), Advanced Driving Systems, Chemnitz
10. Valasek, M. (1986) Synthesis of Optimal Trajectory of Industrial Robots, Kybernetika 22 (1986), pp. 409-424
11. Bastl, P. (2002) Optimal Control of Multibody Systems Described by Redundant Coordinates, PhD Thesis, Czech Technical University in Prague.
12. Joos, H.D. (1992) Informationstechnische Behandlung des mehrzieligen optimierungsgestutzten regelungstechnischen Entwurfs, PhD Thesis, Univ. of Stuttgart.
13. Valasek, M. et al. (1998) Development of Semi-Active Road-Friendly Truck Suspensions, *Control Engineering Practice* 6(1998), pp. 735-744
14. Banks, S.P., Mhana, K.J. (1992) Optimal control and stabilization for nonlinear systems, *SIAM Journal of Mathematical Control and Information*, 9(1992), pp. 179-196
15. Cloutier, J.R. (1997) *State-Dependent Riccati Equation Techniques: An Overview*, Proc. of 1997 American Control Conference, Albuquerque 1997, pp. 932-936
16. Langson, W., Alleyne, A. (1997) Inifinte horizon optimal control of a class of nonlinear systems, *Proc. of the IEEE American Control Conference* 1997, pp. 3017-3022
17. Valasek, M., Steinbauer, P. (1999) Nonlinear Control of Multibody Systems, In: Ambrosio, J., Schiehlen, W. (eds.): *Proc. of Euromech Colloquium 404, Advances in Computational Multibody Dynamics*, Lisabon, pp. 437-444
18. Cloutier, J. R. et al. (1999) On the recoverability of nonlinear state feedback laws by extended linearization control techniques, *Proc. of the 1999 American Control Conference,* San Diego, pp. 1515-1519.
19. Erdem, E.B., Alleyne, A. (1999) Globally stabilizing second order nonlinear systems by SDRE control, *Proc. of the 1999 American Control Conference,* San Diego, pp. 2501-2505.
20. Valasek, M., Kejval, J. (2000) New Direct Synthesis of Nonlinear Optimal Control of Semi-Active Suspensions, In: Proc. of Advanced Vehicle Control AVEC 2000, Ann Arbor, pp. 691-697.
21. Roubicek, T., Valasek, M. (in print) Optimal Controil of Causal Differential-Algebaric Systems, *JMAA*
22. Valasek, M. (1993) Control of Elastic Industrial Robots by Nonlinear Dynamic Compensation, *Acta Polytechnica* 33, 1, (1993), pp. 15-30.

CONTROL AND OPTIMIZATION OF SEMI-PASSIVELY ACTUATED MULTIBODY SYSTEMS

V. E. BERBYUK
Department of Machine and Vehicle Systems
Chalmers University of Technology
412 96 Gothenburg, Sweden
viktor.berbyuk@me.chalmers.se

Abstract

The controlled multibody systems are under consideration. Several questions are addressed about the role of inherent dynamics, and how much multibody system should be governed by external powered drives and how much by the system's inherent dynamics. Mathematical statement of the optimal control problem that is suitable for modelling of controlled motion and optimization of semi-passively actuated multibody systems has been proposed. The methodology and numerical algorithms have been described for solving the control and optimization problems for semi-passively actuated multibody systems. Special emphasis is put on the study of controlled multibody systems having different degrees and types of actuation (underactuated and overactuated systems, external powered drives, unpowered spring-damper like drives, etc.). The solutions of energy-optimal control problems have been presented for different kinds of semi-passively actuated multibody systems (closed-loop chain semi-passively actuated robot, multibody system modelled the human locomotor apparatus with above-knee prosthesis and others).

1. Introduction

Today our knowledge in mechanics, control engineering, electronics and computer sciences is actively integrated into a new interdisciplinary science – *mechatronics* [1]. One of the primary goals of mechatronics is to gain as many advantages as possible from the optimal interaction between mechanical, control, electronic and computer subsystems. This requires more fundamental research on a number of topics of controlled multibody systems, e.g. control-structure interaction, parameter identification and optimal design, contact and impact problems, large deformation problems, etc., [2-3]. The research in the above areas can help to improve performance characteristics of modern mechatronic products.

The important and relevant characteristics of interaction between inherent dynamics and control of any mechanical system are its degree and type of actuation. Most technical systems, e.g. industrial robots, have been designed based on the

W. Schiehlen and M. Valášek (eds.), Virtual Nonlinear Multibody Systems, 279–295.

commonsense rule of minimum complexity of structure. The industrial robots have usually the same number of actuators as degrees of freedom of their mechanical subsystems, i.e. they belong to the class of fully actuated mechanical systems. A lot of research has already been done in the area of control and optimization of fully actuated robotic systems that successfully supported industrial robotics.

If multibody system (MBS) has less actuators than joints or more precisely if the dimension of the configuration space exceeds that of the control input space, the system is called *underactuated.* Examples of underactuated MBS are a car with *n* trailers having spring-damper-like joints, manipulator robots with failed actuators, free-flying space manipulators without jets or momentum wheels, manipulator robots with flexible links, legged robots with passive joints, etc. The general advantages of using underactuated mechanical systems reside in the fact that their weight is lower, and they consume less energy than their fully actuated counterparts. For hyper-redundant robots or multi-legged mobile robots, where large kinematic redundancy is available for dexterity and specific task completion, underactuation allows a more compact design and simpler control schemes. The analysis of dynamics and control of underactuated MBS is significantly more complex than that for fully actuated ones. A survey of papers in the above area has shown that the dynamics and control problems of underactuated mechanical systems have actively been studied for the last decade [4-10].

The next generation of robots must be autonomous and dexterous [11]. Dexterity implies the mechanical ability to carry out various kinds of tasks in various situations. Robots must have many sensors and more actuators than degrees of freedom, i.e. being the controlled mechanical system with sensing and actuation redundancies. To carry out optimally the complex tasks in various situations it can be desirable to change a number of actively controlled degrees of freedom of robotic system. It can easily be done, for instance, by locking or unlocking some of the actuators of robot during its performance of a specific subtask of a given complex task. From the point of view of control it means that robot can be considered as over, -fully, or underactuated mechanical system during its performance of the complex task. Obviously, the type of actuators used can also be different depending on the task of robot [12-17].

The analysis of the literature and the above-mentioned fundamental aspects shows the importance of studying dynamics, control and optimization problems of MBS with different degree and type of actuation and the robotic systems, in particular. This research is of a great challenge.

In the paper a controlled MBS of rigid bodies is under the study. External controlling forces and moments can be applied directly to arbitrary points in the system. These controlling stimuli are generated by external (powered) drives. It is assumed that displacement and velocity dependent internal controlling forces and torques can also be applied to the system. These controlling stimuli are generated by internal (unpowered) drives, e.g. spring-damper actuators located between arbitrary points and described by linear and angular stiffness and damping parameters. MBS including both external (powered) and internal (unpowered) drives we shall term a semi-passively actuated MBS.

In the paper we tackle optimization problems for controlled MBS having unpowered actuators. The reasons of this study are as follows. To incorporate spring-damper actuators into the structure of MBS and to design optimally their parameters can give several advantages, e.g. to decrease a number of external drives and, as a consequence, to decrease the weight of moving links and the energy consumption of the system. It can give great advantages to use different passive compliance elements to control some of the degrees of freedom of manipulator robots and legged mechanisms for their performance of working tasks with periodic laws of motion [12-16, 18, 19]. We study a fundamental question about optimal interaction between the controlling stimuli generated by the external drives and the proportional-differential internal forces described by linear and angular stiffness and damping parameters. A range of questions is also addressed about the role of inherent dynamics in controlled motion, and how much MBS should be governed by the external drives and how much by the system's inherent dynamics. We are in particular investigating semi-passively actuated manipulator robots and bipedal locomotion systems having spring-damper actuators.

2. Statement of the Problem

Consider a MBS the controlled motion of which can be described by the following equations:

$$\dot{x} = f(x,u,w(t,\xi)), \qquad g(x,w(t,\xi)) = 0, \qquad t \in [0,T] \tag{1}$$

Here $x = (x_1, x_2, ..., x_n)$ is a state vector, $u = (u_1, u_2, ..., u_m)$ is a vector of controlling stimuli (forces, torques) generated by the external (powered) drives of the MBS, $w = (w_1, w_2, ..., w_r)$ is a vector of the controlling stimuli of the internal (unpowered) drives, and T is the duration of the controlled motion of MBS. Vector functions f and g are determined by the structures of MBS and unpowered drives, respectively, ξ is a vector of design parameters of the unpowered drives.

Constraints and restrictions are imposed on the state vector $x(t)$, the controlling stimuli of the unpowered drives $w(t,\xi)$, and the external control laws $u(t)$ of the system. These restrictions can be written in the following way:

$$\{x(t)\} \in Q, \qquad t \in [0,T] \tag{2}$$

$$w(t,\xi) \in W, \qquad t \in [0,T] \tag{3}$$

$$u(t) \in U, \qquad t \in [0,T] \tag{4}$$

In formulas (2) - (4), Q and U are given domains in the state and control spaces of the system, respectively; W is a set of addmissible controlling stimuli determined by the structure of the unpowered drives.

The differential equations (1) together with the restrictions (2)-(4) are called the mathematical model of the semi-passively actuated MBS. This model can be used for many applications, e.g. to study fundamental questions about the role of inherent dynamics in controlled motion, and how much MBS should be governed by the external drives and how much by the system's inherent dynamics, to analyze power demand of the actively controlled MBS [19], to solve the design problems of lower limb prostheses and to study control strategies for the stable motion of bipedal locomotion systems with compliance elements at the joints [6, 13], for computer simulation of the energy-optimal motion of closed-loop chain manipulator robots with unpowered drives [15], etc.

Assume that there exists a non-empty set of vector-functions $\{x(t), u(t), w(t,\xi), \ t \in [0,T]\}$ which satisfy the equations (1) and the constraints (2)-(4). The following optimal control problem can be formulated.

Problem A. Given a MBS the controlled motion of which is described by equations (1). It is required to determine the vector-function $w_*(t,\xi)$, the motion of the system $x_*(t)$ and the external controlling stimuli $u_*(t, x_*, w_*)$ which alltogether satisfy the equations (1), the restrictions (2)-(4), and which minimize the given objective functional $\Phi[u]$.

As a result of the solution of *Problem A* the optimal structure of MBS having both powered and unpowered drives is designed. The external controlling stimuli for the system are also found which minimize the given objective functional.

One of the primary goals for the incorporation of unpowered drives into the structure of MBS is an improvement of their control processes. It means that the validity of the following inequality is expected: $\Phi[u_*(t, x_*, w_*)] < \Phi[u_{0*}(t, x_{0*})]$, where $x_{0*}(t)$, $u_{0*}(t)$ are the optimal motion and the respective controlling stimuli of MBS without the unpowered drives obtained under the restrictions (2), (4). In this sense the solution of *Problem A* could help to estimate the limiting possibility of improvement of the external control strategies for MBS due to incorporation into their structure different unpowered drives determined by the constraints (3).

3. Methodology

We have formulated the optimal control problem for the semi-passively actuated MBS. The key feature of the proposed mathematical statement of the problem is direct utilization of the equations describing the inherent dynamics of internal unpowered drives together with all other constraints that are imposed on the state vector and the controlling stimuli of the system. It leads to the non-uniqueness of the solution of the direct and the inverse dynamics problems as well as makes it possible to design optimal unpowered actuators for MBS.

In general case for MBS with many degrees-of-freedom powerful numerical algorithms are needed to solve *Problem A*. Futhermore, during the calculation of optimal control law for MBS it is necessary to design at the same time the optimal structure of the unpowered drives taking into account the restriction (3). This can significantly increase the complexity of the computation.

The numerical method has been developed [6, 13, 15] for the solution of *Problem A* for MBS, which models semi-passively actuated manipulator robots and bipedal locomotion systems with unpowered drives at their joints. The method is based on a special procedure to convert the initial optimal control problem (*Problem A*) into a standard nonlinear programming problem: $F(C) \Rightarrow \min_C$, $g(C) \leq 0$, where C is a vector of varying parameters. This is made by an approximation of the independently varying functions $q(t)$ by a combination of the fifth order polynomial and Fourier series, i.e. by using the following expressions:

$$q(t)=\sum_{j=0}^{5} C_{qj}(t-t_0)^j+\sum_{k=1}^{N_q}[a_{qk}\cos(k\omega(t-t_0))+b_{qk}\sin(k\omega(t-t_0))].$$

Here $\omega = 2\pi / T$, and N_q are given positive integers.
Taking into account the restriction (2), the list of independently varying parameters can be determined. For instance, in the case of two-points boundary conditions imposed on the state vector of the system, from the above formula follow that the parameters $C_{q4}, C_{q5}, a_{qk}, b_{qk}, \ k=1,2,...,N_q$ can serve as independently varying parameters.

To solve the nonlinear programming problem different algorithms have been used that are based on the Rozenbrock's method [20], and the sequential quadratic programming method [21] implemented in the software package TOMLAB [22].

The key features of the proposed method for solving *Problem A* are its high numerical efficiency and the possibility to satisfy a lot of restrictions imposed on the phase coordinates of the system automatically and accurately. The efficiency of the developed method has been illustrated by solution of the design problems of the energy-optimal above-knee prostheses with several types of unpowered knee mechanisms and by computer simulations of the energy-optimal motions of closed-loop chain semi-passively actuated manipulator robot, and the bipedal walking robot [6, 13, 15].

In the next two paragraphs the results of solution of *Problem A* are presented for some particular examples of controlled MBS. These results have been obtained analytically within the frame of assumption that the motion of considered MBS is prescribed.

4. Optimal Passive Drives for Given Motion of MBS

Consider a MBS having n degrees-of-freedom. Let the equations of its controlled motion be as follows:

$$A(q)\ddot{q}+B(q,\dot{q})=u(t), \qquad t\in[0,T] \tag{5}$$

Here $q=(q_1,q_2,...,q_n)$ is a vector of the generelized coordinates, $u=(u_1,u_2,...,u_n)$ is a vector of the controlling stimuli (forces, torques) generated by powered drives of the MBS, $A(q)$, $B(q,\dot{q})$ are given matrices.

At the same time, assume that MBS has additional unpowered (passive) drives, namely non-linear visco-elastic spring-damper-like actuators in its structure. The mathematical model of the semi-passively actuated MBS can be written as follows:

$$A(q)\ddot{q}+B(q,\dot{q})=u(t)+w(q,\dot{q}), \ w(q,\dot{q})+kf(q,\dot{q})=0, \qquad t\in[0,T] \tag{6}$$

where $w = (w_1, w_2, ..., w_n)$ is a vector of the controlling stimuli of the passive drives of MBS, $f(q, \dot{q})$ determines the inherent dynamics of the passive drives under the restriction (3) and k is a "damper coefficient".

To estimate the quality of the control processes the following objective functional is exploited

$$\Phi[u(t)] = \int_0^T \|u(t)\|^2 dt, \quad \|u(t)\| = (u_1^2(t) + ... + u_n^2(t))^{1/2} \tag{7}$$

Let $\{q_0(t), \ u_0(t), \ t \in [0,T]\}$ be any pair of functions that satisfy equations (5).
It is assumed that the motion $\{q_0(t), \ t \in [0,T]\}$ can also be realised by the considered semi-passively actuated MBS. Using the equations (6), the external controlling stimuli needed for the motion are written as follows

$$u_{w0}(t) = u_0(t) + kf(q_0, \dot{q}_0), \text{ where } u_0(t) = A(q_0)\ddot{q}_0(t) + B(q_0, \dot{q}_0) \tag{8}$$

For the control law (8) the objective functional (7) will be equal to

$$\Phi[u_{w0}(t)] = \int_0^T \|u_0(t) + kf[q_0(t), \dot{q}_0(t)]\|^2 dt = \Phi[u_0(t)] + ak^2 + 2bk, \tag{9}$$

$$a = \int_0^T \|f(q_0, \dot{q}_0)\|^2 dt, \quad b = \int_0^T \langle u_0(t), f(q_0, \dot{q}_0)\rangle dt, \tag{10}$$

$$\langle u_0(t), f(q_0, \dot{q}_0)\rangle = (u_{01} f_1 + ... + u_{0n} f_n).$$

It can be shown that the function $\Phi[u_{w0}(t)]$ has a global minimum with respect to the damper coefficient k. The value of this minimum is equal to $\Phi_{\min} = \Phi[u_0(t)] - b^2 / a$ for the following optimal value of the damper coefficient $k_* = -b / a$.

The above mentioned makes it possible to conclude that for the considered MBS with n degrees-of-freedom and for any admissible motion $\{q_0(t), \ t \in [0,T]\}$ the energy-optimal non-linear visco-elastic spring-damper-like actuators are determined by the formulas

$$w_*(q_0, \dot{q}_0) + k_* f(q_0, \dot{q}_0) = 0, \ t \in [0,T]$$

As follows from above due to the incorporation of the optimal spring-damper-like actuators into MBS structure the decrease in energy consumption is equal to

$$\Phi[u_0(t)] - \Phi_{\min} = b / \int_0^T \|f(q_0, \dot{q}_0)\|^2 dt.$$

This value depends only on the given motion $\{q_0(t),\ t\in[0,T]\}$ and the function $f(q,\dot{q})$ determining the inherent dynamics of the passive drives.

Usually some restrictions are imposed on the external controlling stimuli $u(t)$. In this case the function $f(q,\dot{q})$ can not be chosen arbitrarily. Indeed, let us assume that the external controlling stimuli $u(t)$ are restricted by the constraint $\|u(t)\|\le u_{\max},\ t\in[0,T]$ with given positive number $u_{\max}$. Then the function $f(q,\dot{q})$ must satisfy not only the restriction (3) but also the inequality $\|u_0(t)-bf(q_0,\dot{q}_0)/a\|\le u_{\max},\ t\in[0,T]$, where $u_0(t)$, a and b are determined by the expressions (8) and (10).

5. Optimization of Controlled Motion of the Semi-Passively Actuated Manipulator Robot

Here we present several results of optimization of controlled motion of the semi-passively actuated closed-loop chain SCARA-like robot. The sketch of the robot is shown in Figure 1.

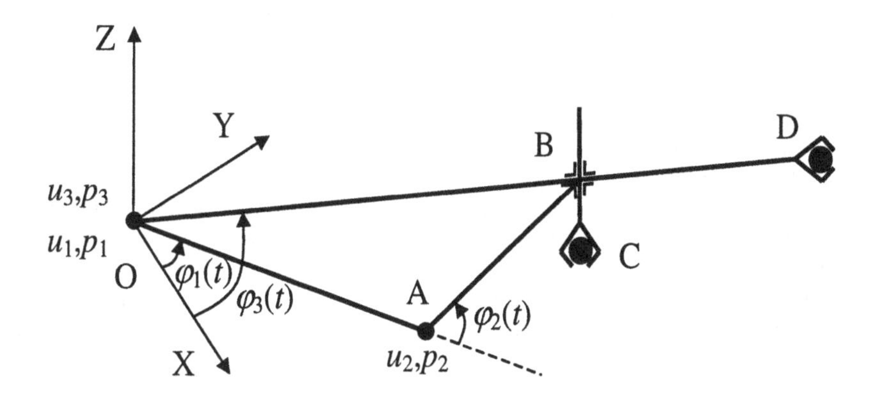

Figure 1. Sketch of the Semi-Passively Actuated Closed-Loop Chain SCARA-Like Robot.

Proposed robotic system has the following new features in comparison with the well-known SCARA robot [23]. In addition to powered drives it comprises several unpowered (passive) spring-damper-like drives. An additional link has also been incorporated into the structure that gives the possibility to obtain the semi-passively actuated closed-loop chain robot. Emphasis is put on the study of the interaction between the controlling stimuli of the powered drives and the torques exerted by the unpowered drives needed to provide the optimal motion of the robots with different degrees of actuation.

The robot depicted in *Figure 1* comprises four links that are modelled by the rigid bodies OA, AB, OD and BC. There are one degree-of-freedom rotational joints at the points O and A, and translational joints at the point B. All joints are considered frictionless.

It is assumed that the robot's links OA, AB and OD move in the horizontal plane OXY under the action of the torques $u_1(t)$, $u_2(t)$ and $u_3(t)$ applied to the links OA, AB, and OD, respectively. Under the action of the force $F(t)$ the link BC moves along the direction of the axes OZ. The controlling stimuli $u_i(t)$, $i = 1, 2, 3$ and $F(t)$ are exerted by the powered drives of the robot. The robotic system also comprises spring-damper actuators at joints O and A. The torques exerted by these actuators p_1 p_2 and p_3 act on the links OA, AB and OD, respectively. They will be treated as the controlling stimuli of unpowered (passive) drives of the robot.

Here we study motions of the robot in the horizontal plane OXY only. Using ϕ_1 and ϕ_2 as the generalized coordinates the equations of motion of the considered mechanical system can be derived using the Lagrange formalism [15].

The equations of the plane motion of the robot can be written as follows:

$$f_1(\phi_i,\dot{\phi}_i,\ddot{\phi}_i) = u_1 + p_1 + u_3 + p_3, \quad f_2(\phi_i,\dot{\phi}_i,\ddot{\phi}_i) = u_2 + p_2 + b(\phi_i)(u_3 + p_3) \tag{11}$$

The functions f_1 and f_2 are determined by means of the Lagrange operator [15].

The torques of the unpowered drives are modelled by formulas

$$p_i = -k_i(\varphi_i - \varphi_{i0k}) - c_i\dot{\varphi}_i \quad i = 1,2,3 \tag{12}$$

where k_i are the stiffness coefficients, c_i are the damping coefficients, φ_{i0k} are the no-load angles of the torsional spring.

The considered robot is an overactuated mechanical system. This makes it possible to optimize the controlling stimuli of powered drives for an arbitrary given motion of the robot.

Problem A.1. Assume that arbitrary motion of the robot and control torques of unpowered drives are given, i.e. the functions $\phi_i(t), p_i(t)$ are specified. It is required to find the control stimuli $u = (u_1, u_2, u_3)$ which minimize the functional

$$E[u(t)] = \int_0^T (u_1^2(t) + u_2^2(t) + u_3^2(t))dt \tag{13}$$

subject to the differential constraints (11).

It can be shown that the solution of *Problem A.1* is given by the following formulas

$$u_3^*(t) = (g_1 + bg_2)/(2 + b^2), \quad u_1^*(t) = g_1 - u_3^*(t), \quad u_2^*(t) = g_2 - bu_3^*(t) \tag{14}$$

Here the functions g_1 and g_2 have the expressions:

$$g_1 = f_1 - p_1 - p_3, \quad g_2 = f_2 - p_2 - bp_3 \tag{15}$$

The obtained controlling stimuli (14) provide execution of an arbitrary given motion of the overactuated robot with minimal energy consumption E^*.

The simplest way to reduce the overactuation of the considered robot is to exclude one of the powered drives. For instance, assuming that

$$u_3(t) = 0, \quad t \in [0, T] \tag{16}$$

the unique solution of *Problem A.1* for the functions $u_1(t)$, $u_2(t)$ can be obtained from the equations (11). In this case the functional (13) is

$$E^0 = \int_0^T (g_1^2(t) + g_2^2(t))dt \tag{17}$$

where the functions $g_1(t)$, $g_2(t)$ are given by the formulas (15).

Comparing the value E^0 with the value of the functional (13) for the obtained optimal controlling stimuli $u_i^*(t)$ it is easy to show the validity of the following expression

$$E^0 - E^* = \int_0^T (g_1(t) + bg_2(t))^2 / (2 + b^2)dt \tag{18}$$

The formula (18) shows that the energy consumption needed to execute an arbitrary given motion by the considered overactuated robot with obtained optimal controlling stimuli (14), (15) is less than the energy consumption of the same robot but without powered drive acting on the link OD.

Early, dynamics, control and optimization problems of the semi-passively actuated closed-loop chain SCARA-like robot (*Figure 1*) that performs a pick-and place operation have been studied numerically. The results have been obtained by using the methodology described in paragraph 3, and can be found in [15, 24].

6. Optimization of Controlled Motion of the Semi-Passively Actuated Bipedal Locomotion Systems

Here the application of methodology of optimization of semi-passively actuated MBS described in paragraph 3 is demonstrated for solving the design problem of lower limb prostheses.

There is an important difference between the dynamics of an intact limb and a prosthetic limb of an amputee. In the paper the mathematical modelling of a human gait of an amputee with above-knee prosthesis is considered based on an assumption that the force moments at the knee and at the ankle joints of the prosthetic leg are passive ones. The values of these moments depend not only on the gait pattern, but also on the prosthesis construction.

The sketch of the amputee locomotor system (ALS) with above-knee prosthesis is depicted in *Figure 2.* The system is modeled as the mechanical system of seven rigid bodies connected by ideal cylindrical hinges. The bodies HG, HK_i, and K_iA_i (i=1,2), which model the torso, thighs, and shins respectively, are assumed to have weight and inertia, and the bodies $A_iT_iH_i$ (the feet) are weightless and inertialess.

In addition to the weights of the trunk, thighs and the shanks, the external forces acting on the ALS include the interaction forces between the feet and the ground, which are replaced by resultant forces $\boldsymbol{R}_i$, (i=1,2). It is also assumed that the control torques $q_i(t), u_i(t), p_i(t)$ acting at the hip (point H), knee (point K_i) and the ankle (point A_i) joints, respectively.

As generalized coordinates that jointly determined the position of the given mechanical system we chose the following: x and y, the Cartesian coordinates of the point of attachment of the legs (the point H); $\psi, \alpha_i, \beta_i, \gamma_i$, the angles of deviation of the link HG, HK_i, K_iA_i, and $A_iT_iH_i$, (i=1,2) respectively from vertical (*Figure 2*).

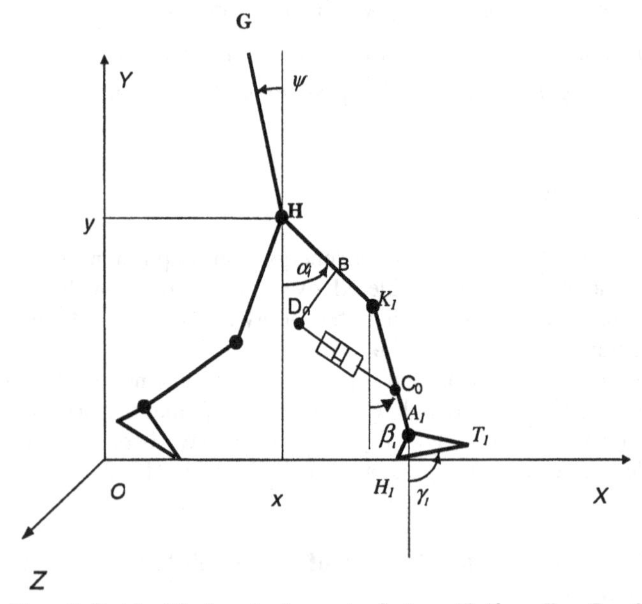

Figure 2. Sketch of the Amputee Locomotor System with Above-Knee Prosthesis.

The above-knee prosthesis comprises the linear-viscoelastic ankle mechanism and the hydraulic or the pneumatic knee mechanism that are assumed to be weightless and inertialess.

During locomotion of ALS with above-knee prosthesis the control torques

$$p_1(t) = C(\beta_1 - \gamma_1 + \pi/2) + K(\dot{\beta}_1 - \dot{\gamma}_1) + D, \tag{19}$$

$$u_1(t) = (P_2 - P_1) S_p d_2 (d_1^2 + l_0^2)^{1/2} \sin(\alpha_1 - \beta_1 + \eta) / l_1$$

are generated at the ankle and at the knee joints of the prosthetic leg, respectively.

Here C, K are the torsion spring and the damping coefficients of the ankle mechanism; D is determined by the free angle of the spring and torsion spring coefficients; P_1, P_2 are the chamber pressures of the hydraulic or the pneumatic actuator that can be calculated by using the equations of dynamics of the knee mechanism of the prosthesis [25], S_p is the cylinder piston cross-area,

$$l_1 = (d_1^2 + d_2^2 + l_0^2 + 2d_2(d_1^2 + l_0^2)^{1/2} \cos(\alpha_1 - \beta_1 + \eta))^{1/2}, \quad \eta = a\tan(l_0 / d_1), \tag{20}$$

$$d_1 = |BK_1|, \quad d_2 = |K_1C_0|, \quad l_0 = |BD_0|.$$

The detailed description of the considered model of ALS can be found in [6, 13, 25].

The design problem of the above-knee prosthesis can be formulated in the same way as *Problem A*. It should be taken into account that the considered semi-passively actuated MBS has the state vector $\left\{x, \dot{x}, y, \dot{y}, \psi, \dot{\psi}, \alpha_i, \dot{\alpha}_i, \beta_i, \dot{\beta}_i, \gamma_i, \dot{\gamma}_i, i = 1,2\right\}$, the vector of controlling stimuli of the powered drives $u(t) = \{q_1, q_2, u_2, p_2\}$, and the vector of the constructive parameters of the unpowered drives $C_p = (C, K, D, d_1, d_2, l_0, S_p, S_0)$.

The controlled motions of MBS that models ALS with above-knee prosthesis are described by the Lagrange equations that can be found in [6, 13], and by the expressions (19), (20). The boundary conditions and other constraints on the phase coordinates of the system have been given on the basis of known experimental data on the human gait [26-28].

The following functional

$$E = \frac{1}{2L} \int_0^T \{\sum_{i=1}^{2} |q_i(\dot{\psi} - \dot{\alpha}_i)| + |u_2(\dot{\alpha}_2 - \dot{\beta}_2)| + |p_2(\dot{\beta}_2 - \dot{\gamma}_2)|\}dt \tag{21}$$

is used for solving *Problem A*. The objective functional (21) estimates the energy expenditure per unit of distance traveling of ALS [27-29]. The same approach as described in paragraph 3 has been used for solving the problem of design energy-optimal above-knee prostheses. Due to the dynamic constraints (19) the procedure of converting the *Problem A* into the standard nonlinear programming problem includes the solution of the semi-inverse dynamics problems for the controlled mechanical system that models ALS with above-knee prosthesis. It sufficiently increases the time consumption of the numerical algorithm for designing the energy-optimal above-knee prosthesis.

We now present the individual results of mathematical modeling of motion of ALS with energy-optimal above-knee prosthesis obtained in the context of the proposed formulation of the optimal control problem (*Problem A)* and the methodology for solving it numerically.

The computations were carried out for ALS of a person of height 1.76 m, and mass 73.2 kg. The respective values of linear and mass-inertia characteristics of particular links of ALS were calculated on the basis of known experimental data and can be found in [30].

Problem A has been solved numerically for two types of the prostheses: the above-knee prosthesis with the hydraulic actuator at the knee and the prosthesis with the pneumatic knee mechanism. For both of these prostheses three types of human gait have been studied, characterized by different values for the duration of a double step T, velocity V, and length of step L: slow walking with T_S=1.383 s, V_S=0.998 m/s, L_S=0.69 m; walking at a normal pace T_N=1.1396 s, V_N=1.325 m/s, L_N=0.755 m; fast walking at T_F=0.9733 s, V_F=1.685 m/s, L_F=0.82 m [26].

The analysis of the solutions obtained has shown that the kinematic, dynamic, and energetic characteristics of controlled motion of ALS are strongly sensitive to the essential prosthesis' parameters. For a given individual and pace of a gait there exist optimal values of constructive parameters of the prosthesis' knee and ankle mechanisms $C_p = (C, K, D, d_1, d_2, l_0, S_p, S_0)$. These parameters give minimum energy expended per unit of distance traveled. For above mentioned types of human walking we obtained the following minimal values for the energy consumption for slow, normal and fast paces of motion respectively: E_S=117 J/m, E_N=114 J/m, E_F=147 J/m (for pneumatic knee mechanism), and E_S=103 J/m, E_N=96 J/m, E_F=125 J/m (for hydraulic knee mechanism). Comparison of these data shows that the normal pace of the ALS gait gives a minimum to the energy expended per unit of distance traveled comparing to the amount of energy needed for the slow or fast gaits. This is valued for both energy-optimal pneumatic and hydraulic knee mechanisms.

Some kinematic and dynamic characteristics of the energy-optimal motion of ALS with optimal structure of the above-knee prosthesis obtained by the numerical solution of *Problem A* for the gait with normal pace are shown in *Figures 3 - 6* (solid thin curves correspond to the prosthesis with the hydraulic actuator at the knee, dashed curves - to the prosthesis with the pneumatic knee mechanism).

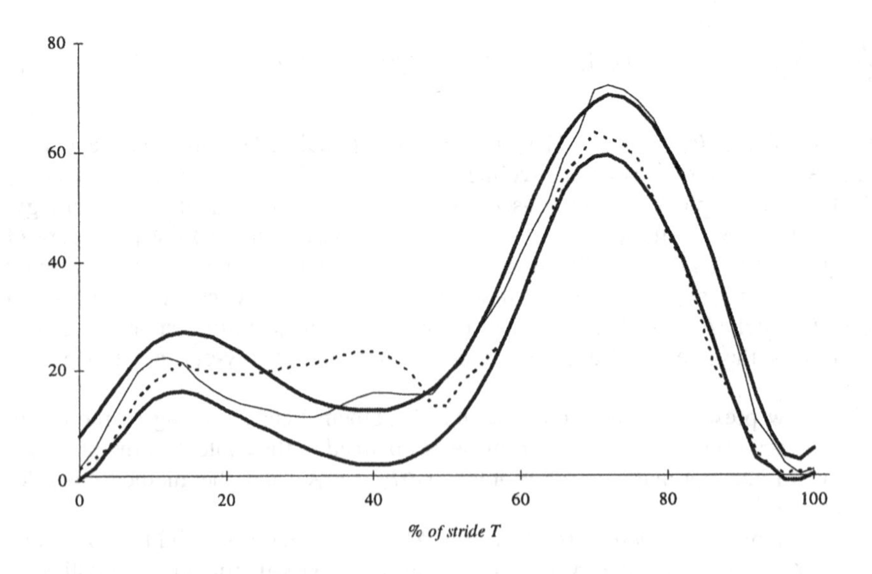

Figure 3. Knee Angle of the Prosthetic Leg, ($\alpha_1 - \beta_1$), in degrees.

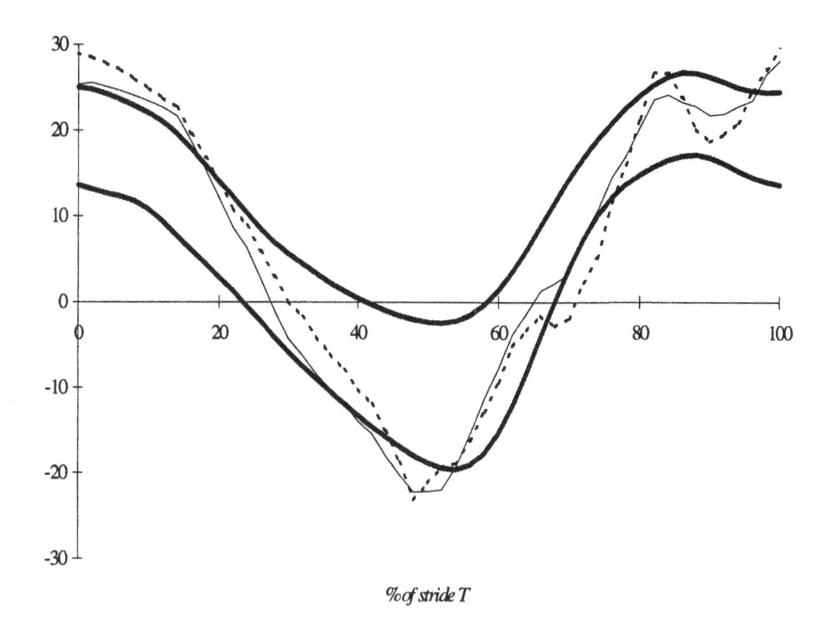

Figure 4. Hip Angle of the Prosthetic Leg, ($\alpha_1 - \psi$), in degrees.

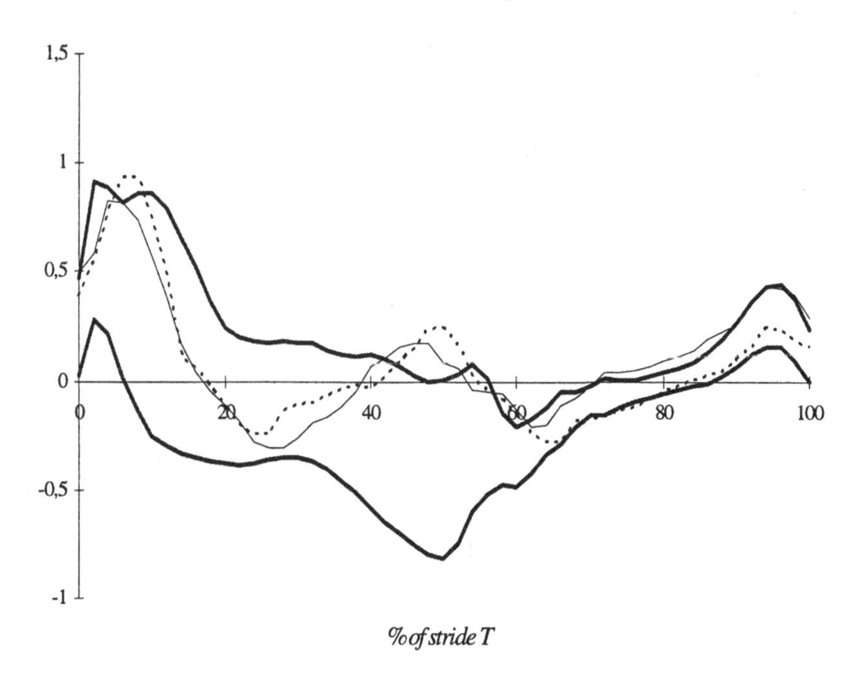

Figure 5. Hip Torque of the Prosthetic Leg, ($q_1(t)/M$), in Nm/kg.

Knee angle ($\alpha_1 - \beta_1$) and hip angle ($\alpha_1 - \psi$) of the prosthetic leg are depicted in *Figures 3-4*, respectively. Hip torque of the prosthetic leg, ($q_1(t)/M$), and knee torque of the healthy leg, ($u_2(t)/M$) are presented in *Figure 5-6*, respectively. For the comparison purposes in *Figures 3-6* the domains of the values of the respective kinematic and dynamic characteristics obtained by the biomechanical experiments for a human normal gait [26] are depicted by heavy solid curves.

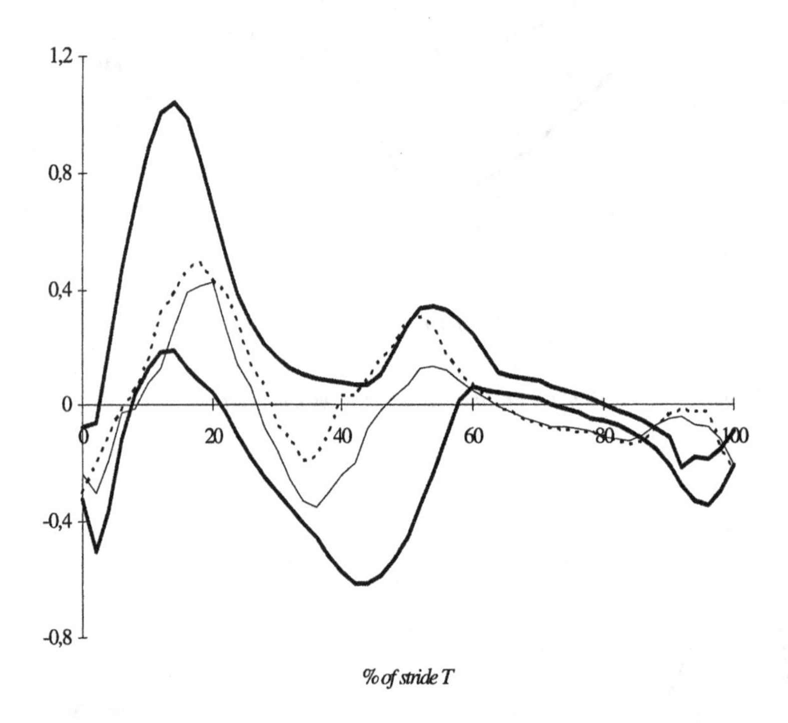

Figure 6. Knee Torque of the Healthy Leg, ($u_2(t)/M$), in Nm/kg.

Analysis of the plots that depicted in Figure 3-6 shows that the kinematic and dynamic characteristics of the motion of ALS with obtained energy-optimal structures of above-knee prostheses are within reasonable proximity to the respective characteristics of a human normal gait [26].

7. Conclusion

We have formulated the optimal control problem for semi-passively actuated MBS (*Problem A*). The key feature of the proposed mathematical statement of the problem is the direct utilization of the differential equations describing the inherent dynamics of the passive actuators (unpowered drives) together with all other constraints imposed on the state vector and the controlling stimuli of the system. It leads to the non-uniqueness of the solution of the direct and inverse dynamics problems and makes it possible to design optimally both structure (passive actuators) and external control of MBS.

Closed-form solution of *Problem A* has been obtained for arbitrary prescribed motion of semi-passively actuated MBS with n degrees-of-freedom. The analysis of the obtained solution shows that in several cases the incorporation of passive drives into the structure of controlled MBS can decrease the energy consumption needed for the prescribed motion of the system.

The problem of energy-optimal control of overactuated closed-loop chain SCARA-like robot having both powered and unpowered drives has also been solved analytically. It has been shown that optimally designed overactuation can decrease the energy consumption needed for arbitrary prescribed motion of the considered semi-passively controlled robot. Moreover, the previous study [15, 24] demonstrated that incorporation of the optimal passive linear spring-damper actuators into the structure of the considered closed-loop chain robot leads to a significant reduction of the energy consumption of the robot for cyclic pick and place operations.

For solving optimization problems of general type of semi-passively actuated MBS the numerical method has been presented. Efficiency of the proposed method is illustrated by the solution of design problem of the energy-optimal above-knee prostheses with two types of passively controlled knee mechanisms. Analysis of the numerical results obtained has shown that during the optimal motion of the considered MBS there is a strong interaction between the gravity force, the external ontrol torque exerted by the actively controlled drives and the internal torque exerted by the passive linear spring-damper actuators. The kinematic, dynamic, and energetic characteristics of controlled motion of MBS that models human locomotor system with above-knee prosthesis are strongly sensitive to the essential parameters of the passive actuators of the prosthesis. For a given individual and pace of a gait there exist optimal values of the spring and damper parameters of the prosthesis's ankle and knee mechanisms. These parameters give minimum energy expended per unit of distance travelled.

Results obtained give some insight into the study of questions about the role of inherent dynamics in controlled motion, and how much MBS should be governed by the external drives and how much by the system's inherent dynamics. They can help to design simpler control systems of manipulator robots and autonomous legged mechanisms having less weight and energy consumption, and can also be use to design energy efficient passively controlled mechanisms of the lower limb prostheses.

8. Acknowledgements

I am grateful to M. Lidberg and N. Nishchenko who have joined with me in the work described in this paper.

The support of the Volvo Research Foundation, TFR (Swedish Research Council for Engineering Sciences), and National Academy of Sciences of Ukraine is gratefully acknowledge.

9. References

1. Wikander, J., Torngren M., and Hanson M. (2001) The science and education of mechatronics engineering, *IEEE Robot Autom Mag* **8** (2), 20-26.
2. Schiehlen, W. (1997) Multibody system dynamics: Roots and perspectives. *J. Multibody System Dynamics* **1,** No 2, 149-188.
3. Shabana, A. A. (1997) Flexible multibody dynamics: Review of past and recent developments, *J. Multibody System Dynamics* **1,** No 2, 189-222.
4. De Luca, A., Mattone, R. and Oriolo, G. (1996) Control of underactuated mechanical systems: Application to the planar 2R robot, *Proceedings of the 35th Conference on Decision and Control,* Kobe, Japan, December 1996, 1455-1460.
5. Suzuki, T., Koinuma, M. and Nakamura, Y. (1996) Chaos and nonlinear control of nonholonomic free-joint manipulator, *Proceedings of the 1996 IEEE International Conference on Robotics and Automation,* 2668-2675.
6. Berbyuk, V. (1996) Multibody systems modeling and optimization problems of lower limb prostheses, in D. Bestle and W. Schiehlen (eds.), *IUTAM Symposium on Optimization of Mechanical Systems,* Kluwer Academic Publishers, pp.25-32.
7. Bullo, F. and Lynch, K. (2001) Kinematic controllability for decoupled trajectory planning in underactuated mechanical systems, *IEEE Transactions on Robotics and Automation,* **8**, (4), 402-412.
8. Nakamura, Y., Chung, W., and Sørdalen, O.J. (2001) Design and control of the nonholonomic manipulator, *IEEE Transactions on Robotics and Automation,* **17**, (1), 48-59.
9. De Luca, A., and Oriolo, G. (2000) Motion planning and trajectory control of an underactuated three-link robot via dynamic feedback linearization, *Proceedings of the IEEE International Conference on Robotics & Automation,* San Francisco, CA, April 2000, 2789-2795.
10. De Luca, A., Iannitti, S., Mattone, R. and Oriolo, G. (2001) Control problems in underactuated manipulators, *Proceedings of the 2001 IEEE/ASME International Conference on Advanced Intelligent Mechatronics,* Como, Italy, 8-12 July 2001, 855-861.
11. Nakamura, Y. (1991) *Advanced Robotics: Redundancy and Optimization,* Addison-Wesley Publishing Company.
12. Berbyuk, V., Peterson, B. and Nishchenko, N. (1998) Linear viscoelastic actuator-based control system of a bipedal walking robot, in J. Adolfsson and J. Karlsen (eds.), *MECHATRONICS'98,* Elsevier Science Ltd., pp.379-384.
13. Berbyuk, V., (1999) Dynamic simulation of human gait and design problems of lower limb prostheses, *Biomechanics Seminar,* (ISSN 1100-2247), **12,** pp.1-20.
14. Berbyuk, V., Boström, A. and Peterson, B. (2000) Modelling an design of robotic systems having spring-damper actuators, in *Proceedings of the 7th Mechatronics Forum International Conference,* 6-8 September 2000, Atlanta, Georgia, USA, (ISBN 0 08 043703 6), PERGAMON.
15. Lidberg, M. and Berbyuk, V. (2000) Modeling of controlled motion of semi-passively actuated SCARA-like robot, in *Proceedings of the 7th Mechatronics Forum International Conference,* 6-8 September 2000, Atlanta, Georgia, USA, (ISBN 0 08 043703 6), PERGAMON.
16. Berbyuk, V. and Boström, A. (2001) Optimization problems of controlled multibody systems having spring-damper actuators, *International Applied Mechanics,* **37**, No. 7, pp.935-940.
17. Okada, M., Nakamura, Y. and Ban, S. (2001) Design of programmable passive compliance shoulder mechanism. In: *Proceedings of the IEEE International Conference on Robotics and Automation,* Seoul.
18. Gruber, S. and Schiehlen, W. (2000) Low-energy biped locomotion, in *Proceedings of the Ro.Man.Sy.'2000,* Zakopane, Poland, 03-06 July 2000, Springer-Verlag, Wien, 459-466.
19. Schiehlen, W. and Guse, N. (2001) Power demand of the actively controlled multibody systems, in *Proceedings of DETC'01 ASME 2001 Design Engineering Technical Conference and Computers and Information in Engineering Conference,* Pittsburghg, PA, September 9-12, 2001, 1-10.
20. Bazara, S. and Shetty, C. (1979) *Nonlinear Programming. Theory and Algorithm,* John Wiley and Sons, New York.

21. Gill, P. E. , Murray, W., Saunders, M. A. and Wright, M. H., (1998) *User's Guide for NPSOL 5.0: A Fortran Packege for Nonlinear Programming.* Technical Report SOL 86-1, Dept. Operations Research, Stanford University, Stanford.
22. Holmström, K. (1999) The TOMLAB optimization environment in Matlab". *Advanced Modeling and Optimization,* **1,** (1), 47-69.
23. Stadler, W. (1995) *Analytical Robotics and Mechatronics,* McGraw-Hill, New York.
24. Lidberg M. and Berbyuk V. (2002) Energy-optimal control of semi-passively actuated SCARA-like robot, In: Peter Maisser and Peter Tenberge (eds.), *Proceedings of the First International Symposium on Mechatronics,* March 21-22, 2002, Chemnitz, Cermany, (ISBN 3-00-007504-6), pp.302-311.
25. Berbyuk, V. and Nishchenko, N. (1998) Mathematical design of the energy-optimal above-knee prostheses, *Mathematical Methods and Physicomechanical Fields,* **41**, (4), 110-117.
26. Winter, D. (1991) *The Biomechanics and Motor Control of Human Gait,* University of Waterloo Press, Canada.
27. Beletskii, V. V. (1984) *Bipedal Walking: Model Dynamic and Control Problems.* Nauka, Moscow.
28. Berbyuk, V. and Lytwyn, B. (1998) The mathematical modelling of the human motions based on the optimization of the biodynamic system's controlled processes, *Mathematical Methods and Physicomechanical Fields,* **41**, (3), 153-161.
29. Becket, R. and Chang, K.(1968) An evaluation of kinematics of the gait by minimum energy, *J. Biomechanics,* **1,** 147-159.
30. Berbyuk, V. E., Crasyuk, G. V. and Nishchenko, N. I. (1999) Mathematical modelling of the dynamics of the human gait in the saggital plane, *Journal of Mathematical Scinces,* **96,** No. 2, 3047-3056.

OPTIMIZATION OF PASSIVE AND ACTIVE DYNAMIC SYSTEMS

D. BESTLE and M. GLORA

Brandenburg University of Technology, Department of Machine Dynamics, P.O. Box 10 13 44, 03013 Cottbus, Germany

1. Introduction

The multibody system approach is widely accepted as an analysis tool for kinematic and kinetic investigation of rigid body systems connected by rigid frictionless bearings and massless coupling elements. Due to the specific structure of the equations of motion, they may easily be coupled with additional differential equations describing the dynamics of active coupling elements like pneumatic or hydraulic actuators, electric motors, or general control laws. Especially in the context of active system design, a coupling of the multibody system approach with high-level simulation software like Matlab/Simulink [1] shows quite some overall design capability although the performance of a single simulation run may be less than for specific numerical integration algorithms. However, the use of such a model in the context of virtual prototyping is often restricted to user-driven parameter studies whereas the design capability for a more systematic synthesis is overlooked.

A reason for this state of the art use of multibody system modeling in practice certainly is the disillusion about most results of classical nonlinear programming tools. Even if we assume perfect convergence to a final optimum, the solution often misses practical requirements or is obviously inferior to solutions resulting from practical experience. This, however, is not a matter of optimization in principle, but a result of the classical way of dealing with optimization on the basis of a single objective only. Technical design has to take into account several conflicting criteria and the final solution has to be a compromise between these requirements. Multicriterion or vector optimization can deal with such conflicting objectives. It is a mean for strolling through the design space like experience guided parameter studies for finding compromise solutions, but with the advantage of releasing the designer from sub-optimal solutions.

W. Schiehlen and M. Valášek (eds.), Virtual Nonlinear Multibody Systems, 297–316.

Virtual prototyping is an ideal environment for optimization: all model parameters can be varied and thus used as potential design variables to improve the design. Further, the full state of the dynamic system is accessible and thus may be introduced to criterion functions evaluating the design performance with respect to kinematic or dynamic aspects.

Virtual prototyping will finally optimize the model instead of the original dynamic system which can be quite close to reality for good models. However, in most cases field testing has to be the last step of a design process in order to perform the fine-tuning. Optimization methods may also be applied to such an experimental design environment, if care is taken of its specific properties like low accuracy of design evaluation, measurement noise and thus non-repeatable performance numbers, safety aspects, and high evaluation costs.

As mentioned already, the result of virtual prototyping will be only transferable to the real system, if the agreement of model and real system is good enough with respect to the major dynamic properties. Especially the dynamics of coupling elements or analog controllers is often hard to model just from basic physical laws, since many of the parameters cannot be measured directly. In this case, the components or parts of the system have to be put on a test rig and excited by forces, motion or control. Optimization may than be used in order to force best correspondence between measurement and simulation. Due to the different influence of the system parameters on different parts of the system behavior, a multicriterion formulation can be equally valuable for this identification process like for optimization.

The paper will show some of these aspects on the basis of two examples: a tuned mass damper being designed by both simulation and by a HiL-experiment, and a controlled tool axis of a machine tool being identified by vector optimization.

2. Virtual Prototyping via Multicriterion Optimization

Cabriolets have a major problem with torsional vibrations due to the low stiffness of the open-topped chassis. Passive or active tuned mass dampers may be used to suppress such vibrations. As a rough model for studying the effect, a frame structure like the one shown in Figure 1a may be used. The springs represent the vehicle suspension where excitation $u(t)$ is applied to one vertex only, and additional concentrated masses are used in order to lower the eigenfrequencies [2].

The steel frame is modelled by 6 beam elements for the long edges and 8 beam elements for the short edges, respectively, resulting in 28×6 DOFs (3 translational and 3 rotational degrees of freedom for each node). The masses at the vertices are taken into account as rigid bodies. Assuming small displacements, this results in a linear system of differential equations being written as

$$\boldsymbol{M}\,\ddot{\boldsymbol{y}} + \boldsymbol{K}\,\boldsymbol{y} = \boldsymbol{0}\,. \tag{1}$$

The interesting torsional mode and some of the additional bending modes which have been computed from the eigenvalue problem $(-\boldsymbol{M}\omega^2 + \boldsymbol{K})\hat{\boldsymbol{y}} = \boldsymbol{0}$, are shown in Figure 1. They have proven to be close to results obtained from experimental modal analysis of a real frame which will be used later on in the HiL-approach.

In order to suppress torsional vibrations, an actively controlled mass m is attached to one of the frame vertices, Figure 1b, where the relative position Δv is controlled by feed-back of torsion angle and angular velocity, i.e.

$$\Delta v = -K_\alpha\,\alpha - K_{\dot{\alpha}}\,\dot{\alpha}\,. \tag{2}$$

This will result in a force

$$F = m\left(\ddot{z}_1 + \Delta\ddot{v}\right) \tag{3}$$

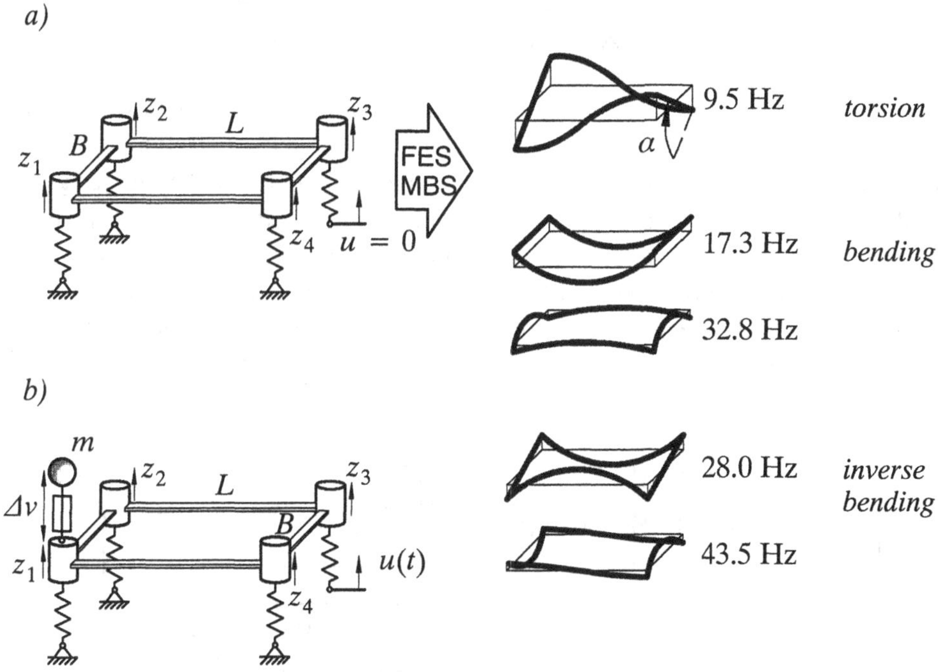

Figure 1. Frame structure with torsion and bending modes (a) and additional active mass damper (b)

acting on the left front corner of the frame. Together with excitation $u(t)$ this will change the system equation of the frame to

$$\boldsymbol{M}\,\ddot{\boldsymbol{y}} + \boldsymbol{K}\,\boldsymbol{y} = \boldsymbol{B}\,\boldsymbol{u} \tag{4}$$

where $\boldsymbol{u} = [u(t) \quad F]^T$ and $\boldsymbol{B}$ results from geometry.

In practical applications, the torsion angle used in the control law (2) cannot be measured directly but has to be computed from

$$\alpha = \frac{(z_4 - z_3) - (z_1 - z_2)}{B} \tag{5}$$

where the frame node positions z_i themselves have to computed by stabilized integration filters from measured accelerations $\ddot{z}_i$ as inputs. Further the actuator dynamics has to be taken into account which can be described sufficiently by the differential equation

$$\Delta\ddot{v} + 2\delta\,\Delta\dot{v} + \omega_0^2\,\Delta v = \omega_0^2\,\Delta v_{target} \tag{6}$$

where $\delta = 75.4\ \mathrm{s}^{-1}$ and $\omega_0 = 471$ rad/s have been identified for the hydraulic actuator used in the HiL-experiment later on. Altogether, the dynamics of the system can be summarized in a control loop as shown in Figure 2.

The control gains K_α, $K_{\dot{\alpha}}$ of Eq. (2) can be designed either by classical controller design methods like pole placement and Riccati equations, or by optimization. In the latter case, they will be interpreted as design variables $\boldsymbol{p} = \left[K_\alpha \quad K_{\dot{\alpha}}\right]^T$ and suitable design criteria have to be defined. The major goal of the mass damper is to suppress torsional vibrations in a given frequency range $\Omega = [\omega_l, \omega_u]$. If $G(s) = A(s)/U(s)$ denotes the transfer function of the controlled loop with the Laplace transformations $A(s) = \mathcal{L}\{\alpha(t)\}$ and $U(s) = \mathcal{L}\{u(t)\}$, the maximum response should be minimized. Thus, a potential objective would be the maximum value of the response function for harmonic excitation

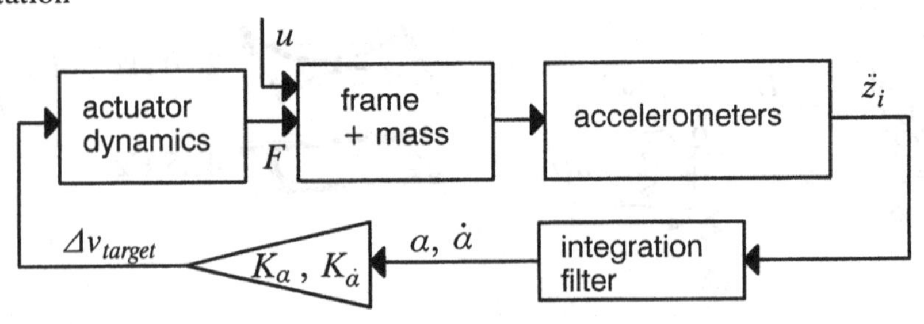

Figure 2. Control loop for frame structure with active mass damper

$$f_1 := \max_{\omega \in \Omega} |G(i\omega)| \,. \tag{7}$$

If this is used as the only objective, it turns out that high amplitudes of the tuned mass damper are required. In practise, however, the space for such a device will be limited. Thus, as a second criterion one may define the maximum of the response function for the amplitude of the mass point, i.e.

$$f_2 := \max_{\omega \in \Omega} |G_T(i\omega)| \tag{8}$$

where $G_T(s) = V(s)/U(s)$ and $V(s) = \mathcal{L}\{\Delta v(t)\}$. In principle, this criterion could serve as an inequality constraint, but in that case the designer has to assign a value to the maximum space prior to optimization without any idea about potential compromise solutions balancing both criteria.

Thus, a better way of dealing with such a problem is to define a multicriterion or vector optimization problem

$$\underset{p \in \mathcal{P}}{optimize}\ \boldsymbol{f}(\boldsymbol{p}) \,, \quad \mathcal{P} = \left\{ \boldsymbol{p} \in I\!R^h \,|\, \boldsymbol{g}(\boldsymbol{p}) = \boldsymbol{0},\ \boldsymbol{h}(\boldsymbol{p}) \leq \boldsymbol{0} \right\} , \tag{9}$$

see e.g. [3]. In our case the criterion vector $\boldsymbol{f}(\boldsymbol{p})$ consists of the two objectives (7) and (8) to be minimized and the set of feasible designs is chosen to be

$$\mathcal{P} = \left\{ p = \begin{bmatrix} K_a \\ K_{\dot{a}} \end{bmatrix} \,\middle|\, -3 \leq K_a \leq 0,\ 0 \leq K_{\dot{a}} \leq 0.05 \right\} . \tag{10}$$

According to the low dimensions of the design and criterion space, the criterion functions may be evaluated explicitly to obtain a contour plot in the design space, Figure 3. With some experience it becomes obvious that there is a design conflict where torsion criterion f_1 forces the gains to the upper left corner of the design space whereas amplitude criterion f_2 pushes the design to the lower right corner, however, it is hard to figure out the compromise solutions. Better insight in the problem can be gained in the design space where the vector criterion function is used as a mapping of the feasible designs to their criterion values. In this figure the design goal clearly is the lower left corner with minimum torsion angles and minimum mass amplitudes. The limit of simultaneous minimization of both criterion values is given by the boundary line of accessible designs called Pareto-optimal or more precisely Edgeworth-Pareto-optimal (EP-optimal) solutions. Each point of this set is an optimal compromise in the sense that lower torsion amplitudes can be gained only at the prize of higher mass amplitudes and vice versa. The corresponding design points $\boldsymbol{p}^*$ have to be identified as origins of EP-optimal images $\boldsymbol{f}(\boldsymbol{p}^*)$.

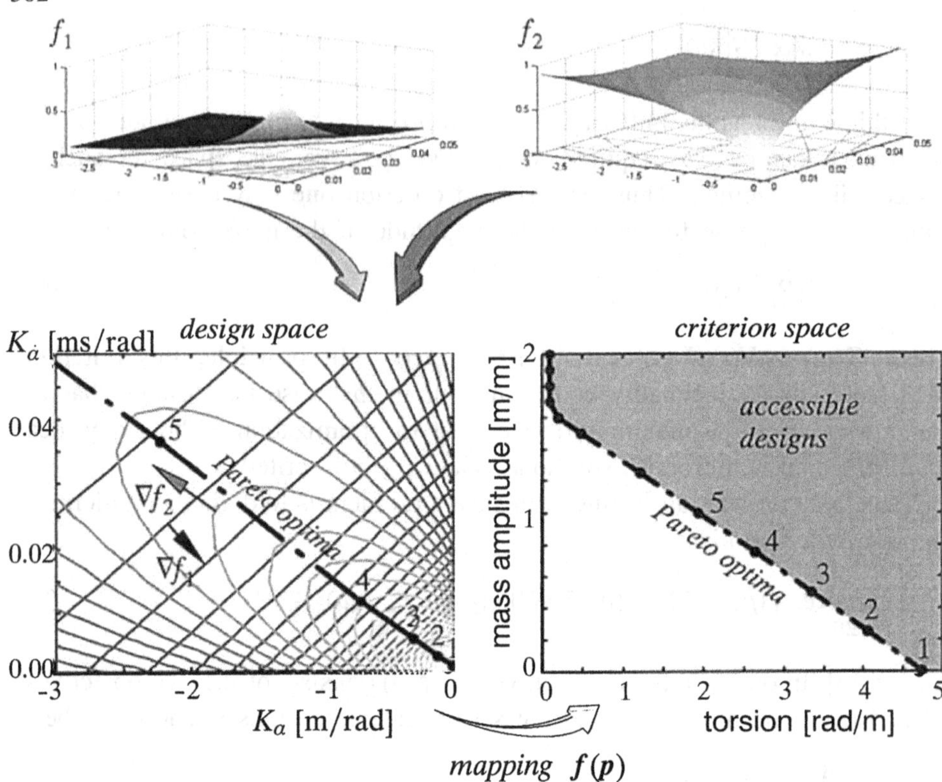

Figure 3. Optimization problem for active tuned mass damper

In more general cases, it won't be possible to find the whole set of EP-optimal designs by graphical means. In such cases, numerical methods have to be applied which basically reduce the multicriterion optimization problem to one or more single-criterion problems and solve them by standard nonlinear programming algorithms [4]. The choice of a proper reduction principle has to be part of the design process and is essential for a suitable imagination of potential trade-offs, e.g. [5], [6].

One of the most popular reduction principles is the weighted objectives method where an utility function

$$u(\boldsymbol{p}) = \sum w_i f_i(\boldsymbol{p}), \qquad w_i \geq 0, \tag{11}$$

is minimized instead of the criterion vector. For a specific choice of weighting coefficients w_i, where at least one has to be non-zero, only one single solution out of the EP-optimal set will be obtained. In the design space of our two-dimensional problem, the contour lines of the utility function are straight lines with slope $(-w_1/w_2)$ and the solution will be the boundary point touching the contour line with minimum value of u, Figure 4a. In order to find a representative subset of the EP-optimal solution set, the weighting coefficients have to be varied which will change the slope of the contour lines and thus the resulting trade-off solution. However, it is obvious that in this example for a reasonable choice for the w_i only the segment between points A and B will be found, and there will be no choice for finding trade-offs between B and C, because as the slope decreases the solution will jump from B to C. These nonlinearities and the incompleteness of EP-optimal solutions are the major drawbacks of the weighted objectives approach, and therefore it cannot really be recommended.

A better approach e.g. is the hierarchical method where we first introduce a lexicographical order by assigning a level of importance l_i to the criteria (low number $\hat{=}$ high priority). Then we successively minimize the objectives in this sequence as follows:

$$
\begin{aligned}
for\ j = 1,2,\ldots:\ \bar{f}_i &= \min_{p \in \mathcal{P}^{j-1}} f_i(p) \quad \text{where} \quad i:\ l_i = j, \\
\mathcal{P}^j &:= \left\{ p \in \mathcal{P}^{j-1} \,|\, f_i(p) \le (1+\varepsilon_i)\bar{f}_i \right\}, \\
\mathcal{P}^0 &:= \mathcal{P}\,. \qquad (12)
\end{aligned}
$$

We start with the most important criterion $(l_i = 1)$ to find its best value $\bar{f}_i$ which provides an idea about what is achievable. Based on this value we can define an upper bound by releasing the criterion from its best value by a user-

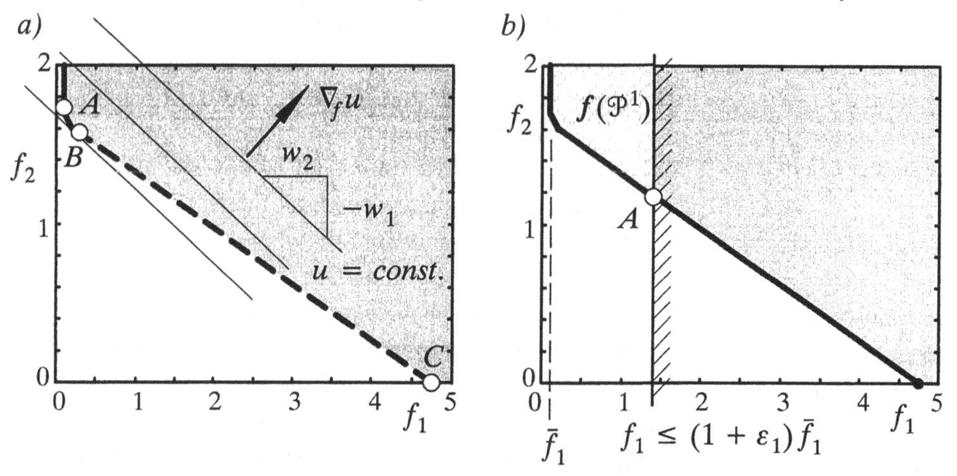

Figure 4. Scalar substitute problems for the tuned mass damper: (a) weighted objectives method, (b) hierarchical optimization

defined relative tolerance $\varepsilon_i > 0$ which restricts the set of feasible designs to $\mathcal{P}^1$. We then proceed with the next important criterion ($l_i = 2$) which in our case will already yield EP-optimum A, Figure 4b. By changing the user-defined tolerances the constraint bound is moved, and thus the designer may get some imagination about the whole EP-optimal set.

Beside these two examples there are many other heuristic multicriterion optimization approaches around which may be classified either as a scalarization or hierarchization principle, or a combination of it. Important is to accept it as part of the design process yielding only single solution points of non-comparable optimal trade-offs, leaving the final decision for a specific compromise solution to the designer. Such a final decision may be based on additional non-formal goals and properties or just a preference of some criteria over others.

3. Hardware-in-the-Loop Optimization

The process of modelling always requires simplifications and idealizations resulting in differences between the real system and the model. Thus, optimization on the basis of virtual prototyping always will result in an optimized model instead of an optimized technical system. For this reason, field testing and fine-tuning on a real prototype will always be the final step of the development process of a new product, and it would be desirable to use the same systematic multicriterion optimization approach as during the phase of virtual prototyping.

In order to find out about differences and problems when applying optimization to real systems instead of mathematical models, a hardware-in-the-loop setup for the above described active mass damper was set up, Figure 5. The dynamic system is realized as a steel frame (1.2×0.8 m, cross section 5×1 cm) with masses (3 kg) at its vertices being supported by soft springs analogously to Figure 1. The upper node of the spring at the left rear corner in Figure 5 is excited by an hydraulic actuator whereas the mass at the right front corner is substituted by a second hydraulic actuator of equal total mass with a movable piston of 0.2 kg weight used as a controlled mass damper. The vertical corner accelerations are measured by accelerometers and transmitted to a real-time dSpace computer [7] being connected to a PC. The real-time computer controls the excitation and the position Δv of the mass damper in real time, whereas the PC automatically organizes the setting of proper feedback gains, starting of an experiment, evaluation of objectives based on the measurements and the execution of optimization steps.

The objectives are the same as defined in Eqns. (7) and (8), however, their values cannot be found analytically anymore. Assuming a control loop analo-

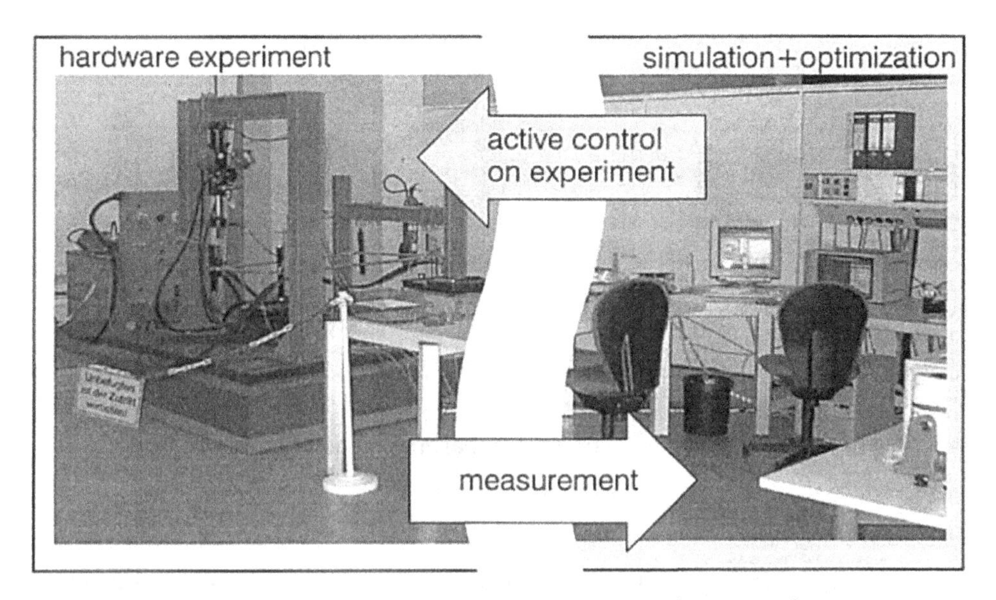

Figure 5. Hardware-in-the-loop setup for an active mass damper

gously to Figure 2, where now the upper blocks are substituted by hardware components, measurements can be performed and stored for excitation $u(t)$, torsion angle $\alpha(t)$ and relative mass displacement $\Delta v(t)$ at a discrete time rate. In our case, the data were sampled over $T_s = 61.5$ s at a rate of $f_s = 100$ Hz. The transfer functions were obtained from averaging three FFTs with a block length of 2048, the maximum was computed for the frequency range $\Omega = [10\pi\ ,\ 30\pi]$.

Starting with a function evaluation similar to Figure 3 already shows the major differences between criterion evaluation via models and experiments. The results of Figure 3 are rather smooth compared to the experimental results in Figure 6, where the function graphs are crumpled and have many local minima due to the far more limited measurement accuracy compared to numerical analysis of differential equations. Even if one would try to perform two different experiments with exactly the same feedback gains, the results would be different due to measurement noise and unsure initial conditions. Thus, use of conventional optimization algorithms like SQP methods and computation of gradients by finite differences fails due to the rough criterion estimates. However, finding the optima of the reduced nonlinear programming problems and not getting stuck in local minima is also essential for the overall success of multicriterion optimization and finding EP-optimal solutions.

A second important point regards efficiency. Applying optimization to dynamic systems, and especially multicriterion optimization, requires many time-

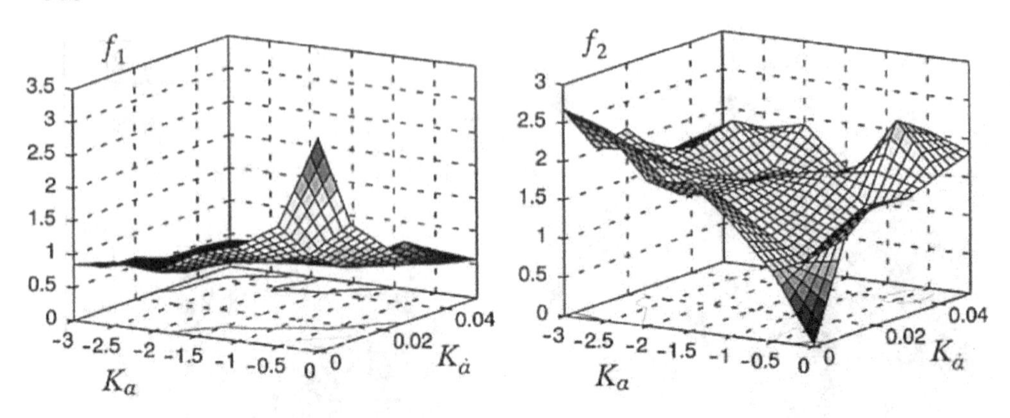

Figure 6. Criterion evaluation for experimental active mass damper

consuming analysis steps with modified design parameter values. Application of fast converging algorithms like the SQP approach reduces the number of optimization steps, however, such optimization routines require a lot more analysis steps just for computing gradient information e.g. from finite differences with the already mentioned accuracy problems.

In order to overcome both problems, low accuracy and high computational costs, the process of function evaluation is decoupled from the optimization process, Figure 7. Let us denote the design points proposed by an iterative optimization algorithm as $\boldsymbol{p}^{(i-2)}$, $\boldsymbol{p}^{(i-1)}$, ... with the actual design point being $\boldsymbol{p}^{(i)}$. Then the next optimization step with a first order optimization algorithm requires the function value $f^{(i)} = f(\boldsymbol{p}^{(i)})$ and the gradient $\nabla f^{(i)} = \partial f / \partial \boldsymbol{p}|_{\boldsymbol{p}^{(i)}}$, where the lower index indicating a specific objective or constraint of the multicriterion optimization problem is suppressed for simplicity. They may be estimated from $n \geq h + 1$ experiments started with design points $\boldsymbol{p}_j$, $j = 1(1)n$, which in general will be different from $\boldsymbol{p}^{(i)}$. As a result we obtain n function values $f_j = f(\boldsymbol{p}_j)$ from the measurements which in our case are the corner accelerations of the steel frame, the excitation and the mass displacement. These values may be interpreted as a first order Taylor-expansion about $\boldsymbol{p}^{(i)}$, i.e.

$$f_j = f(\boldsymbol{p}_j) \approx f(\boldsymbol{p}^{(i)}) + \nabla f^{(i)T} \left(\boldsymbol{p}_j - \boldsymbol{p}^{(i)}\right), \qquad j = 1(1)n\,, \tag{13}$$

which can be summarized in a system of linear equations:

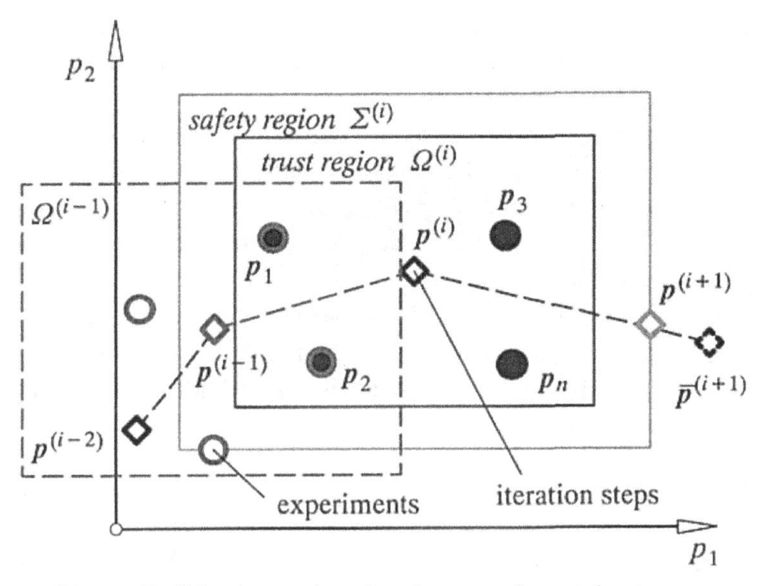

Figure 7. Criterion estimation for experimental setups

$$\begin{bmatrix} f_1 \\ \vdots \\ f_n \end{bmatrix} = \begin{bmatrix} 1 & \left(p_1 - p^{(i)}\right)^T \\ \vdots & \vdots \\ 1 & \left(p_n - p^{(i)}\right)^T \end{bmatrix} \begin{bmatrix} f^{(i)} \\ \vdots \\ \nabla f^{(i)} \end{bmatrix}, \qquad j = 1(1)n\,. \tag{14}$$

In case of $n = h + 1$ the solution will be unique, in case of $n > h + 1$ the system of equations is overdetermined and the solution may be obtained by least square methods like Householder transformations [8]. The accuracy of the estimates for the function value $f^{(i)}$ and the gradient $\nabla f^{(i)}$ can be controlled by the number of experiments involved and the size of the trust region where the experimental design points are taken from, Figure 7. In practise the trust regions of successive iteration steps will overlap, and it turns out that a new iteration step does not necessarily require n additional experiments, but also experimental results from former iterations may be taken into account cutting down experimental effort. This will be especially important in the final stage of convergence where design changes will be rather small. The initial trust region is chosen to cover 10% of the design space defined by its lower and upper bounds, and it is reduced by a constant factor in each iteration step.

If a general purpose SQP algorithm is applied to the problem, the rules for design changes are rather sophisticated, and in case of convergence problems such algorithms explore the boundaries of the design space resulting in rather large design changes. What may happen then, e.g. in control design via optimization, is that a design point with unstable dynamic behavior will be proposed endangering the experimental setup. Therefore, a second so-called safety region is defined which restricts the maximum design changes, Figure 7. If the objectives and constraint functions involve quantities becoming large in case of instability, that will help to detect stability problems early and keep away from dangerous designs.

Applied to our mass damper problem, we get e.g. an iteration history as shown in Figure 8. A hierarchical type method was used where the vibration amplitude is minimized according to objective $f_1(\boldsymbol{p})$ and the mass amplitude is bounded by $f_2(\boldsymbol{p}) \leq \hat{f}_2$:

$$\min_{\boldsymbol{p} \in \mathcal{P}^1} f_1(\boldsymbol{p}) \quad \text{where} \quad \mathcal{P}^1 := \left\{ \boldsymbol{p} \in \mathcal{P} \,|\, f_2(\boldsymbol{p}) \leq \hat{f}_2 \right\}. \tag{15}$$

Choosing $\hat{f}_2 = 1$ and starting from the initial design $\boldsymbol{p}^{(0)} = \boldsymbol{0}$ yields the desired EP-optimum $\boldsymbol{p}_1^*$ with the transfer functions shown in Figure 9. If the achieved reduction in the torsion amplitude is not sufficient, one may just release the bound on the mass amplitude by setting $\hat{f}_2 = 1.5$ and will gain another optimal trade-off with reduced torsional vibrations according to design $\boldsymbol{p}_2^*$. A comparison to Figure 3 shows slight differences according to differences between model and real system emphasizing the need of HiL optimization methods.

4. Parameter Identification by Multicriterion Optimization

System modelling is a creative, not clearly defined procedure driven by the kinematic and kinetic properties of a given or virtual technical system, but also by the analysis goals. Decisions have to be made with respect to rigidity or compliance of bodies, the number of degrees of freedom, and choice of coordinates. Simplifications are necessary in order to reduce model complexity and idealizations with respect to the force behavior of coupling elements have to be performed. Thus, differences between dynamic behavior of the model and the original technical system will be observed, and model parameters may loose their physical meaning defying direct measurement.

The goal of identification is to find parameter values such that the dynamic behavior of the model will be close to measurements on the real system. Classical approaches try to minimize the differences of the simulated behavior and some given measurement in an integral sense. However, different model param-

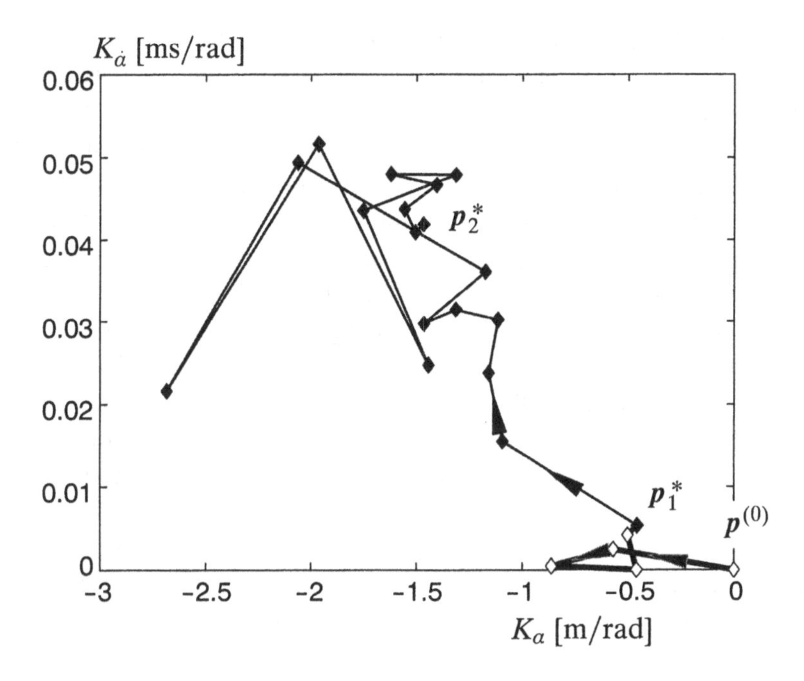

Figure 8. Iteration history for HiL mass damper design

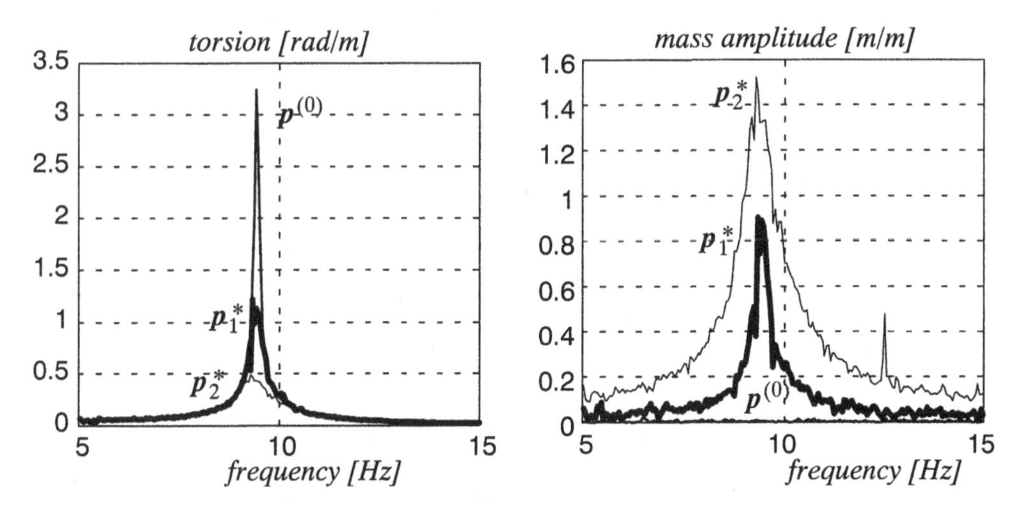

Figure 9. Transfer functions of steel frame with active mass damper for initial design $p^{(0)}$ and EP-optimal designs p_1^* (a) and p_2^* (b), respectively

eters will have different influence on specific parts of the trajectory, e.g. inertia errors will become most visible for non-stationary motion whereas the integration part of a controller will influence the decay of deviations from the stationary equilibrium position. Thus, the experimental maneuvers and the criteria for differences between real and model behavior should be devoted to the specific identification problem resulting in a multicriterion identification problem.

As an application example let us consider the hardware-in-the-loop simulation of a CNC-machine, Figure 10. On the left side the CNC is connected with the real machine tool axis whereas on the right side the CNC is coupled with a real-time computer simulating the dynamic behavior of the electronic amplifier and the tool axis including electric motor, belt drive, gears, and sensor dynamics. The goal is to provide a virtual test environment for CNCs with realistic feedback, and thus the model behavior has to be in good agreement with the real dynamics.

A first modal analysis showed that the eigenfrequencies of the belt drive are very high and may be neglected. Thus, the model reduces to a 1 DOF system described by the angular velocity ω of the motor:

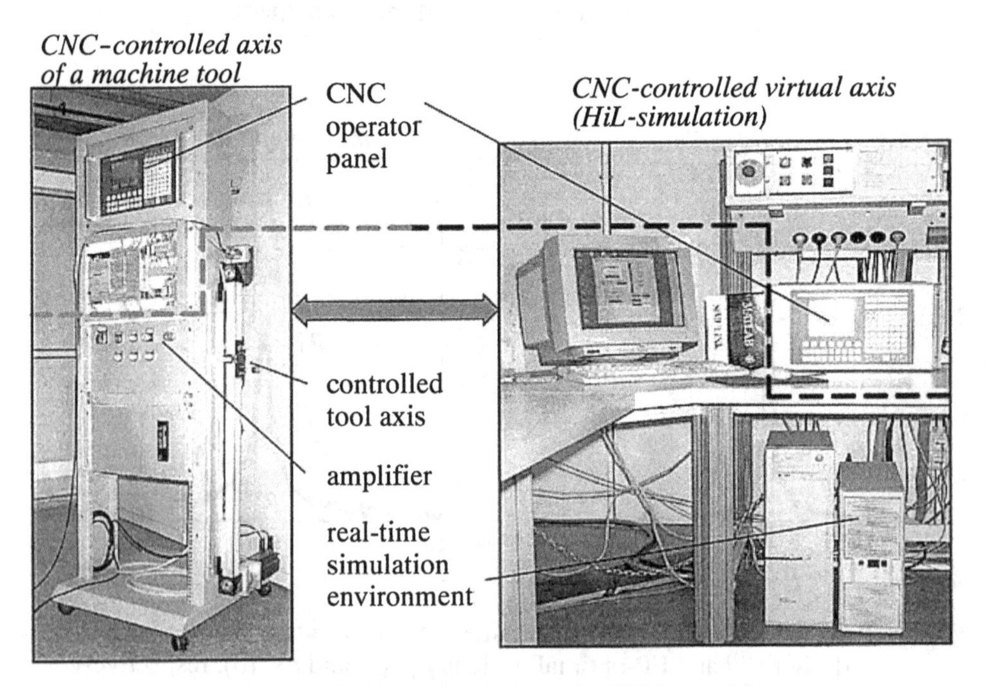

Figure 10. HiL-simulation of a CNC-machine

$$I\dot{\omega} = M_M - M_R - M_L(t) \tag{16}$$

where the unknown moment of inertia I summarizes the inertia terms of motor, gears and tool support. The motor moment can be considered as proportional to the current, i.e. $M_M = K_T\, i$ where the factor K_T can be taken from the motor manual. The friction moment M_R summarizes resisting forces and moments of the bearings, gears and belt drive and has to be identified with respect to both quality and quantity. The load moment resulting from forces acting on the tool support can be neglected for the identification process, i.e. $M_L = 0$.

The amplifier involves a hierarchical structure of analog control loops for current and angular speed. Since the dynamic response of the current control loop is high compared to the overlying speed control loop, it can be neglected. Thus, the amplifier's dynamics could finally be modelled as PI-controller where the state of the integral part is bounded by $\pm\,\hat{i}$:

$$i = P\,\Delta\omega + \left[\frac{1}{T_I}\int \Delta\omega\, dt\right]_{-\hat{i}}^{\hat{i}}. \tag{17}$$

The proportional factor P, the time constant T_I of the integral part and limit $\hat{i}$ for the state of the integral part are unknown and have to be identified.

The resulting flow chart of the model is shown in Figure 11. The real system is accessible only by imposing target speed signals which are generated by the already mentioned dSpace computer. Measurements have been performed with a sampling rate of $f_s = 1$ kHz on the motor current and the angular velocity.

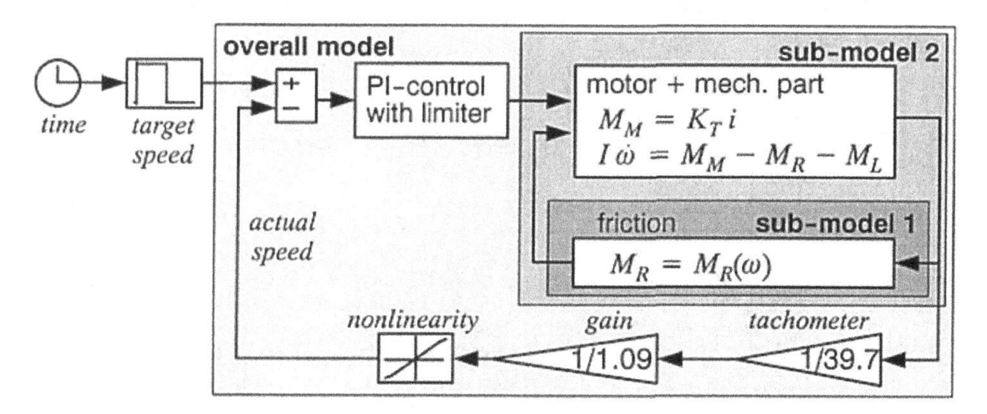

Figure 11. Model for the numerically controlled machine tool axis

In order to decouple effects for identification purposes, the overall model is split into sub-models according to the hierarchical structure of the loops. The most inner sub-model 1 consists of the resistance moment $M_R(\omega)$ which can be best identified for constant angular velocities. From Eq. (16) we find for $M_L = 0$ and $\omega \approx const.$

$$M_R \approx M_M = K_T\, i \tag{18}$$

which can be directly found from current measurements. Thus, we speed up the tool support to prescribed velocities ω_j and keep them constant, as far as the bounds on the stroke will allow. During that time the current will be almost constant yielding the moment M_j according to Eq. (18) and resulting in Figure 12. A proper functional description is

$$M_R = c_1\, \omega^{c_2} \tag{19}$$

with unknown constants c_1 and c_2. Thus, a first unconstrained optimization problem arises from minimization of the differences between measurements and approximation function for the resistance moment:

$$\min_{c_1, c_2} f_1 \qquad \text{where} \quad f_1 = \sum_j \left(M_j - c_1\, \omega_j^{c_2}\right)^2 . \tag{20}$$

In a second step, the unknown moment of inertia I is identified from sub-model 2. Since inertia terms become most visible for high accelerations, a step-function with the target velocity ω_j is imposed on the system and the current $i_j(t)$ and the time $T_{j,M}$ for reaching the target speed for the first time are measured, Figure 13. On the other hand, the measured current is used as input for the simulation of sub-model 2 also resulting in a speed-up time $T_{j,S}$. By comparing the two results we define a second optimization problem for the unknown inertia:

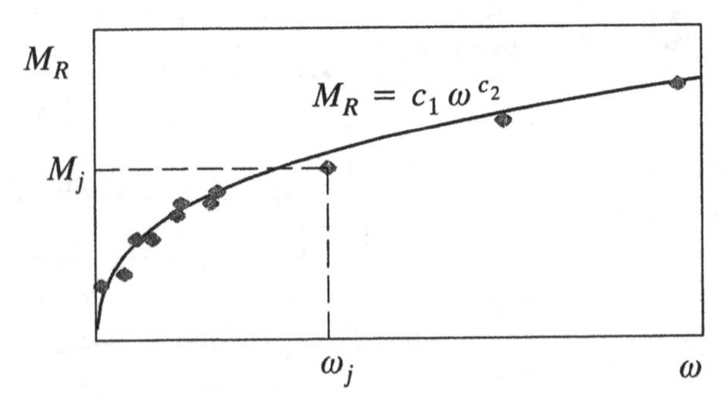

Figure 12. Resistance moment of the machine tool axis

$$\min_{I} f_2 \quad \text{where} \quad f_2 = \sum_j \left(T_{j,M} - T_{j,S}\right)^2 . \tag{21}$$

Based on these results a final identification problem can be set up for all unknown parameters $\boldsymbol{p} = [P, T_I, \hat{i}, c_1, c_2, I]^T$ of the overall model where the values from the optimization problems described above are taken as initial design. In order to obtain full information on both the controller and the mechanical part, complete speeding-up and breaking maneuvers with different target speeds ω_j are carried out, Figure 14. Applied to the real system we may find the current $i_{j,M}(t)$ from measurements which can be compared to the current $i_{j,S}(t)$ obtained from the simulation model:

$$\min_{p \in \mathcal{P}} f_3 \quad \text{where} \quad f_3 = \sum_j \int_0^T \left(i_{j,M}(t) - i_{j,S}(t)\right)^2 dt ,$$
$$\mathcal{P} = \left\{\boldsymbol{p} \in \mathbb{R}^6 \mid f_1(\boldsymbol{p}) \le \hat{f}_1 , f_2(\boldsymbol{p}) \le \hat{f}_2\right\} . \tag{22}$$

The quasi-static and highly instationary dynamic behavior expressed by criteria (20) and (21) is taken into account by inequality constraints according to the final step of the hierarchical multicriterion optimization method, where the bounds are user-defined based on the information obtained from the first two optimization tasks.

A close view to the current trajectories showed some differences between measurements and simulation during the phases of high current changes. Therefore, a fourth criterion was defined weighting the current differences by current changes:

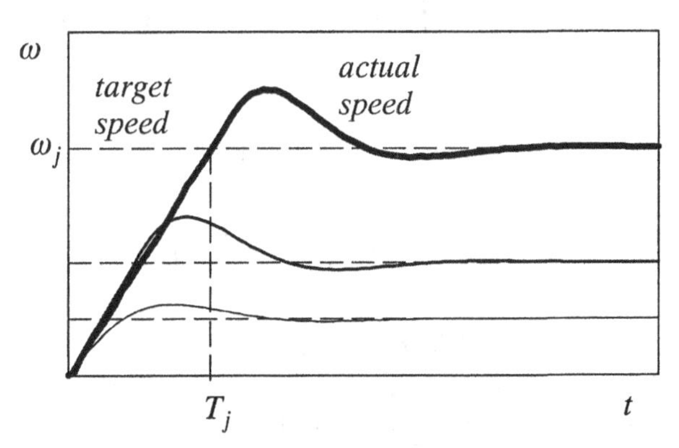

Figure 13. Step-resonse of the machine tool axis

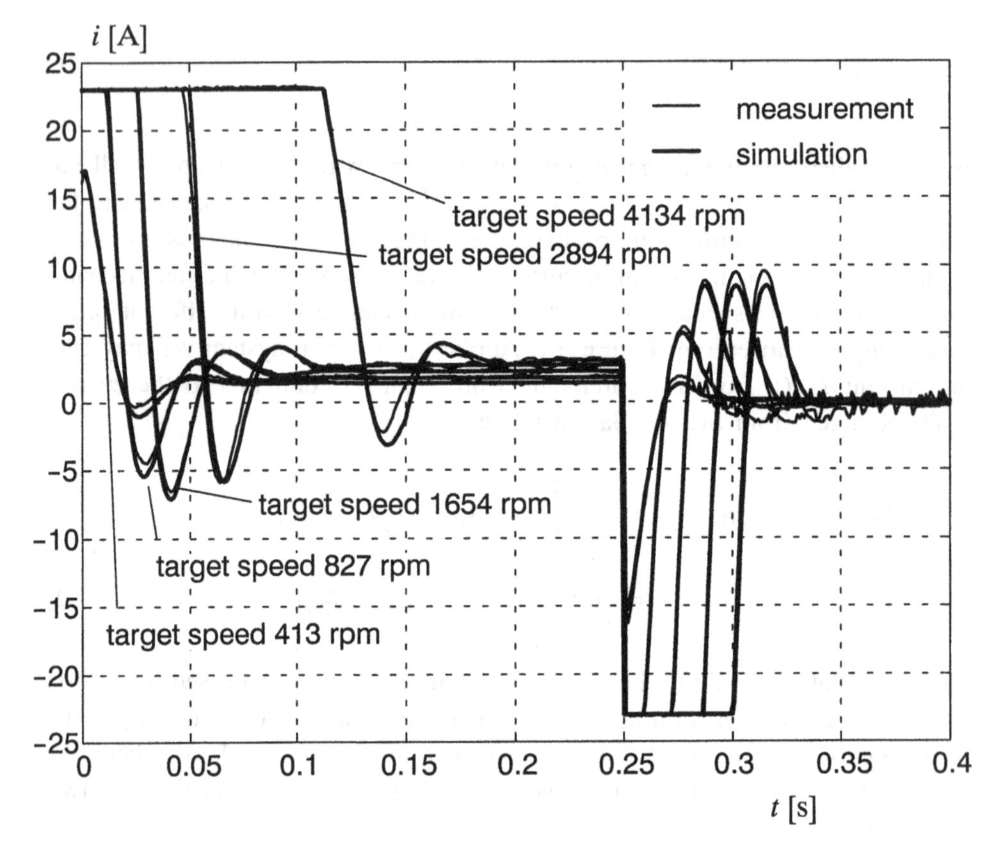

Figure 14. Speeding-up and breaking maneuvers of the machine tool axis

$$\min_{p \in \mathcal{P}} f_4 \quad \text{where} \quad f_4 = \sum_j \int_0^T w(t) \left(i_{j,M}(t) - i_{j,S}(t) \right)^2 dt \,,$$

$$w(t) = \frac{1}{2\Delta t} \int_{t-\Delta t}^{t+\Delta t} \left| \frac{di(\tau)}{d\tau} \right| d\tau \,,$$

$$\mathcal{P} = \left\{ p \in \mathbb{R}^6 \,|\, f_1(p) \leq \hat{f}_1 \,,\, f_2(p) \leq \hat{f}_2 \right\} . \tag{23}$$

For small windows Δt the amplification will be very locally, whereas large Δt will result in a uniform weighting. The definition of such a weighting function relies on the experience of the designer and other weighting functions are supposable, see ref. [2].

A comparison of the final model with the measurements in Figure 14 shows acceptable agreement for maneuvers with different target speeds, and also the coupling with a real CNC according to Figure 10 was very convincing showing the potential of multicriterion identification.

5. Conclusions

The modelling process has been developed over many centuries and has been broken down to clearly described procedures which can be highly automatized by computer-aided formalisms for Finite Element and Multibody Systems. Optimization is certainly still far away from that stage; objectives have to be designed via trial and error, and especially in the case where they should reflect the rating of humans it is hard to match the expectations. However, that is no reason for overlooking the high potential especially of multicriterion optimization methods in the design process. They provide a clear concept for handling conflicting criteria, reflect the experience of good designs being compromise solutions and release the designer from non-optimal solutions.

The paper shows applications with respect to three different tasks: design optimization on the basis of virtual prototyping, fine-tuning of an experimental setup, and parameter identification. In the first case, if a simulation model is available already, it takes only little effort to introduce automatic design analysis and optimization. The advantage of virtual prototype based design is that all model parameters can be varied and used as potential design variables, and criterion functions can be based on the full state information. Thus, it is only a matter of experience to define well qualified criterion functions and get started with a systematic design synthesis. The advantages over user-driven design improvements are time-savings and a clear classification of the kind of optimality and not just a feeling that it may be best with respect to whatever. Available ingredients for such a concept are existing analysis tools like MATLAB [1] for criterion evaluation, finite differences, symbolic differentiation [9], semi-analytic methods [3] or automatic differentiation [10] for sensitivity analysis, multicriterion optimization methods [5] and nonlinear programming algorithms [4].

Application of optimization algorithms to experimental setups requires more effort due to measurement needs and takes more caution about problems in criterion evaluation and sensitivity analysis due to measurement noise. The proposed approach of decoupling experimental design points from optimization steps has shown to be promising. Accuracy can be increased by estimating function values and gradients from more sampling points and experimental effort can be decreased by multiple use of experimental results in several iteration steps of the optimization process.

Finally, multicriterion optimization methods can also improve the identification task by allowing for specified maneuvers devoted to single design parameters. If unsatisfactory differences occur in parts of the dynamic behavior only, additional criteria can be defined and included in the identification process without loosing the correlation effect of former criteria and without the need of a super-criterion including all effects. This was demonstrated for a rather simple model, but will work equally well for more complex machines [2].

References

1. *MATLAB, The Language of Technical Computing* (2000). The MathWorks Inc, Natick.
2. Glora, M. (2001) *Ein Beitrag zur Mehrkriterienoptimierung von Simulationsmodellen und Experimenten,* Shaker, Aachen.
3. Bestle, D. (1994) *Analyse und Optimierung von Mehrkörpersystemen,* Springer, Berlin.
4. Fletcher, R. (1987) *Practical methods of optimization,* Wiley, Chichester.
5. Osyczka, A. (1984) *Multicriteria optimization in engineering,* Ellis Horwood, New York.
6. Hwang, Ch.-L.; Masud, A.S. (1979) *Multiple objective decision making - Methods and applications,* Lect. Notes in Economics and Math. Sys., Vol. 164, Springer, Berlin.
7. *ControlDesk - Experiment guide* (2002), dSpace GmbH, Paderborn.
8. Golub, G.H.; van Loan, F. (1986) *Matrix computations,* North Oxford Academic Publ., London.
9. *Maple 8* (2002) Waterloo Maple Inc., Waterloo.
10. Griewank, A. (1991) Evaluating Derivatives, SIAM, Philadelphia.

SYMOFROS: A VIRTUAL ENVIRONMENT FOR MODELING, SIMULATION AND REAL-TIME IMPLEMENTATION OF MULTIBODY SYSTEM DYNAMICS AND CONTROL

JEAN-CLAUDE PIEDBŒUF, JÓZSEF KÖVECSES, BRIAN MOORE†, RÉGENT L'ARCHEVÊQUE
Space Technologies, Canadian Space Agency,
6767 Route de l'Aéroport, St-Hubert, Québec, Canada, J3Y 8Y9
Jean-Claude.Piedboeuf@space.gc.ca, Jozsef.Kovecses@space.gc.ca,
Brian.Moore@space.gc.ca, Regent.Larcheveque@space.gc.ca

Abstract. This paper briefly describes Symofros, the modeling, simulation and control environment developed and used at the Canadian Space Agency for multibody and robotic systems. This environment is based on a symbolic modeling and code generation engine supported by Maple, and the Matlab/Simulink environment. Symofros serves two main purposes: control and real-time implementation, and analysis and design. Applications of the Symofros environment in space robotics will also be demonstrated in this paper.

1. Introduction

Multibody dynamics is of central importance in design and analysis of mechanical systems and their controllers. In space systems, multibody modeling and analysis is fundamental in developing and operating systems and technologies. Simulations (both non-real-time and real-time) are required for space robotics and space systems in general. The Canadian Space Agency's (CSA) in-house multibody dynamics software package Symofros has been developed since 1994. Symofros permits modeling, simulation and real-time control of multibody systems. The software architecture of Symofros is based on the Maple symbolic modeling engine and the Matlab-Simulink environment. Symofros is used for various projects in robotics both inside and outside CSA.
This paper describes the integrated virtual environment provided by Symofros. This environment allows the user to efficiently model, simulate in non-real-time and in real-time, and then do the implementation on a real hardware. The paper details the modeling environment based on XML, Maple and on a server system. We will then discuss the generation of the functions used for the simulation and the controller development. We will describe how a system can be simulated using the libraries built into Symofros. The next stage is the generation of a real-time simulation. As it will be discussed in the

† OPAL-RT TECHNOLOGIES ON SECONDMENT TO CANADIAN SPACE AGENCY

W. Schiehlen and M. Valášek (eds.), Virtual Nonlinear Multibody Systems, 317–324.

following, the Symofros architecture provides a very flexible environment that allows users to perform rapid prototyping.

2. Modeling

Symofros multibody dynamics engine is based on a formulation relying on Jourdain's principle. Jourdain's principle provides a physically clear framework for multibody analysis for both holonomic and nonholonomic systems. The various parts of Symofros' modeling engine have been extensively validated by experiments, analytical examples and simulations.

Complex systems (e.g. closed-loop multibody systems, parallel robots) can be split to sub-systems, and the system model can then be assembled by employing constraints between the various sub-systems. Open-loop systems and sub-systems are modeled using a generic recursive formulation, which can consider both rigid and flexible elements in the system (Piedbœuf, 1998).

In general, the consideration of the system constraints is a key issue in multibody dynamics. Symofros is able to handle both holonomic and nonholonomic constraints based on the Lagrangian multiplier technique with Baumgarte stabilization, and the use of projection and decomposition techniques.

Flexible beams are implemented for flexible body modeling with various choices of shape functions. Besides the traditional assumed modes approximations, a characteristic modeling approach employed is the advanced use of the assumed modes method, where the discretization is carried out in a way similar to the finite element method, i.e. interpolation functions are generated locally for an element, but the shape functions are represented globally as in the traditional assumed modes method. Besides body flexibility, the finite stiffness of the mechanical structure of the connecting joints is also a dominant effect in multibody systems. Symofros is capable of modeling joint flexibility using discrete stiffness models.

For contact mechanics modeling, Symofros currently uses the Contact Dynamics Toolkit developed by MacDonald Dettwiler Space and Advanced Robotics Ltd. Work is in progress to extend the contact-impact modeling capabilities of the Symofros environment with special attention to the real-time aspects. There are two main approaches being investigated in contact dynamics modeling: the local compliance based models, and the rigid body models based on unilateral constraints.

Dynamic parameter identification is an important area in multibody systems simulations, analysis and control. This area is currently being looked at to develop an identification toolbox for Symofros The two main purposes of the identification toolbox is to facilitate the optimum generation of experimental data for identification, and to process the measured data to determine the required parameters. This work involves the formulation and analysis of the dynamic equations in the form suitable for identification, and the solution techniques of these equations for the parameters.

Symofros also includes a control system toolbox comprising a library of Simulink blocks of various control algorithms (e.g. model based control with PD compensation). These can be easily linked and tested with the dynamic model of a multibody system to form the model of a controlled system. Also, new control algorithms can be readily built from the existing primitives.

3. Symofros software architecture

Symofros is based on commercial tools and is composed of three main modules for mechanical system description, modeling and simulation (see Fig. 1).

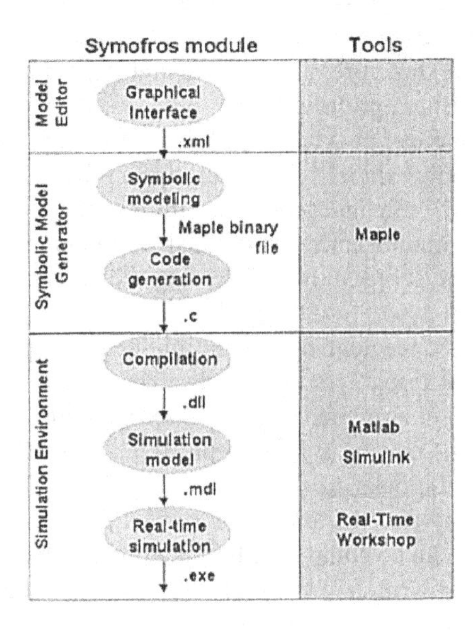

Figure 1. Overview of Symofros modules

Creating a model of a mechanical system consists of describing the bodies, the joints and the topology of the system. This model description is based on the XML language[1], a standardized language used to describe any kind of data and used for many applications. For mechanical system description, this language is also used by researchers in Spain (Rodríguez et al., 2001).
The Symbolic Model Generator (SMG) comprises modules written in the Maple language to perform the symbolic modeling. The input of the module is an XML or Maple file describing the properties of the mechanical system. This file is used by the module to compute the kinematic and dynamic quantities of the bodies and the joints. From the input file, the topology of the mechanical system is analyzed to generate a graph model. Using the topology with the body and joint data, the SMG develops the kinematic equations. Using the kinematic formulation, the SMG builds the dynamic equations in various forms, for simulation (forward dynamics), control (inverse dynamics), and parameter identification (currently in development). Special kinematic quantities are also

[1] www.w3.org

generated for parallel mechanisms based on the approach proposed in (Monsarrat and Gosselin, 2002). The SMG is normally used as an automatic model generator, but it is also a powerful tool to analyze the dynamic equations and to develop models on-line. More details on the symbolic modeling part of Symofros can be found in (Piedbœuf, 1996) and (Moore et al., 2002).

For simulation and real-time implementation, the SMG generates C code to represent the multibody system. The code generation requires optimization tools to break the complex expressions down to smaller expressions and to avoid redundant computation. The C functions are the links between the modeling part of Symofros, and the simulation/real-time implementation parts. Therefore, using the model in an advanced simulation or in the real-time environment is straightforward.

The Symofros SMG module can also be called using a server. The user has to connect to the server and send the mechanical system description files. These files are then processed by the SMG and the C file and processing information are sent back to the user. This approach reduces the maintenance required, since the upgrades and modifications have to be carried out on the server only. Also, using the server reduces the load on the user's computer resources.

To allow an efficient and convenient use of the mathematical model derived, and to enable the numerical simulation, Symofros is directly linked to the Matlab/Simulink environment. The Simulink environment allows to create complex models and generate complex simulation systems in only a few simple steps without the need of advanced programming skills. Special blocks are available in the library in order to call the functions generated symbolically and written in the .c file. As an example, Figure 2 shows how the forward dynamics can be computed. In this example, the dark blocks (*Mnl*, *gnl*) are used to call the functions written in the .c file. Then, using standard Simulink blocks, the system of equations is solved to obtain the accelerations, and integrated to obtain the generalized velocities and generalized coordinates. This block (Forward Dynamics) can then be found in the Symofros library and re-used with other models.

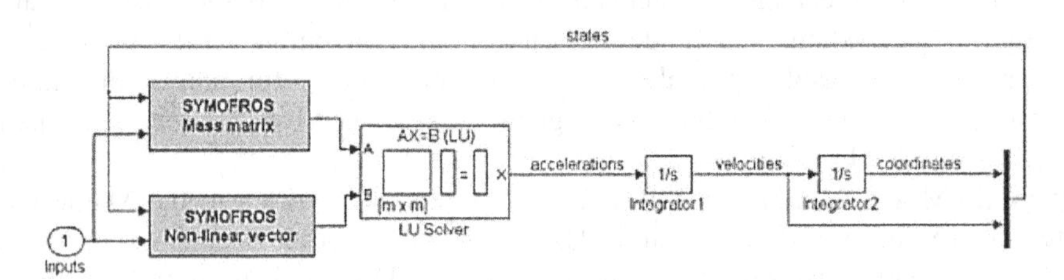

Figure 2. Example of a simple simulation model within Simulink

Real-time simulation and hardware-in-the-loop simulation can be achieved by using complementary tools like the Real-Time Workshop and RT-Lab for generating real-time simulation code and distributing the computations on several computers. More details on this topic can be found in (L'Archevêque et al., 2000), (Lambert et al., 2001) and (Piedbœuf et al., 2001).

4. Applications

Canada's contribution to the International Space Station (ISS) is the Mobile Servicing System (MSS) which is composed of the Mobile Transporter, the Space Station Remote Manipulator (SSRMS) and the Special Purpose Dexterous Manipulator (SPDM). An important aspect of a typical SPDM task is the insertion/extraction of payloads. To support the MSS, CSA has developed the STVF[2] and the SMP[3]. These two systems demonstrate the two main application areas of Symofros: model based control, and simulation and analysis.

4.1. STVF: HARDWARE-IN-THE LOOP SIMULATOR

Due to the complexity of an SPDM task, a verification of the operation must be performed on the ground for each payload manipulation. The main difficulty in this validation is verifying the part of the task for which the SPDM end-effector or payload undergoes contact with the environment. This part is verified using a hardware-in-the-loop simulation (HLS) to generate the real contact force using a mockup of the payload that needs to be manipulated.
The STVF Manipulator Testbed (SMT) (Aghili et al., 1999) is used to perform the HLS. The output of a real-time simulator representing a space robot is used as the input to the ground robot controller. The real contact forces are measured and fed back to the simulator.
Figure 3 shows the hardware architecture required for the test-bed. The real-time simulation is performed using the MSS Operation and Training Simulator (MOTS). The simulator includes the dynamics of the mobile base and the SSRMS in addition to the two arms of the SPDM. The full model has more than 50 degrees of freedom. The dynamic engine (SMT-SIM) is running at 1 kHz on an Origin 200 machine with four processors. The visualization is running at 25 Hz on a four processor ONYX machine. The real-time control of the robot is achieved using a cluster of six Pentium processors running QNX, and using Simulink Real-Time Workshop with Opal-RT RT-LAB for the code generation and multi CPU management. The graphical user interface on the SMT-CS is developed using Labview. The models required for the controller and for the simulation on the cluster are generated using Symofros. The robot controller in the HLS mode is based on a Cartesian feedback linearization (de Carufel et al., 2000). For the design and the tune-up phases, we developed an equivalent model of the robot using Symofros. This model reproduces exactly the same interfaces (in terms of inputs and outputs) what the SMT robot has. Therefore, we can choose between the real robot and the simulated robot simply by clicking a switch. For the same reason, a simplified model with a reduced number of degrees of freedom has been developed for the space robot using Symofros. This model uses exactly the same interface as the SMT-SIM. This flexibility is critical for the development since this allows the engineers to develop the

[2] SPDM Task Verification Facility
[3] System for Maintaining, Monitoring MSS Robotic Operator (MRO) Performance on board the ISS

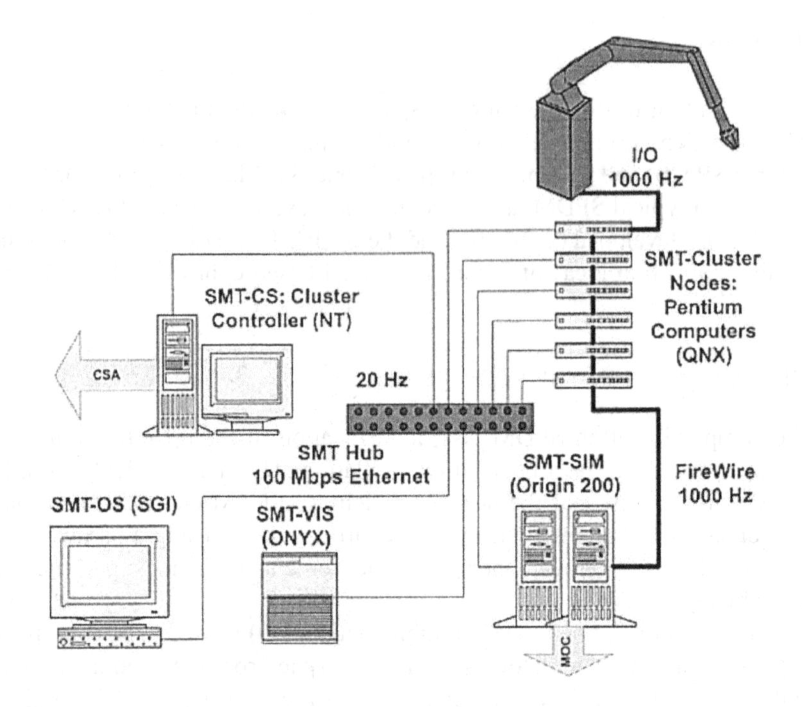

Figure 3. Computer Architecture for the HLS

overall software architecture in their offices and then download the code on the real-time system. There is no re-coding necessary between the pure simulation phase and the HLS phase.

4.2. SMP SIMULATOR

Experimental tests and analysis have shown that the capture of free-flyers is the most complicated task to be performed by a robotic operator on board of the ISS. The dexterity and accuracy of the astronauts decrease over time if they are not trained on-board. It is an obvious choice to have a simulator on-orbit to keep the skills of the astronauts at the required level. In order to support the on-orbit training, the SMP [4] simulator is being developed. The main objective of the simulator is to determine if an astronaut is ready to perform an operation with the real SSRMS. The training scenario, implemented in the SMP, consists of capturing a free-flyer with the SSRMS.

The simulator is composed of four modules, the Graphical User Interface (GUI), the Analysis Module, the Visual Renderer (VR) and the Dynamic Simulator (SIM). It has the same architecture as the Basic Operations Robotic Instructional System (BORIS) simulator used to provide generic robotic training to the astronauts (L'Archevêque et al., 2001).

[4] System for Maintaining, Monitoring MRO Performance on board the ISS

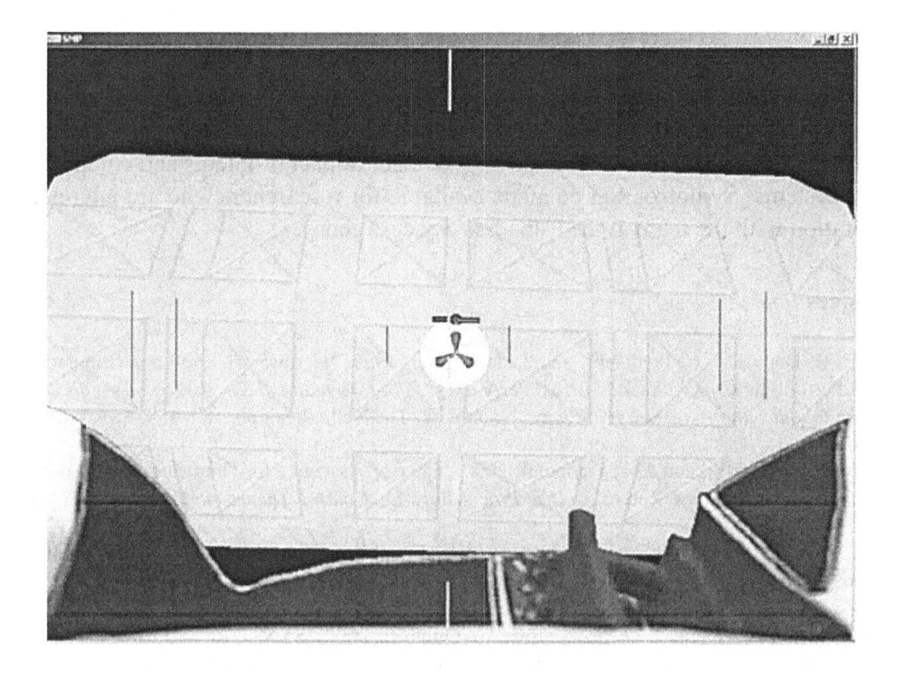

Figure 4. SMP Visual Renderer

The GUI has been developed with Labview 6 and runs on Windows operating systems. During a training session, the operator is firstly prompted to log into the system. Then, he has the choice to start a simulation session, a session analysis or a trend analysis. The session analysis provides information such as the hand-controller rates, the relative position and velocity between the end-effector and the free-flyer, and the capture status. Operational criteria and heuristics are used to provide a score, which allows the astronaut to have a good picture of his personal progress over time using trend analysis. The astronaut can then determine if he needs more training or not. Figure 4 shows the VR model of a generic free-flyer as viewed by the SSRMS end-effector camera. The VR module has been developed with OpenGL toolboxes.

Two Symofros models have been used to represent the SSRMS and the free-flyer. The SSRMS model has been configured and tuned using real flight data (data gathered during SSRMS operations) in order to obtain a realistic model. Generic parameters have been established to configure the dynamic behavior of the free-flyer. A Simulink diagram, using Symofros toolboxes, performs the simulation of the SSRMS, models the attitude control system of the free-flyer, interprets the hand-controller input values, handles the capture sequence, and gathers session data. The SMP simulator running in soft real-time on Windows 2000 has been generated using Real-Time Workshop. The experimental system will be launched in February 2003 and will be used by several astronauts and cosmonauts.

5. Conclusion

The virtual environment of Symofros has been used in several projects related to space applications. Future developments will include modules for parameter identification of multibody systems, advanced techniques for contact/impact problems and complex constrained systems. Symofros can be made available for researchers who are interested in collaboration with the team at the Canadian Space Agency.

References

Aghili, F., E. Dupuis, J.-C. Piedbœuf, and J. de Carufel: 1999, 'Hardware-in-the-Loop Simulations of Robots Performing Contact Tasks'. In: M. Perry (ed.): *Fifth International Symposium on Artificial Intelligence, Robotics and Automation in Space*. Noordwijk, The Netherland, pp. 583–588, ESA Publication Division.

de Carufel, J., E. Martin, and J.-C. Piedbœuf: 2000, 'Control Strategies for Hardware-in-the-Loop Simulation of Flexible Space Robots'. *IEE Proceedings-D: Control Theory and Applications* **147**(6), 569–579.

Lambert, M., B. Moore, and M. Ahmadi: 2001, 'Essential Real-Time and Modeling Tools for Robot Rapid Prototyping'. In: *The 6th International Symposium on Artificial Intelligence and Robotics & Automation in Space: i-SAIRAS 2001*.

L'Archevêque, R., M. Doyon, J.-C. Piedbœuf, and Y. Gonthier: 2000, 'SYMOFROS: Sofware Architecture and Real Time Issues'. In: *DASIA 2000 - Data Systems in Aerospace*. Montreal, Canada.

L'Archevêque, R., Z. Joukakelian, and P. Allard: 2001, 'BORIS: A Simulator for Generic Robotic Training'. In: *The 6th International Symposium on Artificial Intelligence and Robotics & Automation in Space: i-SAIRAS 2001*.

Monsarrat, B. and C. Gosselin: 2002, 'Jacobian Matrix of General Parallel and Hybrid Mechanisms with Rigid and Flexible Links: A Software-Oriented Approach'. In: *Proceedings of the 2002 ASME Design Engineering Technical Conferences*.

Moore, B., J.-C. Piedbœuf, and L. Bernardin: 2002, 'Maple as an automatic code generator ?'. In: *Maple Summer Workshop*.

Piedbœuf, J.-C.: 1996, 'Modelling Flexible Robots with Maple'. *Maple Tech: The Maple Technical Newsletter* **3**(1), 38–47.

Piedbœuf, J.-C.: 1998, 'Recursive Modelling of Flexible Manipulators'. *The Journal of Astronautical Sciences* **46**(1).

Piedbœuf, J.-C., F. Aghili, M. Doyon, Y. Gonthier, E. Martin, and W.-H. Zhu: 2001, 'Emulation of Space Robot Through Hardware-in-the-Loop Simulation'. In: *The 6th International Symposium on Artificial Intelligence and Robotics & Automation in Space: i-SAIRAS 2001*. Canadian Space Agency, St-Hubert, Quebec, Canada.

Rodríguez, J. I., J. M. Jimńez, F. J. Funes, and J. Garcià de Jalón: 2001, 'Dynamic Simulation of Multi-Body Systems on Internet using Corba, Java and XML'. In: *USACM, Sixth U.S. National Congress on Computational Mechanics*. p. 398.

SIMULATION OF AN ACTIVE CONTROL SYSTEM IN A HOT-DIP GALVANIZING LINE

O. BRULS AND J.-C. GOLINVAL
Département Aérospatiale, Mécanique et Matériaux,
University of Liège
Chemin des Chevreuils, 1 (B52/3), 4000 Liège, Belgium

Abstract. This paper concerns the modeling and the integrated numerical simulation of a flexible mechanism subject to the action of a digital control system. A general method is proposed, based on the formalism of flexible multibody systems using the Finite Element Method. The numerical simulation tool is applied to design an active control system in a hot-dip galvanizing line, which aims at reducing the vibrations of the steel strip.

1. Introduction

Lately, numerous investigations appeared in control of flexible mechanisms such as flexible manipulators, high precision machine tools, vehicles and foldable structures [1, 7, 8]. In order to estimate the performances before experimental testing, designers would like to simulate the whole mechatronic system, including the structure, the controller, the sensors and the actuators, using the most rigorous dynamical model as possible. Many standard simulation tools are available in both fields of flexible multibody systems and control systems. These software packages are usually not able to consider simultaneously the structural behaviour and the control system without lost of generality. In this paper, a general method for the simulation of mechatronic systems is presented and applied to design an active vibration control system in a hot-dip galvanizing line. The number of actuators and their configuration are defined on the basis of the simulation results.

The time-integration algorithm is presented in section 2 and the design of the active control system for the galvanizing line is described in section 3.

W. Schiehlen and M. Valášek (eds.), Virtual Nonlinear Multibody Systems, 325–332.

2. Simulation of mechatronic systems

2.1. MULTIBODY DYNAMICS

The purpose of this section is to recall some concepts of multibody dynamics [4]. The Finite Element methodology is adopted so that the motion is directly referred to the inertial frame. This formalism, implemented in the MECANO computer code [6], accounts for non-linear structural flexibility and large displacements.

Applying the Lagrangian multipliers method with the Hamilton principle leads to a system of Differential Algebraic Equations (DAE) of general form:

$$\begin{cases} M(q)\ddot{q} + B^T\lambda - g(q,\dot{q},t) = 0 \\ \Phi(q,t) = 0 \end{cases} \tag{1}$$

where one defines t, the time; M, the mass matrix describing the inertia terms proportional to acceleration; q, the generalized degrees of freedom of the system; g, the sum of external, internal and complementary inertia forces; Φ, the set of holonomic kinematic constraints; λ, the set of Lagrangian multipliers and B, the matrix of constraint gradients.

The first set of equations describes the dynamic equilibrium of the system and the second one represents the holonomic kinematic constraints.

The equations (1) may be integrated numerically through time-domain with the well known Newmark α-family of implicit algorithms. Assuming that the solution is known at time t_n, the unknowns of the problem are q_{n+1}, $\dot{q}_{n+1}$, $\ddot{q}_{n+1}$ and λ_{n+1} at time $t_{n+1} = t_n + h$, where h is the time step. The algebraic system consisting of the non-linear equations (1) at time t_{n+1} together with the Newmark equations may be solved with a Newton-Raphson iterative procedure.

2.2. SIMULATION OF MECHATRONIC SYSTEMS

A sampled-data control system with a sampling period T can be modelled by the following state equations:

$$\begin{aligned} x_{i+1} &= f_u(x_i, u_i, t_i) \\ f_{i+1} &= f_o(x_i, u_i, t_i) \end{aligned} \tag{2}$$

where the subscript i denotes the ith sampling instant, u_i the vector of the inputs, x_{i+1} the state vector of the control system, and f_{i+1} the vector of the outputs applied to the mechanism during the time interval $[t_i, t_{i+1}]$. f_u and f_o are respectively the update and output functions. In our case, the input data are measured on the mechanical system, and thus are related with its generalized coordinates: $u = u(q, \dot{q}, \ddot{q})$. The action of the control

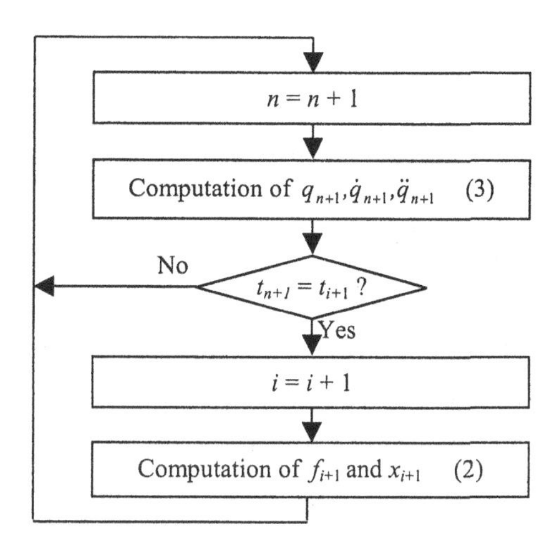

Figure 1. Numerical integration algorithm for a mechanical system subject to the action of a digital controller.

system on the structure modifies its dynamic equilibrium:

$$\begin{cases} M(q)\ddot{q} + B^T\lambda - g(q,\dot{q},t) = Df_{i+1} & \forall t \in [t_i, t_{i+1}] \\ \Phi(q,t) = 0 & \end{cases} \tag{3}$$

where D is the influence matrix of the control forces on the generalized coordinates, which is assumed to be constant. The integration algorithm of those equations is illustrated in Figure 1. The time step h is a divisor of the sampling period T. Inside each sampling period, the time integration of the mechanical equations (3) is performed taking into account a constant vector f. At the sampling instants, a FORTRAN routine, describing the state equations (2), updates the control forces as well as the state variables. In most cases, the dynamics of the control system is faster than the dynamics of the mechanism and a reasonable choice for the time step is $h = T$.

3. Application of the simulation tool

3.1. GALVANIZING PROCESS

Figure 2(a) illustrates a continuous hot-dip galvanizing line. The steel strip, of the order of 1 m wide by 1 mm thick, is preheated and passed at the speed of about 1 m/s through a pot of molten zinc. A zinc film is entrained onto the strip as it emerges from the pot. The deposited film solidifies while the strip moves vertically upwards. After the top roller, the finished product is guided to a delivery section where it is coiled and cut. The distance from the stabilizing roller to the top roller is of the order of 50 m.

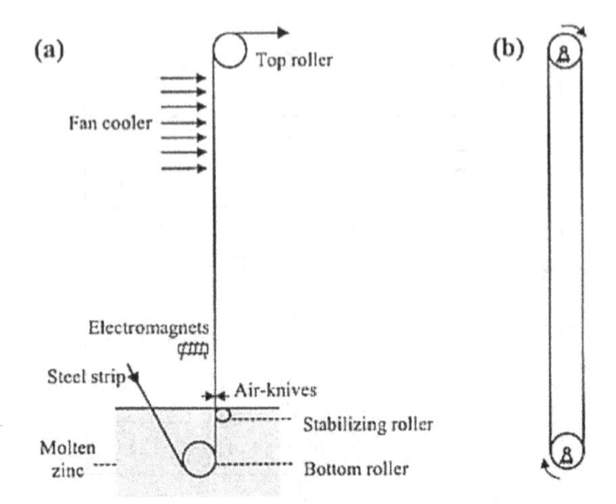

Figure 2. (a) Galvanizing line, (b)Two-dimensional Finite Element model of the moving steel strip.

Accurate control of the amount of solidified deposit has a great commercial issue: overdeposition results in excessive use of zinc which increases the production costs; underdeposition results in an unsatisfactory product. Air-knives, consisting of a pair of nozzles, regulate the zinc thickness. However, the vibration movement of the steel strip in front of the air-knives leads to variations in the amount of deposit.

Our purpose is to design a collocated active control system able to reduce those vibrations. Based on the information received by a sensor, a digital controller drives an electromagnet acting on the strip. Several independent sensor-actuator pairs can thus be installed.

In the following, a mechanical model is first established. Then the design and the modeling of the control system is described. Finally, the simulation tool which has been presented above is used to estimate the performances of the system.

3.2. MECHANICAL MODELING

Although the steel strip is prestressed, the structure remains very flexible and the mechanical excitation induced by the fan cooler causes a high vibration level. This section aims at constructing a reliable mechanical model able to capture all the significant effects.

3.2.1. *Basic Model*

The steel strip may be assumed to be fixed at the stabilizing roller and at the top roller. It is modeled with shell finite elements which allows to take the

TABLE 1. Natural frequencies.

Flexion (Hz)	Torsion (Hz)
0.55	0.56
1.10	1.11
1.66	1.67
2.21	2.23
...	...

prestressing effect and the gravity field into account. As the strip bends, its extension produces modifications of the stresses inside the structure, which influences its stiffness. This non-linear phenomenon, well known as geometrical stiffening, was considered in a preliminary model. But the results showed that the stress modifications remain small, so that the geometrical stiffening can be neglected and a linear model is sufficient to describe the dynamic behaviour of the steel strip. The first natural frequencies of the structure are listed in Table 1.

The pressure field produced by the fan cooler is modeled as a white noise excitation in the frequency range from 0.2 Hz to 10 Hz, as suggested by experimental data. This excitation appears to be spatially uncorrelated. However, to avoid the definition of a time-domain excitation function at each node of the Finite Element model, the excitation zone may be decomposed into a few independent zones in which the nodes are simultaneously excited.

3.2.2. *Speed of the Steel Strip*

The vertical motion of the steel strip during the process may affect the vibrations. A two-dimensional Finite Element model of the moving steel strip has been developed to study this phenomenon. The galvanizing line has been replaced by a line enclosing the stabilizing roller and the top roller as shown in Figure 2(b). Despite numerical difficulties encountered in the elaboration of this model, the natural frequencies have been computed for increasing values of the vertical speed. For speed values up to 15 m/s, they remain almost unaffected.

As a conclusion, we can assume that the steel strip is motionless and that both the stabilizing roller and the top roller are fixed.

3.2.3. *Model Reduction*

All non-linear effects were found to be negligible in the structure. Thus, assuming a linear behaviour, the Craig-Bampton substructuring method can be used to build a reduced model of the steel strip [3]. This method requires the partitioning of the initial degrees of freedom into two groups: the boundary degrees of freedom, which will be retained, and the internal degrees of freedom which will not appear explicitly in the reduced model and are considered as free. The movement of the structure is described as the superposition of constrained modes describing the static behaviour of the boundaries and a few clamped vibration modes obtained when fixing the boundary.

The degrees of freedom situated on the rollers, in the excitation zone, in front of the electromagnets and in front of the air-knives are defined as boundary degrees of freedom. To cover the frequency range of the excitation (0.2 Hz - 10 Hz), 50 clamped vibration modes have been kept. As the initial model contains 2400 degrees of freedom, the reduced one contains only 300 degrees of freedom so that the computation time decreases by a factor of 3.

3.3. ACTIVE CONTROL SYSTEM

This paragraph concerns the design and the modeling of the active control system when a single actuator acts on the structure (single input - single output system). If several independent sensor/actuator pairs are present, a similar procedure has to be followed for each of them.

3.3.1. *Control Law*

Design methods for active control systems are extensively described in the reference [5]. This paragraph presents the results of the design procedure.

A collocated configuration of the actuator and the sensor is chosen in order to maximise the robustness. The active damping control law is a direct velocity feedback :

$$f^d = -g\dot{q} \tag{4}$$

The desired force f^d is proportional and opposite to the measured velocity $\dot{q}$, which guarantees energy dissipation and unconditional stability. This control law is thus stabilizing for any flexible structure. No matter the dimensions of the steel strip, all vibration modes will be damped. The gain g, which defines the impact of the control system on the structure, has to be carefully optimized.

3.3.2. *Actuator Placement*

The performance of the active control depends on the position of the actuator on the structure. A method developed by Gawronski [2] has been applied to find the best location of the actuator. The detailed description of this method is beyond the scope of this paper. In brief, accounting for technological constraints, the optimization yields the following conclusion:

Displacement on the edge, electromagnet side :

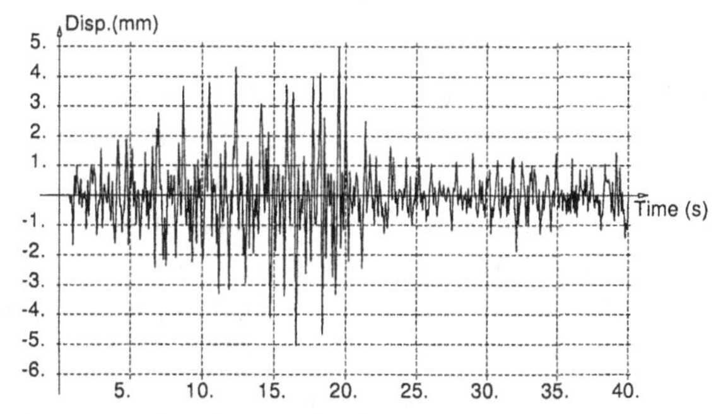

Displacement on the edge, opposite side :

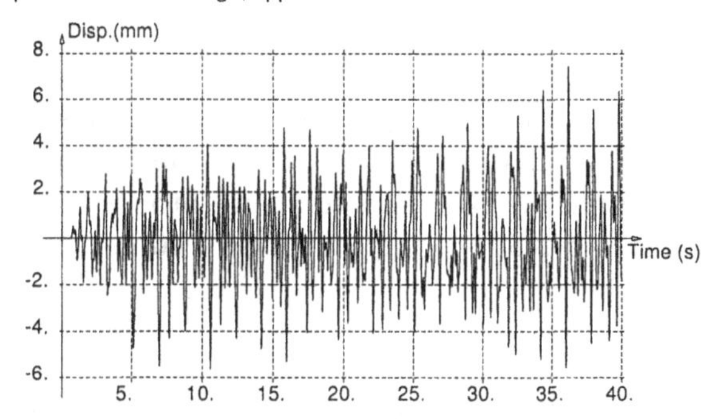

Figure 3. Displacements in front of the air-knives when a single actuator is located on one edge of the steel strip; the active control system is turned on at time $t = 20s$ (sampling period : $T = 0.5$ ms, gain : $g = 100$).

the actuator should be placed 3.5 m above the stabilizing roller, on the edge of the strip in order to control both flexion and torsion modes.

3.3.3. *Actuator Modeling*

The controller drives the electrical current i in the electromagnet. But, the relation between the electrical current and the force f applied on the steel strip is highly non-linear and dependent on the air gap e. An analytical expression of the relation $f(i, e)$ has been established to fit experimental data. This relation is the non-linear model of the actuator.

3.3.4. *Description of the Control Routine*

The control routine receives the input vector u containing two components: q and $\dot{q}$. First, it computes f^d according to the control law (4). Then, the value of q is used to compute the air gap e and the desired electrical current i is adequately estimated. Finally, the actuator model $f(i, e)$ defines the force f applied on the structure.

3.4. RESULTS

A parametric study has been led for several configurations of the actuators and several dimensions of the steel strip. For the sake of conciseness, the detailed results are not presented here. Figure 3 illustrates the results obtained with mean dimensions of the strip and a single actuator placed on the edge, 3.5 m above the stabilizing roller. After 20 seconds, the control system is turned on and the vibrations are efficiently attenuated on the actuator side, but not on the other side. Better performances are observed with more actuators: three actuators are able to reduce efficiently the vibration level in front of the air-knives.

4. Conclusion

The method presented for the simulation of mechatronic systems turned out to be really helpful for the design of an active control system in a hot-dip galvanizing line. It may be extended to many other kinds of applications such as machine tools, flexible manipulators, foldable structures, vehicles...

Acknowledgements

M. Brüls is supported by a grant from the Belgian National Fund for Scientific Research (FNRS) which is gratefully acknowledged. This work also presents research results of the Belgian programme on Inter-University Poles of Attraction initiated by the Belgian state, Prime Minister's office, Science Policy Programming. The scientific responsibility is assumed by its authors.

References

1. Cannon, H. R. and Schmitz, E. (1984) Initial Experiments on the End-Point Control of a Flexible One-Link Robot, *Int. J. Robot. Res.*, **Vol. 3 no. 3**, pp. 62–75.
2. Gawronski, W.K. (1998) *Dynamics and Control of Structures - A Modal Approach*, Springer.
3. Géradin, M. and Rixen, D. (1997) *Mechanical Vibrations: Theory and Application to Structural Dynamics*, John Wiley & Sons.
4. Géradin, M. and Cardona, A. (2001) *Flexible Multibody Dynamics: A Finite Element Approach*, John Wiley & Sons.
5. Preumont, A. (1997) *Vibration Control of Active Structures, An Introduction*, Kluwer.
6. SAMTECH (1999) *SAMCEF Users Manual - v8.0*, Liège, Belgium.
7. Van Brussel, H., Sas, P., Németh, I., De Fonseca, P. and Van den Braembussche, P. (2001) Towards a Mechatronic Compiler, *IEEE/ASME Transactions on Mechatronics*, **Vol. 6 no. 1**, pp. 90–105.
8. Wang, H.-L., Eischen, J.W. and Silverberg, L.M. (1998) On Control and Optimization of Elastic Multilink Mechanisms, *Computers and Structures*, **Vol. 67**, pp. 483–502.

MULTIBODY SYSTEM DYNAMICS IN OCEAN ENGINEERING

KATRIN ELLERMANN (ellermann@tu-harburg.de) and
EDWIN KREUZER (kreuzer@tu-harburg.de)
Technical University Hamburg-Harburg,
Mechanics and Ocean Engineering

Abstract. This paper deals with the dynamics of marine structures under wave excitation. The modeling of floating bodies is presented in detail. As one example ships with a forward speed of the vessel are considered and the stability of motion until the occurence of capsizing is investigated. In this case the vessel is floating freely in waves, in other cases the motion of vessels or marine structures is bounded by mooring systems. The dynamics of mooring cables and the lifting of a catenary from the ocean floor cause nonlinear restoring forces on the moored vessel. Thus, as a second example, moored floating cranes are considered. These vessels are known to exhibit large amplitude motions caused by resonance conditions. Operating ranges, in which the motion of the vessel or the load is highly dependent on the initial conditions and disturbances or in which small changes of parameter values cause large qualitative changes in the dynamics, are of special interest.

Key words: Multibody systems, nonlinear dynamics, ship stability, mooring systems, floating cranes

1. Introduction

Marine technology holds a number of challenging engineering problems. Due to the economic meaning of oil exploration or transport systems and the risk of accidents endangering human lifes and the sensitive environment, the prediction and characterization the dynamic behavior of these systems is essential. Determining or even increasing operation ranges is generally a non-trivial task. Excited by external forces due to wind, waves or currents, the dynamics of offshore systems exhibit different nonlinear phenomena ranging from subharmonic response to chaotic motion. While in many cases the nonlinearities arise from the kinematic coupling of different rigid bodies, they can also be caused by the fluid-structure interaction.

There are numerous examples for multibody systems in ocean engineering: platforms consist of many different parts which are either modeled as stiff or flexible bodies, underwater robots and remotely operated vehicles (ROV) used for deep sea exploration and assembly tasks, or different types of floating vessels which again can be modeled as multibody systems, e. g. catamarans. For a long time, the dynamics of these systems was either neglected completely or only considered by

W. Schiehlen and M. Valášek (eds.), Virtual Nonlinear Multibody Systems, 333–342.

simple linear techniques. Nonlinear dynamics provides several tools to perform the analysis of the dynamics of these systems more precisely and thus allow for a more reliable prediction and characterization of the dynamical behavior. In this process, the quality of the underlying mathematical model determines the accuracy of the analysis. A crucial part in the modeling of offshore systems is the description of the fluid-body interaction. In some cases, linear wave theory allows for a simplified description of the fluid-body interaction in frequency domain. Other problems require the analysis in time domain due to the complex structure of the underlying nonlinear effects, even though the modeling of the fluid forces is still based on a linear technique.

As in most technical systems, analytically exact solutions to the equations of motion can only rarely be obtained. On the other hand, a number of tools is available for generating approximate solutions either by numerical or by analytical methods: Numerical simulations provide insight in the dynamics of even complicated systems and computer animations are used for visualizing the results. Other techniques reveal the influence of different parameters, e. g. the numerical bifurcation analysis based on path following techniques or analytical methods based on perturbation theory.

In this article, the nonlinear dynamics of floating structures are considered. First, the problem of capsizing is investigated. Conventional approaches use the concepts of the metacentric height or the righting lever, which both neglect the dynamical aspects of the capsizing process. Here, the problem of ship stability is addressed using a nonlinear dynamical model. Additional nonlinearities can arise from the interaction of several connected bodies. Not only the multi-link catenary systems can be seen as a multibody system, but also the so-called single point mooring systems (spm), which consist of a moored buoy to which a ship is connected, can be seen as a multibody system. Finally, as a specific example of a moored vessel, the dynamics of a floating crane are considered.

2. Hydrodynamics of Marine Structures

The description of the fluid-structure interaction is the key point in the modeling process. Detailed models, using time-stepping methods for the calculation of the fluid forces are very precise but also very time consuming. These methods allow for the simulation of the dynamics of a system with a defined set of parameters. When investigating the influence or specific parameters on the dynamics of a floating body, i. e. when allowing certain values to vary in the calculation, a detailed but also easy to evaluate description of the fluid-structure interaction is needed. A standard technique to determine fluid-forces is based on potential theory. The total velocity potential Φ is then divided into parts corresponding to the potential of the incoming wave Φ_0, the radiation potentials $\Phi_{1...6}$, corresponding to the six degrees of freedom of the floating body, and the scattering or diffraction potential Φ_7. Software packages for the calculation of the hydrodynamic forces often combine the potentials Φ_0 and Φ_7 to give the total wave excitation forces. The so-called added mass and added damping coefficients a_{ij} and b_{ij} are derived from the radiation potentials

$$a_{ij} = -\rho\Re\left(\int\limits_S \phi_j n_i dS\right) \quad \text{and} \quad b_{ij} = -\rho\omega\Im\left(\int\limits_S \phi_j n_i dS\right) \quad , \tag{1}$$

with ρ being the density of the water, S is the submerged body surface and n_i is the projection of the surface normal vector on the ith coordinate direction. In some cases, the added mass and damping coefficients enter the equations of motion directly:

$$\left(\mathbf{M} + \mathbf{A}(\omega)\right)\ddot{\mathbf{u}} + \mathbf{B}(\omega)\dot{\mathbf{u}} + \mathbf{C}(\omega)\mathbf{u} = \mathbf{F}(\omega). \tag{2}$$

Here, $\mathbf{M}$ is the inertia matrix, $\mathbf{A}$ and $\mathbf{B}$ are the frequency dependent matrices of added mass and linear damping, respectively. The stiffness matrix $\mathbf{C}$ is usually obtained from hydrostatic considerations and the external forces $\mathbf{F}$ include the effects of the incident wave and diffraction. Additional forces, e. g., due to wind or mooring systems can also be included but generally cause nonlinearities in the equation. The validity of (2) is limited: Not only that in deriving this equation the resulting oscillation $u_i = \Re(\hat{u}_i e^{i\omega t})$ is assumed to be harmonic, it also has to be noted that the coefficient matrices not only depend on the wave frequency but also on the angle of the vessel relative to the wave direction. Different computer programs are available for determining the hydrodynamic coefficients in (2). They are based on either panel or strip methods as described in [1].

For the investigation of nonlinear systems a formulation of the equations of motion in time domain is required. A frequency or a spectrum of frequencies, the wave excitation can easily be determined in time domain. For the transformation of the radiation forces there exist different methods: Jiang [6] applies the so-called state space model. The frequency dependent added mass and damping coefficients $\mathbf{A}$ and $\mathbf{B}$ are approximated using rational polynomials

$$\left(\sum_{k=0}^{K} \mathbf{Q}_k (i\omega)^k\right)^{-1} \left(\sum_{k=0}^{K} \mathbf{R}_k (i\omega)^k\right) \approx \mathbf{A}(\omega) + \frac{i}{\omega}\mathbf{B}(\omega) \quad , \tag{3}$$

where K is the order and $\mathbf{R}$, $\mathbf{Q}$ are the coefficients of the approximation. A different formulation of the radiation forces is given by Cummins [3]: The coefficients

$$K_{ij}(\tau) = \frac{2}{pi} \int\limits_0^\infty b_{ij} cos(\omega\tau) d\omega \tag{4}$$

are used for the equations of motion. This approach results in an integro-differential equation

$$\sum_{j=1}^{6} \left((m_{ij} + a_{ij})\ddot{u}_j + \int_{-\infty}^{t} K_{ij}(t-\tau)d\tau + k_{ij} \right) = q_j \quad . \tag{5}$$

When evaluating (5) numerically, the integral is only evaluated for a short period of time, since the value of the K_{ij} approaches zero after only few wave cycles, see Figure 1.

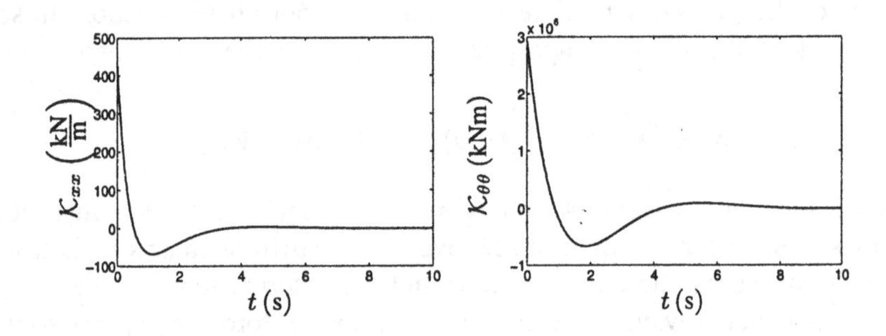

Figure 1. Examples for the coefficients K_{ij}

3. Ship stability

When a ship is moving forward in calm water at a constant speed, a stationary wave field develops. This wave field determines the ship's resistance in water, see Figure 2. Like the radiation and diffraction forces, the resulting resistance forces can be determined from potential theory. The potential of this stationary wave field is constant and superposed to the other potentials gives the total potential. To determine the stability of the upward position of a ship, the vessel is modeled as a single rigid body with six degrees of freedom, see Figure 3. Collecting the degrees of freedom in $\mathbf{u} = [x, y, z, \varphi, \theta, \psi]^T$, the equations of motion are given by

$$\mathbf{M}\ddot{\mathbf{u}} + \mathbf{k}(\mathbf{u}, \dot{\mathbf{u}}, t) = \mathbf{q}(\mathbf{u}, \dot{\mathbf{u}}, t) \tag{6}$$

with the mass matrix $\mathbf{M}$, the vectors of internal and external forces $\mathbf{k}$ and $\mathbf{q}$, respectively.

When investigating ship stability, the influence of the different parameters such as the amplitude of the encounter frequency of the waves is of major concern. One way of examining this dependence is by means of path following techniques as described in [8]. These methods allow for a determination of the solution of implicit algebraic equations. Requiring that the motion of a vessel is periodic,

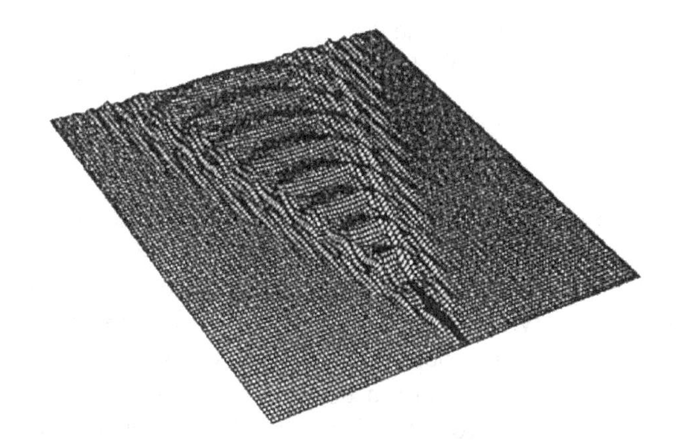

Figure 2. Stationary wave field of a ship moving in calm water,[5]

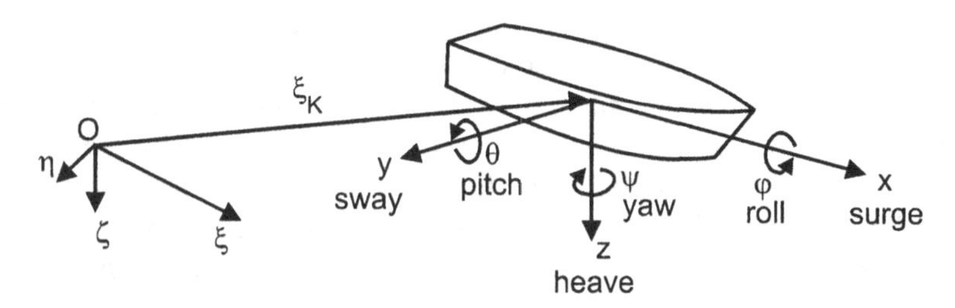

Figure 3. Coordinate systems for the investigation of capsizing

i. e., $P^{\ell}(\mathbf{u}, \lambda) - \mathbf{u} = 0$, where P^{ℓ} is the Poincaré map and describes the state of the system after ℓ times the reference period, the effect of different parameters λ on the dynamics can be studied. Even though the reference period is sometimes known in advance, for many dynamical systems, this is not the case for the floating ship investigated here. Depending on the speed of the ship, the encounter frequency and thus the reference period changes. To overcome this difficulty, the unknown period T is considered as an additional state of the system. The period is fixed by requiring that

$$H_i(t_0 + 2T) - H_i(t_0 + T) = 0 \quad , \tag{7}$$

where H_i is an arbitrary component of the algebraic equation. As an example of this procedure [9] considers the dynamics of two container ships. It is revealed that the coupling between the yaw and the roll motion contributes to the stability of the motion. It is observed from simulations that for ships in following seas or almost following seas, capsizing takes place after a significant yaw motion of the vessel. This scenario which also occurs in experiments can only be observed when all six degrees of freedom are taken into account. For increasing amplitudes it is observed that just before the critical amplitude for capsizing is reached, a bifurcation occurs. This bifurcation can be considered as an indicator for capsizing.

4. Mooring systems

In Offshore Engineering, mooring systems are used to hold buoys, ships or platforms in position during certain operations. This dominantly refers to the three translational degrees of freedom. These mooring systems can spread out over many kilometers. They typically consist of anchors, heavy chains and nylon ropes or steel cables which are attached to the moored object. A displacement of the vessel with respect to the equilibrium position causes lowering or lifting of the heavy chain links from the ground and thus results in a changed restoring force. The force characteristics of this system shows a significantly nonlinear behavior: As the displacement of the vessel increases, the catenary system shows stiffening of the restoring forces. When modeling the dynamics of mooring lines, this multibody system can be split into two parts: The mooring lines can be considered as hydrodynamically transparent since they show significant distortion of the wave field. In contrast to that, the floating vessel can be described by the radiation-diffraction model as given in section 2. The fluid forces on the transparent mooring lines can be calculated from the modified Morison's equation, which gives the change in the normal force dF_nin a part of the mooring line of length ds:

$$dF_n = \left(\rho \frac{\pi D^2}{4} \frac{\partial v_n}{\partial t} + C_a \rho \frac{\pi D^2}{4} \frac{\partial u_{rn}}{\partial t} + C_d \frac{\rho D}{2} |u_{rn}| u_{rn} \right) ds \tag{8}$$

with the normalized acceleration of the fluid $\frac{\partial v_n}{\partial t}$, the relative normalized acceleration between the fluid and the structure $\frac{\partial u_{rn}}{\partial t}$ and the normalized relative velocity between fluid and structure u_{rn}. Considering the dynamics of mooring lines as a multibody system leads to a large set of differential equations with hundreds of degrees of freedom. Dividing the catenary and the cable into several subsystems as shown in Figure 4 and including the appropriate boundary conditions for each subsystem can be advantageous for the integration of the equations of motion, compare [7]. While in many publications on mooring lines only static aspects of the mooring line forces are considered, this multibody approach also reveals the dynamic part of the forces: Simulations show that instead of just following the motion of the moored vessel, some parts of the mooring system might rest or even move in the opposite direction as compared to the vessel.

5. Floating cranes

One example of a moored multibody system is a floating crane. In addition to the mooring line forces and the fluid-structure interaction, its dynamics is influenced

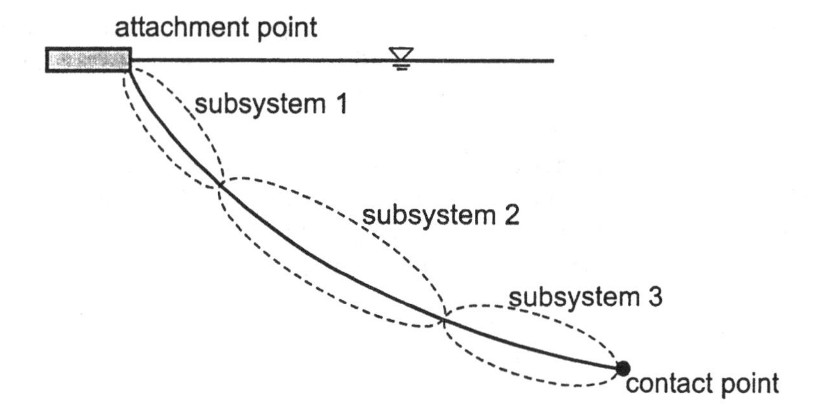

Figure 4. Subsystems in a catenary mooring system

by the coupling between the vessel and the swinging load. When the payload is small compared to the mass of the ship the influence of the motion of the load on the motion of the vessel may be neglected. Here, the floating crane is modeled as indicated in Figure 5. A system of two connected rigid bodies under the in-

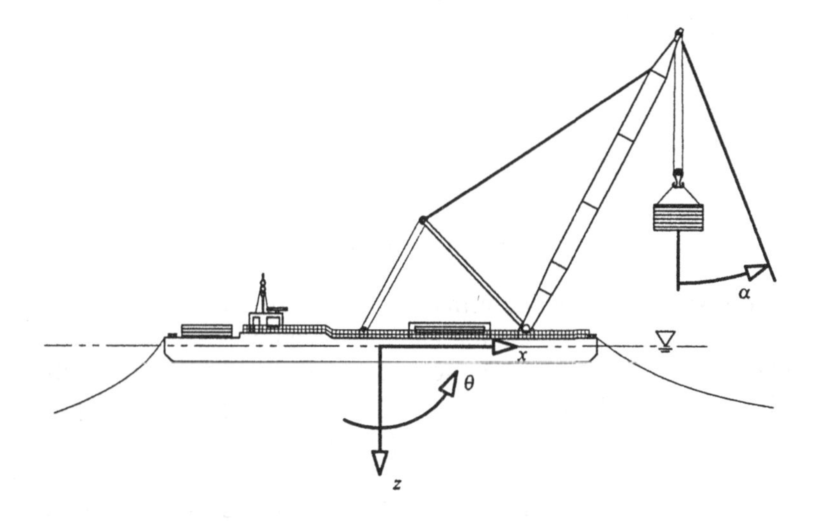

Figure 5. Model of a floating crane

fluence of periodic wave forces. The forces of the mooring system are simplified by using a polynomial approximation of the static mooring line curve. For the case of excitation in the plane indicated in Figure 5, this gives a system with four degrees of freedom with $\mathbf{u} = [x, \theta, z, \alpha]^T$. Experiments with floating cranes in a wave tank have shown that this system can show large amplitude subharmonic motion, see [2]. This subharmonic motion was observed to become particularly obvious from the surge motion. To investigate this phenomenon mathematically, two different techniques are applied, see [4]: One, the multiple scales method, is a tool that comes from perturbation theory. After some assumptions on the order of magnitude of the different parameters in the equations of motion and the relation between the forcing frequency and the resonance frequencies of the system, this method yields an analytical approximation for the solution. The advantage of this tool is that it can easily be evaluated for any set of parameters. Figure 6 gives an example of a solution obtained by the multiple scales method. The solid curves

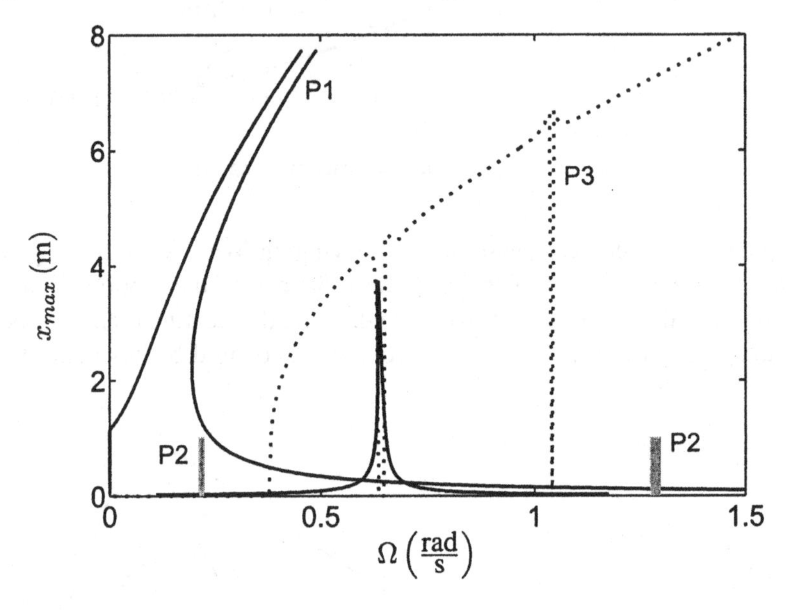

Figure 6. Results of the multiple scales analysis

indicate the first and the second primary resonance. The shaded areas give the range of the subharmonic P2-motion and the dotted curve shows the amplitude of the subharmonic P3-solution. The second technique, the numerical bifurcation analysis based on path following methods is applied to give a more precise solution for the individual motions and the position of the bifurcation points. The procedure for the path following is as given in section 3, except that this time, the fundamental period, the forcing period of the waves, is known in advance. Figure 7 shows an example for a numerically determined bifurcation diagram. The parameters correspond to those used in Figure 6. The bifurcation diagram clearly shows the same two peaks for the first two primary resonances, it gives the P2-motion at the frequencies predicted by the multiple scales method and it also reveals the strong bending of the curve for the P3-motion. The difficulty when applying the path following technique is that isolated solutions such as the P2 motion at 0.35 rad/S cannot be found directly. Only those solutions which result from a bifurcation can be traced systematically. By using different free parameters in the bifurcation analysis and possibly follow periodic solutions beyond the range of validity of the model, some of these different solutions can be found. A detailed discussion and many interesting results of the crane ship example are presented in [4].

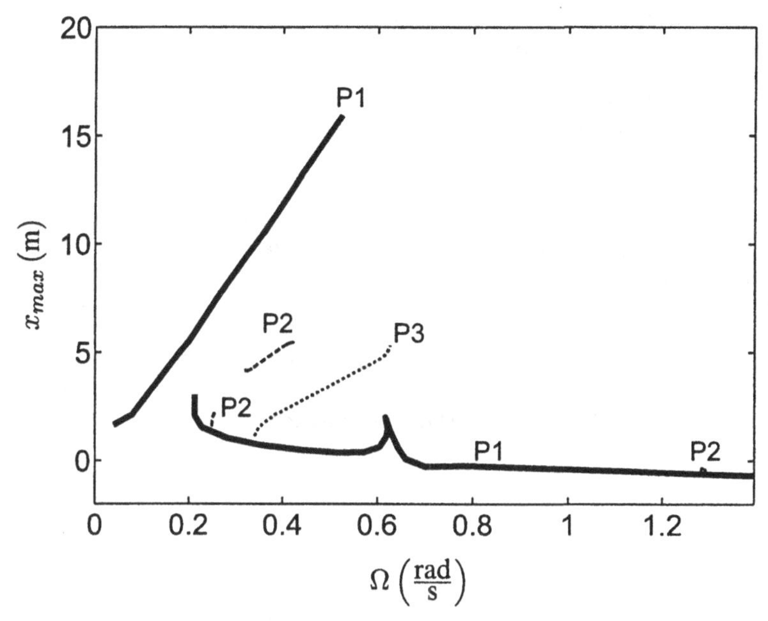

Figure 7. Results of the numerical bifurcation analysis

6. Conclusions

Even though the advances in computer technology and in the development of mathematical tools to describe nonlinear phenomena now allow for a more detailed investigation of offshore systems, the analysis of these systems will remain a challenging engineering task. The complexity of offshore systems is characterized by the interaction between the structure and the surrounding fluid as well as the uncertainty of parameters describing the operating conditions such as wind or waves. The examples given in this paper show that a systematic analysis of ocean engineering systems is possible but the mathematical analysis cannot be completely decoupled from the modeling process. Considering the pros and cons of simple mathematical models is helpful when applying certain mathematical tools and a verification of the results by more precise models is in some cases necessary.

References

1. BAUMGARTEN, R.; KREUZER, E.; WENDT, M. : Nonlinear Dynamics in Marine Technology. In: *Europ. J. of Mechanics A/Solids* 16, 1997, pp 25 – 44.
2. CLAUSS, G.; VANNAHME, M.; ELLERMANN, K.; KREUZER, E.: Subharmonic Oscillations of Moored Floating Cranes. In: *Proc. of the Offshore Technology Conference* (Housten, Texas, USA, May 1 – 4, 2000). OTC 11953.
3. CUMMINS, W.E.: The Impulse Response Function and Ship Motion. In: *Schiffstechnik* 9, No. 47, (1962), pp 101-109.
4. ELLERMANN, K.; KREUZER, E.; MARKIEWICZ, M.: Nonlinear Dynamics of Floating Cranes. In: *Nonlinear Dynamics* 27, (2002), pp 107-183.
5. FALTINSEN, O.M.: *Sea Loads on Ships and Offshore Structures.* Ocean Technology Series. Cambridge: University Press, 1990.
6. JIANG, T.: *Untersuchung nichtlinearer Schiffsdynamik mit Auftreten von Instabilitäten und Chaos an Beispielen aus der Offshoretechnik.* Hamburg, Institut für Schiffbau der Universität Hamburg, Dissertation, 1991.
7. KREUZER, E.; WILKE, U.: Dynamics of Mooring Systems in Ocean Engineering. In: *Archive of Applied Mechanics.*
8. SEYDEL, R.: *Practical Bifurcation and Stability Analysis.* New York/..., Springer, 1994.
9. WENDT, M.: *Zur nichtlinearen Dynamik des Kenterns intakter Schiffe im Seegang.* Düsseldorf : VDI Verlag GmbH, 2000 (Fortschritt-Berichte VDI, Reihe 12, Nr. 433).

SNAKE-LIKE LOCOMOTIONS OF MULTILINK SYSTEMS

F.L. CHERNOUSKO
Institute for Problems in Mechanics
of the Russian Academy of Sciences,
pr. Vernadskogo 101-1, Moscow 119526, Russia

KEYWORDS/ABSTRACT: modelling / control / multibody system / biomechanics / dry friction / periodic motions / snake-like locomotion / optimization / simulation / mobile robots

Motions of multilink mechanisms with two, three, and many links along a horizontal plane are investigated in the presence of dry friction between the mechanism and the plane. Control torques are created by actuators placed at the joints. It is shown that the system can perform various periodic and wavelike motions and reach any position and configuration in the plane. The speed of the motion and required torques are estimated. Optimal geometrical and mechanical parameters are determined which maximize the average speed. The results of computer simulation and experiments confirm the theoretical results.

1. Introduction

The crawling motions of snakes and other limbless animals have always been of great interest for specialists in mechanics and biomechanics. By contrast to walking and running creatures who alternate their supporting legs, snakes mostly keep the permanent contact between their bodies and the ground. Though the friction force acting upon each moving segment of the body is directed against the velocity of the segment, the resultant of the friction forces, i.e., the thrust, should be directed along the velocity of the center of mass of the body. To explain this phenomenon, various models of snake-like locomotions have been proposed.

The motion of a snake inside a curved tube has been considered in [1]. It was shown that the required thrust can be created by the normal reactions of the tube. Locomotions of snakes have been described and classified into

W. Schiehlen and M. Valášek (eds.), Virtual Nonlinear Multibody Systems, 343–362.

three classes in [2]. One of them, a rectilinear one, requires the displacement of the snake's mass along its body, whereas the other two classes (in which the snake twists its body) are possible only in the presence of vertical walls or other vertical or inclined objects. By pressing its body against these objects such as stones, grass, sand slopes, etc., the snake creates horizontal components of reactions along the direction of motion. Snakes always try to use vertical or inclined obstacles and avoid flat surfaces. Biomechanical aspects of snake-like locomotions have been discussed in [3]. Mechanisms of snake-like locomotions in the presence of obstacles have been considered in [4,5].

Nonholonomic multilink snake-like robots have been designed and investigated in [6,7]. These mechanisms consist of many elements equipped with passive wheels and connected by joints. By twisting, these robots can perform snake-like locomotions. In these motions, the wheels exert forces directed along their axes and thus produce the desirable forces in the direction of motion. In fact, the wheels here play the role similar to that of vertical walls for snakes. Kinematics of snake-like locomotions of such nonholonomic mechanisms have been considered in [8-10].

In this paper, we consider plane multibody systems (linkages) without wheels which can move along a horizontal plane in the presence of dry friction between the linkage and the plane. Control torques are created by actuators installed at the joints of the linkage. It is shown that the linkage can perform various motions along the plane. For three-link and two-link mechanisms, periodic motions consisting of slow and fast phases are designed. The three-member linkage is considered in more detail. Displacements, the average speed of motions, and the required magnitude of control torques are estimated. The geometrical and mechanical parameters of the linkage as well as the parameters of its motion are calculated which maximize the average speed. For the multilink system with more than four links, a slow wavelike motion is possible. The obtained theoretical results are confirmed by the results of computer simulation as well as by the experimental data. The paper is based on the resulllts obtained in [11-16].

2. Mechanical Model

Consider first a plane three-member linkage $O_1C_1C_2O_2$ moving along a fixed horizontal plane Oxy (Fig. 1). For the sake of simplicity, we assume that the links O_1C_1, C_1C_2, and C_2O_2 are rigid and massless bars, and the entire mass of the linkage is concentrated at the end points O_1, O_2 and joints C_1, C_2 which can slide along the plane. The mass of each of the end points is denoted by m_0, and the mass of each of the joints is denoted by m_1. Thus, the total mass of the linkage is $m = 2(m_0 + m_1)$. The length of the central

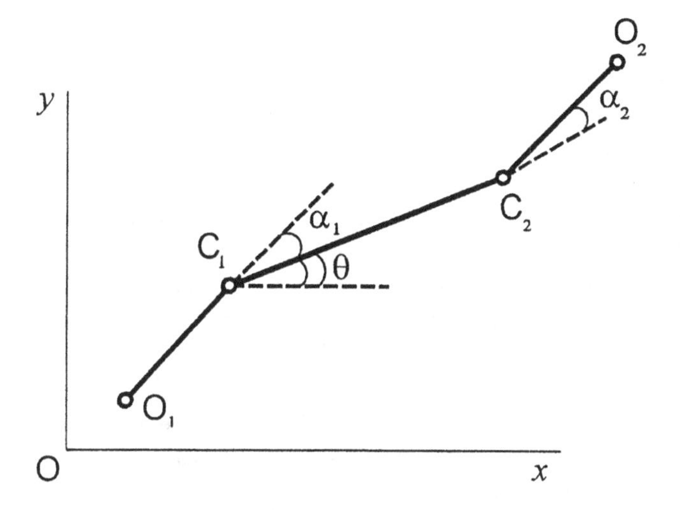

Figure 1. Three-member linkage

link C_1C_2 is denoted by $2a$, and the lengths of the end links are equal to l.

Denote by x, y the Cartesian coordinates of the middle of the central link, by θ the angle between this link and the x-axis, and by α_i the angles between the central link C_1C_2 and the end links O_iC_i, $i = 1, 2$ (Fig.1).

We assume that the dry friction forces acting between the masses O_i, C_i, $i = 1, 2$, and the horizontal plane obeys Coulomb's law. If a point mass m_0 (or m_1) moves, the friction force is directed opposite to the point velocity and equal to its weight m_0g (or m_1g) multiplied by the friction coefficient k_0 (or k_1). If the point mass m_0 (or m_1) is at rest, the friction force does not exceed m_0gk_0 (or m_1gk_1), and its direction can be arbitrary.

The control torques M_1 and M_2 about the vertical axes are created by the actuators installed at the joints C_1 and C_2. We assume that these torques can produce some prescribed time-history of angles $\alpha_i(t)$, $i = 1, 2$.

3. Elementary Motions

We will construct various motions of the linkage as a combination of more simple motions, which we call elementary [11]. All elementary motions begin and end at the states of rest of the linkage. In each elementary motion, the angles $\alpha_i(t)$, $i = 1, 2$, change within the interval $(-\pi, \pi)$ between the prescribed initial value α_i^0 and terminal value α_i^1; the time-histories $\alpha_i(t)$ can be more or less arbitrary. Either one or both angles α_1, α_2 can change in the elementary motion. In the latter case, they must change synchronously so that $\dot{\alpha}_2(t) = \pm\dot{\alpha}_1(t)$. Thus, the end links can rotate either in the same direction, or in the opposite directions.

Elementary motions are divided into slow and fast ones.

In *slow* motions, the values of the angular velocity $\omega(t) = |\dot{\alpha}_i(t)|$ and angular acceleration $\varepsilon(t) = \dot{\omega}(t)$ are small enough, so that the central link stays at rest during these motions. To derive conditions under which the central link does not move in slow motions, one should write down the equations of motion of the end links and determine forces and torques with which these links act upon the central one. Then one should consider the equations of equilibrium for the central link C_1C_2. This link will stay at rest, if friction forces acting upon the points C_1 and C_2 exist which satisfy Coulomb's law and ensure the equilibrium of the central link. According to this plan, sufficient conditions were derived which guarantee that the central link does not move during the slow motions [11, 13, 16]. To write down these conditions, we denote by ω_0 and ε_0, respectively, the maximal values of the angular velocity and acceleration of the end links during the slow motion:

$$\omega_0 = \max|\dot{\alpha}_i(t)|, \quad \varepsilon_0 = \max|\ddot{\alpha}_i(t)|. \tag{1}$$

Here, the maxima are taken along the whole slow motion. Since $\dot{\alpha}_2 = \pm\dot{\alpha}_1$, they do not depend on $i = 1, 2$.

If the both end links rotate in the same direction in the slow motion, then the sufficient condition, which ensures that the central link stays at rest, can be expressed as follows [13,16]:

$$m_0 l\left\{\left[\omega_0^4 + (\varepsilon_0 + gk_0 l^{-1})^2\right]^{1/2} + (\varepsilon_0 + gk_0 l^{-1})la^{-1}\right\} \leq m_1 gk_1. \tag{2}$$

This condition is also true for the case where only one end link rotates during the slow motion.

Note that condition (2) holds, if the motion is slow enough (i.e., if ω_0 and ε_0 in (1) are sufficiently small) and $m_0k_0(a + l) < m_1k_1a$. The latter inequality can be easily satisfied by choice of lengths a, l and masses m_0, m_1.

If the end links rotate in the opposite directions, condition (2) can be replaced by a weaker one [13,16]:

$$m_0 l\left[\omega_0^4 + (\varepsilon_0 + gk_0 l^{-1})^2\right]^{1/2} \leq m_1 gk_1. \tag{3}$$

This condition holds, if ω_0 and ε_0 in (1) are small enough and $m_0k_0 \leq m_1k_1$.

In *fast* motions, the angular velocities and accelerations of the end links are sufficiently high, and the duration τ of this motion is much less than the duration T of the slow motion: $\tau \ll T$. The magnitudes of the control torques M_1 and M_2 during the fast motion are high compared to

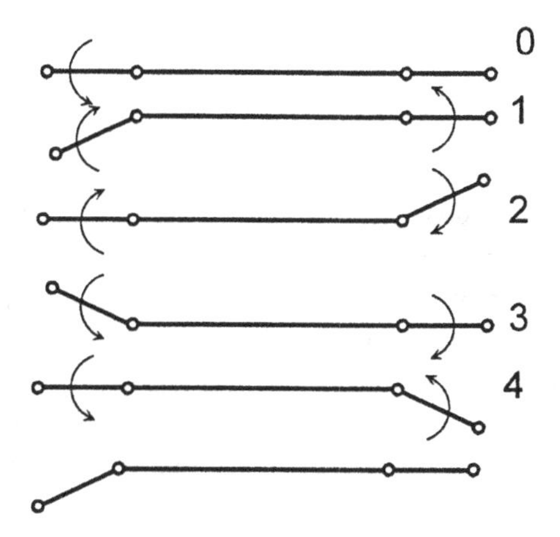

Figure 2. Longitudinal motion

the torques due to the friction forces:

$$|M_i| \gg m^* g k^* l^*, \quad i = 1, 2, \quad m^* = \max(m_0, m_1),$$
$$k^* = \max(k_0, k_1), \quad l^* = \max(a, l). \tag{4}$$

Hence, the friction can be neglected during the fast motion, and the conservation laws for the momentum and angular momentum hold in this motion. Therefore, the center of mass C of the linkage stays at rest, and its angular momentum is zero during the fast motion. Using these conservation laws, one can evaluate the terminal state of the linkage after the fast motion.

4. Periodic Locomotions

Let us show how longitudinal (or lengthwise), lateral (or sideways), and rotational motions of the three-member linkage can be composed from elementary ones. Suppose that at the initial instant of time the linkage is at rest, and all its links are parallel to the x-axis. We have $\theta = \alpha_1 = \alpha_2 = 0$ in this state.

For the sake of brevity, we denote slow and fast motions by the capital letters S and F, respectively. Let us indicate the initial and terminal values of the angles α_1, $i = 1, 2$, in each elementary motion by the respective superscripts ^{0} and ^{1}. Thus, we will describe the limits between which the angle α_i changes in an elementary motion as follows α_i: $\alpha_i^0 \to \alpha_i^1$, $i = 1, 2$.

Longitudinal motion (Fig. 2).

First, the auxiliary slow motion is carried out: $S, \alpha_1 : 0 \to \gamma, \alpha_2(t) \equiv 0$ where $\gamma \in (-\pi, \pi)$ is a fixed angle. Then, the following four motions are performed:

$$1)F, \alpha_1 : \gamma \to 0, \alpha_2 : 0 \to \gamma; \qquad 2)S, \alpha_1 : 0 \to -\gamma, \alpha_2 : \gamma \to 0;$$
$$3)F, \alpha_1 : -\gamma \to 0, \alpha_2 : 0 \to -\gamma; \quad 4)S, \alpha_1 : 0 \to \gamma, \alpha_2 : -\gamma \to 0.$$

After stage 4, the configuration of the linkage coincides with its configuration before stage 1, so that stages 1–4 can be repeated any number of times. Thus, we obtain a periodic motion consisting of two fast and two slow phases. To return the linkage to its initial rectilinear configuration $\alpha_1 = \alpha_2 = 0$, one should perform the motion $S, \alpha_1 : \gamma \to 0, \alpha_2 \equiv 0$.

Let us evaluate the displacements of the linkage during the cycle of the periodic motion described above. During the slow phases, the central link stays at rest, whereas the center of mass C of the linkage moves. One can see from Fig.1 that the x-displacements of the both end masses are negative in the both slow phases, whereas their y-displacements are of the opposite signs. Hence, the x-coordinate x_C of the center of mass increases in the both slow phases, whereas its y- coordinate y_C does not change. During the fast phases, the center of mass C stays at rest, while the central link moves. Since the x-displacements of the both end masses are negative in the both fast phases, the central link moves forward along the x-axis in these phases. The y-displacements of the both end masses are positive in the first fast phase and negative in the other one. As a result, the total displacement of the central link along the y-axis during the whole cycle is zero. Since the both end links rotate in the opposite directions in the both fast phases, the central link, by virtue of the conservation law for the angular momentum, does not rotate at all.

The considerations presented above and the calculations performed in [11,16] give the following total increments for the coordinates x, y of the middle of the central link and for the angle θ during the cycle of the longitudinal motion:

$$\Delta x = 8m_0 m^{-1} l \sin^2(\gamma/2), \quad \Delta y = 0, \quad \Delta\theta = 0. \tag{5}$$

Since the duration of the fast phases τ is negligible compared to that of the slow ones, the average speed v_1 of the longitudinal motion is evaluated by $v_1 = \Delta x(2T)^{-}1$.

Lateral motion (Fig. 3).

First, the auxiliary slow motion S, $\alpha_1 : 0 \to -\gamma$, $\alpha_2 : 0 \to \gamma$ is performed. Then, the following two motions are carried out:

$1)F, \alpha_1 : -\gamma \to \gamma, \alpha_2 : \gamma \to -\gamma;$

$2)S, \alpha_1 : \gamma \to -\gamma, \alpha_2 : -\gamma \to \gamma.$

Note that the configuration of the linkage after stage 2 coincides with its configuration before stage 1. To return to the initial configuration, it is sufficient to perform the motion $S, \alpha_1 : -\gamma \to 0, \alpha_2 : \gamma \to 0$.

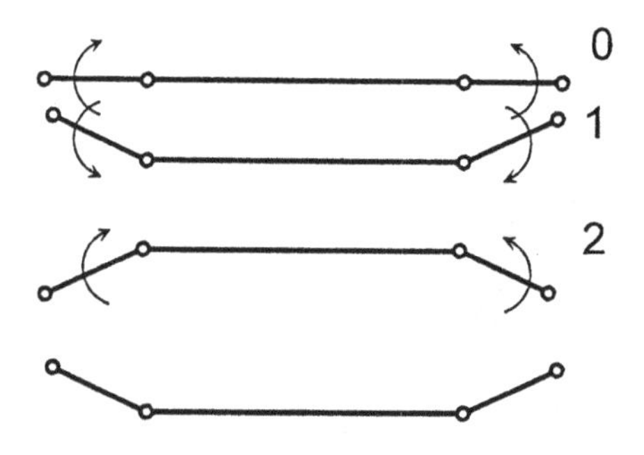

Figure 3. Lateral motion

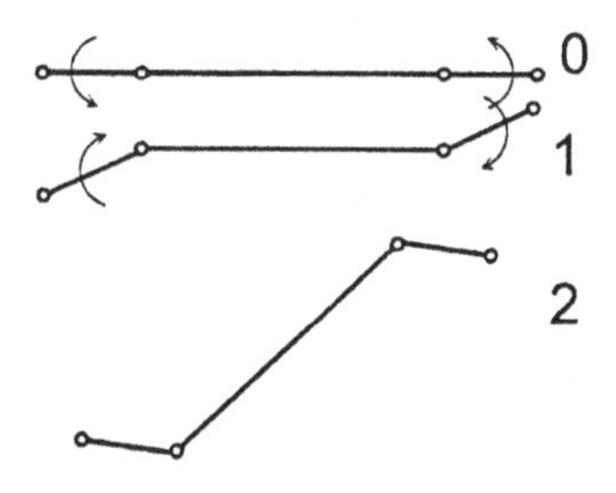

Figure 4. Rotation

Due to the symmetry of the motions of the end links, the displacement of the central link along the y-axis and its angle of rotation stay equal to zero during the whole cycle of motion. The total displacement of the middle of the central link along the y-axis and the average speed of the lateral motion are given by [11]:

$$\Delta y = 4m_0 m^{-1} l \sin\gamma, \quad v_2 = \Delta y T^{-1}. \tag{6}$$

Rotation (Fig. 4).

First, the auxiliary motion $S, \alpha_1 : 0 \to \gamma, \alpha_2 : 0 \to \gamma$ is carried out. Then, the following two motions are performed:

1) $F, \alpha_1 : \gamma \to -\gamma, \quad \alpha_2 : \gamma \to -\gamma;$

2) $S, \alpha_1 : -\gamma \to \gamma, \quad \alpha_2 : -\gamma \to \gamma.$

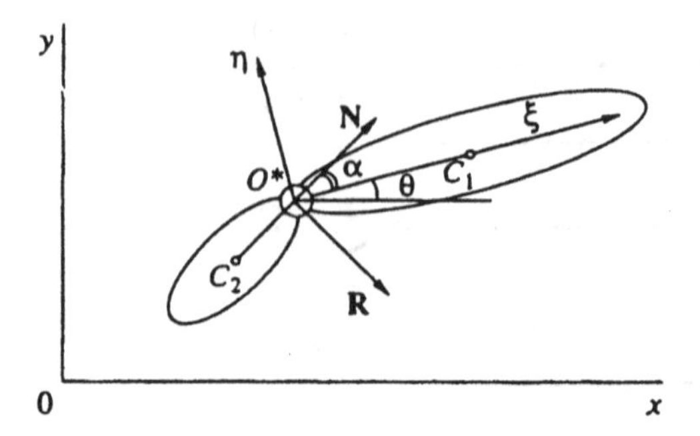

Figure 5. Two-member linkage

Stages 1, 2 can be repeated any number of times. To bring the linkage to its initial configuration, one should perform the motion $S, \alpha_1 : \gamma \to 0, \alpha_2 : \gamma \to 0$.

Using the conservation laws, we can calculate the displacements of the linkage during the cycle of the rotational motion. We obtain [16]:

$$\Delta x = \Delta y = 0,$$

$$\Delta\theta = \gamma + 2\frac{[m_0 l^2 - (m_0 + m_1)a^2]}{R} \arctan\left(\frac{R \tan(\gamma/2)}{m_0(a+l)^2 + m_1 a^2}\right), \tag{7}$$

$$R = [m_0^2(l^2 - a^2)^2 + m_1 a^2(2m_0 l^2 + 2m_0 a^2 + m_1 a^2)]^{1/2}.$$

Thus, the displacements and the average speed of the longitudinal, lateral, and rotational motions are evaluated by (5)–(7). Combining these three types of motions, the linkage can move from any initial position and configuration in the horizontal plane to any prescribed terminal position and configuration in this plane.

5. Two-Member Linkage

Consider now a system of two rigid bodies connected by a joint O^* which can move along the horizontal plane. Denote the masses of the bodies by m_1 and m_2, their centers of mass by C_1 and C_2, their moments of inertia about the vertical axis passing through the point O^* by J_1 and J_2, the distances O^*C_1 and O^*C_2 by a_1 and a_2, and the friction coefficients for these bodies by k_1 and k_2, respectively. The body with index 1 will be referred to as the body, whereas the body with index 2 will be called the tail (Fig. 5).

The joint O^* is treated as a point mass m_0 with the friction coefficient k_0.

We again consider the motion of the linkage along the horizontal plane Oxy. The actuator is installed at the point O^* and creates the torque M. The coordinates of the joint O^* are denoted by x, y, the angle between O^*C_1 and the x-axis by θ, and the angle between C_2O^* and O^*C_1 by α (Fig.5).

Again, we introduce the notion of slow and fast motions. These motions begin and end at the states of rest of the linkage, and the angle $\alpha(t)$ changes between α^0 and α^1 in each of these motions, where α^0, $\alpha^1 \in (-\pi, \pi)$. The duration of slow and fast motions are denoted by τ and T, respectively, $\tau \ll T$. In the slow motions, the tail rotates slowly enough, so that the body stays at rest. In the fast motions, the friction can be neglected, so that the conservation laws for the momentum and angular momentum of the linkage hold.

Denote by ω_0 and ε_0, respectively, the maximal values of the angular velocity and acceleration of the tail during the slow motion:

$$\omega_0 = \max |\dot{\alpha}(t)|, \quad \varepsilon_0 = \max |\ddot{\alpha}(t)|. \tag{8}$$

Here, the maxima are taken along the whole slow motion.

As in the case of the three-member linkage, we are to obtain sufficient conditions which ensure that the body does not move during the slow motion. Here, however, we do not assume that the mass of the system is concentrated at certain points. Therefore, the distribution of the normal reactions and, hence, of the friction forces is not known a priori. In other words, we are to deal with the statically indeterminate case. For this case, the conditions of equilibrium for the rigid body in the presence of dry friction forces were analyzed in [17] where the notion of the guaranteed equilibrium conditions was introduced. The guaranteed equilibrium conditions ensure that the equilibrium holds under any admissible distribution of the normal reactions.

For our two-member linkage, the following two inequalities make up a sufficient condition which ensures that the body stays at rest during the slow motions of the tail [14]:

$$\begin{aligned} &J_2\varepsilon_0 + m_2gk_2a_2 \le m_1gk_1a_1, \\ &J_2\varepsilon_0 + m_2gk_2a_2 + m_2a_1a_2[\omega_0^4 + (\varepsilon_0 + gk_2a_2^{-1})^2]^{1/2} \le m_0gk_0a_1. \end{aligned} \tag{9}$$

These inequalities hold, if ω_0 and ε_0 in (8) are sufficiently small and the following two simpler inequalities are true:

$$m_2k_2a_2 < m_1k_1a_1, \quad m_2k_2(a_1 + a_2) < m_0k_0a_1. \tag{10}$$

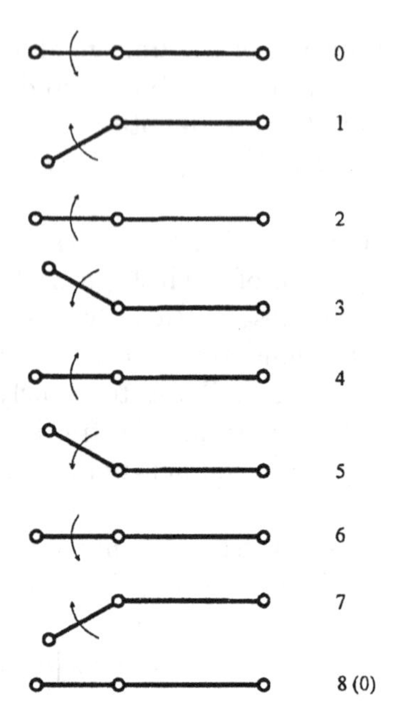

Figure 6. Motion of the two-member linkage

The magnitude of the control torque M during the fast motion must satisfy the following condition similar to (4):

$$|M| \gg m^* g k^* a^*, \qquad m^* = \max(m_0, m_1, m_2),$$
$$k^* = \max(k_0, k_1, k_2), \quad a^* = \max(a_1, a_2). \tag{11}$$

The following sequence of slow (S) and fast (F) motions form the longitudinal motion of the two-member linkage (Fig. 6):

$$\begin{array}{ll} 1) S, \alpha : 0 \to \beta; & 2) F, \alpha : \beta \to 0; \\ 3) S, \alpha : 0 \to -\beta; & 4) F, \alpha : -\beta \to 0; \\ 5) S, \alpha : 0 \to -\beta; & 6) F, \alpha : -\beta \to 0; \\ 7) S, \alpha : 0 \to \beta; & 8) F, \alpha : \beta \to 0. \end{array}$$

Here, β is a fixed angle, $\beta \in (-\pi, \pi)$. As a result of this sequence of motions, the linkage moves along itself by the distance [14]:

$$\Delta x = 8 m_2 m^{-1} a_2 \sin(\beta/2) \cos(\gamma/2) \sin[(\beta - \gamma)/2]. \tag{12}$$

The total lateral displacement and rotation are equal to zero. In (12), the following denotations are used:

$$m = m_0 + m_1 + m_2, \quad \gamma = \beta/2 + A_0 A_1^{-1} A_2^{-1} \arctan\left[A_1 A_2^{-1} \tan(\beta/2)\right],$$

$$A_0 = m(J_2 - J_1) + m_1^2 a_1^2 - m_2^2 a_2^2, \tag{13}$$

$$A_{1,2} = [m(J_1 + J_2) - (m_1 a_1 \pm m_2 a_2)^2]^{1/2}.$$

The average speed of the longitudinal motion is $v_3 = \Delta x/(4T)$. It is shown [14] that the two-member linkage satisfying conditions (10) can, starting from any given position and configuration in the horizontal plane and combining slow and fast motions, reach any prescribed position and configuration.

6. Optimization

The average speed of motion of the linkages depends on their geometrical and mechanical parameters such as the lengths and masses of links, coefficients of friction, etc., as well as on the parameters of the motion itself. This dependence was analyzed in [15,16] where also the parametric optimization of the average speed with respect to the parameters of the linkages and their motions was performed.

To specify the problem, let us consider the longitudinal motions of the three-member linkage. We assume that the angular velocity $\omega(t)$ of the end links in slow motions first increase linearly from 0 to its maximal value ω_0 and then decrease linearly from ω_0 to 0. Then we have for the slow stages of the longitudinal motion:

$$\omega(t) = \varepsilon_0 t, \quad t \in [0, T/2],$$

$$\omega(t) = \varepsilon_0 (T - t), \quad t \in [T/2, T], \tag{14}$$

$$\omega_0 = |\dot{\alpha}_i(t)|, \quad i = 1, 2, \quad \omega_0 = 2\gamma T^{-1}, \quad \varepsilon_0 = 4\gamma T^{-2}.$$

The sufficient condition for the longitudinal motion (2) for our case takes the form:

$$m_0 l \left\{ \left[(2\gamma T^{-1})^4 + P^2\right]^{1/2} + P l a^{-1} \right\} \le m_1 g k_1, \quad P = 4\gamma T^{-2} + g k_0 l^{-1}. \tag{15}$$

Let us fix the mass m_1 of the joints, the length $2a$ of the central link, and the angle of rotation γ. The mass m_0, the length l of the end links, the duration T of the slow motion, and the friction coefficients k_0 and k_1 are to be chosen in order to maximize the speed $v_1 = \Delta x (2T)^{-1}$, see equation

(5). The imposed constraints comprise the inequality (15) and the bounds on the friction coefficients:

$$k^- \leq k_0 \leq k^+, \quad k^- \leq k_1 \leq k^+. \tag{16}$$

Here, k^- and k^+ are given positive constants.

It is shown [15, 16] that the desired maximum of v_1 is reached in case of the equality sign in (15), and the optimal values of k_0 and k_1 are equal to k^- and k^+, respectively. Let us introduce the characteristic speed and time given by

$$v_0 = (gak^+)^{1/2}, \quad T_0 = a^{1/2}(gk^+)^{-1/2}, \tag{17}$$

and the following dimensionless quantities:

$$\lambda = l/a, \quad \mu = m_1/m_0, \quad T_1 = T/T_0,$$
$$\chi = k^-/k^+, \quad V_1 = v_1/v_0. \tag{18}$$

Then our optimization problem is reduced to maximizing the dimensionless speed V_1 given by

$$V_1 = 2\lambda(\mu+1)^{-1}T_1^{-1}\sin^2(\gamma/2) \tag{19}$$

over positive λ, μ, and T_1 under the constraint

$$[16\gamma^4 T_1^{-4}\lambda^2 + (4\gamma T_1^{-2}\lambda + \chi)^2]^{1/2} + (4\gamma T_1^{-2}\lambda + \chi)\lambda = \mu. \tag{20}$$

Equation (19) stems from (5), whereas constraint (20) follows from (15) where the inequality sign is replaced by the equality. Also, equations (14), (17), and (18) are taken into account.

The optimization problem stated above was solved numerically in [15, 16]. Some results are presented in Figs. 7 - 10 where the maximal speed V_1 as well as optimal values of λ, μ, and T_1 are shown as functions of the angle γ given in radians. These data correspond to the following values of $\chi = k^-/k^+ = 0;\ 0.2;\ 0.5;\ 1$ which are shown in the figures.

The obtained numerical results reveal the following properties of the optimal solutions. The maximal speed and the duration T_1 depend significantly on the angle of rotation γ and increase with γ. The saturation occurs as $\gamma \to 2$. The optimal value of λ also increases with γ, whereas the optimal value of μ does not depend significantly on γ .

All optimal parameters depend considerably on the ratio $\chi = k^-/k^+$ of the friction coefficients: the speed V_1 as well as the optimal values of λ and T_1 decrease with χ, whereas μ increases with χ.

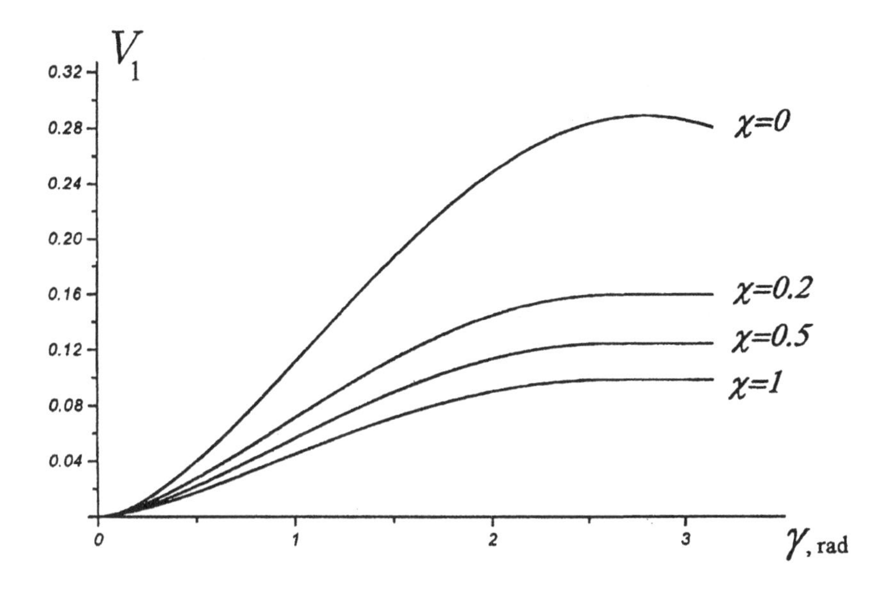

Figure 7. Maximal speed

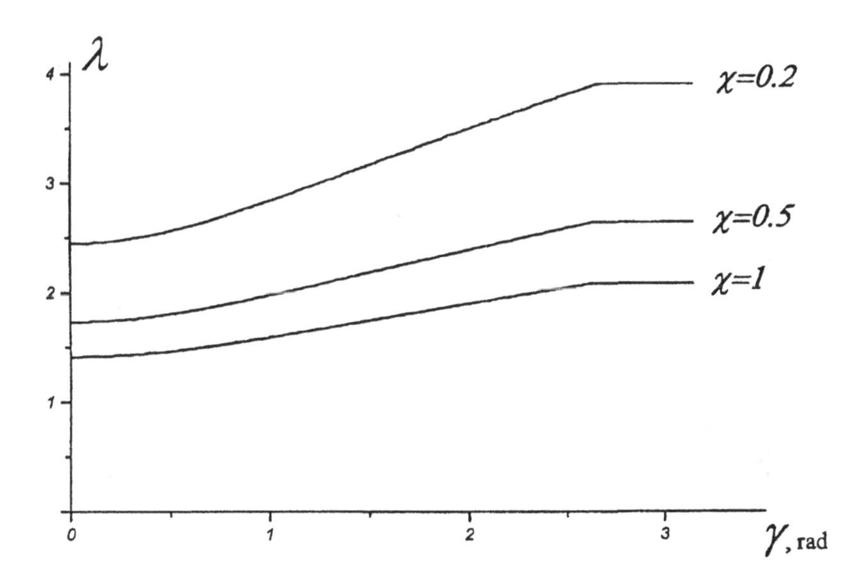

Figure 8. Optimal λ

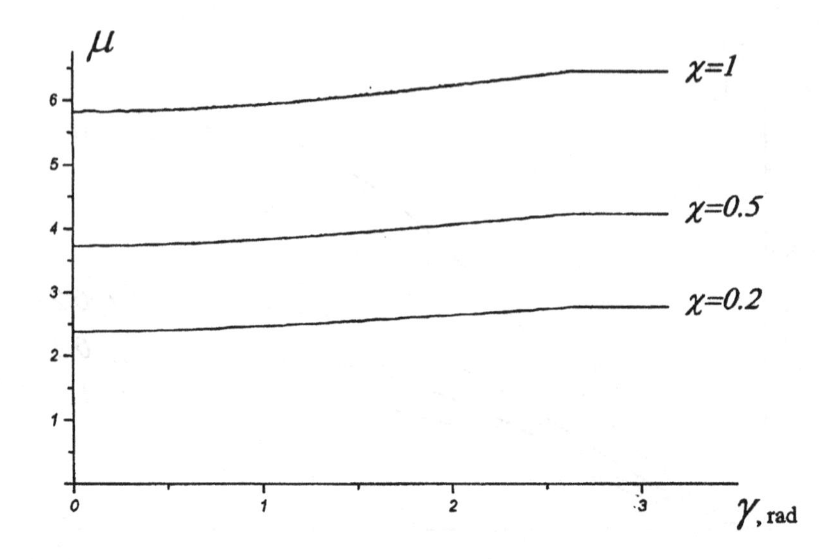

Figure 9. Optimal μ

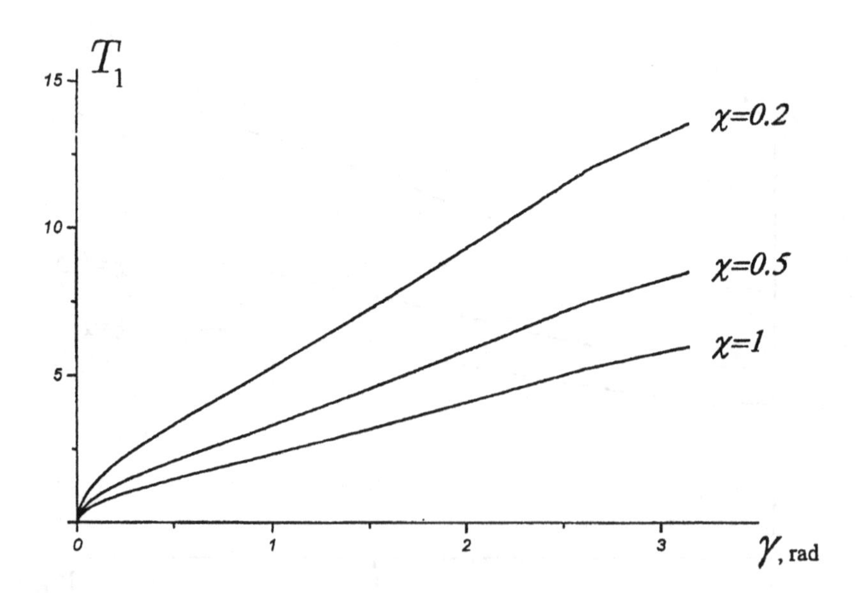

Figure 10. Optimal T_1

The required magnitude of the control torques, according to (4), can be now estimated as follows:

$$|M| \sim 10 m_1 g k^+ l. \tag{21}$$

Note that we used the sufficient condition (2) as the constraint in our optimization problem. Strictly speaking, we should use instead the necessary and sufficient condition, which is not available in the explicit form for our nonlinear system. However, the results of computer simulations based on the numerical integration of the complete set of nonlinear differential equations show that our sufficient condition is rather close to the necessary one for a wide range of parameters. Therefore, the obtained results are close to the optimal ones.

As an example, we consider the three-member linkage having the parameters $a = 0.2$m, $m_1 = 1$kg. Optimal dimensional parameters of the longitudinal motion for some values of the friction coefficients and angle γ are presented in Table I.

TABLE 1. Optimal parameters for the longitudinal motion

k^-/k^+	0.2/0.2	0.1/0.5	0.2/1	0.5/1	0/1
γ, deg	60	60	90	60	90
l, m	0.32	0.57	0.64	0.4	∞
m_0, kg	0.17	0.40	0.39	0.26	1
$m = 2(m_0 + m_1)$, kg	2.34	2.80	2.78	2.52	4
T_1, s	0.77	1.10	1.08	0.49	∞
V_1, m/s	0.030	0.075	0.167	0.084	0.28

Similar results are obtained also for the optimal lateral motion of the three-member linkage as well as for the two-member linkage [15, 16].

Note some differences between the three-member and two-member optimal linkages. For the three-member linkage, the optimal length l of the end links is greater than the half-length a of the central link: $l > a$, see Fig.8. By contrast, the optimal length a_2 of the tail of the two-member linkage is smaller than the half-length of the body: $a_2 < a_1/2$. The gain in speed due to the optimization of lengths and masses is rather essential, up to 50% and more.

It follows from the obtained results that the maximal longitudinal speed of the three-member linkage can be estimated by the formula:

$$v_1 \approx 0.1 (g a k^+)^{1/2}. \tag{22}$$

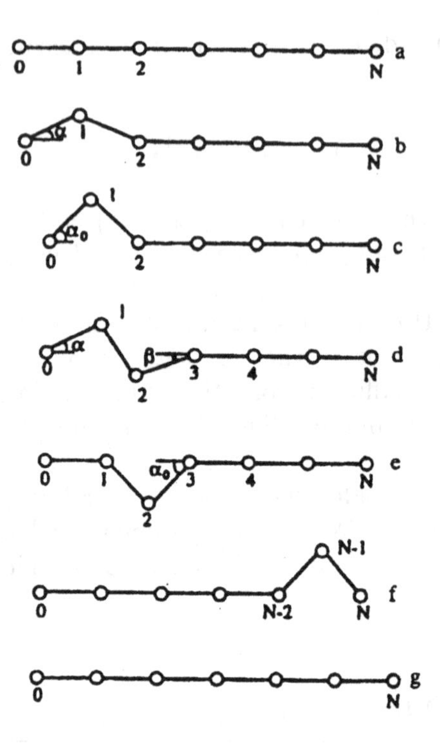

Figure 11. Wavelike motion with three moving links

The maximal speed of the lateral motion for this linkage is several times higher.

As for the two-member linkage, its maximal longitudinal speed is approximately two times smaller than v_1 from (22). This difference can be easily explained by the fact that the three-member linkage is equipped with two actuators, whereas the two-member linkage has only one.

7. Multilink Systems

Consider a multilink system consisting of N identical links assumed to be rigid straight rods of length a. The mass of the system is concentrated at the joints and end points P_i, $i = 0, 1, ..., N$ which are equal point masses m. The actuators installed at the joints $P_1, .., P_{N-1}$ can create torques about vertical axes.

Two types of wavelike quasi-static motions of the system (with three and four moving links) along the horizontal plane Oxy are proposed [12].

At the beginning and at the end of the cycle of motions, the linkage is straight and placed along the x-axis. Its successive configurations for two

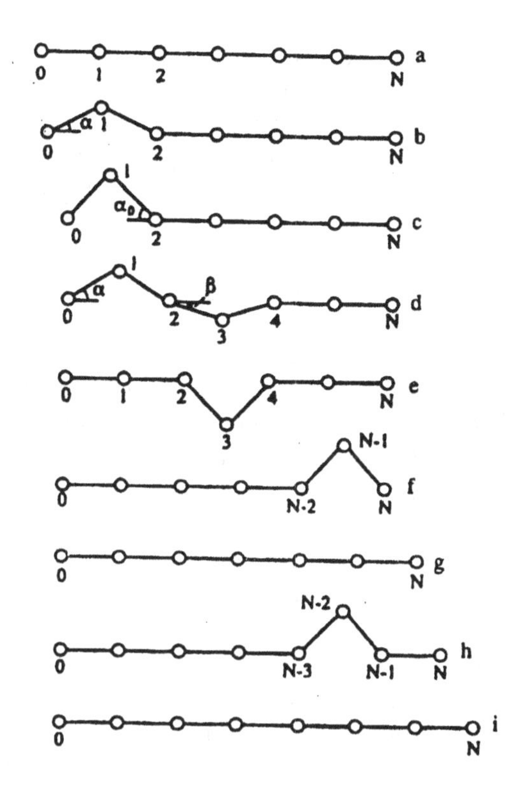

Figure 12. Wavelike motion with four moving links

types of motion are shown in Figs. 11 and 12.

In the first stage of the motion with *three moving links* (Fig. 11), the end mass P_0 advances along the x-axis while the points P_i, $i \geq 2$, remain fixed. The angle α between the x-axis and link P_0P_1 grows monotonically from zero to a certain given value α_0, see states a, b, c in Fig. 11. At the next stage, links P_0P_1, P_1P_2, and P_2P_3 are moving. The angle α decreases monotonically from α_0 to zero while the angle β between link P_2P_3 and the x-axis grows from O to α_0, see state d in Fig. 11. At the end of this stage, the system will be in state e in which links P_1P_2 and P_2P_3 form an isosceles triangle congruent to the triangle $P_0P_1P_2$ in state c but with its apex pointing in the opposite direction. Here, all points except P_2 lie on the x-axis. Next, the motion involves links P_1P_2, P_2P_3, and P_3P_4, and is identical with the preceding stage, apart from a displacement along the x-axis and a mirror reflection in the axis. Continuing this process, we see that after each stage all joints except one lie on the x-axis, and that one joint is the apex of the isosceles triangle with angle α_0 at the base. The triangle gradually moves towards the right. Finally, the point P_{N-1} becomes the apex of such a triangle, see state f in Fig.11. At the last stage of the motion, the point P_N advances to the right along the x-axis. The angle at the base of the triangle $P_{N-2}P_{N-1}P_N$ decreases from α_0 to zero, and the linkage becomes straight again, see state g in Fig. 11.

As a result of the entire cycle of motions, the linkage advances along the x-axis for a distance

$$L = 2a(1 - \cos\alpha_0). \tag{23}$$

In the motion with *four moving links*, the first stage proceeds exactly as in the previous case (states a–c are the same in Figs. 11 and 12). At the next stage, links P_0P_1, P_1P_2, P_2P_3, and P_3P_4 are involved. The point P_2 moves to the right along the x-axis. As this happens, the angle α at the base of the isosceles triangle $P_0P_1P_2$ decreases from α_0 to zero, while the angle β at the base of the triangle $P_2P_3P_4$ grows from zero to α_0, see state d in Fig. 12. At the end of this stage, all joints of the linkage except P_3 lie on the x-axis, see state e in Fig. 12. Continuing this process, we come to the right end of the linkage. The final stages are somewhat different for the cases of even and odd N, see states f–i in Fig. 12. The resultant displacement of the linkage is again given by formula (23).

Comparing the two types of wavelike motions, we see that the motion with three moving links is somewhat simpler. However, it requires larger angles between links, or more intensive twisting, for the same values of α and L [12].

The wavelike motions whose kinematics is described above were analyzed in a quasi-static formulation [12]. We assume that all velocities and accelerations are extremely small, and hence the external forces applied to the system must balance out. In other words, the friction forces must satisfy three equilibrium conditions (two for the forces and one for the moments). The friction forces applied to the moving points are readily evaluated, whereas for the points at rest they are unknown but bounded by the inequalities. Our equilibrium problem is statically indeterminate and can have nonunique solution. The simplest distributions of the friction forces are found [12] for which the equilibrium is attained with the participation of the least possible number of points adjacent to the moving ones. It is shown that the wavelike motion with three moving links is feasible, if the linkage has at least five links ($N \geq 5$), whereas the motion with four moving links is possible, if $N \geq 6$. Also, the magnitude M of the required control torques at the joints of the linkage is estimated [12]. It is shown that $M \leq 2mgka$ where k is the friction coefficient. Comparing this result to (4), we see that the requirements imposed on the control torques for the wavelike quasi-static motions are more moderate than those for the fast phases of the motions of the three-member and two-member linkages.

8. Conclusions

A plane linkage consisting of two, three, or more bodies can move along a horizontal surface in different directions using internal control torques created by the actuators placed at the joints. The mechanisms have a simple structure and can use, in fact, only one actuator. Various modes of motions are described, and sufficient conditions are derived which ensure the possibility of the locomotions. Displacements, speed, and the required control torques are estimated. Optimization of the average speed of linkages with respect to their geometrical and mechanical parameters is carried out. Computer simulation of the motion of linkages confirms theoretical considerations and estimates. The proposed principle of snake-like locomotions can be useful for mobile robots, especially for small ones .

The experiments performed at the Technical University of Munich by F. Pfeiffer, M. Gienger, and J. Mayr, and at the Institute for Problems in Mechanics of the Russian Academy of Sciences by V.G. Gradetsky, V.B. Veshnikov, and L.N. Kravchuk show that the proposed snake-like motions can be implemented.

The work was supported by the Russian Foundation for Basic Research, Project 02-01-00201.

References

1. Lavrentyev, M.A. and Lavrentyev, M.M. (1962) On one principle of creating the thrust force in motion, *Journal of Applied Mechanics and Technical Physics* (4), 6–9.
2. Gray, J. (1968) *Animal Locomotion*, Weidenfeld & Nicolson, London.
3. Dobrolyubov, A.I. (1987) *Travelling Waves of Deformation*, Nauka i Tekhnika, Minsk.
4. Bayraktaroglu, Z.Y., Butel, F., Pasqui, V., and Blazevic, P. (1999) A geometrical approach to the trajectory planning of a snake-like robot, in G.C. Virk, M. Randall, and D. Howard (eds.), *Proceedings of the Second International Conference on Climbing and Walking Robots CLAWAR 1999*, Portsmouth, pp. 851–856.
5. Bayraktaroglu, Z,Y. and Blazevic, P. (2000) Snake-like locomotion with a minimal mechanism, in M. Armada and P. Gonzalez de Santos (eds.), *Proceedings of the Third International Conference on Climbing and Walking Robots CLAWAR 2000*, Madrid, pp. 201–207.
6. Hirose, S. and Morishima, A. (1990) Design and control of a mobile robot with an articulated body, *International Journal of Robotics Research* **9**(2), 99–114.
7. Hirose, S. (1993) *Biologically Inspired Robots: Snake-like Locomotors and Manipulators*, Oxford University Press, Oxford.
8. Chirikjan, G.S. and Burdick, J.W. (1991) Kinematics of hyper-redundant robot locomotion with applications to grasping, in *Proceedings of the IEEE International Conference on Robotics and Automation* **1**, Sacramento, pp. 720–725.
9. Burdick, J.W., Radford, J., and Chirikjan, G.S. (1993) A 'sidewinding' locomotion gait for hyper-redundant robots, in *Proceedings of the IEEE International Conference on Robotics and Automation* **3**, Atlanta, pp. 101–106.
10. Ostrowski, J. and Burdick, J. (1996) Gait kinematics for a serpentine robot, in *Proceedings of the IEEE International Conference on Robotics and Automation* **2**,

Minneapolis, pp. 1294–1300.
11. Chernousko, F.L. (2000) The motion of a multilink system along a horizontal plane, *Journal of Applied Mathematics and Mechanics* **64** (1), 5–15.
12. Chernousko, F.L. (2000) The wavelike motion of a multilink system on a horizontal plane, *Journal of Applied Mathematics and Mechanics* **64** (4), 497–508.
13. Chernousko, F.L. (2001) On the motion of a three-member linkage along a plane, *Journal of Applied Mathematics and Mechanics* **65** (1), 13–18.
14. Chernousko, F.L. (2001) Controllable motions of a two-link mechanism along a horizontal plane, *Journal of Applied Mathematics and Mechanics* **65** (4), 565-577.
15. Smyshlyaev, A.S. and Chernousko, F.L. (2001) Optimization of the motion of multilink robots on a horizontal plane, *Journal of Computer and Systems Sciences International* **40**(2), 340-348.
16. Chernousko, F.L. (in press) Snake-like locomotions of multilink mechanisms, *Journal of Vibration and Control.*
17. Chernousko, F.L. (1988) Equilibrium conditions for a solid on a rough plane, *Mechanics of Solids* **23** (6), 1-12.

OPTIMIZATION OF RIGID AND FLEXIBLE MULTIBODY SYSTEMS WITH APPLICATION TO VEHICLE DYNAMICS AND CRASHWORTHINESS

M. S. Pereira and J. P. Dias
IDMEC - Instituto de Mecânica - Polo IST
Instituto Superior Técnico.
Lisbon, Portugal.

Abstract

The optimal design of vehicle structures is a challenging task and is usually carried out using finite elements programs. However, design tools for multibody systems can be an alternative to the finite elements codes, especially in the first stages of the project and when crashworthiness and dynamic requirements are to be considered.

In this paper a framework for the dynamics, sensitivity analysis and optimization of rigid and flexible multibody systems is presented.

The formulation for the dynamic analysis is based in a moving frame approach to describe the kinematics of the deformable bodies where the large rigid body motion is described using Cartesian coordinates and the flexibility is introduced using the finite element method.

Analytic sensitivities obtained by the direct differentiation method and finite differences sensitivities are used in the optimization process. The optimization process includes constrained and multiload optimization methodologies.

Several applications are presented, including the design of train structures and energy absorption devices in crashworthiness and design of automotive chassis under dynamic loads.

1. Introduction

The design of vehicle structures is usually carried out using finite elements programs. In order to optimize parameters or characteristics of the vehicle structure these programs lead to several drawbacks: the time necessary to develop the models is large and in general these programs do not include general optimization procedures.

In the first stages of the design project, when the detailed characteristics of the structure are not yet known or defined, simplified models that allow simulation and optimization can be used with great advantage. Multibody dynamics formulations together with design tools fulfill these requirements. The design of vehicle structures for crashworthiness and the elastic design of vehicle chassis under dynamic loads are

W. Schiehlen and M. Valášek (eds.), Virtual Nonlinear Multibody Systems, 363–382.

examples of these application scenarios.

Several types of simplified models have been used for crashworthiness problems. Mass-spring models have been used for the crash simulation of automotive vehicles [1, 2] and for the crash simulation of train sets [3, 4]. Models based on multibody dynamics formulations, with rigid and flexible bodies 2D and 3D, have also been used successfully in the crash simulation of train structures and validated with experimental results [5, 6].

During the different stages of structural design different design tools can be explored in order to achieve different design goals. Train crashworthiness is one of the industrial applications where this design framework can be applied. One-dimensional rigid models can be used for the analysis of the energy distribution, overall deformations, forces and accelerations along the train during a collision. These models together with optimization methodologies can optimize the characteristics of the energy absorption devices. Two-dimensional models (rigid or flexible) are suitable for the analysis of the structural behavior of the end-underframe of the train cars or the overriding phenomena and to the optimal design of the structural elements. Three-dimensional models can be used for detailed analysis and design of the structures and to study the entire car shells as also to design the structural elements. In all these models optimal design plays an important role, so the sensitivity analysis and the optimization of multibody systems are important topics in the development of these tools.

Optimal design is currently a topic of intensive research in the field of multibody systems [7]. Multibody dynamics formulations have seen considerable developments in the last two decades and several commercial codes for the analysis of such systems are now available. However there is still a lack in the development of optimal design tools especially for flexible systems.

In this work, formulations and methodologies for the analysis and optimal design of rigid and flexible multibody systems are presented within the framework of multibody dynamics, sensitivity analysis and optimization. This framework is schematically presented in figure 1. It includes a range of simulation tools including 1D models suitable for the simulation of train collisions, 2D rigid and flexible simulation tools suitable for planar mechanisms and structures for crashworthiness, and 3D tools for vehicle dynamics.

Multibody dynamics formulations are developed using a moving frame approach to describe the kinematics of the deformable bodies. The large rigid motion is described using Cartesian coordinates and the flexibility is introduced using the finite element method. To reduce the number of elastic degrees of freedom, the component mode synthesis is used. The mean axis condition method is applied to the reference conditions to reduce the dynamic coupling between rigid and flexible coordinates.

For applications in crashworthiness, special topics are necessary to be included into the simulation tools, such as plastic hinges to model the plastic deformations, contact-impact-models based on a Hertzian formulation to model the contact between bodies and bodies and barriers.

In order to develop the design tools, optimization algorithms are necessary. These algorithms can be divided in two groups: those that do not use sensitivity information as for example genetic algorithms and those that require sensitivity information, which are more suitable for problems where the simulation time is large as in the case of crashworthiness.

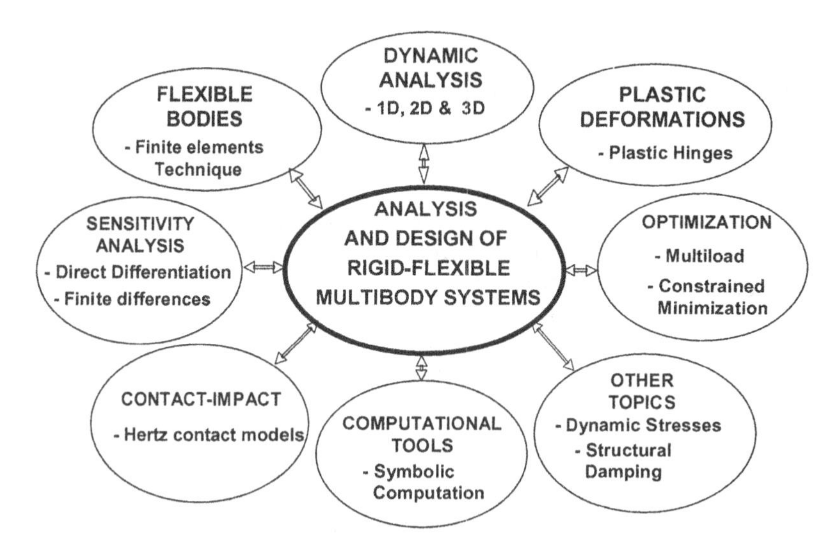

Figure 1. Framework for the optimal design of rigid-flexible multibody systems.

The method of finite differences for sensitivity analysis [8] has the advantage, of being easily integrated with an analysis program with minor implementation efforts. However this technique presents accuracy problems, in particular for nonlinear systems and is computationally expensive as an evaluation of the perturbed system is necessary for each design variable. The use of analytical algorithms can overcome these disadvantages but the effort in formulation and development is much larger. Basically two methods are available for obtaining analytic sensitivities for multibody systems, direct differentiation [9, 10, 11] and adjoint variable [12, 13]. Each one has its advantages and its drawbacks. The adjoint method is more appropriate for problems having a large number of design variables and a small number of design constraints. On the other hand the direct differentiation method is more suitable for problems having a large number of design constraints and a small number of design variables.

For the optimization procedure the multibody dynamic analysis code is linked with general optimization algorithms included in the package DOT/DOC [14]. The optimization procedure requires the sensitivities of the objective function and design constraints. For this purpose, numerical sensitivities using the finite difference method and analytic sensitivities from the direct differentiation method are used.

The proposed methodology for the design of rigid and flexible multibody systems is applied in different design problems, including the optimization of train structures in crashworthiness, design of automotive chassis using dynamic loads and other mechanisms related with vehicles.

2. Dynamic Analysis

The configuration of a flexible body in a multibody system can be described by a set of reference coordinates $\mathbf{q}_r^i$ and a set of finite element elastic coordinates $\mathbf{q}_f^i$. As indicated in figure 2, the position of a flexible body in an inertial frame XYZ is defined by the spatial location $\mathbf{R}^i$ of the local frame $X^iY^iZ^i$ and by its orientation parameters θ^i summarized in the vector

$$\mathbf{q}_r^{i^{\mathrm{T}}} = [\mathbf{R}^{i^{\mathrm{T}}}, \mathbf{q}^{i^{\mathrm{T}}}] \tag{1}$$

For a planar system, each flexible body has (3+*nf*) coordinates, which correspond to the 3 reference coordinates, and *nf* elastic coordinates. For a spatial system, 3 translational coordinates and 4 Euler parameters are used. The vector of generalized coordinates for body *i* is written as

$$\mathbf{q}^i = [\mathbf{q}_r^{i^{\mathrm{T}}}, \mathbf{q}_f^{i^{\mathrm{T}}}]^{\mathrm{T}} \tag{2}$$

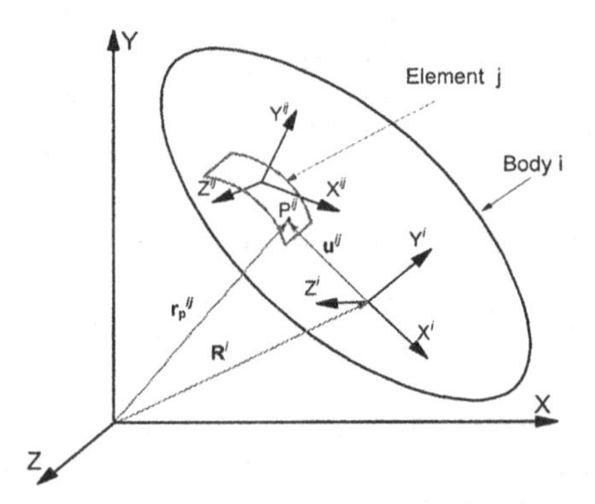

Figure 2. Reference generalized coordinates.

2.1 EQUATIONS OF MOTION IN TERMS OF ELASTIC COORDINATES

The equations of motion for a constrained multibody system, and for each flexible body *i* can be obtained from a variational approach [15]

$$\mathbf{M}^i(\mathbf{q}^i)\ddot{\mathbf{q}}^i + \mathbf{C}^i\dot{\mathbf{q}}^i + \mathbf{K}^i\mathbf{q}^i = \mathbf{g}^i(\dot{\mathbf{q}}^i, \mathbf{q}^i) + \mathbf{f}^i(\dot{\mathbf{q}}^i, \mathbf{q}^i, t) - \mathbf{\Phi}_{\mathbf{q}}^{k^{\mathrm{T}}}\boldsymbol{\lambda} \tag{3}$$

Where $\mathbf{M}^i$ is the mass matrix, $\mathbf{K}^i$ is the stiffness matrix, $\mathbf{C}^i$ is the structural damping matrix, $\mathbf{f}^i$ and $\mathbf{g}^i$ are the external and quadratic force vectors and $\mathbf{\Phi}_{\mathbf{q}}^{k^{\mathrm{T}}}\boldsymbol{\lambda}$ is the vector of reaction forces.

The kinematic constraint equations of the complete system can be expressed in the form

$$\mathbf{\Phi}(\mathbf{q},t) = \mathbf{0} \tag{4}$$

Differentiating (4) twice with respect to time, the constraint velocity and acceleration equations are obtained as

$$\dot{\mathbf{\Phi}} \equiv \mathbf{\Phi}_{\mathbf{q}}\dot{\mathbf{q}} + \mathbf{\Phi}_{t} = \mathbf{0} \tag{5}$$

$$\ddot{\mathbf{\Phi}} \equiv \mathbf{\Phi}_{\mathbf{q}}\ddot{\mathbf{q}} + (\mathbf{\Phi}_{\mathbf{q}}\dot{\mathbf{q}})_{\mathbf{q}}\dot{\mathbf{q}} + \mathbf{\Phi}_{tt} + 2\mathbf{\Phi}_{\mathbf{q}t}\dot{\mathbf{q}} = \mathbf{0} \tag{6}$$

For rigid-flexible systems, equation (6) can be written also in the form

$$\mathbf{\Phi}_{\mathbf{q}_r}\ddot{\mathbf{q}}_r + \mathbf{\Phi}_{\mathbf{q}_f}\ddot{\mathbf{q}}_f = \gamma \tag{7}$$

where

$$\gamma = -(\mathbf{\Phi}_{\mathbf{q}}\dot{\mathbf{q}})_{\mathbf{q}}\dot{\mathbf{q}} - \mathbf{\Phi}_{tt} - 2\mathbf{\Phi}_{\mathbf{q}t}\dot{\mathbf{q}} \tag{8}$$

The equations of motion (3) and the acceleration constraint equation (8) form a set of differential-algebraic equations. Separating rigid and elastic coordinates they can be written as

$$\begin{bmatrix} \mathbf{M}_{rr} & \mathbf{M}_{rf} & \mathbf{\Phi}_{\mathbf{q}_r}^{\mathrm{T}} \\ \mathbf{M}_{fr} & \mathbf{M}_{ff} & \mathbf{\Phi}_{\mathbf{q}_f}^{\mathrm{T}} \\ \mathbf{\Phi}_{\mathbf{q}_r} & \mathbf{\Phi}_{\mathbf{q}_f} & \mathbf{0} \end{bmatrix} \begin{Bmatrix} \ddot{\mathbf{q}}_r \\ \ddot{\mathbf{q}}_f \\ \boldsymbol{\lambda} \end{Bmatrix} = \begin{Bmatrix} \mathbf{f}_r + \mathbf{g}_r \\ \mathbf{f}_f + \mathbf{g}_f - \mathbf{K}_{ff}\mathbf{q}_f - \mathbf{C}_{ff}\dot{\mathbf{q}}_f \\ \gamma \end{Bmatrix} \tag{9}$$

Where $\mathbf{M}_{rr}$ contains the rigid body inertia, $\mathbf{M}_{ff}$ is the standard finite element mass matrix, $\mathbf{M}_{rf}$ is the inertia coupling between rigid and flexible coordinates, $\boldsymbol{\lambda}$ is the Lagrange multiplier vector, $\mathbf{f}_r$ and $\mathbf{f}_f$ are the generalized external forces, $\mathbf{g}_r$ and $\mathbf{g}_f$ are the quadratic forces including gyroscopic, Coriolis forces and other terms associated with kinetic energy. The coupling between elastic deformation and rigid body motion is represented by the term $\mathbf{K}_{ff}\mathbf{q}_f$ in the right-hand side of equation (9). Concerning the damping matrix a proportional damping scheme based on the Rayleigh model is considered as

$$\mathbf{C}_{ff} = \alpha\mathbf{K}_{ff} + \beta\mathbf{M}_{ff} \tag{10}$$

where the constants α and β are calculated from the vibration frequencies of the structure, and β is limited by 0.1 [15].

2.2. EQUATIONS OF MOTION IN TERMS OF A REDUCED SET OF MODAL COORDINATES

The solution of the dynamic and sensitivity equations with elastic coordinates is time consuming. The number of elastic coordinates can be drastically reduced, if the elastic deformation is represented by a set of eigenmodes of the system. The component mode synthesis is used for such purpose.

If a flexible body i its assumed to vibrate freely about its reference configuration, and assuming small damping, equation (3) yields

$$\mathbf{M}^{i}_{ff}\ddot{\mathbf{q}}^{i}_{f} + \mathbf{K}^{i}_{ff}\mathbf{q}^{i}_{f} = \mathbf{0} \tag{11}$$

where $\mathbf{K}^{i}_{ff}$ is positive definite once proper reference conditions are imposed. The solution of equation (11) has the form

$$\mathbf{q}^{i}_{fk} = \mathbf{a}^{i}_{k} e^{j\upsilon_k t} \tag{12}$$

Substituting the state vectors in (11) by equation (12) gives

$$\mathbf{K}^{i}_{ff}\mathbf{a}^{i}_{k} = (\upsilon_k)^2 \mathbf{M}^{i}_{ff}\mathbf{a}^{i}_{k} \tag{13}$$

Equation (13) is the standard eigenvalue problem, which is solved for a set of eigenvalues $(\Omega_i)^2$ and the respective eigenvectors $\mathbf{u}^i$. These eigenvalues are useful also for the calculation of the structural damping. The coordinate transformation from the modal coordinates to the physical elastic coordinates can be written as

$$\mathbf{q}^{i}_{f} = \mathbf{U}^{i}_{nm}\mathbf{p}^{i}_{f} \tag{14}$$

where $\mathbf{U}^{i}_{nm}$ is the modal transformation matrix whose columns are the *nm* low-frequency eigenvectors and $\mathbf{p}^{i}_{f}$ is the vector of modal coordinates. Generically for all the bodies of the system, equation (14) is written as

$$\mathbf{q}_f = \mathbf{U}\,\mathbf{p}_f \tag{15}$$

The equations of motion (9) are now written in terms of the modal coordinates, as [10]

$$\begin{bmatrix} \overline{\mathbf{M}}_{rr} & \overline{\mathbf{M}}_{rf} & \overline{\mathbf{\Phi}}^{T}_{\mathbf{q}_r} \\ \overline{\mathbf{M}}_{fr} & \overline{\mathbf{M}}_{ff} & \overline{\mathbf{\Phi}}^{T}_{\mathbf{p}_f} \\ \overline{\mathbf{\Phi}}_{\mathbf{q}_r} & \overline{\mathbf{\Phi}}_{\mathbf{p}_f} & \mathbf{0} \end{bmatrix} \begin{Bmatrix} \ddot{\mathbf{q}}_r \\ \ddot{\mathbf{p}}_f \\ \boldsymbol{\lambda} \end{Bmatrix} = \begin{Bmatrix} \bar{\mathbf{f}}_r + \bar{\mathbf{g}}_r \\ \bar{\mathbf{f}}_f + \bar{\mathbf{g}}_f - \overline{\mathbf{K}}_{ff}\mathbf{p}_f - \overline{\mathbf{C}}_{ff}\dot{\mathbf{p}}_f \\ \bar{\gamma} \end{Bmatrix} \tag{16}$$

The transformations from nodal to modal coordinates can be found in ref. [16] Equation (16) with the initial conditions

$$\left\{ \mathbf{q}_r(t^0) = \mathbf{q}^0_r \;,\; \mathbf{p}_f(t^0) = \mathbf{p}^0_f \;,\dot{\mathbf{q}}_r(t^0) = \dot{\mathbf{q}}^0_r \;,\; \dot{\mathbf{p}}_f(t^0) = \dot{\mathbf{p}}^0_f \right\} \tag{17}$$

can be integrated in time in order to obtain all the state variables of the system, i. e. positions, velocities, accelerations and Lagrange multipliers.

The dynamic analysis tools developed include several features necessary to solve specific problems related with crashworthiness or mechanism and chassis design such as, local plastic hinges models, dynamic stress analysis and Hertz contact-impact models.

For the simulation of train set collisions, 1D unconstrained rigid body models are to be used. In this case, equation (3) is simplified as

$$\mathbf{M}\,\ddot{\mathbf{q}} = \mathbf{f} \tag{18}$$

where $\mathbf{M}$ is the mass matrix containing the masses of the bodies, $\ddot{\mathbf{q}}$ the acceleration vector and $\mathbf{f}$ the forces vector.

A variable order, variable step size algorithm [17] has been used. This kind of algorithm is particularly adequate for crash simulation where the force vector changes very quickly as the result of the contact between the different components of the structure that occurs for instance in crashworthiness applications.

3. Sensitivity Analysis

The analytic sensitivity equations, obtained using the direct differentiation method are here summarily presented in the modal form. The sensitivity equations can be written in a compact form matrix form identical to the equations of motion, as [10]

$$\begin{bmatrix} \overline{\mathbf{M}}_{rr} & \overline{\mathbf{M}}_{rf} & \overline{\boldsymbol{\Phi}}_{\mathbf{q}_r}^{\mathrm{T}} \\ \overline{\mathbf{M}}_{fr} & \overline{\mathbf{M}}_{ff} & \overline{\boldsymbol{\Phi}}_{\mathbf{p}_f}^{\mathrm{T}} \\ \overline{\boldsymbol{\Phi}}_{\mathbf{q}_r} & \overline{\boldsymbol{\Phi}}_{\mathbf{p}_f} & \mathbf{0} \end{bmatrix} \begin{bmatrix} \ddot{\mathbf{q}}_{r_{\mathrm{b}}} \\ \ddot{\mathbf{p}}_{f_{\mathrm{b}}} \\ \boldsymbol{\lambda}_{\mathrm{b}} \end{bmatrix} = \begin{bmatrix} \overline{\mathbf{R}} \\ \overline{\mathbf{F}} \\ \overline{\gamma}_{\mathrm{b}} \end{bmatrix} \tag{19}$$

where $\overline{\mathbf{R}}$, $\overline{\mathbf{F}}$ and $\overline{\gamma}_{\mathrm{b}}$ are

$$\begin{aligned} \overline{\mathbf{R}} = & -(\overline{\mathbf{M}}_{rr}\hat{\ddot{\mathbf{q}}}_r)_{\mathbf{q}_r}\mathbf{q}_{r_{\mathrm{b}}} - (\overline{\mathbf{M}}_{rr}\hat{\ddot{\mathbf{q}}}_r)_{\mathbf{p}_f}\mathbf{p}_{f_{\mathrm{b}}} - (\overline{\mathbf{M}}_{rr}\hat{\ddot{\mathbf{q}}}_r)_{\mathrm{b}} \\ & - (\overline{\mathbf{M}}_{rf}\hat{\ddot{\mathbf{p}}}_f)_{\mathbf{q}_r}\mathbf{q}_{r_{\mathrm{b}}} - (\overline{\mathbf{M}}_{rf}\hat{\ddot{\mathbf{p}}}_f)_{\mathbf{p}_f}\mathbf{p}_{f_{\mathrm{b}}} - (\overline{\mathbf{M}}_{rf}\hat{\ddot{\mathbf{p}}}_f)_{\mathrm{b}} \\ & + \overline{\mathbf{f}}_{r\mathbf{q}_r}\mathbf{q}_{r_{\mathrm{b}}} + \overline{\mathbf{f}}_{r\dot{\mathbf{q}}_r}\dot{\mathbf{q}}_{r_{\mathrm{b}}} + \overline{\mathbf{f}}_{r\mathbf{p}_f}\mathbf{p}_{f_{\mathrm{b}}} + \overline{\mathbf{f}}_{r\dot{\mathbf{p}}_f}\dot{\mathbf{p}}_{f_{\mathrm{b}}} + \overline{\mathbf{f}}_{r\mathrm{b}} \\ & + \overline{\mathbf{g}}_{r\mathbf{q}_r}\mathbf{q}_{r_{\mathrm{b}}} + \overline{\mathbf{g}}_{r\dot{\mathbf{q}}_r}\dot{\mathbf{q}}_{r_{\mathrm{b}}} + \overline{\mathbf{g}}_{r\mathbf{p}_f}\mathbf{p}_{f_{\mathrm{b}}} + \overline{\mathbf{g}}_{r\dot{\mathbf{p}}_f}\dot{\mathbf{p}}_{f_{\mathrm{b}}} + \overline{\mathbf{g}}_{r\mathrm{b}} \\ & - (\overline{\boldsymbol{\Phi}}_{\mathbf{q}_r}^{\mathrm{T}}\hat{\boldsymbol{\lambda}})_{\mathbf{q}_r}\mathbf{q}_{r_{\mathrm{b}}} - (\overline{\boldsymbol{\Phi}}_{\mathbf{q}_r}^{\mathrm{T}}\hat{\boldsymbol{\lambda}})_{\mathbf{p}_f}\mathbf{p}_{f_{\mathrm{b}}} - (\overline{\boldsymbol{\Phi}}_{\mathbf{q}_r}^{\mathrm{T}}\hat{\boldsymbol{\lambda}})_{\mathrm{b}} \end{aligned} \tag{20}$$

$$\begin{aligned} \overline{\mathbf{F}} = & -(\overline{\mathbf{M}}_{fr}\hat{\ddot{\mathbf{q}}}_r)_{\mathbf{q}_r}\mathbf{q}_{r_{\mathrm{b}}} - (\overline{\mathbf{M}}_{fr}\hat{\ddot{\mathbf{q}}}_r)_{\mathbf{p}_f}\mathbf{p}_{f_{\mathrm{b}}} - (\overline{\mathbf{M}}_{fr}\hat{\ddot{\mathbf{q}}}_r)_{\mathrm{b}} - (\overline{\mathbf{M}}_{ff}\hat{\ddot{\mathbf{p}}}_f)_{\mathrm{b}} \\ & + \overline{\mathbf{f}}_{f\mathbf{q}_r}\mathbf{q}_{r_{\mathrm{b}}} + \overline{\mathbf{f}}_{f\mathbf{p}_f}\mathbf{p}_{f_{\mathrm{b}}} + \overline{\mathbf{f}}_{f\dot{\mathbf{q}}_r}\dot{\mathbf{q}}_{r_{\mathrm{b}}} + \overline{\mathbf{f}}_{f\dot{\mathbf{p}}_f}\dot{\mathbf{p}}_{f_{\mathrm{b}}} + \overline{\mathbf{f}}_{f\mathrm{b}} \\ & + \overline{\mathbf{g}}_{f\mathbf{p}_f}\mathbf{p}_{f_{\mathrm{b}}} + \overline{\mathbf{g}}_{f\dot{\mathbf{q}}_r}\dot{\mathbf{q}}_{r_{\mathrm{b}}} + \overline{\mathbf{g}}_{f\dot{\mathbf{p}}_f}\dot{\mathbf{p}}_{f_{\mathrm{b}}} + \overline{\mathbf{g}}_{f\mathrm{b}} \\ & - (\overline{\boldsymbol{\Phi}}_{\mathbf{p}_f}^{\mathrm{T}}\hat{\boldsymbol{\lambda}})_{\mathbf{q}_r}\mathbf{q}_{r_{\mathrm{b}}} - (\overline{\boldsymbol{\Phi}}_{\mathbf{p}_f}^{\mathrm{T}}\hat{\boldsymbol{\lambda}})_{\mathbf{p}_f}\mathbf{p}_{f_{\mathrm{b}}} - (\overline{\boldsymbol{\Phi}}_{\mathbf{p}_f}^{\mathrm{T}}\hat{\boldsymbol{\lambda}})_{\mathrm{b}} \\ & - (\overline{\mathbf{K}}_{ff}\hat{\mathbf{p}}_f)_{\mathrm{b}} - \overline{\mathbf{K}}_{ff}\mathbf{p}_{f_{\mathrm{b}}} - (\overline{\mathbf{C}}_{ff}\hat{\dot{\mathbf{p}}}_f)_{\mathrm{b}} - \overline{\mathbf{C}}_{ff}\dot{\mathbf{p}}_{f_{\mathrm{b}}} \end{aligned} \tag{21}$$

$$\overline{\gamma}_{\mathrm{b}} = -2\dot{\overline{\boldsymbol{\Phi}}}_{\mathbf{q}_r}\dot{\mathbf{q}}_{r\mathrm{b}} - 2\dot{\overline{\boldsymbol{\Phi}}}_{\mathbf{p}_f}\dot{\mathbf{p}}_{f\mathrm{b}} - \ddot{\overline{\boldsymbol{\Phi}}}_{\mathbf{q}_r}\mathbf{q}_{r\mathrm{b}} - \ddot{\overline{\boldsymbol{\Phi}}}_{\mathbf{p}_f}\mathbf{p}_{f\mathrm{b}} - \ddot{\overline{\boldsymbol{\Phi}}}_{\mathrm{b}} \tag{22}$$

Equation (19) can be integrated in time provided the initial conditions

$$\left\{\mathbf{q}_{r\mathrm{b}}(t^0) = \mathbf{q}_{r_{\mathrm{b}}}^0,\ \mathbf{p}_{f\mathrm{b}}(t^0) = \mathbf{p}_{f_{\mathrm{b}}}^0,\ \dot{\mathbf{q}}_{r_{\mathrm{b}}}(t^0) = \dot{\mathbf{q}}_{r_{\mathrm{b}}}^0,\ \dot{\mathbf{p}}_{f_{\mathrm{b}}}(t^0) = \dot{\mathbf{p}}_{f_{\mathrm{b}}}^0\right\} \tag{23}$$

From the integration of the equations (19-23) the modal sensitivities of the state variables, are obtained. Since the objective function and constraints are defined in terms of nodal coordinates, the modal sensitivities are transformed to nodal sensitivities using equation

$$\mathbf{q}_{f_{b^k}} = \mathbf{U}_{b^k}\mathbf{p}_f + \mathbf{U}\,\mathbf{p}_{f_{b^k}} \tag{24}$$

Two different cases can be considered, an updated-mode approach where equation (24) is evaluated and a fixed-mode approach where the derivatives of vibration modes are neglected. However, in general the updated-mode approach is more accurate. Herein a Nelson's scheme [19] is used to evaluate the derivatives of the vibration modes.

Different types of design variables related to flexible elements such as area of the sections, Young Modulus, moment of inertia, and design variables involving element actuators and kinematic constraints may be considered.

The calculation of the terms in the right-hand side of sensitivity equations (20-22) represents a major computational effort. The symbolic computation program MAPLE is used to obtain these terms for a given set of design variables. For the differentiation of the finite element matrices, an exact method is used once the analytical expressions for the mass, stiffness and other matrices involving the flexible terms, are given in closed form for specific finite elements.

As in the case of rigid body sensitivities, the left-hand side of the dynamic and sensitivity equations is the same. Both equations can be integrated simultaneously in time. In order to minimize the constraint violations for dynamic and sensitivity equations, Baumgarte's method [10] is applied.

4. Examples

In this section, five examples corresponding to the application of the different design tools are presented.

4.1. SIMULATION OF TRAIN COLLISIONS WITH 1D MODELS

In figure 3 a typical front of a train car is illustrated showing the different energy absorption devices. Figure 4 shows the corresponding force-displacement curve for the axial crushing of this car front. Figure 5 illustrates the model for the analysis of a head-on collision between two equal trains composed by 3 cars. One of the trains travels at 55 km/h and collides with the other train stopped and braked on the line. This speed has been established within the European project SAFETRAIN [20] and covers 90% of all the real train collisions in this collision scenario (front-front). In figure 6 an example of the results obtained from the simulation is presented. The simulation results include crush displacements, velocities of individual cars, sustained accelerations, energy distribution along the train and dissipated energy on each energy absorption device.

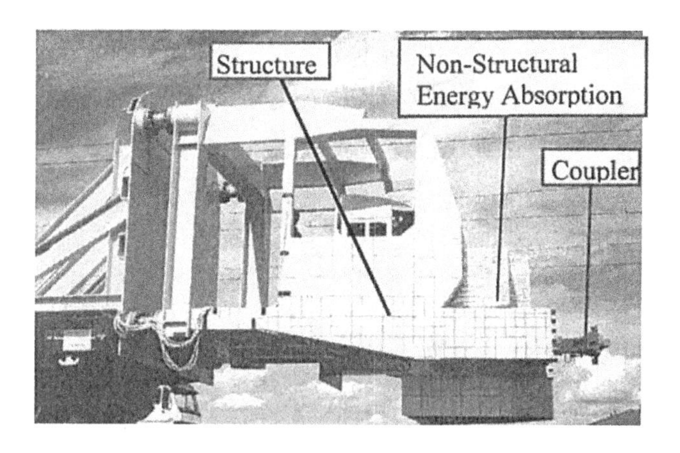

Figure 3. Location of energy absorption devices in a train car.

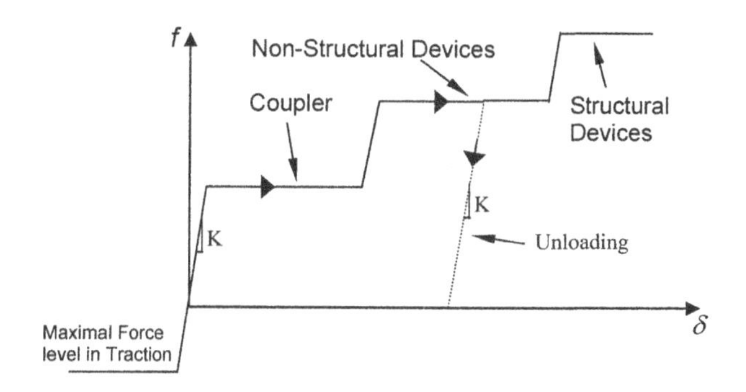

Figure 4. Typical force-displacement curve representing the front of a train car.

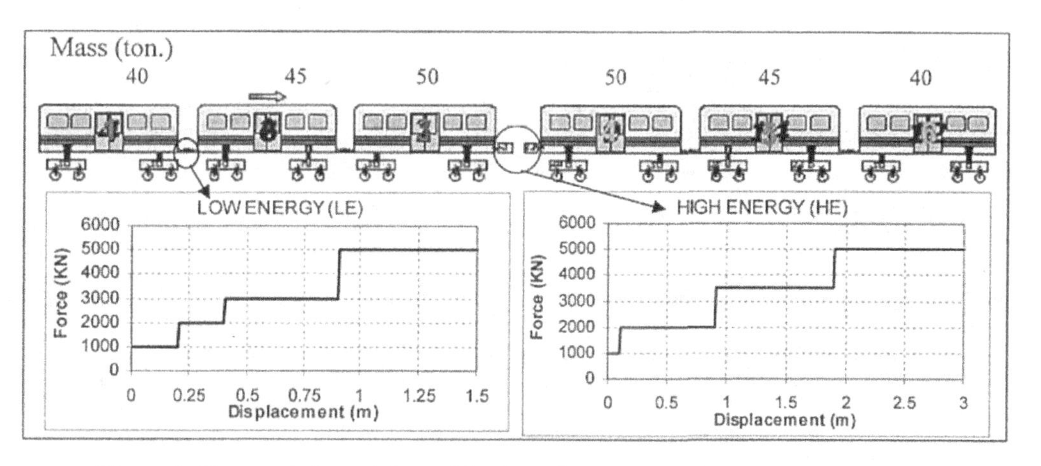

Figure 5. Characteristics and data of the front-front collision between two 3-car trains.

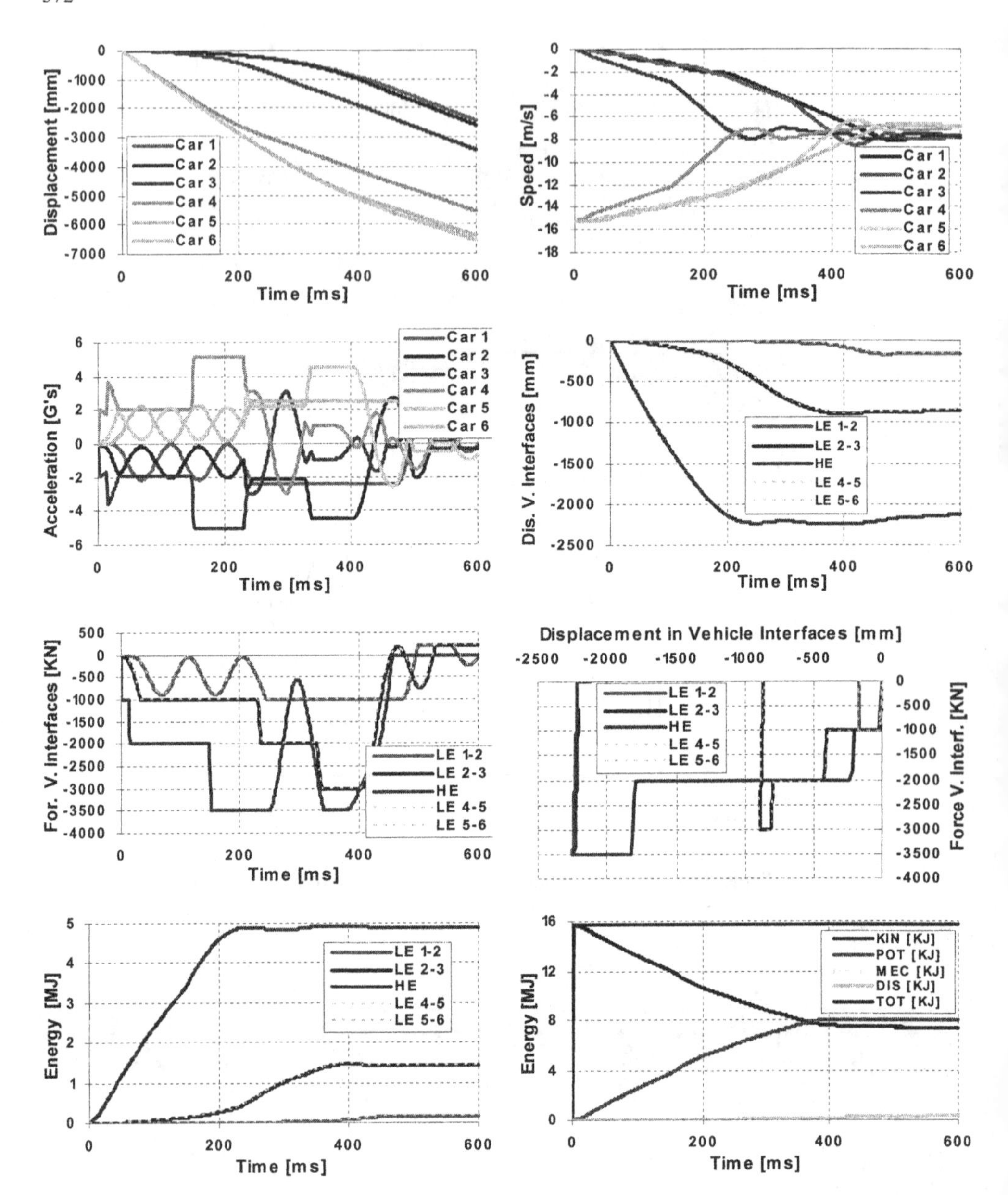

Figure 6. Simulation results for the front-front collision between two 3-car trains.

From the analysis of figure 6, it can be observed that accelerations levels are below 5 G that is the value usually accepted as the limit allowed. In fact the accelerations are one of the most important parameters to be considered into the project, as they are directly related with vehicle occupant injuries. Another important crashworthiness design requirement is the survivability space, which implies that no deformation occurs in the passenger's compartment. As can be observed in the results, the occupant compartment integrity is verified since the maximal deformation is within the deformation space available in the front of the cars.

4.2. OPTIMIZATION OF THE ENERGY ABSORPTION DEVICES

The characteristics of energy absorption devices located in the front of the cars establish the overall behavior of the trains and consequently the occupant's safety. The characteristics of these devices are represented in the models by force-displacement curves, as shown in figure 5, and can be optimized in order to increase the safety of the vehicles.

Preliminary studies have shown that passive safety improves as the amount of energy absorbed in the front of the trains increases. Another important aspect is that the space available for energy absorption between cars should be minimized by functional reasons, as less space required for deformation allows more space for the passenger compartment. A design function has been chosen as the energy absorbed between cars (LE-Low energy interfaces). The minimization of this design function will cause a transfer of energy absorption from the LE interfaces to the HE Interfaces, so a design constraint should be specified for the energy absorbed in fronts (HE-High Energy Interfaces). The design variables are related with the force levels of the energy absorption devices. In this example the energy absorption devices are optimized for the head-on collision scenario at 55 km/h. The optimization problem is summarized in table I.

In this example, a 5-car train is considered. The optimization problem has 6 design variables and 36 design constraints. This constrained minimization problem is solved using a quadratic programming algorithm. The initial and optimal plastic levels are presented in table II. The design function that corresponds to the energy absorbed in the Low-Energy interfaces has been reduced; from 8.88 MJ to 7.03 MJ that represents a decrease of 22%.

In figure 7 the results for the accelerations, forces, deformations and energies are presented for the initial design and for the optimal design. As can be observed, reductions in the sustained acceleration peaks have been obtained. These reductions that correspond to an increase in passive safety of the vehicle are achieved with higher deformations in the HE-Front. In fact, due to the design function selected, an energy absorption transfer from the LE interfaces to the HE Front has occurred.

TABLE I. Optimization problem of the energy absorption devices in a train

Design Function	$\min \sum_{i=1}^{n_{Le}} \mathrm{LE}_i$
Design Constraints	$-\mathrm{acc}_i < 5g \quad i=1, \ldots, n_{\mathrm{cars}}$ $-\ \mathrm{LE}_i < 1\ \mathrm{MJ}\ i=1, \ldots, n_{\mathrm{LE}};\ \ \mathrm{HE}_i < 10\ \mathrm{MJ}\ i=1, \ldots, n_{\mathrm{HE}}$ $-\mathrm{Def}_i < x_1 \quad i=1, \ldots, n_{\mathrm{Le}} + n_{\mathrm{He}};\ L_{xe_i} > L_{xe_{i-1}} + 500KN$
Design Variables	$\left[\{L_{Le1}, L_{Le2}, L_{Le3}\};\{L_{He1}, L_{He2}, L_{He3}\}\right]$

With:

n_{cars} Total number of cars; n_{LE} - Number of Low-Energy Interfaces (LE);

n_{HE} - Number of High-Energy Interfaces (HE);

LE_i - Energy Absorbed in Low Energy interface *i*;

acc_i - Maximal acceleration level in car *i;*

Def_i - Maximal allowed deformation in interface i;

$L_{LE1}, L_{LE2}, L_{LE3}$ - Low-Energy plastic limits; $L_{HE1}, L_{HE2}, L_{HE3}$ - High energy plastic limits.

TABLE II. Optimization results of the energy absorption devices in a train

		Low Energy Devices			High Energy Devices		
Force Level ID		L_{LE1}	L_{LE2}	L_{LE3}	H_{HE1}	H_{HE2}	H_{HE3}
Force	Initial	1000	2000	3500	1000	2000	3000
	Optimal	1022	2245	3234	1343	2126	2929
Displacement	Both	100	800	1000	200	200	500

From the optimal specifications for the energy absorption devices obtained, and verified the safety of the train in other collision scenarios, the devices can be then designed, using more detailed simulation tools such as non-linear finite element programs. The requirement for the detailed models is that the crashworthiness behavior of the designed structure and energy absorption devices must satisfy the force levels specified.

4.3. DESIGN OF EXPERIMENTAL TESTS USING OPTIMIZATION TOOLS

The design tools presented previously can also be used for the specification of the conditions of the experimental crash tests.

The present example corresponds, to a full-scale test carried out within the European Project Safetrain [20], which is shown in figure 8. In this experimental set-up a wagon B collides with two wagons (A and C) (Figure 9). Wagons A and C are connected by a Low-Energy device (LE), which was to be tested and validated. A honeycomb structure (Figure 10), simulating a High-Energy (HE), is mounted in the front of wagon C.

The previous design tools are used in the evaluation of the conditions of the experimental test (masses and velocity of the impact wagon), in order to absorb in the LE extremity 1.4 MJ. In table III the optimization problem is presented.

	Optimal Design	Initial Design
Vel [km/h]	55	55
Maximum load [KN]		
car 1, 2	1281	1000
car 2, 3	1343	2000
car 3, 4	2126	3000
car 4, 5	2929	3000
car 5, 6	3233	3500
car 6, 7	2929	3000
car 7, 8	2126	3000
car 8, 9	1343	2000
car 9, 10	1271	1000

	Optimal Design	Initial Design
Vel [km/h]	55	55
Maximum deformation - d [mm]		
car 1, 2	10	19
car 2, 3	192	575
car 3, 4	792	932
car 4, 5	978	997
car 5, 6	3724	3158
car 6, 7	979	996
car 7, 8	789	935
car 8, 9	199	577
car 9, 10	9	18

	Optimal Design	Initial Design
Vel [km/h]	55	55
Peak accelerations - G's & S.I.		
1	2.4	1.9
2	2.5	2.6
3	2.9	4.2
4	3.3	4.2
5	2.5	3.0
6	2.6	3.1
7	3.4	4.3
8	3.1	4.3
9	2.6	2.7
10	2.5	2.0

	Optimal Design	Initial Design
Vel [km/h]	55	55
Energy distribution [MJ]		
car 1, 2	0.01	0.01
car 2, 3	0.25	0.74
car 3, 4	1.36	1.58
car 4, 5	1.89	1.78
car 5, 6	10.00	8.13
car 6, 7	1.90	1.77
car 7, 8	1.35	1.59
car 8, 9	0.26	0.74
car 9, 10	0.01	0.01
Total	17.02	16.36

	Optimal Design	Initial Design
Vel [km/h]	55	55
Initial kinetic energy [MJ]		
	32.27	32.27

Figure 7. Initial and optimal results for the 5-cars train.

Figure 8. Experimental test: Physical set-up.

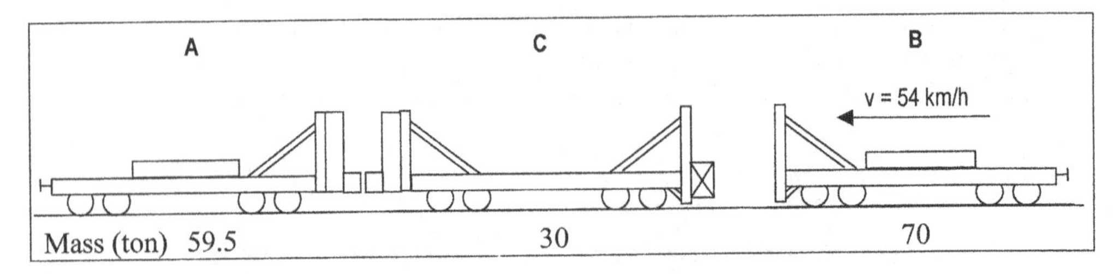

Figure 9. Experimental test: Collision scenario and conditions.

Figure 10. Experimental test: HE Front Extremity.

TABLE III. Optimization problem for the determination of the experimental test set-up

Design Function	$\psi_0 = (LE-1.4)^2$
Design Constraints	$\psi_1 = HE < 3MJ$
Design Variables	$b = [M_B, V_B, M_C, M_A, L_1]^T$
With: M_B: Mass of the impact wagon (Car B) V_B: Velocity of the impact wagon M_A and M_C: Mass of the wagons (Car A and C) L_1: Force level for the Honeycomb.	

The specified design constraints and the limits for the design variables are related with structural and manufacturing constraints. The optimization results and the limits considered for the design variables are presented in table IV.

TABLE IV. Optimization results for the determination of the experimental test set-up

	M_B (Ton.)	M_C (Ton.)	M_A (Ton.)	V_B (Km/h)	L_1 (KN)	ψ_1 HE (MJ)	ψ_0 LE (MJ)
Lower Limits	30	30	30	0	0	-	-
Upper Limits	70	50	60	72	3000	-	-
Optimal	70	30	59.5	53.7	3000	3	1.41

The experimental crash test has been performed using the test conditions obtained in the optimization process. In figure 11 the velocities of the wagons obtained experimentally are compared with the simulation velocities.

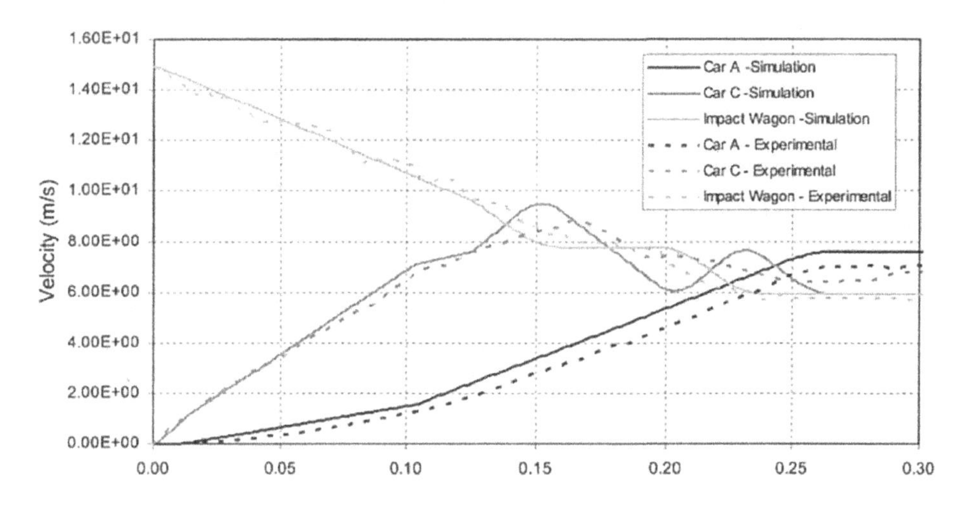

Figure 11. Correlation between velocities of the experimental test and simulation.

From the analysis of figure 11 it can be observed a good correlation between the simulation and experimental results. This demonstrates that the simulation model can accurately predict the energy absorption mechanism.

4.4. MULTI-LOAD DESIGN FOR CRASHWORTHINESS OF A TRAIN WAGON [18]

This example corresponds to the design of the end-underframe of a train and illustrates the application of the 2D rigid-flexible multibody system methodologies together with multiload optimization procedures and analytical sensitivities. The objective function is associated with the acceleration levels in the compartment for a typical impact velocity (30 km/h) against a rigid wall. Service loads (5 km/h) are included as design constraints. The optimization problem is indicated in table V and the optimization results are presented in table VI. More information about the model can be found in [18]. In figure 12 the train carshell and multibody model are shown schematically. The end underframe is modeled with 13 rigid bodies and 10 flexible bodies, connected by 26 plastic hinges.

TABLE V. End-underframe optimization problem

Design Function	$VCSI = \frac{1}{T}\int \ddot{a}^2 \, dt \Big\vert_{V = 30 \text{ km/h}}$
Design Constraints	$M_{p_i} < M_{\max_i} \Big\vert_{V = 5 \text{ km/h}}$ (i=1, .., 7)
Design variables	Width of the I beams (Figure 12)

TABLE VI. End-underframe optimization results

		Initial	Optimal
Design Function		20.2	13.6
Design variables	b_1 (mm)	120	97.36
	b_2 (mm)	100	75.3
	b_3 (mm)	150	130.38
Critical Design Constraint		$4.71\text{X}10^{-1}$	$-9.06\text{X}10^{-3}$

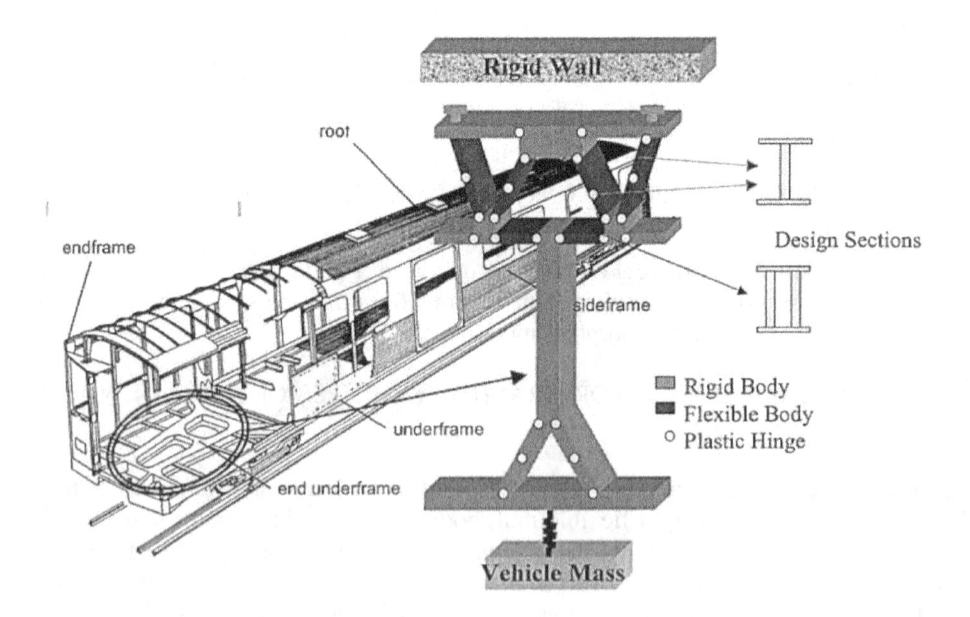

Figure 12. Train Carshell and model of the end underframe.

In the initial design, all the constraints are satisfied which means that the structure for the low speed simulation can allow higher levels of deformation. In the optimization process the acceleration levels for the simulation at 30 km/h are reduced with the increase in deformation levels for the low velocity simulation (5 km/h). At the optimal design all the design constraints are satisfied so no plastic deformation occurs at 5 km/h. This is achieved by reducing the dimensions of the sections of the beams. The initial and optimal values for VCSI are quite low. This is a result of the location where the VCSI is evaluated, which in this case as been considered at the middle of the carshell where the remaining mass of the vehicle is located.

4.5. DESIGN OF A 3D VEHICLE CHASSIS UNDER DYNAMIC LOADS

The design of a vehicle chassis under the action of dynamic loads, using the 3D tools, is presented in this example. The vehicle model comprises four wheels connected to the flexible chassis by a simplified suspension system. A simplified unidirectional tire model is used in the simulation. For the optimization of the chassis, two bump conditions are considered in order to induce structural bending and torsion. The mass of the chassis is chosen as the design function to be minimized. The chassis is divided in three different zones, identified in figure 13. For each zone a design variable is considered, corresponding to the thickness of the hollow circular cross sections of the beams of the chassis.

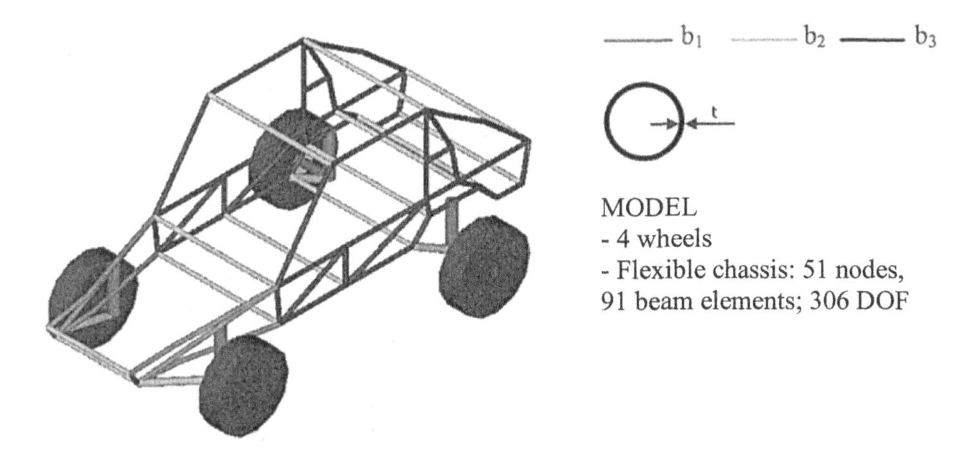

Figure 13. 3D Model of the vehicle chassis.

The optimal design problem of the vehicle chassis is presented in table VII. The results are presented in table VIII.

TABLE VII. Optimal design problem of the vehicle chassis

Design Function	Total Mass of the Vehicle
Design Constraints	$\sigma_i \leq \sigma_{adm}$ i=1,..,2*N_{LC}*N_{ele}*N_{cc}
Number of load cases (N_{LC})	2
Number of elements (N_{ele})	91
Maximal admissible stress σ_{adm} (Mpa)	200
Number of critical stress constraints per element (N_{cc})	2
Total number of design constraints	364
Design Variables	$b^k = t^k$
Number of design variables	3
Initial design vector ($\mathbf{b}^I$) (mm)	$\{5 \quad 5 \quad 5\}$
Lower limits vector ($\mathbf{b}^L$) (mm)	$\{0.5 \quad 0.5 \quad 0.5\}$
Upper limits vector ($\mathbf{b}^U$) (mm)	$\{50 \quad 50 \quad 50\}$

TABLE VIII. Optimization results of the Vehicle Chassis

		Initial	Optimal
Objective Function (Mass)		262.1	63.4
Design Variables	b_1 (mm)	5.0	3.49
	b_2 (mm)	5.0	1.33
	b_3 (mm)	5.0	0.97
Critical Design Constraint		-0.567	0.0

For the initial design, the section of the beams of the chassis are clearly oversized, being the value of the maximum stress in the chassis (86 Mpa) considerably below the allowable maximum stress (200 MPa). The optimization process drives to a global reduction of the thickness of the beam sections, with the consequent decrease of the weight of the chassis and the increase of the stresses. For the optimal design, it is observed that the largest thickness of the beam sections corresponds to the frontal zone (variable b_1) and the zone of smaller thickness is located in the lateral panels. For the optimal design, two of the design constraints are active and the remaining ones are verified. The active design constraints correspond to the traversal beams of the roof in the zone of the passenger's compartment. These active design constraints occur for the torsion ride, which shows that the critical loads correspond to torsion conditions.

5. Conclusions

General design methodologies for multibody systems have been presented. These methodologies that are based on rigid and flexible multibody dynamics formulations have been coupled with general optimization algorithms and include numerical and analytical sensitivities. These methodologies have been applied in the design of vehicle crashworthiness. 1D models have been used to simulate collisions between trains, where overall deformations, accelerations, forces and absorbed energies are obtained. 1D design tools have been applied to the specification of the energy absorption devices. These methodologies have been applied to the simulation and optimization of the energy absorption devices of real trains. 2D models, where the plastic deformations are modeled using plastic hinges, have been used to simulate end-underframes of trains under impact. Multiload design concepts that allow simulating impacts at different velocities have been used in the design process, and applied to the design of an end underframe of a train. The optimized structures allow a reduction of the accelerations without any plastic deformations for the service loads in shunting operations. 1D and 2D models provide computationally efficient and sufficiently accurate tools, which can be used in the earlier design stages.

Finally these design methodologies have been applied for the design of a 3D vehicle chassis. When compared with the classical structural design techniques that use only static load conditions, the proposed methodology allows the design of the vehicle chassis under dynamic loads, which result from transposition of obstacles or vehicle trajectories.

6. Acknowledgements

The support by the European Commission under BRITE/EURAM project BE-3092-89 and Fundação para a Ciência e Tecnologia under Project POCTI/EME/34340/99-00 are gratefully acknowledged.

7. References

1. Lust, R. V. (1992) Structural optimization with crashworthiness constraints, *Structural Optimization*, **4**, 85-89.
2. Kim, C.H. Mijar, A.R. and Arora, J.S. (2001) Development of simplified models for design and optimization of automotive structures for crashworthiness, *Struct. Multidisc. Optim.*, **22-4**, 307-321.
3. Oyan C. (1998) Dynamic simulation of Taipei EMU Train, *Vehicle System Dynamics*, **30**, 143-167.
4. Yamakawa, H., Tsutsui, Z. Takemae, K. and Ujita, Y. (1999) Structural optimization for improvement of train crashworthiness in conceptual and preliminary designs, In Proceedings of the *WCSMO-3, Third World Congress of Structural and Multidisciplinary Optimization*, Buffalo, New York.
5. Ambrósio, J. C., Pereira M. S. and Dias, J. P. (1996) Distributed and discrete nonlinear deformations on multibody dynamics, *Nonlinear Dynamics*, **10**, 359-379.

6. Pereira, M. S. Ambrósio, J. A. and Dias, J. P. (1997) Crashworthiness analysis and design using rigid-flexible multibody dynamics with application to train vehicles, *International Journal for Numerical Methods in Engineering*, **40**, 655-687.
7. Bestle, D. and Schiehlen, W., Editors, (1996), *IUTAM Symposium on Optimization of Mechanical Systems*, Kluwer Academic Publishers, Dordrecht.
8. Haftka, R. T. and Gürdal, Z. (1992) *Elements of Structural Optimization*, Kluwer Academic Publishers, Dordrecht.
9. Chang, C. O. and Nikravesh, P. E., (1985) Optimal Design of Mechanical Systems with Constraint Violation Stabilization Method, *ASME J. Mech., Trans., and Auto. in Design*, **107**, 493-498.
10. Dias, J. P. and Pereira, M. S., (1997), Sensitivity Analysis of Rigid Flexible Multibody Systems, *Multibody System Dynamics*, **1**, 303-322.
11. Ashrafiuon, H. and Mani, N. K. (1990) Analysis and Optimal Design of Spatial Mechanical Systems, *ASME Journal of Mechanical Design*, **112**, 200-207.
12. Bestle, D. and Eberhard, P., (1996) Multi-Criteria Multi-Model Design Optimization, in D. Bestle and W. Schiehlen (Eds.) *IUTAM Symposium on Optimization of Mechanical Systems*, Kluwer Academic Publishers, Dordrecht, pp. 33-40.
13. Mani, N. K. and Haug, E. J. (1985) Singular Value Decomposition for Dynamic System Design Sensitivity Analysis, *Engineering with Computers*, **1**, 103-109.
14. Vanderplaats, G. N. (1992) *DOT- Design Optimization Tools*, Version 3.0, VMA Engineering, Colorado Springs.
15. Dias, J. P. and Pereira, M. S. (1995) Dynamics of flexible mechanical systems with contact-impact and plastic deformations, *Nonlinear Dynamics*, **8**, 491–512.
16. Pereira, M. S. and Proença, P. L. (1991) Dynamic Analysis of Spatial Flexible Multibody Systems Using Joint Co-Ordinates", *International Journal of Numerical Methods in Engineering*, **32** 1799-1812.
17. Brown, P. N., Hindmarsh, A. C. and Petzold, L. R. (1994), Using Krylov Methods in the Solution of Large-Scale Differential-Algebraic Systems, *SIAM J. Sci. Comp.*, **15**, 1467-1488.
18. Dias, J. P. and Pereira, M. S. (2002) Optimal Design of Train Structures for Crashworthiness using a Multi-load Approach, *I. J. of Crashworthiness,* **7-3**, 331-343.
19. Haftka, R. T. and Gürdal Z. (1992). *Elements of Structural Optimization*, Kluwer Academic Publishers, Dordrecht.
20. http://europa.eu.int/comm/research/growth/gcc/projects/safe-train.html

MULTIBODY SYSTEM CONTACT DYNAMICS SIMULATION

E. ZAHARIEV
Institute of Mechanics
Bulgarian Academy of Sciences
Acad. G. Bonchev, bl. 4
1113 Sofia, Bulgaria
evtimvz@bas.bg

ABSTRACT: A novel numerical approach for modeling of impact and unilateral contact constraints of rigid and flexible multibody systems is presented. Constraint dynamic equations and Lagrange's multipliers are applied. Coulomb friction is taken into account. Impact hypotheses of Newton and Poisson are regarded. The effects stiction, sliding, stick – slip are discussed. Numerical analysis of the velocities, as well as, the reactions and impulses in the contact points is achieved. In case of impact, a new definition "dominant jump of the tangential velocity" is defined analyzing the transition event sliding - stiction. Transformation of the dynamic equations and impulse – momentum equations is implemented. In case of contact, the reactions and/or the impulses in the contact points are parameters of the dynamic equations, while the number of the equations is constant. The method of the applied contact forces is compared to the approach presented. Simultaneous contact in multiple points is regarded.

1. Introduction

In the article multibody systems are considered mechanical systems built of rigid and flexible bodies. The possible motions are subject of kinematic and force constraints that are imposed by joints, springs, prescribed motions and etc. The main subjects are unilateral constraints and impact, i.e.: the contact that occurs as a result of the system motion and is not guarantied by the design scheme. The discontinuity of this motion and of the dynamic equations makes the simulation procedure complicated problem that requires specific approaches for its solution.

Mainly two approaches have been used in the scientific literature. These are:

- classical approach of the constrained dynamics and Lagrange's multipliers;
- approach of estimation of the contact forces as a result of possible penetration of the contact surfaces.

The distance at the very beginning or end of the contact defines the normal direction and the tangential plane in the contact points. In the classical theory existence of contact means additional kinematic constraints [1]. The addition or deletion of constraints that correspond to contact phenomena causes transformation of the dynamic equations. The signs of the Lagrange's multipliers define the directions of the contact forces and,

W. Schiehlen and M. Valášek (eds.), Virtual Nonlinear Multibody Systems, 383–402.

respectively, existence of contact or separation of the contacting points. The contact problem solution becomes more complicated if nonlinear friction in the tangential plane is assumed. One unilateral constraint could have four different states: separation of the contact; sliding, stiction; transition stiction - sliding. The stiction phenomenon and possible transition stiction - sliding causes branching of the calculations and causes different transformations of the dynamic equations. This branching implies the necessity of solution of combinatorial task. Another complicated problem is the definition of the sliding direction after stiction. The aforementioned problems are discussed in the investigations of Glocker and Pfeiffer [2, 3], Seyfferth and Pfeiffer [4]], Wosle and Pfeiffer [5].

The classical theory of the impact analysis uses the Impulse – Momentum Equations (IME) and the so-called "coefficient of restitution". The most frequently used definitions of the coefficient of restitution are based on (Hunt and Grossley [6] , Keller [7], Brach [8], Wang and Mason [9]): the Newton's theory of the ratio between the relative velocities of the contact points after and before the impact; the Poisson's theory of the ratio of the impulses in the contact point during the phases of restitution and compression.The coefficient of restitution provides one step solution of the velocity problem and is widely applied for analysis of impact. The basic hypotheses have been used in many treatments. Glocker and Pfeiffer [10] presented a two-dimensional impact model based on Poisson's hypothesis and Coulomb friction. In their recent book, Pfeiffer and Glocker [11] developed this methodology in case of impact of multibody systems with unilateral constraints. Schiehlen [12] regarded different approaches of multibody contact dynamics and multi-rate integration methods are proposed to enhance the efficiency of the calculation procedures. Lankarani and Pereira [13] made a detailed review of the studies concerning this topic. The authors solved the rigid multibody systems frictional impact and described seven phases of possible motion.

In the second approach of contact and impact simulation the distance (or the gap) between the contact surfaces plays significant role. For modeling of impact the contact points are loaded by forces [14, 15]. The parameters of the forces correspond to the stiffness properties of the body materials and shapes, their values being calculated using contact Hertz' theory [16]. Using the contact forces as external ones, the dynamic equations of motion are numerically integrated. Similar approach is this of the penalty functions [17 - 19]. In [19] the convergence of the penalty function method and the reduced – integration penalty methods are discussed. The main part of this approach is the procedure for calculation of the contact forces, not the dynamic model.

In the article, a novel numerical approach of modified dynamic model using Lagrange's equations is suggested. The possible motions in the normal directions and in the tangential planes of the contact points are considered active coordinates. If no contact exists these motions appear as parameters or, in case of contact, transformation of the dynamic equations is implemented and the corresponding reaction forces become parameters. Using this approach the size and number of the parameters is constant. Multiple contact points (simultaneous unilateral contact an impact points) are included in the dynamic equations.

The algorithm is consequently applied, if rigid bodies are considered, to the two phases of the impact - compression and restitution. In every phase the possible events sliding, reverse sliding, stiction and sliding-stiction are analyzed satisfying the impulse-momentum equations. The algorithm is developed in case of Newton's and Poisson's hypotheses for the coefficient of restitution. A new definition of "dominant tangential velocity jump" is applied for impact analysis and of the stick-slip process simulation. The same numerical procedure is applied in case of flexible systems discretized using finite element theory. The jumps of the velocities are estimated as a result of deletion of the coordinates that define the distances between the colliding nodes and elements. The resulting global deflections of the flexible system are computed integrating the dynamic equations with the initial velocities so obtained. The reaction forces in the contact points and nodes are calculated. The numerical algorithm is applied solving examples of flexible multibody systems in case of contact and impact of many elements and nodes. The method of the applied contact forces is also used, the numerical results so obtained being compared

2. Theoretical Background

The analytical form of a multibody system dynamic equations with kinematic constraints are presented by second order Differential Algebraic Equations (DAE)

$$\boldsymbol{M} \cdot \ddot{\boldsymbol{Q}} = \boldsymbol{G}(\boldsymbol{Q}, \dot{\boldsymbol{Q}}) \tag{1}$$

$$\boldsymbol{\Phi}(\boldsymbol{Q}) = \boldsymbol{0}_m \tag{2}$$

with respect to $n \times 1$ matrix - column $\ddot{\boldsymbol{Q}}$ of the second time derivatives of the coordinates of motion $\boldsymbol{Q} = [q_1 \quad q_2 \quad \cdots \quad q_n]^{\mathrm{T}}$. For a common case the analytical solution cannot be provided. The numerical solution requires discretization of the equations (1, 2). Using the Lagrange's equations and after transformations one obtains the linear $(n+m) \times (n+m)$ equation system with respect to the unknown $\ddot{\boldsymbol{Q}}$ and Lagrange's multipliers, m sized matrix - column $\boldsymbol{\Lambda}$, i. e.:

$$\boldsymbol{M} \cdot \ddot{\boldsymbol{Q}} - \boldsymbol{\Phi}_Q^{\mathrm{T}} \cdot \boldsymbol{\Lambda} = \boldsymbol{G}(\boldsymbol{Q}, \dot{\boldsymbol{Q}}) \tag{3}$$

$$\boldsymbol{\Phi}_Q(\boldsymbol{Q}) \cdot \ddot{\boldsymbol{Q}} + \dot{\boldsymbol{\Phi}}_Q(\boldsymbol{Q}) \cdot \dot{\boldsymbol{Q}} = \boldsymbol{0}_m \tag{4}$$

Equations (3, 4) provide solution of the initial value problem (with given $\boldsymbol{Q}$ and $\dot{\boldsymbol{Q}}$) that should satisfy the discretized position constraint equations, as well as, their time derivatives, i. e.:

$$\boldsymbol{\Phi}(\boldsymbol{Q}) = \boldsymbol{0}_m \tag{5}$$

$$\boldsymbol{\Phi}_Q(\boldsymbol{Q}) \cdot \dot{\boldsymbol{Q}} = \boldsymbol{0}_m \tag{6}$$

In equations (1 – 6) $\boldsymbol{M}$ is $n \times n$ mass-matrix; $\boldsymbol{G}$ is n sized matrix - column of the applied and velocity dependent inertia forces; $\boldsymbol{\Phi}_Q = \dfrac{\partial \boldsymbol{\Phi}}{\partial \boldsymbol{Q}}$ is $m \times n$ matrix of the partial derivatives of the constraint equations; $\boldsymbol{0}_m$ is m sized zero matrix-column ($\boldsymbol{0}^n$ and $\boldsymbol{0}_m^n$ are $1 \times n$ sized zero matrix - row and $m \times n$ zero matrix, respectively).

For the constraint dynamics the IME in case of impact in a point are as follows [1]:

$$\boldsymbol{M} \cdot \Delta\dot{\boldsymbol{Q}} - \boldsymbol{\Phi}_{\boldsymbol{Q}}^{\mathrm{T}} \cdot \boldsymbol{I}_\lambda = \frac{\partial s}{\partial \boldsymbol{Q}}^{\mathrm{T}} \cdot I_s \tag{7}$$

$$\boldsymbol{\Phi}_{\boldsymbol{Q}} \cdot \Delta\dot{\boldsymbol{Q}} = \boldsymbol{0}_m \tag{8}$$

$$s_{\boldsymbol{Q}} \cdot \Delta\dot{\boldsymbol{Q}} = -(1+e) \cdot s_{\boldsymbol{Q}} \cdot \dot{\boldsymbol{Q}}^{(-)} \tag{9}$$

where $s = s(\boldsymbol{Q})$ is the distance between the colliding surfaces at the time of impact; $s_{\boldsymbol{Q}} = \frac{\partial s}{\partial \boldsymbol{Q}}$; $\Delta\dot{\boldsymbol{Q}} = \dot{\boldsymbol{Q}}^{(+)} - \dot{\boldsymbol{Q}}^{(-)}$ is the jump of the velocities after (superscript +) and before (-) the impact; $\boldsymbol{I}_\lambda$ is m sized matrix – column of the impulses caused by the constraints; and e is the coefficient of restitution. Here the coefficient of restitution is defined according to the Newton's hypothesis. Existence of impact adds to equations (7, 8) one constraint (9) and one parameter – the impulse I_s in the point of impact. So, the IME (7 - 9) are $n+m+1$ linear equation system with respect to $n+m+1$ unknown parameters - velocities jumps $\Delta\dot{\boldsymbol{Q}}$, impulses of the constraints $\boldsymbol{I}_\lambda$ and impulse in the contact point I_s.

The Poisson's hypothesis could be applied for the restitution phase only, calculating first the impulse during the compression phase applying the Newton's hypothesis with coefficient of restitution $e = 0$. For the restitution phase the impulse I_s in equation (7) is known parameter, i. e., $I_s = I_s^r = e \cdot I_s^c$, since I_s^c is computed at the compression stage. Then the solution for the velocity jumps $\Delta\dot{\boldsymbol{Q}} = \dot{\boldsymbol{Q}}^{(+)}$ and the impulses of the constraints $\boldsymbol{I}_\lambda^r$ at the restitution phase are obtained solving the linear system of equations (7, 8).

3. Kinematics of Contact

In Fig. 1 (a) a multibody system is shown at the time of possible contact of two bodies. The problem of contact point detection [20] is not subject of this paper. The both tangential planes in the colliding points, respectively, their normal directions coincide. If contact occurs as a result of active kinematic restrictions the motion in the normal direction of the contact tangential plane vanishes and the number of the coordinates decreases with one. If stiction occurs the restrictions become two more (in case of three-dimensional motion) and two more coordinates become dependent. The possible small translations s_{τ_1}, s_{τ_2} in the tangential plane and in its normal direction, as well as, the possible small rotations θ_{τ_1}, θ_{τ_2}, θ_n are also presented in the figure. In the article, the friction forces imposed on the possible rotations are not regarded. In case of stiction reaction forces appear in the normal direction and both directions of the tangential plane. Friction forces appear in the tangential plane if sliding occurs.

In Fig. 1 (b) two flexible bodies of a multibody system at the moment of incoming impact are shown. The axes defined from the tangential plane and the normal direction in the point of contact are shown. Similarly of Fig. 1 (a), the three possible translations

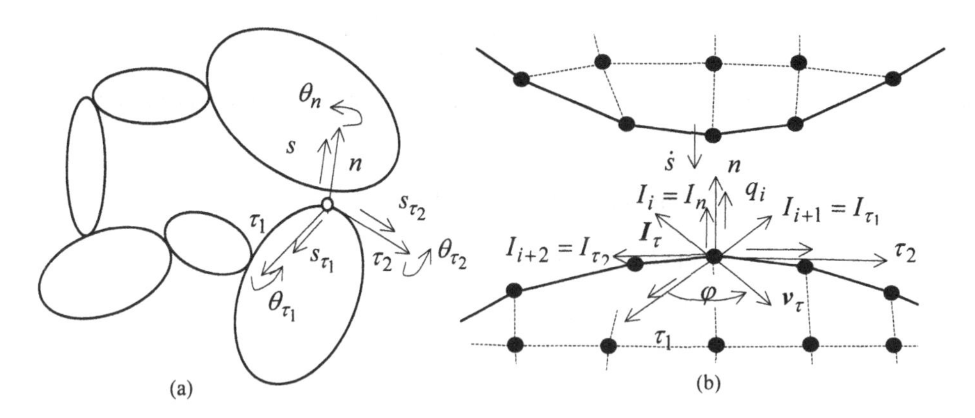

Figure 1. Kinematic parameters of contact and impact points.

are q_i, q_{i+1}, q_{i+2}, that coincide with the translations s, s_{τ_1}, s_{τ_2}, respectively. The impulses in the contact points are I_i, I_{i+1}, I_{i+2}, where I_i corresponds to the normal impulse ($I_i = I_n$) while the other are components of the impulse $\boldsymbol{I}_\tau = \begin{bmatrix} I_{i+1} & I_{i+2} \end{bmatrix}^{\mathrm{T}}$ in the tangential plane. In case of sliding, the frictional impulse is in the opposite direction of the tangential velocity $\boldsymbol{v}_\tau = \begin{bmatrix} \dot{q}_{i+1} & \dot{q}_{i+2} \end{bmatrix}^{\mathrm{T}}$ while its magnitude depends on the friction coefficient and the impulse in the normal direction. The new notations of these possible translations are to point out that we include them as elements of matrix $\boldsymbol{Q}$ and the indexes show the number of the elements. As well as, the impulses in the contact points are stored in a matrix – column of the contact impulses $\boldsymbol{I}$ that differs from the impulses $\boldsymbol{I}_\lambda$ of the constraints in equation (7). If contact occurs the possible translations in the normal direction and in the tangential plane are stored in matrix $\boldsymbol{Q}$ with indexes presented above. The same order will be used for the impulses in the contact point.

The kinematic constraints, equations (4 – 6), are assumed invariant in numbers and forms. They do not present the constraints as a result of possible contact. Existence of contact adds additional constraints to the dynamic equations (3, 4) that, for a single contact point for which the parameter q_i is the motion along the normal, are as follows:

$$\Phi_{m+1} = q_i = a_i = const \qquad (10)$$

$$\dot{\Phi}_{m+1} = \dot{q}_i = 0 \qquad (11)$$

$$\ddot{\Phi}_{m+1} = \ddot{q}_i = 0 \qquad (12)$$

The matrix – row of the partial derivatives of the additional constraint Φ_{m+1} is

$\frac{\partial \Phi_{m+1}}{\partial \boldsymbol{Q}} = \frac{\partial q_i}{\partial \boldsymbol{Q}} = [0 \quad 0 \quad \cdots \quad 0 \quad 1 \quad 0 \quad \cdots \quad 0]$, where the place of the unity coincides with the place of the corresponding coordinate q_i of matrix $\boldsymbol{Q}$. If impact occurs the distance s between the colliding points coincides with the corresponding coordinate (motion in the normal direction) that is the ith element of matrix $\boldsymbol{Q}$. The partial derivatives of s are stored in matrix – row $\frac{\partial s}{\partial \boldsymbol{Q}} = \frac{\partial q_i}{\partial \boldsymbol{Q}} = [0 \quad 0 \quad \cdots \quad 0 \quad 1 \quad 0 \quad \cdots \quad 0]$.

The jump of the velocity $\Delta \dot{s} = \dot{s}^{(+)} - \dot{s}^{(-)} = \Delta \dot{q}_i$ is also the ith element of matrix $\Delta \dot{\boldsymbol{Q}}$.

4. Transformation of the Contact Dynamic Equations

Dynamic equations (3, 4) could be rewritten in matrix form as follows:

$$\begin{bmatrix} \boldsymbol{M} & -\boldsymbol{\Phi}_{\boldsymbol{Q}}^{\mathrm{T}} \\ \boldsymbol{\Phi}_{\boldsymbol{Q}} & \boldsymbol{0}_m \end{bmatrix} \cdot \begin{bmatrix} \ddot{\boldsymbol{Q}} \\ \boldsymbol{\Lambda} \end{bmatrix} = \begin{bmatrix} \boldsymbol{G} \\ -\dot{\boldsymbol{\Phi}}_{\boldsymbol{Q}} \cdot \dot{\boldsymbol{Q}} \end{bmatrix} = \boldsymbol{F}(\boldsymbol{Q}, \dot{\boldsymbol{Q}}) \tag{13}$$

Substituting equation (12) in equation (13) one obtains the dynamic equations in case of contact along the coordinate i, i. e.:

$$\begin{bmatrix} \underline{\boldsymbol{M}} & -\boldsymbol{\Phi}_{\boldsymbol{Q}}^{\mathrm{T}} \\ \underline{\boldsymbol{\Phi}}_{\boldsymbol{Q}} & \boldsymbol{0}_m \end{bmatrix} \cdot \begin{bmatrix} \underline{\ddot{\boldsymbol{Q}}} \\ \boldsymbol{\Lambda} \end{bmatrix} = \boldsymbol{F} \tag{14}$$

where matrix $\underline{\boldsymbol{\Phi}}_{\boldsymbol{Q}}$ is composed of matrix $\boldsymbol{\Phi}_{\boldsymbol{Q}}$ for which all elements of row i are zero;

$$\underline{\boldsymbol{M}} = \begin{bmatrix} m_{1,1} & m_{1,2} & \cdots & 0 & m_{1,i+1} & \cdots & m_{1,n} \\ m_{2,1} & m_{2,2} & \cdots & 0 & m_{2,i+1} & \cdots & m_{2,n} \\ \cdots & \cdots & \cdots & \cdots & \cdots & \cdots & \cdots \\ m_{i,1} & m_{i,2} & \cdots & -1 & m_{i,i+1} & \cdots & m_{i,n} \\ \cdots & \cdots & \cdots & \cdots & \cdots & \cdots & \cdots \\ m_{n,1} & m_{n,2} & \cdots & 0 & m_{n,i+1} & \cdots & m_{n,n} \end{bmatrix} \tag{15}$$

is composed of matrix $\boldsymbol{M}$ for which the ith diagonal element is equal 1 and all other elements of ith column are 0; for matrix $\underline{\ddot{\boldsymbol{Q}}}$, $\underline{\ddot{\boldsymbol{Q}}} = [\ddot{q}_1 \quad \ddot{q}_2 \quad \cdots \quad \lambda_{m+1} \quad \ddot{q}_{i+1} \quad \cdots \quad \ddot{q}_n]^{\mathrm{T}}$ the element $\ddot{q}_i$ is substituted by λ_{m+1}. This simple substitution gives the possibility, in case of contact and if the conditions (10 – 12) are fulfilled, the dynamic equations to be solved directly for the accelerations and for the normal reaction λ_{m+1} in the contact point using the latter as a parameter instead of $\ddot{q}_i$. So, the size of the differential equations and the number of the parameters are not changed. The sign of λ_{m+1} defines the direction of the reaction in the contact point and if it is negative the reaction coincides with the positive direction of the normal, which means that the solution of the transformed dynamic equations is true. If the reaction is negative the solution is not real. Separation of the contact points follows and the corresponding constraint must be excluded from the DAE.

If impact occurs and using the motion in the normal direction as a coordinate (in this case the ith element of matrix $\boldsymbol{Q}$) one can rewrite the IME (7 – 9) as follows:

$$\begin{bmatrix} \boldsymbol{M} & -\boldsymbol{\Phi}_{\boldsymbol{Q}}^{\mathrm{T}} \\ \boldsymbol{\Phi}_{\boldsymbol{Q}} & \boldsymbol{0}_m \end{bmatrix} \cdot \begin{bmatrix} \Delta\dot{\boldsymbol{Q}} \\ \boldsymbol{I}_{\lambda} \end{bmatrix} - \begin{bmatrix} \dfrac{\partial s}{\partial \boldsymbol{Q}}^{\mathrm{T}} \\ \boldsymbol{0}_m \end{bmatrix} \cdot I_i = \boldsymbol{0}_{n+m} \tag{16}$$

$$\Delta\dot{q}_i = -(1+e)\cdot \dot{q}_i^{(-)} \tag{17}$$

which means that, with given coefficient of restitution, the velocity jump of the ith parameter $\Delta\dot{q}_i$ is calculated explicitly from (17) and could be excluded from equation (16). In equation (16) I_i is the impulse in the normal direction. The resulting IME are:

$$\begin{bmatrix} \underline{\boldsymbol{M}} & -\boldsymbol{\Phi}_{\boldsymbol{Q}}^{\mathrm{T}} \\ \underline{\boldsymbol{\Phi}}_{\boldsymbol{Q}} & \boldsymbol{0}_m \end{bmatrix} \cdot \begin{bmatrix} \Delta\underline{\dot{\boldsymbol{Q}}} \\ \boldsymbol{I}_{\lambda} \end{bmatrix} + \begin{bmatrix} \boldsymbol{M}_i \\ \boldsymbol{0}_m \end{bmatrix} \cdot \Delta\dot{q}_i = \boldsymbol{0}_{n+m} \tag{18}$$

In equation (18) the matrices $\underline{\boldsymbol{\Phi}}_{\boldsymbol{Q}}$ and $\underline{\boldsymbol{M}}$ are the same matrices as described above in equation (14, 15); $\boldsymbol{M}_i$ is the ith column of matrix $\boldsymbol{M}$. In matrix $\Delta\underline{\dot{\boldsymbol{Q}}}$, similarly to matrix $\underline{\ddot{\boldsymbol{Q}}}$, the ith element is substituted by I_i, $\Delta\underline{\dot{\boldsymbol{Q}}} = [\Delta\dot{q}_1 \quad \Delta\dot{q}_2 \quad \cdots \quad I_i \quad \Delta\dot{q}_{i+1} \quad \cdots \quad \Delta\dot{q}_n]^{\mathrm{T}}$. This selection of the coordinates results in transformation of the mass matrix and the matrix of the partial derivatives modifying only the columns that correspond to the coordinates (motions) in the contact point. The transformation is independent for every contact point and could be achieved simultaneously for all possible contact points.

5. Frictional Contact of Rigid Multibody Systems

5.1. FRICTIONAL UNILATERAL CONTACT

As the magnitude of the reaction force in the contact point is equal to the coefficient λ_{m+1}, we denote this coefficient as a contact reaction – force $R_i = \lambda_{m+1}$ that corresponds to coordinate i. The friction forces in the tangential plane are denoted F_{i+1}, F_{i+2}. The values and directions of the friction forces are calculated as follows:

$$F_{i+m} = -\mu \cdot \frac{v_{\tau_m}}{\|v_\tau\|} . R_i = -\mu \cdot \frac{\dot{q}_{i+m}}{\sqrt{\dot{q}_{i+1}^2 + \dot{q}_{i+2}^2}} \cdot R_i = -\mu_{i+m} \cdot R_i \,; m = 1, 2 \tag{19}$$

where μ is the coefficient of friction, while with μ_{i+1}, μ_{i+2} directional frictional coefficients are pointed out. The transformation of the DAE (14), respectively of matrix $\boldsymbol{M}$, taking into account the friction in the contact point is as follows:

$$\underline{\boldsymbol{M}}\langle i_{sl} \rangle = \begin{bmatrix} m_{1,1} & m_{1,2} & \cdots & 0 & m_{1,i+1} & \cdots & m_{1,n} \\ m_{2,1} & m_{2,2} & \cdots & 0 & m_{2,i+1} & \cdots & m_{2,n} \\ \cdots & \cdots & \cdots & \cdots & \cdots & \cdots & \cdots \\ m_{i,1} & m_{i,2} & \cdots & -1 & m_{i,i+1} & \cdots & m_{i,n} \\ m_{i+1,1} & m_{i+1,2} & \cdots & \mu_{i+1} & m_{i+1,i+1} & \cdots & m_{i+1,n} \\ m_{i+2,1} & m_{i+2,2} & \cdots & \mu_{2+1} & m_{i+2,i+1} & \cdots & m_{i+2,n} \\ \cdots & \cdots & \cdots & 0 & \cdots & \cdots & \cdots \\ m_{n,1} & m_{n,2} & \cdots & 0 & m_{n,i+1} & \cdots & m_{n,n} \end{bmatrix} \tag{20}$$

where the notation $\underline{\boldsymbol{M}}\langle i_{sl} \rangle$ points out that column i of matrix $\boldsymbol{M}$ is transformed according to the conditions for sliding in case of contact for i^{th} coordinate (element) of $\boldsymbol{Q}$. The modified matrix of the accelerations is

$$\underline{\ddot{\boldsymbol{Q}}}\langle i_{sl} \rangle = \begin{bmatrix} \ddot{q}_1 & \ddot{q}_2 & \cdots & R_i & \ddot{q}_{i+1} & \cdots & \ddot{q}_n \end{bmatrix}^{\mathrm{T}} \tag{21}$$

In the modified matrix $\underline{\boldsymbol{\Phi}}_{\boldsymbol{Q}}\langle i_{sl} \rangle$ of matrix $\boldsymbol{\Phi}_{\boldsymbol{Q}}$ all elements of column i are equal zero.

If stiction occurs constraint in the form of the equations (10 – 12) are also imposed on the motions q_{i+1} and q_{i+2}. Then, matrix $\boldsymbol{M}$ is modified as follows:

$$\underline{\boldsymbol{M}}\langle i_{st} \rangle = \begin{bmatrix} m_{1,1} & m_{1,2} & \cdots & 0 & 0 & 0 & m_{1,i+3} & \cdots & m_{1,n} \\ m_{2,1} & m_{2,2} & \cdots & 0 & 0 & 0 & m_{2,i+3} & \cdots & m_{2,n} \\ \cdots & \cdots & \cdots & \cdots & \cdots & \cdots & \cdots & \cdots & \cdots \\ m_{i,1} & m_{i,2} & \cdots & -1 & 0 & 0 & m_{i,i+3} & \cdots & m_{i,n} \\ m_{i+1,1} & m_{i+1,2} & \cdots & 0 & -1 & 0 & m_{i+1,i+3} & \cdots & m_{i+1,n} \\ m_{i+2,1} & m_{i+2,2} & \cdots & 0 & 0 & -1 & m_{i+2,i+3} & \cdots & m_{i+2,n} \\ \cdots & \cdots & \cdots & \cdots & \cdots & \cdots & \cdots & \cdots & \cdots \\ m_{n,1} & m_{n,2} & \cdots & 0 & 0 & 0 & m_{n,i+3} & \cdots & m_{n,n} \end{bmatrix} \tag{22}$$

where the modified matrix – column of the accelerations is

$$\underline{\ddot{\boldsymbol{Q}}}\langle i_{st} \rangle = \begin{bmatrix} \ddot{q}_1 & \ddot{q}_2 & \cdots & R_i & R_{i+1} & R_{i+2} & \ddot{q}_{i+3} & \cdots & \ddot{q}_n \end{bmatrix}^{\mathrm{T}} \tag{23}$$

In (23) R_{i+1}, R_{i+2} are the reactions in the tangential plane, because of the existing stiction. In matrix $\underline{\Phi}_Q\langle i_{st}\rangle$ all elements of columns i, i+1, i+2 are equal zero. If the tangential velocity in the point of contact is zero, then there exist two possibilities – sliding or stiction. In this case, three-step procedure should be applied.

Step A. Using the modified matrix $\underline{\boldsymbol{M}}\langle i_{st}\rangle$ solution of the DAE (14) with respect to the modified matrix $\underline{\ddot{\boldsymbol{Q}}}\langle i_{st}\rangle$ that includes the reactions R_i, R_{i+1}, R_{i+2}.

Step B. With the coefficient of dry friction μ_0 checking the condition for stiction, i.e.: if the inequality

$$\mu_0 \cdot R_i \geq \sqrt{R_{i+1}^2 + R_{i+2}^2} \tag{24}$$

is fulfilled the solution for $\underline{\ddot{\boldsymbol{Q}}}\langle i_{st}\rangle$ is the true solution; if not, transition stiction - sliding must be considered where the direction of the friction force should be estimated.

Step C. Definition of the friction force directions and solution of the DAE (14) in case of sliding. With the values of the reactions R_i, R_{i+1}, R_{i+2} so calculated the direction of friction forces in the tangential plane are estimated from the equation

$$F_{i+m} = -\mu_0 \cdot \frac{-R_{i+m}}{\sqrt{R_{i+1}^2 + R_{i+2}^2}} \cdot R_i = -\mu_{0\,i+m} \cdot R_i \,;\, m = 1, 2 \tag{25}$$

The algorithm proposed in the article, in case of contact, provides transformation of the DAE without changing the number of the differential equations and parameters. The approach is applicable for multiple contact points. The matrices $\boldsymbol{M}$, $\boldsymbol{\Phi}_Q$, $\ddot{\boldsymbol{Q}}$ are modified using simple substitution of the elements of the columns that correspond to the coordinates describing the motions in the points of contact. For multiple contact points a combinatorial task of all the possible contact events (sliding and stiction) in every contact point should be compiled and every combination should be checked for fulfilling of the contact conditions.

5.2. FRICTIONAL IMPACT

In terms of the mathematics the IME do not regard the kind of the velocity jump, i.e. compression or restitution – the IME are identical for these two phases. The separate investigation of the two phases of the impact process is applied in several treatments [10, 13]. Here, according to the positive direction of the normal in the contact point (Fig. 1), the normal velocity at the stage of compression is negative. The tangential velocity could be negative or positive. The normal impulse for both the stages is positive and the tangential impulse depends on the direction of the tangential velocity. The impact laws for the compression and restitution phases are identical and the procedures for solution of the IME are similar for both stages.

5.2.1. *Stiction with zero and non-zero approach tangential velocity*

5.2.1.1. Hypothesis of Newton. At the beginning of impact the configuration of a multibody system and the coefficient of friction impose conditions that prevent any kind of sliding, i.e. instant stiction occurs. With given normal $\Delta\dot{q}_i^c = 0 - \dot{q}_i^{(-)}$ and tangential $\Delta\dot{q}_{i+1}^c = 0 - \dot{q}_{i+1}^{(-)}$, $\Delta\dot{q}_{i+2}^c = 0 - \dot{q}_{i+2}^{(-)}$ jumps of the velocities at the compression (superscript "c") phase the impulses I_i^c, I_{i+1}^c, I_{i+2}^c and the other jumps of the velocities are obtained solving the modified linear equation system of the IME. The superscripts for compression and restitution "r" will be used if it is needed for the explanations. The IME in case of stiction for the ith coordinate are:

$$\begin{bmatrix} \underline{\boldsymbol{M}}\langle i_{st}\rangle & -\boldsymbol{\Phi}_{\boldsymbol{Q}}^{\mathrm{T}} \\ \underline{\boldsymbol{\Phi}}_{\boldsymbol{Q}}\langle i_{st}\rangle & \boldsymbol{0}_m \end{bmatrix} \cdot \begin{bmatrix} \Delta\underline{\dot{\boldsymbol{Q}}}\langle i_{st}\rangle \\ \boldsymbol{I}_\lambda \end{bmatrix} + \begin{bmatrix} \boldsymbol{M}_i & \boldsymbol{M}_{i+1} & \boldsymbol{M}_{i+2} \\ \boldsymbol{0}_m & \boldsymbol{0}_m & \boldsymbol{0}_m \end{bmatrix} \cdot \begin{bmatrix} \Delta\dot{q}_i \\ \Delta\dot{q}_{i+1} \\ \Delta\dot{q}_{i+2} \end{bmatrix} = \boldsymbol{0}_{n+m} \tag{26}$$

The modified matrices $\underline{\boldsymbol{M}}\langle i_{st}\rangle$ and $\underline{\boldsymbol{\Phi}}_{\boldsymbol{Q}}\langle i_{st}\rangle$ coincides with the same matrices in Section 5.1. The matrix of the velocity jumps $\Delta\dot{\boldsymbol{Q}}$ is transformed to matrix of the unknown parameters that include the impulses in the point of impact, i. e.:

$$\Delta\underline{\dot{\boldsymbol{Q}}}\langle i_{st}\rangle = \begin{bmatrix} \Delta\dot{q}_1 & \Delta\dot{q}_2 & \dots & I_i & I_{i+1} & I_{i+2} & \Delta\dot{q}_{i+3} & \dots & \Delta\dot{q}_n \end{bmatrix}^{\mathrm{T}} \tag{27}$$

The free part of the linear equation system (26) is obtained by simple multiplication of columns *i*, *i*+1, *i*+2 of matrix $\boldsymbol{M}$ by the known jumps of the velocities in the contact point. The inertia and mass parameters of the system are the reason for tangential impulse I_τ,

$$I_\tau = \sqrt{I_{i+1}^2 + I_{i+2}^2} \tag{28}$$

calculated from equation (26). The conditions for stiction are:

$$\left|I_\tau\right| \le \mu_0 \cdot \left|I_i\right|,\ \left\|v_\tau\right\| = 0 \tag{29}$$

$$\left|I_\tau\right| \le \mu \cdot \left|I_i\right|,\ \left\|v_\tau\right\| > 0 \tag{30}$$

Using the Newton's coefficient of restitution the velocity jumps $\Delta\dot{q}_i^r = e \cdot \Delta\dot{q}_i^c$, $\Delta\dot{q}_{i+1}^r = e \cdot \Delta\dot{q}_{i+1}^c$, $\Delta\dot{q}_{i+2}^r = e \cdot \Delta\dot{q}_{i+2}^c$ are calculated and the same IME (26) is solved for the restitution phase.

5.2.1.2. Hypothesis of Poisson. The hypothesis of Poisson defines the coefficient of restitution by the ratio $e_p = I_i^r / I_i^c$ of the accumulated normal impulses I_i^r, I_i^c at the phases of restitution and compression, respectively. Obviously, both hypotheses demand the solution of common modified IME (24) for the compression phase. Solving equation (26) for the compression phase we obtain the impulses I_i^c, I_{i+1}^c, I_{i+2}^c and using the data for the Poisson's coefficient of restitution we obtain the impulses $I_i^r = e_p \cdot I_i^c$, $I_{i+1}^r = e_p \cdot I_{i+1}^c$, $I_{i+2}^r = e_p \cdot I_{i+2}^c$ for the restitution phase. The impulses in the point of contact that should be reversed at the restitution phase build the free part of the following IME

$$\begin{bmatrix} \boldsymbol{M} & -\boldsymbol{\Phi}_{\boldsymbol{Q}}^{\mathrm{T}} \\ \boldsymbol{\Phi}_{\boldsymbol{Q}} & \boldsymbol{0}_m \end{bmatrix} \cdot \begin{bmatrix} \Delta\dot{\boldsymbol{Q}} \\ \boldsymbol{I}_\lambda \end{bmatrix} - \begin{bmatrix} \boldsymbol{0}_{i-1} \\ I_i \\ I_{i+1} \\ I_{i+2} \\ \boldsymbol{0}_{n+m-i-2} \end{bmatrix} = \boldsymbol{0}_{n+m} \tag{31}$$

for the unknown velocity jumps $\Delta\dot{\boldsymbol{Q}}$ and impulses $\boldsymbol{I}_\lambda$ of the constraints.

5.2.2. Sliding

5.2.2.1. Hypothesis of Newton. If sliding occurs, the tangential frictional impulses either at the compression or at the restitution stage depend on the normal impulse I_i and the kinetic coefficient of friction μ. The directions of the frictional impulses I_{i+1}, I_{i+2} depend on the relative tangential velocity and the kinetic coefficient of friction, i. e.:

$$I_{i+m} = -\mu \cdot \frac{v_{\tau_m}}{\|v_\tau\|} . I_i = -\mu \cdot \frac{\dot{q}_{i+m}}{\sqrt{\dot{q}_{i+1}^2 + \dot{q}_{i+2}^2}} \cdot I_i = -\mu_{i+m} \cdot I_i ;\ m = 1, 2 \tag{32}$$

With know jump $\Delta\dot{q}_i^c$ of the normal velocity at the compression phase the IME are modified as follows:

$$\begin{bmatrix} \underline{\boldsymbol{M}}\langle i_{sl}\rangle & -\boldsymbol{\Phi}_{\boldsymbol{Q}}^{\mathrm{T}} \\ \underline{\boldsymbol{\Phi}}_{\boldsymbol{Q}}\langle i_{sl}\rangle & \boldsymbol{0}_m \end{bmatrix} \cdot \begin{bmatrix} \Delta\underline{\dot{\boldsymbol{Q}}}\langle i_{sl}\rangle \\ \boldsymbol{I}_\lambda \end{bmatrix} + \begin{bmatrix} \boldsymbol{M}_i \\ \boldsymbol{0}_m \end{bmatrix} \cdot \Delta\dot{q}_i = \boldsymbol{0}_{n+m} \tag{33}$$

The modified matrices $\underline{\boldsymbol{M}}\langle i_{sl}\rangle$ and $\underline{\boldsymbol{\Phi}}_{\boldsymbol{Q}}\langle i_{sl}\rangle$ are the same defined in Section 5.1. The modified matrix of the parameters is:

$$\Delta\underline{\dot{Q}}\langle i_{sl}\rangle = [\Delta\dot{q}_1 \quad \Delta\dot{q}_2 \quad \cdots \quad I_i \quad \Delta\dot{q}_{i+1} \quad \cdots \quad \Delta\dot{q}_n]^{\mathrm{T}} \tag{34}$$

The solution of equation (33) gives the velocity jumps of the coordinates, the normal impulse I_i^c at the compression phase, and the impulses of the constraints. Using the coefficient of restitution the jump of the normal velocity $\Delta\dot{q}_i^r$ at the restitution phase is calculated and used as known parameter in the same linear equation system (33) which is to be solved for $\Delta\underline{\dot{Q}}^r$ of the unknown velocity jumps and normal impulse I_i^r at the restitution phase.

5.2.2.2. Hypothesis of Poisson. The normal impulse I_i^c obtained solving the linear system (33) and the coefficient of restitution of Poisson are used for calculation of the normal impulse I_i^r at the time of the restitution, which impulse is in the free part of the IME

$$\begin{bmatrix} \boldsymbol{M} & -\boldsymbol{\Phi}_{\boldsymbol{Q}}^{\mathrm{T}} \\ \boldsymbol{\Phi}_{\boldsymbol{Q}} & \boldsymbol{0}_m \end{bmatrix} \cdot \begin{bmatrix} \Delta\dot{\boldsymbol{Q}}^r \\ \boldsymbol{I}_\lambda \end{bmatrix} - \begin{bmatrix} \boldsymbol{0}_{i-1} \\ I_i \\ \boldsymbol{0}_{n+m-i} \end{bmatrix} = \boldsymbol{0}_{n+m} \tag{35}$$

that should be solved for the unknown velocity jumps $\Delta\dot{Q}^r$.

5.2.3. Dominant velocity jump in the tangential plane.
A phenomenon we discuss here could appear in case of reverse sliding with non-zero approach tangential velocity. This event is obtained either in the compression or in the restitution phase. The physical nature of the problem could be easily understood if impact is regarded in two-dimensional space, which means that the tangential velocity has only one component along one tangential axis. If this velocity at the beginning of impact is non zero and there is no condition for instant stiction one could obtain resultant tangential velocity with different sign of the initial one. This means that the tangential velocity jump is greater than the initial tangential velocity and with different sign, i.e. during the impact the tangential velocity becomes zero and changes it sign afterwards. For example, if compression phase is analyzed, this means that the tangential velocity goes through its zero value before the normal velocity. In this case, the IME should be solved for the known jump of the tangential velocity till its zero value, i.e. for "dominant jump of the tangential velocity".

If dominant tangential velocity jump should be considered, the modified IME are as follows:

$$\begin{bmatrix} \underline{\boldsymbol{M}}\langle i_{tj}\rangle & -\boldsymbol{\Phi}_{\boldsymbol{Q}}^{\mathrm{T}} \\ \underline{\boldsymbol{\Phi}}_{\boldsymbol{Q}}\langle i_{tj}\rangle & \boldsymbol{0}_m \end{bmatrix} \cdot \begin{bmatrix} \Delta\underline{\dot{\boldsymbol{Q}}}\langle i_{tj}\rangle \\ \boldsymbol{I}_\lambda \end{bmatrix} + \begin{bmatrix} \boldsymbol{M}_{i+1} \\ \boldsymbol{0}_m \end{bmatrix} \cdot \Delta\dot{q}_{i+1} = \boldsymbol{0}_{n+m} \tag{36}$$

where $\underline{M}\langle i_{tj} \rangle$ is the mass matrix modified according to the known jump for the tangential velocity. Solving the IME (36) we obtain only a part of the whole normal velocity jump (Newton's hypothesis) or of the whole normal impulse (Poisson's hypothesis). The remaining part of the normal velocity jump (or the normal impulse) should be realized checking, because of zero tangential velocity so obtained, for stiction and if not then solving the IME in case of sliding.

5.3. NUMERICAL ALGORITHM FOR IMPACT PROBLEM SOLUTION

The compression and restitution phases have its own initial and final velocities. The initial velocities for the restitution phase are the final velocities for the compression phase. The solution of the rigid-body system impact problem will be achieved using one and the same numerical algorithm for the two phases, compression and restitution, of the impact event. Three-step numerical procedure is developed.

Step A - with the initial normal and tangential velocities (Newton's hypothesis) or normal impulse (Poisson's hypothesis) the modified IME are solved for the case of stiction. If the conditions for stiction are:

- fulfilled, then ***the solution so obtained is the final solution***;
- not fulfilled, then the next step follows.

Step B - with the same initial tangential and normal velocities or normal impulse, the modified IME are solved in case of sliding. If the condition for dominant tangential velocity jump is:

- not fulfilled, then ***the solution so obtained is the final solution***;
- fulfilled, then the modified IME are solved in case of dominant tangential velocity jump and the impact analysis continues with the next step.

Step C - the solution so obtained is used the new values of the initial normal velocity or impulse to be updated and with the tangential velocity equal zero involved in a new cycle of calculations starting with Step A.

By the algorithm presented here all possible impact events are numerically simulated, i. e.:

(a) compression phase
- instant stiction with no relative approach tangential velocity;
- instant stiction with relative approach tangential velocity;
- sliding with no relative approach tangential velocity;
- sliding;
- sliding and reverse sliding;
- sliding and stiction;

(b) restitution phase
- continuos stiction;
- continuous sliding;
- continuous sliding and reverse sliding;
- continuous sliding and stiction.

5.4. DIRECTION OF FRICTIONAL IMPULSE IN CASE OF SPATIAL IMPACT

While the numerical algorithm suggested in the article is in good agreement with the conditions for plane impact simulation, because of existing only one tangential line in the contact point, in case of spatial impact (existing of tangential plane) the sliding process is different. In Fig. 2, a case of spatial impact with sliding of two bodies of a multibody system is shown. In the figure, the initial $v_\tau^{(-)}$ and final $v_\tau^{(+)}$ sliding velocities are also presented. According to the one step approach of impact simulation in case of sliding, in the IME the tangential impulse I_τ is included with its direction defined by the direction of the initial velocity $v_\tau^{(-)}$. But, in general, the initial and final velocities in the tangential plane at the time of spatial impact are not collinear and the direction of impulse I_τ, quite the contrary of the plane impact, is not constant.

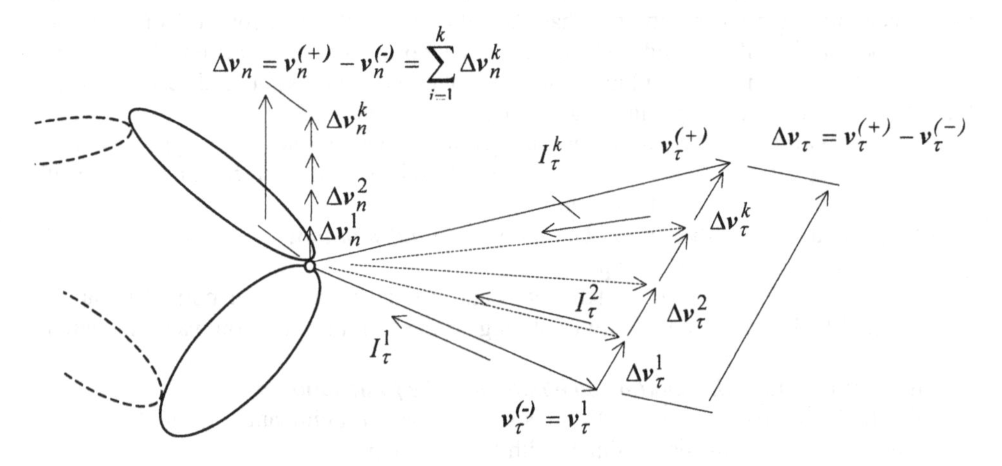

Figure 2. Spatial impact and sliding of two bodies of a myltibody system

Here a numerical procedure is suggested, where the velocity jump Δv_n, respectively Δv_τ (hypothesis of Newton), is discretized on final number of intervals Δv_n^i, $i = 1, 2,\dots k$. For all of this these discrete values of the normal velocity jump the IME are solved with the current direction of the frictional impulse I_τ^i, $i = 1, 2,\dots, k$ that correspond to the current values of the velocity $v_\tau^{i+1} = v_\tau^i + \Delta v_\tau^i$. For these discrete values of the velocity jumps the three-step numerical procedure presented in Section 5.3 is applied and the possible events are analyzed. In case of transition effect, for which the tangential velocity goes through zero (sliding – zero velocity – sliding or stiction), for the sake of higher precision in the transition point, the step of discretization is to be decreased.

The similar algorithm could be used if the hypothesis of Poisson for the coefficient of restitution is applied. In this case the discretization is achieved for the normal impulse while the velocities, respectively the frictional impulses, are obtained step by step as a result of the solution of IME.

6. Frictional Contact of Flexible Multibody Systems

In the finite element theory the flexible bodies are discretized of flexible elements and nodes and their position approximate the shape of the flexible bodies. Although detection of contact is not a subject of this article, some examples of definition of the tangential plane and its normal will be presented. So, the algorithm for rigid body contact dynamics simulation could be applied in terms of the definitions for the flexible systems.

The most simplest case of definition of the contact point and its tangential plane and normal is the contact of flexible body, respectively a node, to a rigid body (Fig. 3 a). The contact point of the rigid body, if detected, defines uniquely the plane and its normal. The nodes of tetrahedron (Fig 3 b) or prismatic elements (Fig. 3 c) that are over the surface of the flexible body define a mash of hyper-surface patches. If the elements are tetrahedrons the surfaces of these patches are planes and the normal direction is one and the same for the all the patch. If the elements are prismatic the patches are nonlinear surfaces defined by the shape function of the corresponding elements. The normal could be obtained differentiating the shape function in the contact point detected. Significant simplification of the procedure for contact detection, as well as, estimation of the tangential plane and normal, could be achieved if the coordinate systems that pertain to the flexible elements are fixed in the nodes that build the body surface (Fig. 3 b, c, d). So, for tetrahedron, prismatic and shell elements, one of the coordinate system plane (in the figures it is *XY* plane) coincides with the patch of the body surface. For prismatic and shell elements, although the patch is nonlinear-surface, it also could be assumed coincident with *XY* plane as the relative flexible deflections of the element nodes are assumed small. In case of beam elements (Fig. 3 e) the possible contact point, in the general case, does not coincide with a node. So, the unity vectors of the beam axes

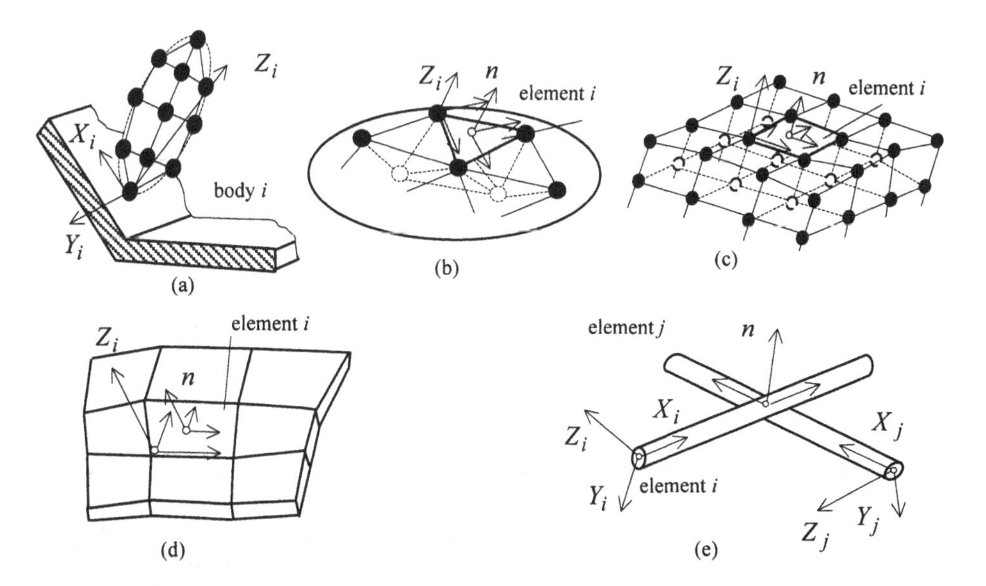

Figure 3. Flexible elements and definition of contact point tangential planes.

define the tangential plane while the normal is defined by their cross product. The readers familiar with the contact problems will conclude that the definitions of the normal penetration and the normal in the tangential plane are very similar. Of course, there are many difficulties and one of it is if a contact point (node) goes from one element to another.

In the finite element theory the deflections of the nodes are with respect to the inertial reference frames. Absolute coordinates are very appropriate if a flexible body comes into contact with a body that coincides with the inertial reference frame. But if contact occurs between two moving bodies, the contact should be described in terms of relative coordinates. The author of this treatment suggested in [21] an approach of relative coordinates for modeling of rigid and flexible multibody system dynamics. This method defines at every time step the position of the body or element coordinate system, respectively the contact tangential plane and its normal. So, the contact of flexible systems could be described as a contact of a node of a flexible body with an element of another body. Respectively, the possible motions of this node are with respect to the coordinate system of the element of the other body (Fig. 4 a) and these motions are included in matrix – vector $\boldsymbol{Q}$ of the coordinates. It could be easily verified that this presentation allows the algorithm suggested here to be directly applied for simulation of contact of flexible bodies. In case of contact of flexible beam elements (Fig. 4 b) the coordinates that describe the possible motions in the contact point are the translations along the beam axes and the common normal.

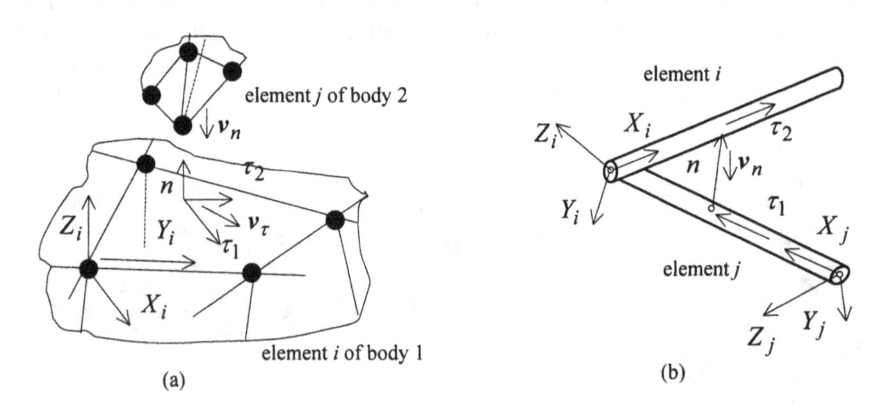

Figure 4. Incoming contact of flexible elements

7. Examples

The author applied [22] the numerical algorithm presented to rigid multibody systems, results being analyzed and compared with experimental investigation [13]. Here, the method is applied to flexible systems and multiple contact points. Two examples of motion simulation of flexible bodies that achieve large deflections are presented. The results are compared to these obtained by the method of the contact forces. One and the same general – purpose computer program [21] for deriving the dynamic equations and the IME is applied.

7.1. FALLING FLEXIBLE BEAM OVER TWO RIGID SURFACES

In Fig. 5, a slender beam and locations of two rigid surfaces imposing constraints of the beam motion are shown. All measures in the examples are in SI UNITS. The beam is falling down with the vertical velocity $v = -1$. The parameters of the beam are: length $L = 15$; modulus of elasticity $E = 2.1 \cdot 10^{11}$; cross section area S = 0.002; moment of inertia of the cross section $I = 1. \cdot 10^{-6}$; mass density ρ = 8690. The beam is divided into 30 elements and 31 nodes and has 93 degree of freedom. Simultaneous impacts of beam nodes over two rigid surfaces are achieved. After the impact, the nodes remain in contact till the conditions for existence of unilateral constraints are fulfilled. The snapshots of the large beam deflections are shown in the figure.

The same example is also solved using the method of the applied forces. In this case, the contact is not clearly defined and oscillations for the nodes 14 and 16 in the normal directions of the surfaces are observed and presented in Fig. 6.

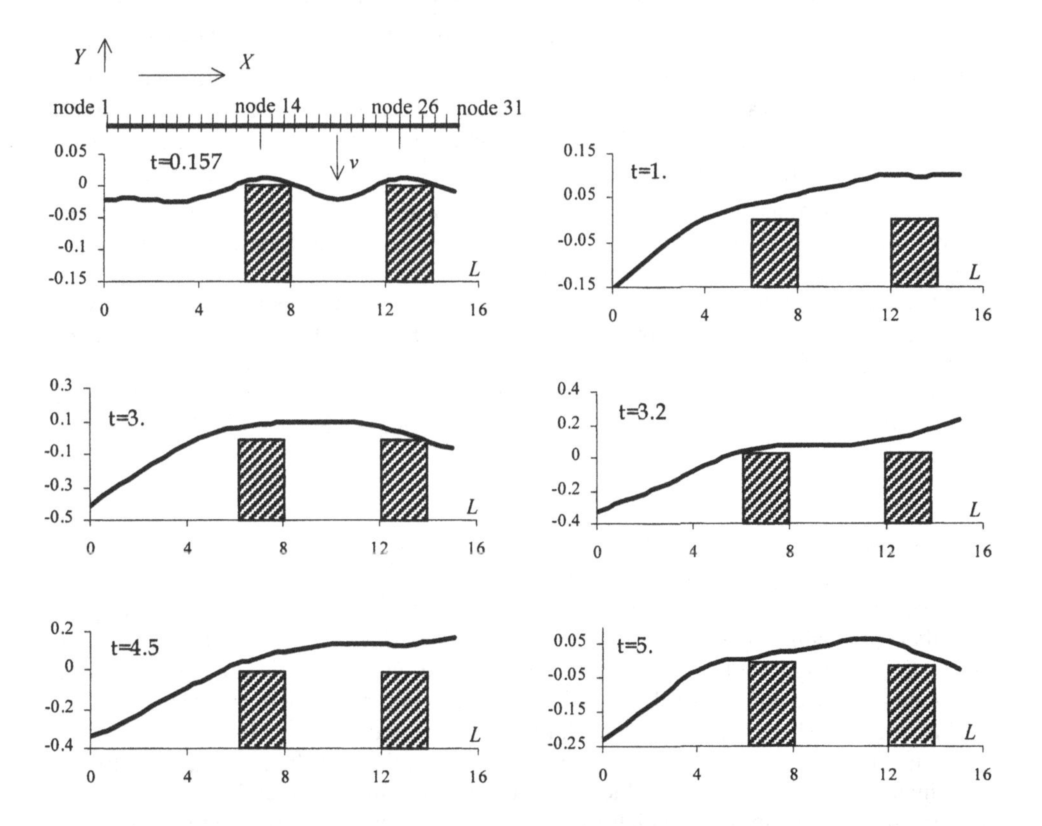

Figure 5. Impact over two rigid surfaces and large deflections of a slender beam

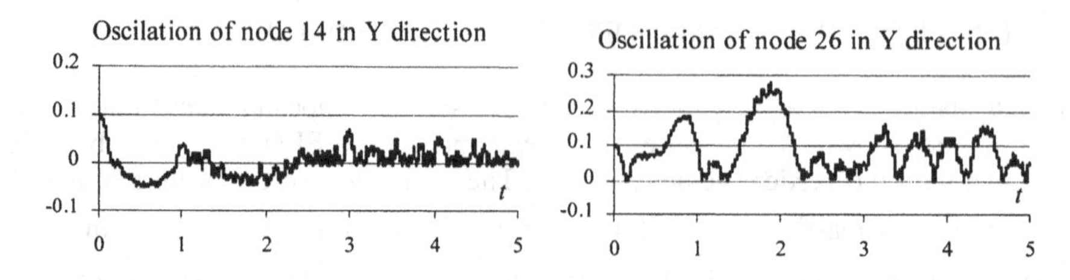

Figure 6. Time history of the oscillations of node 14 and 26 in vertical (Y) direction

7.2. FALLING SLENDER RING

In Fig. 7, a falling ring made of slender rod is shown. The radius of the ring in its not deformed position is 2.3917. The parameters of the rod are: modulus of elasticity $E = 7.\cdot 10^{10}$; cross section area S = 0.0018; moment of inertia of the cross section $I = 1.2151\cdot 10^{-8}$; mass density ρ = 2690. The ring is divided into 30 elements and 30 nodes and has 90 degree of freedom. The initial position of its center at time $t = 0$ is shown in the figure. The ring rotates with angular velocity $\omega = 1$. Friction is taken into account. The friction coefficient is $\mu = 0.6$. In the figure, a small part of the ring is not drawn and its rotation could be easily observed. The coefficient of friction provides instant stiction of the nodes at their impact over the rigid surface. In the figure snapshots of the ring motions are also presented.

The same example is solved using the method of the applied contact forces. There is no visible difference between both solutions. It could be stated that, using this approach, neither contact in the normal direction nor stiction in the tangential plane could be clearly defined. In Fig. 8 the oscillation at time interval 0.6 – 1.5 along the normal, as well as, the oscillation of node 27 in the tangential plane, where stiction should be, are presented.

8. Conclusions

A novel numerical method for simulation of contact dynamics of rigid and flexible multibody systems using the classical approach of DAE and IME with the Lagrange's multipliers is suggested and analyzed. Transformation of the dynamic equations is implemented. The number of the dynamic equations and parameters is constant as it is for the method of the applied contact forces. The method is very effective for flexible systems and finite element discretization, since the motions of the nodes are described by kinematic parameters that coincide with the kinematic parameters of the contact. Many simultaneous existing unilateral constraints and impact points are regarded. The events sliding, stiction, stic-slip are effectively simulated. A new definition for "dominant tangential velocity jump" is suggested with which the impact event of transition sliding – stiction is truly simulated.

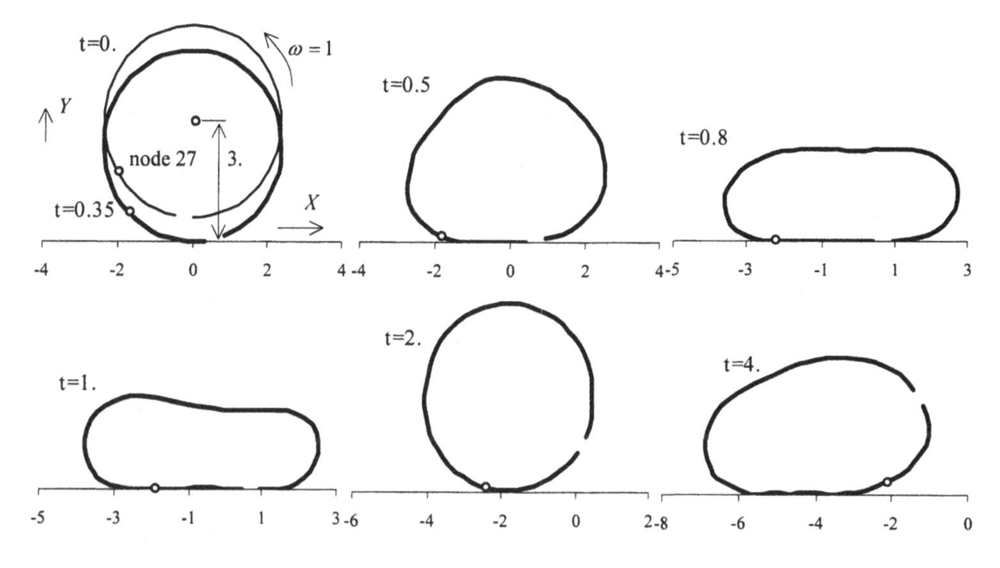

Figure 7. Impact and large deflections of slender ring

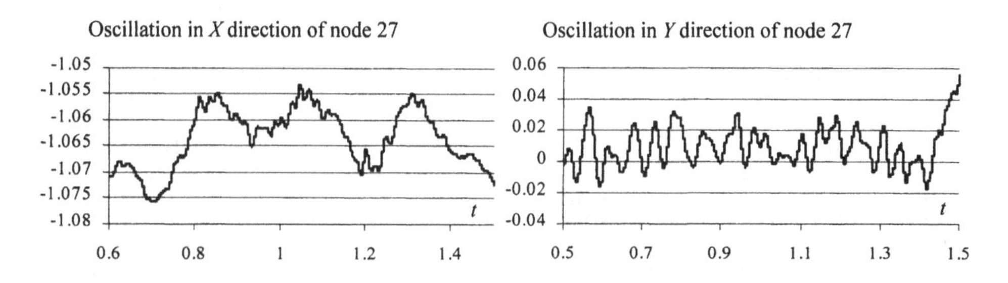

Figure 8. Oscillation of node 27 in the tangential plane (X) and along the normal (Y)

Examples of motion simulation of flexible systems including simultaneous impact and contact in many points are presented, and the results obtained are analyzed. Although the method of the applied contact forces is easy to be implemented, the results obtained using the methods of the Lagrange's multipliers and IME suggested here are more precise and reliable.

Acknowledgments

The research work reported here was made possible under Grant B33/1999 of the German Research Council.

The financial support of Natural Science and Engineering Research Council of Canada (2001) for investigation of flexible multibody beam systems, part of this article, is appreciated.

The author acknowledged the financial support of NATO SEAD for presentation of this lecture.

References

1. Haug, E.J., Wu, S.C. and Yang, S.M. (1986) Dynamics of Mechanical Systems with Coulomb Friction, Stiction, Impact and Constraint Addition-Deletion. Part 1: Theory, *Mechanism and Machine Theory*, **21**, 401-406.
2. Glocker, Ch. and Pfeiffer, F. (1992) Dynamic systems with unilateral contacts, *Nonlinear Dynamics*, **9**, 245 - 259.
3. Glocker, Ch. and Pfeiffer, F. (1994) Stick - slip phenomena and application, in *Nonlinearity and Chaos in Engineering Dynamics, IUTAM Symposium, UCL*, John Wiley, New York, 103 - 113.
4. Seyfferth, W. and Pfeiffer, F. (1994) Modeling of Time - Variant Contact Problem in Multibody systems, *Proceedings of 12th Symposium on Engineering Applications of Mechanics, Montreal*, June 27 - 29, 579 - 588.
5. Wosle, M. and Pfeiffer, F. (1996) Dynamics of Multibody Systems Containing Dependent Unilateral Constraints with Friction, *Journal of Vibration and Control*, **2**, 161 - 192.
6. Hunt, K.H. and Grossley, F.R.E. (1975) Coefficient of Restitution Interpreted as Damping in Vibroimpact, *ASME Journal of Applied Mechanics*, 440-445.
7. Keller, J.B. (1986) Impact with Friction, *ASME Journal of Applied Mechanics*, **53**, 1-4.
8. Brach, R.M. (1989) Rigid body collision, *ASME Journal of Applied Mechanics*, **56**, 133-138.
9. Wang, Y. and Mason, M.T. (1992) Two Dimensional Rigid-Body Collisions with Friction, *ASME Journal of Applied Mechanics*, **59**, 635-642.
10. Glocker, C. and Pfeiffer, F. (1995) Multiple Impacts with Friction in Rigid Multibody Systems, *Nonlinear Dynamics*, **7**, 471-497.
11. Pfeiffer, F. and Glocker, C. (1996) *Multibody Dynamics with Unilateral Contacts*, JOHN WILEY & SONS, INC,.
12. Schiehlen, W. (1998) Unilateral Contacts in Machine Dynamics, *IUTAM Symposium Unilateral Multibody Dynamics*, Munich, August 3-7, Kluwer Academic Publishers, 1-12.
13. Lankarani, H.M. and Pereira, M.F.O.S. (1999) Treatment of Impact with Friction in Multibody Mechanical Systems, in J. A.C. Ambrosio and W. O. Schiehlen (eds.), *Proc. of EUROMECH Colloquium 404 Advances in Computational Multibody Dynamics*, IDMEC/IST, Lisbon, Portugal, September, 20-23.
14. Ambrosio, A.C.J. and Pereira, M.S. (1997) Flexible Multibody Dynamics with Nonlinear Deformations: Vehicle Dynamics and Crashworthiness Application, in J. Angeles and E. Zakhariev (eds.), *Computational Methods in Mechanical Systems: Mechanism Analysis, Synthesis and Optimization*, NATO ASI, Ser. F, **161**, Springer Verlag, Berlin, 382-419.
15. Stronge, W.J. (1994) Planar Impact of Rough Compliant, *Int. J. Impact.Eng.*, **15**, 435 – 450.
16. Stronge, W.J. (1995) Theoretical Coefficient of Restitution for Planar Impact of Rough Elasto-Plastic Bodies, in R.C. Batra, A.K. Mal and G.P. MacSithigh (eds.) *AMD. Impact, Waves, and Fracture*, **205**, Book H00952, 351-362.
17. Kikuchi, N. (1982) A Smoothing Technique for Reduced Integration Penalty Methods in Contact Problems, *Int. Journal of Numerical Methods in Engineering*, **18**, 343 – 350.
18. Oden, J.T. and Kikuchi, N. (1982) Finite Element Methods for Constrained Problems, *Int. Journal of Numerical Methods in Engineering*, **18**, 701 – 725.
19. Yagawa, G. and Kanto, Y. (1993) Finite Element Analysis of Contact Problems using the Penalty Function Method, in M.H. Alibadi and C.A. Brebbia (eds.) *Computational Methods in Contact Mechanics*, Computational Mechanics Publications, Elsevier Applied Science, 127-153.
20. Eberhard, P. and Jiang, S. (1997) Collision Detection for Contact Problems in Mechanics, *Institutsbericht IB - 30*, Institute B of Mechanics, University of Stuttgart.
21. Zahariev, E.V. (2002) Relative Finite Element Coordinates in Multibody System Simulation, *Multibody System Dynamics*, **7**, 51 – 77.
22. Zahariev, E. V. (2001) A Numerical Method for Multibody System Frictional Impact Simulation, *Proc. of DETC'01 ASME 2001 Design Engineering Technical Conferences and Computers and Information in Engineering Conference*, Pittsburgh, Pennsylvania, Sept. 9-12, 1-10.

PARALLEL COMPUTING IN THE CONTEXT OF MULTIBODY SYSTEM DYNAMICS

ANDREAS MÜLLER
Institute of Mechatronics
at the Chemnitz University of Technology
Reichenhainer Str. 88, 09126 Chemnitz, Germany
Andreas.Mueller@ifm.tu-chemnitz.de

1. Distributed simulation of MBS dynamics

The need for high performance simulation of the dynamics of large MBS's is a widely recognized issue stimulated by demands from a variety of different application areas such as interactive real time virtual reality simulation, model based control and of course the design and development process. In particular the development of complex mechatronic systems calls for highly flexible simulation tools which are reconfigurable and model independent. Several interactive tools for the simulation of MBS dynamics exist (Adams, alaska, NewEul, Mobile), which are commonly intended to support the design process of one particular model but not for case studies, model fitting or even MBS optimizations. Approaches to the optimization of complex systems have always been tailor made implementation specific to the problem at hand. A general treatment was not attempted yet.

In general it is left to the engineer to employ novel optimization strategies to their problems at hand. As such he or she shall have an in-depth knowledge in his or her specialized subject but also a broad overview of developments in the field of optimization. It is nowadays very likely that a mechanical engineer must be able to compile computer programs to simulate a particular problem at hand the phenomenology of which he or she presumably is aware of. Indeed many scientists do have this ability but it is in general not very likely for the majority of engineers that he or she is an expert in both fields. In particular when the use of advanced computation facilities is desired one has to carefully divide the different disciplines. In this paper a general approach to simulation/optimization of multibody dynamics on parallel computing facilities is presented. Its hier-

W. Schiehlen and M. Valášek (eds.), Virtual Nonlinear Multibody Systems, 403–410.

archic structure releases the user from dealing with implementation issues and allows to concentrate on the MBS modeling and the straight forward simulation/optimization. This is owed to an adherence of defined interface mechanisms between the modules of the developed system.

The use of parallel computing facilities (PCF) is well established for the numerical simulation of continuum mechanical, fluid mechanical as well as electromagnetic field problems. This is because the large number of degrees of freedom of the mathematical models can be immediately distributed on a parallel computing grid. However, though PCF have not been seriously employed in the context of MBS simulations, PCF are also potentially advantageous in many respects for the MBS dynamics simulation and optimization. The classical single-model/single-processor simulation systems may be extended to evaluate the instantaneous kinematics, kinetics and dynamics exploiting the MBS topology for a distributed evaluation of the motion equations employing very time efficient parallel $O(n)$ algorithms [2]. But the necessary computing resources are not justifiable since it could only speed up the dynamics simulation of one MBS model. On the other hand the entire dynamics of one MBS model can be simulated per processing node. In this way autonomous running MBS models constitute a **task farm**. A combination of both approaches, i.e. model instances on this task farm use parallel $O(n)$ algorithms, is usually not possible with the current state of the art technology.

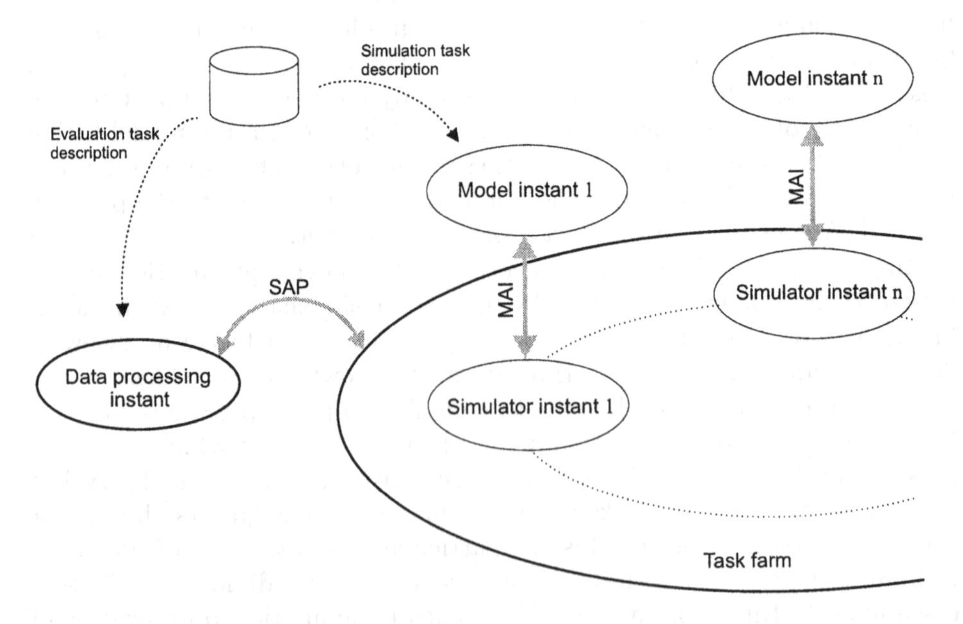

Figure 1. Task farm concept, where several simulator/model instances deliver simulation results to a data processing instance.

Such a task farm plus a superordinated controller/processing instance constitute a **MBS simulation grid**. The aim of the simulation grid is to provide MBS simulation results and incorporate these in data processing tasks. Since each node on the task farm can be considered as a stand-alone simulation tool this methodology shall feature the complete functionality of established simulation packages. Consequently a simulation may be every combination of possible task that a simulation package could perform, e.g. kinematic/dynamic simulation, linear analysis or equilibrium determination. In this way it is possible to perform full simulations of several instances of a parameterized MBS model in parallel.

One single controller/processing instance governs the task farm and serves the model instances on that task farm with necessary parameters. Incoming simulation results are processed by the controller/processing instance, they could simply be stored for later use or model parameters could be optimized to achieve a desired behavior.

A simulation grid for MBS was developed at the Edinburgh Parallel Computing Center (EPCC), Edinburgh, Scotland, UK in cooperation with the University of the Federal Armed Forces, Hamburg, Germany. The MBS modelling is supported by the interactive modelling and simulation system alaska. This system provides MBS models in a suitable form for the task farm.

2. Components of a distributed simulation environment

2.1. CONTROLLER AND PROCESSING INSTANCE (CPI)

The controller/processing instance (CPI) controls the overall simulation grid. It is the only instance of the simulation grid accessible from the external environment. The CPI is the actual data processing unit that provides the task farm with model and simulation parameters. MBS simulation results are obtained by the simulator instances, the data processing is accomplished by the single CPI.

Typical examples for data processing strategies are the parameter variation (collection of simulation results for different parameters) and the optimization of MBS kinematics or dynamics with respect to specified model parameters. From the CPI point of view the task farm entries are simply parameterized input-output relations. Thus the task farm entries may be any instances, not necessary MBS simulators, compliant to the communication framework described below.

2.2. SIMULATOR AND MODEL INSTANCE (SMI)

Each simulator/model instance (SMI) placed on one node of the farm is a composite of the specific MBS model and a simulation engine. Here MBS model means C-code which is generated by an interactive simulation tool. In this way the code fulfills interface specifications in order to ensure model independence. The accompanying simulator can be considered as the simulation kernel of a standard simulation packages so that it is able to carry out the same simulation tasks as a user might do interactively. The problem specific simulation tasks are accomplished by the SMI in batch simulation mode and described by a command file which is common to all SMIs on the farm. As such the SMI has the full simulation functionality of classical stand-alone simulation tools except their interactive modelling capabilities. The SMIs appear to the PCI as black boxes and the CPI strictly has no information about the particular simulations carried out the PCIs.

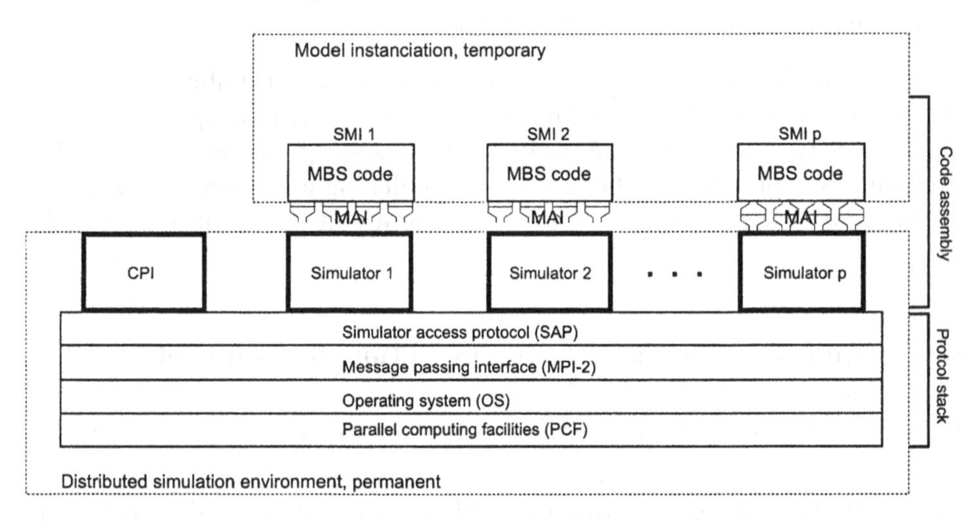

Figure 2. The different implementation levels of the system.

2.3. SIMULATOR ACCESS PROTOCOL (SAP)

One of the main challenges of the developed system is the coordination of SMIs by the PCI. For the sake of flexibility and generality the PCI and SMIs, each being an individual process on a PCF node, communicate via a simulator access protocol (SAP). This SAP is based on the MPI-2 framework for distributed computing systems [8]. The SAP approach ensures a maximum flexibility of the system because the actual PCI as well as the SMIs can be freely substituted as long as they are compliant with the SAP specification. Consequently the data processing of developed simulation grid is not limited in type and complexity, i.e. parameter variation and optimization of MBS are two applications only. It may further cater for the distributed simulation of cooperating systems.

2.4. MODEL ACCESS INTERFACE (MAI)

The SMIs consist of two parts the MBS model and the simulation engine. While the PCI and SMIs are coupled via the SAP (a software protocol) the MBS (the model C-code) must be linked to the simulation engines (C-library). The general conditions for this interconnection are defined by a model access interface (MAI). This is nothing but a predefined set of C-functions with defined calling conventions. Any MBS model fulfilling this MAI convention can be linked to the simulator, which are MAI compliant.

2.5. TASK DESCRIPTION

The consistency of the above instants and interfaces/protocols is essential for the independence with respect to the model and the employed hardware. The control of the grid as well of the simulations that are to be carried out by each simulator/model instance are described by an object oriented description language. That is the grid evaluation strategy (GES) and the model evaluation strategy (MES) are described in form of human-readable syntax. In particular the GES determine the assessment strategy of the grid, i.e. how simulation results are to be appraised and how model parameters are to changed. This is the actual optimization strategy in case of model optimization or the variation strategy in case of parameter variation. The MES is equivalent to the simulation control of the single model simulation tool that describes the simulation tasks to be carried out, such as assembling, integration or linearization.

3. MBS modeling and code generation

Crucial for an easy and straight forward implementation of MBS models on the task farm is the automatic generation of C-code fulfilling the MAI specification. The automatic generation of the model code has several advantages in terms of transparency, modularity and safety. One condition on the model description to facilitate this is a modular, or consequently object oriented, modeling [3,6]. Another condition is that the modelling and simulation tool that engineers use for interactive simulations is able to 'dump' its internal program flow for that MBS at hand in form of portable C-code which is also compliant with the MAI specification. This claim was achieved during the development of the simulation tool box alaska (www.tu-chemnitz.de/ifm).

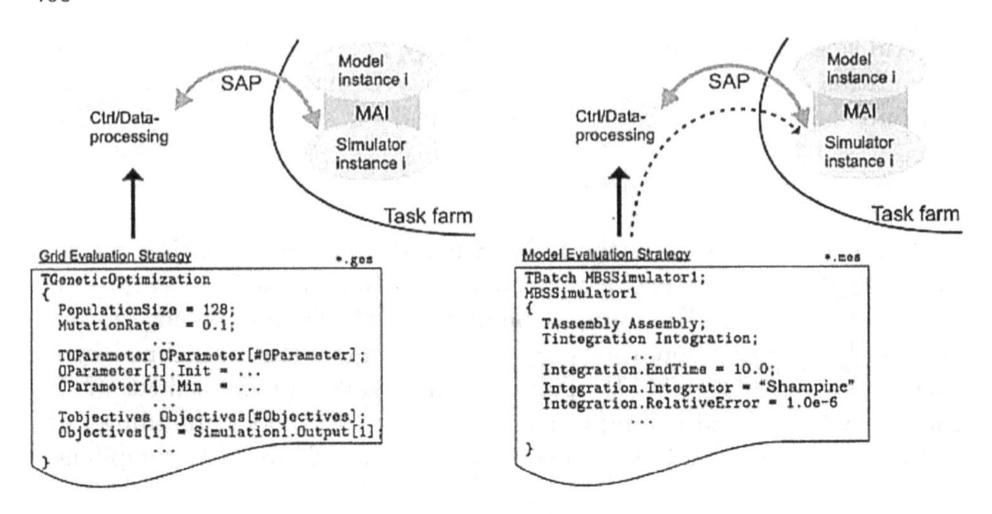

Figure 3. The GES determines the eploitation of the SMI's on the task farm. The set of simulation that are to be accomplished by each SMI is described in the MES.

4. Application: Nanometer coordinate measuring machine

The developed system was employed for the optimization of a fairly complex high precision coordinate measurement device [7] as part of the development process. A nanometer coordinate measuring machine (NCMM) combines the precision of nanometer measuring devices and the large workspace of conventional measuring machines. This is achieved by a novel cascadable setup. The machine is equipped with an atomic force microscope (AFM) as topography sensor. Obviously, the large scan volume contradicts the targeted high precision taking into account the used mechanical components. That is, the NCMM construction demands the use of high quality components to ensure high mechanical precision and the drive control must be able to rapidly reach a target position with very high accuracy. It turned out that the control of the NCMM is crucial and cannot be optimized by trial-and-error. Therefore the positioning system of an existing NCMM prototype was modelled as electromechanical rigid multibody system model with the alaska simulation package as shown in figure 4.

The controller parameter of the (existing) prototype were optimized using a genetic algorithm [4,5]. The typical population consisted of 127 SMIs, i.e. 127+1 processors were in use. The main objective was to minimize the overshooting effect during positioning of the AFM tip. The optimization goal was achieved after less then 40 cycles and the optimal controller parameter constellation now yields 70% less overshooting during the tip approach. Also the scan motion precision could be improved.

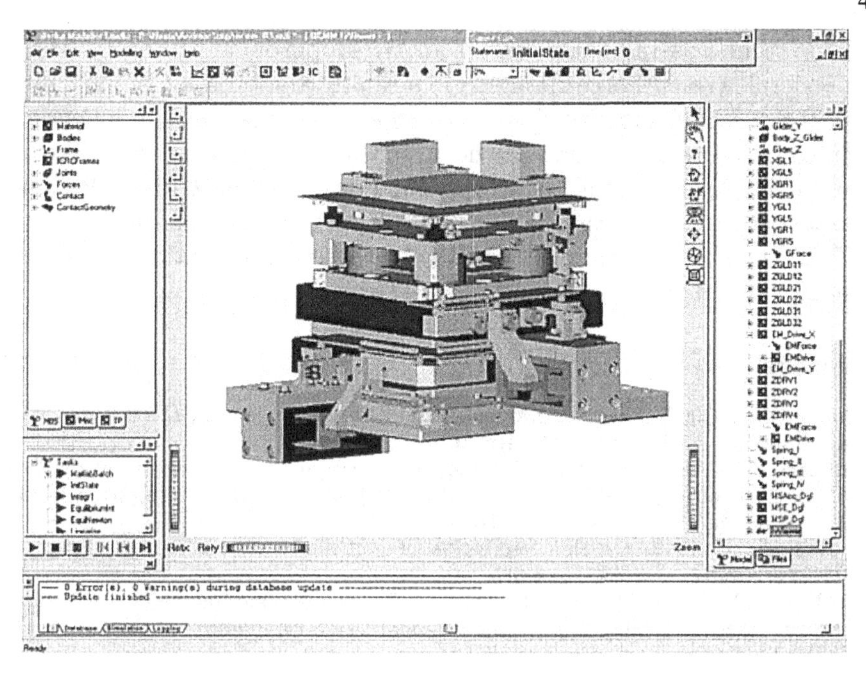

Figure 4. The MBS modell of the NCMM including electromagnetic drives and digital controllers.

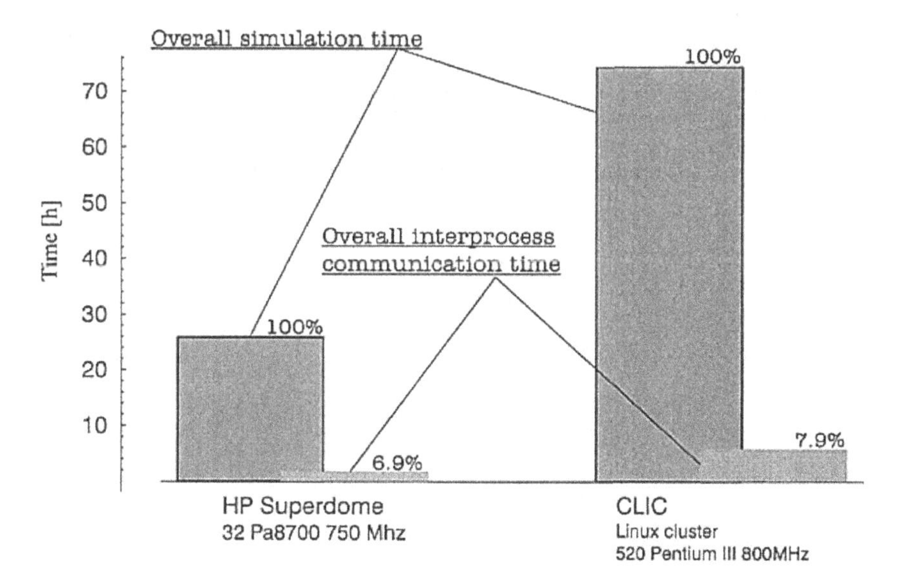

Figure 5. Cumputation time for an optimization run on an HP-Superdome and on the CLIC Chemnitz.

The system was implemented on an HP-Superdome at the University of the Federal Armed Force Hamburg, Germany and on the Cemnitz Linux Cluster (CLIC) in Chemnitz, Germany. The aimed hardware independence is justified in figure 5. It shows the necessary simulation time for an expressive number of runs on the HP-Superdome and on the CLIC. Obviously the performance scales very well with the performance of the respective PCF. That is, the performance loss due to the communication overhead is comparatively small.

References

1. DeJong, K.A. (2000) Adaptive system design: a genetic approach, *IEEE Trans. Syst., Man. and Cyber.*, **Vol. SMC-10 no. 9**
2. S. Duan, K.S. Anderson (2000) Parallel implementation of a low order algorithm for dynamics of multibody systems on a distributed memory computing system, *Engineering with Computers*, **Vol. 16**
3. A. Keil, *et. al* (1999) On the description of multibody system models, *Proc. EUROMECH Colloquium 404, Lisbon*
4. A. Müller, H. Rothe (2001) Optimization of controller parameter of a coordinate measuring machine, *Proc. 6. US national congress on computational mechanics (USNCCM) 2001, Dearborn, Detroit, USA*
5. A. Müller, A. Keil, H. Rothe, R. Petersen (2002) The dynamics of a coordinate measuring machine, *Proc. 1. International symposium on mechatronics (ISOM) 2002, Chemnitz, Germany*
6. U. Neerparsch (1996) Zur Standardisierung der Modellbeschreibung von Mehrkörperformalismen, *VDI-Berichte*, **no. 235**
7. H. Rothe, D. Hüser, and R. Petersen (2000) Simulation of the dynamic behavior of a nanometer coordinate measuring machine, *K. Hasche, W. Mirandé, and G. Wilkening, (ed.), Proceedings of the 4th Seminar on Quantitative Microscopy: Dimensional measurements in the micro- and nanometer range*, **Vol. PTB-F-39**
8. Message Passing Interface Forum (1994), MPI: A Message-Passing Interface Standard, *International Journal of Supercomputer Applications*, **Vol. 8 no. 3/4**
   For MPI related informations refer to http://www.mpi-forum.org

COUPLED MULTIBODY-AERODYNAMIC SIMULATION OF HIGH-SPEED TRAINS MANOEUVRES

A. CARRARINI (antonio.carrarini@dlr.de)
DLR - German Aerospace Center, Vehicle System Dynamics Group
P.O.Box 1116, D-82230 Wessling, Germany

Abstract. In this paper the effects of unsteady aerodynamic loads on the driving dynamics of high speed trains during passing manoeuvres in absence of cross wind have been investigated. To this end a co-simulation MBS/CFD was implemented. A linear aerodynamic model, the panel method, was applied to the computation of the unsteady flow around the driving trailers for the examined manoeuvres. The multibody simulation program SIMPACK simulated the dynamic response of the vehicles to the resulting aerodynamic loads.

Keywords: Multibody dynamics, railway aerodynamics, unsteady aerodynamics, co-simulation

1. Introduction

The most general way to include aerodynamic effects in a multibody system is the coupling of the multibody system code with a solver from computational fluid dynamics (CFD), see [1, 2]. Such *partitioned* approach, which is called *co-simulation* or *simulator coupling* when the coupled codes remain unchanged and completely stand-alone and communicate only through appropriate interfaces at discrete time points, see [3], is capable to describe virtually every *unsteady* aerodynamic phenomenon and to take into account the reciprocal interaction between mechanical and aerodynamical system.

A new application field for the coupled approach is the behavior of ground vehicles invested by transient, irregular flow, due for example to the interaction with other vehicles (*interference*), see [4, 5]. Such problems can not be handled by the conventional approach based on aerodynamic coefficients and present peculiar features like hysteresis in the transient forces. Because of their intrinsic complexity they are usually investigated through experiments, whereby an efficient numerical simulation, also aiming to define simplified modeling solutions, is still an open challenge. A typical case from this field – two high speed trains passing by each other – is presented in this paper.

The interest to this case is justified observing that very high operative speeds together with the cut off of leading car's weight, due to light construction and to the distribution of the traction units along the whole train,

W. Schiehlen and M. Valášek (eds.), Virtual Nonlinear Multibody Systems, 411–418.

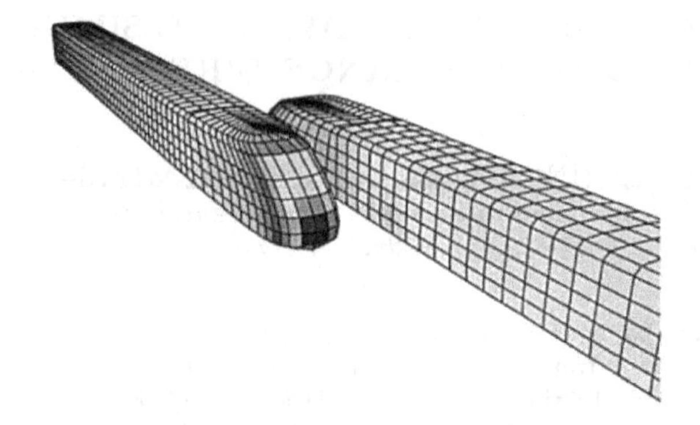

Figure 1. CFD applied to a passing manoeuvre on open track within a co-simulation CFD-MBS.

let today's trains be very sensitive to aerodynamic loads. For example, the driving trailers of many recent high speed trains are precautionary ballasted in order to reduce their aerodynamic sensitivity. The response of vehicles to steady and especially unsteady loads has thus to be carefully investigated to ensure the safety of railway operations under extreme aerodynamic conditions [6].

It must be mentioned that the methods of the multibody dynamics and their implementation in simulation software offer very efficient tools for the analysis of the dynamical behavior of railway vehicles. On the contrary the description of unsteady aerodynamic loads through CFD methods or wind tunnel experiments can be still achieved only with great efforts and high costs, in most cases with poor accuracy.

2. Basic principle of the co-simulation

The modular structure of coupled problems may be adopted in the simulation using for each subsystem its own simulation tool for model setup and time integration [7]. Well established standard software tools are used for the individual subsystems. In this way the subsystems are integrated by *different* time integration methods such that each of these methods can be tailored to the solution behavior of the corresponding subsystem.

The communication between subsystems is restricted to discrete synchronization points T_n. In each subsystem all necessary information from other subsystems can be provided by interpolation or — if data for interpolation are not yet available — by extrapolation from $t \leq T_n$ to the actual *macro step* $T_n \rightarrow T_{n+1}$. But in many cases it is sufficient to keep the value of the coupling variables from the other subsystems constant during the whole macro step $T_n \to T_{n+1}$. The latter is the usual approach adopted by the multibody system tool during the co-simulation and was also used for this study. The validity of this choice for the investigated problem was proved performing additional test computations with halved *macro step size* $H := T_{n+1} - T_n$ and verifying that the results remain sensibly unaltered. This also proved that the macro step size was accurately set for each specific simulation case.

It is important to note that co-simulation techniques are convenient but they may suffer from numerical instability. Furthermore, interpolation and extrapolation introduce additional discretization errors. In most standard applications stability and accuracy is guaranteed if the macro step size is sufficiently small.

For certain classes of coupled problems the instability phenomenon has been analyzed in great detail. Several modifications of the co-simulation techniques help to improve its stability, accuracy and robustness also for larger macro step sizes [8].

3. Formulation of the coupled problem

3.1. Multibody System

The classical topic of interest in multibody dynamics are systems of rigid bodies being connected by joints and force elements like springs and dampers [9]. The equations of motion are given by

$$\mathbf{M}(\mathbf{q})\,\ddot{\mathbf{q}}(t) = \mathbf{f}(t,\mathbf{q},\dot{\mathbf{q}},\lambda) - \mathbf{G}^T(t,\mathbf{q})\,\lambda\,, \tag{1a}$$

$$0 = \mathbf{g}(t,\mathbf{q}) \tag{1b}$$

with $\mathbf{q}$ denoting the position coordinates of all bodies. $\mathbf{M}(\mathbf{q})$ is the generalized mass matrix and $\mathbf{f}$ the vector of applied forces. Joints decrease the number of degrees of freedom in the system and may result in constraints (1b) that are coupled to the dynamical equations (1a) by constraint forces $-\mathbf{G}^T\lambda$ with Lagrange multipliers λ and $\mathbf{G}(t,\mathbf{q}) := (\partial\mathbf{g}/\partial\mathbf{q})(t,\mathbf{q})$. Very efficient numerical methods for the evaluation and for the time integration of (1) have been developed and implemented in industrial multibody simulation tools like ADAMS, SIMPACK or DADS, see [10, 11].

Already in the early days of multibody dynamics these methods have been extended to more general mechanical systems that contain e.g. flexible bodies or force elements with internal dynamics. On the contrary the extension of the simulation scenarios through co-simulation is a recent development which is still in progress.

3.2. Aerodynamic System

The flow around high speed trains in absence of cross wind can be assumed to be inviscid and irrotational, leading to a linear aerodynamic model. Such a flow model is called *potential* flow and is widely used in aircraft aerodynamics but also in railway aerodynamics when cross wind has not to be considered. Its discretized numerical formulations, the *panel methods* [12, 13], lead to small computational effort and other benefits compared to nonlinear aerodynamic models.

A potential flow can be merely described by the Laplace equation by introducing a scalar field function Φ:

$$\nabla^2 \Phi = 0 \tag{2}$$

whereby potential function and velocity field are directly connected:

$$\mathbf{u} = \nabla \Phi \,. \tag{3}$$

Eq. (2) must be completed with some boundary conditions which are the physical interface between multibody and aerodynamical system. Such conditions are presented in the next section.

3.3. Coupled System

The boundary condition for the flow only requires that the normal component of the relative velocity on the vehicles walls $\partial\Omega_V$ vanishes, i.e. that the normal component of the absolute velocity $\mathbf{u}$ is equal to the velocity of the wall $\mathbf{v}$:

$$\nabla \Phi \cdot \mathbf{n} = -\mathbf{v}(\dot{\mathbf{q}}) \cdot \mathbf{n} \qquad \text{on } \partial\Omega_V \tag{4}$$

which shows that the potential Φ must depend on the velocity of the vehicles $\dot{\mathbf{q}}$.

Using Green's formula Eq. (2) can be rearranged to obtain an expression for the potential Φ as integral on the vehicles walls $\partial\Omega_V$ of a *source* distribution σ divided by the module of the position vector $\mathbf{r}$. A *doublet* distribution, which compares in the general formulation, is not necessary for the case of ground vehicles because no special conditions, such as the Kutta-condition, have to be satisfied.

Since $\partial\Omega_V$ depends on the vehicles position, Φ depends on $\mathbf{q}$ as well:

$$\Phi(\mathbf{r}, \mathbf{q}, \dot{\mathbf{q}}) = \frac{1}{4\pi} \int_{\partial\Omega_V} \sigma \, \frac{1}{|\mathbf{r}|} \, \mathrm{d}s \,. \tag{5}$$

The source distribution σ on $\partial\Omega_V$ is unknown and has to be determined using the boundary condition (4). When σ has been computed, Φ and $\mathbf{u}$ can be derived using (5) and (3).

The Bernoulli Equation can now be applied to obtain the pressure field:

$$\frac{\partial \Phi}{\partial t} + \frac{|\mathbf{u}|^2}{2} + \frac{p}{\rho} = \frac{p_\infty}{\rho} = \text{const} \quad \Rightarrow \quad p(\mathbf{r}, \mathbf{q}, \dot{\mathbf{q}}, t)\,. \tag{6}$$

It is finally possible to compute the resulting flow force $\mathbf{L}_f$ and torque $\mathbf{M}_f$ related to the origin O:

$$\mathbf{L}_f(\mathbf{q}, \dot{\mathbf{q}}, t) \;=\; -\int_{\partial\Omega_V} p \cdot \mathbf{n}\,\mathrm{d}s\,, \tag{7a}$$

$$\mathbf{M}_f(\mathbf{q}, \dot{\mathbf{q}}, t) \;=\; -\int_{\partial\Omega_V} \mathbf{r} \times p \cdot \mathbf{n}\,\mathrm{d}s \tag{7b}$$

which couple the flow equations (2) and (4) with the multibody system equations (1).

The used panel method adopts a discretization of the surface integral in (5). The finite surface elements are called *panels* and on each of them the source distribution σ_i is supposed to be constant. The boundary condition (4) leads to an algebraic linear system whose unknown vector is the discrete source distribution σ_i and whose dimension is thus the number of panels. Eq. (6) has also to be discretized: pressure distribution and forces (7) can be finally obtained on a discrete time axis.

In order to minimize the computational effort the number of "aerodynamic" time steps must be minimized, as each of these time steps required for a usual configuration about 15 minutes. The panel method is capable of very large time steps compared to the multibody system part and also to other CFD tools. Furthermore, the flow and driving dynamics are quite weakly coupled. For these reasons following co-simulation technique has been implemented. In each macro step $T_n \rightarrow T_{n+1}$ Eq. (6) is discretized once using the macro step size H as time step. The flow field is thus resolved only at the synchronization points T_n and kept frozen between them. The multibody system part of the coupled problem is integrated by standard techniques from multibody dynamics with step size and order control. In this way about 30 macro steps are necessary for the simulation of a typical manoeuvre.

4. Results

The simulation of a wide range of typical driving manoeuvres (passing on open track and at tunnel entrance, tunnel run-in and run-out, etc.) have been performed. Results can be examined in many ways using different criteria but none of them can be definitively chosen as representative, as each railway company or railway national office uses its own methods to estimate aerodynamic sensitivity, see [6]. In addition to that, the official guidelines – if any is present – don't consider strong unsteady cases, which should be evaluated on the base of dedicated criteria; only idealized unsteady cases (idealized gust) or, more commonly, quasi-steady or even steady cases are taken into account. In the following the attention is focused on the *lateral displacement* of the leading wheelset for the case of two german ICE 2 trains passing by each other at the same speed on open track.

From the simulations emerged that, even if the aerodynamic forces grow to the square of the driving velocity, the response of the system has only a linear dependence on the velocity, see Fig. 2. As a consequence not exclusively very high driving speeds are critical. Results plotted in Fig. 2 refer to trains driving on a perfectly straight and plane track without rail excitations (ideal case); the values are therefore relative small.

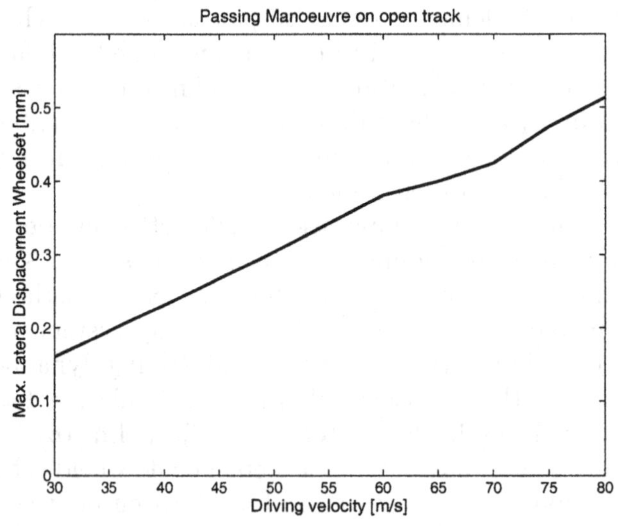

Figure 2. Maximal wheelset's lateral displacement during a passing manoeuvre on open track (ideal case).

The motion of railway vehicles is characterized by very small lateral displacements of the wheelsets relative to the rail, which lead to very large driving forces at the wheel-rail contact. During passing manoeuvres such forces reach large values for very short periods, i.e. very impulsive forces at the wheel-rail interface arise, see Fig. 3. An important question is whether such forces can be borne under any arbitrary real operative condition, e.g. when the wheels are partially unloaded. Actually results show that when little disturbances are present the dynamical response of the vehicles seems to be amplified. Fig. 4 shows a typical situation: a small, low frequency

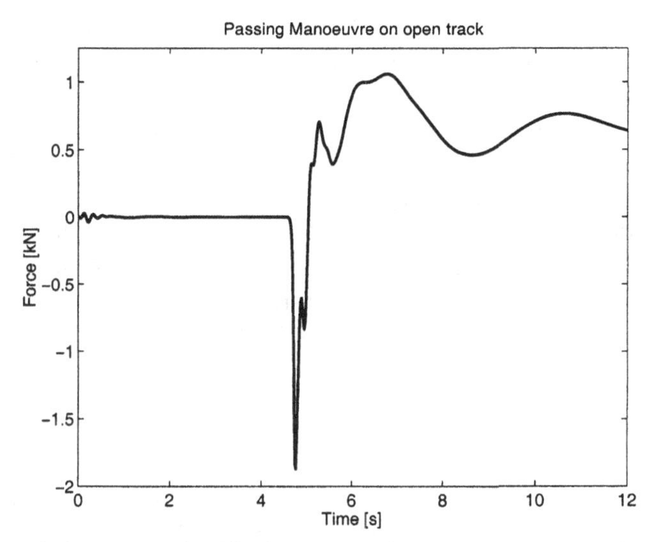

Figure 3. Lateral force at wheel/rail contact during a symmetrical passing manoeuvre on open track at 80 m/s.

perturbation, which could be caused by cross wind or track irregularities in a real environment, lets the displacement of the wheelset (thick line) reach much larger values than in the ideal case (dashed line).

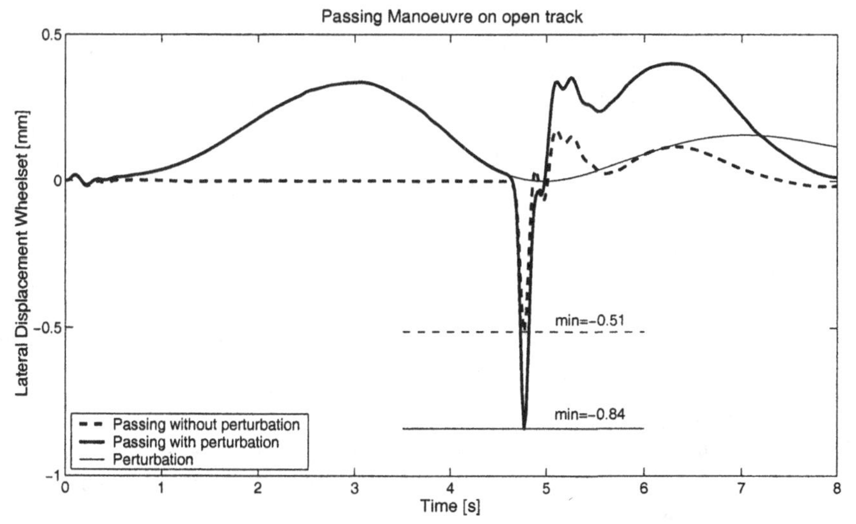

Figure 4. Effect of a small perturbation on wheelset's lateral dynamics during a symmetrical passing manoeuvre on open track at 80 m/s.

Performed co-simulations also pointed out that, even if the unsteady aerodynamic loads can exert a very large influence on the driving dynamics, the effects of the induced vehicle motion on the surrounding flow, i.e. of the multibody variables on the aerodynamic variables, is of some influence only when the fundamental frequency of the transient loads approaches the lowest natural frequencies of the car motion. In the case of symmetrical passing manoeuvres this condition is satisfied at very small driving velocities, which are of no technical interest because at low speed the transient loads have small amplitude. Furthermore such influence causes modifications in the aerodynamic forces which induce in turn an almost insignificant alteration of the mechanical behavior, e.g. of the lateral displacement of the wheelsets. The influence can be thus neglected for the studied cases.

References

1. H.G. Matthies and J. Steindorf. Efficient iteration schemes for nonlinear fluid-structure interaction problems. In B.H.V. Topping, editor, *Computational Mechanics: Techniques and Developments*, pages 263–267. Civil-Comp Press, Edinburgh, 2000.
2. W. Rumold. Modeling and simulation of vehicle carring liquid cargo. *Multibody System Dynamics*, 5:351–374, 2001.
3. A. Veitl, T. Gordon, et al. Methodologies for coupling simulation models and codes in mechatronic system analysis and design. In *16th IAVSD Symposium*, Pretoria, South Africa, 1999. Supplement to Vehicle Sytem Dynamics, Vol. 33, 2000.
4. S. Yamamoto et al. Aerodynamic influenece of a passing vehicle on the stability of the other vehicles. *JSAE Review*, 18:39–44, 1997.
5. K. Pahlke. Application of the standard aeronautical CFD method FLOWer to trains passing on open track. In *TRANSAERO Symposium, Paris*, 1999.
6. S. Lippert. On side wind stability of trains. Technical Report TRITA - FKT Report 1999:38, Royal Insitute of Technology – Railway Technology, Stockholm, 1999.
7. M. Arnold, A. Carrarini, A. Heckmann, and G. Hippmann. Modular dynamical simulation of mechatronic and coupled systems. In H.A. Mang, F.G. Rammerstorfer, and J. Eberhardsteiner, editors, *Proc. of WCCM V, Fifth World Congress on Computational Mechanics, July 7–12, 2002, Vienna, Austria*, 2002.
8. R. Kübler and W. Schiehlen. Two methods of simulator coupling. *Mathematical and Computer Modelling of Dynamical Systems*, 6:93–113, 2000.
9. R.E. Roberson and R. Schwertassek. *Dynamics of Multibody Systems*. Springer-Verlag, Berlin Heidelberg New York, 1988.
10. W.O. Schiehlen, editor. *Multibody Systems Handbook*. Springer–Verlag, Berlin Heidelberg New York, 1990.
11. W. Rulka. *Effiziente Simulation der Dynamik mechatronischer Systeme für industrielle Anwendungen*. PhD thesis, Vienna University of Technology, Department of Mechanical Engineering, 1998.
12. J.L. Hess and A.M.O. Smith. Calculation of potential flow about arbitrary bodies. *Progress in Aeronautical Sciences*, 8, 1985.
13. L.L. Erickson. Panel Methods - An Introduction. Technical Paper 2995, NASA, 1990.

THE DEVELOPMENT OF A REAL-TIME MULTIBODY VEHICLE DYNAMICS AND CONTROL MODEL FOR A LOW COST VIRTUAL REALITY VEHICLE SIMULATOR: AN APPLICATION TO ADAPTIVE CRUISE CONTROL

SUNG-SOO KIM, MOONCHEOL WON
Department of Mechatronics Engineering, Chungnam National University
220 Kung-Dong, Yusong-Ku, Daejeon, 305-764, Korea
E-mail:sookim@cnu.ac.kr

1. Introduction

The concept of the ACC (Adaptive Cruise Control) system is shown in Fig. 1. ACC system controls the throttle valve and the brake master cylinder to maintain a desired space between the preceding vehicle and the ACC vehicle. To evaluate performance of the ACC system and driver's responses to the intelligent vehicles, a vehicle simulator with a virtual reality feature can be a very effective tool. For such a vehicle simulator, a real-time vehicle dynamics and control model is essential.

Real-time models based on multibody vehicle dynamics have been developed with several different ways [1-3]. Among them, the vehicle model with subsystem synthesis method provides most efficient solution method for real-time applications [3]. In the vehicle control area, the researches on intelligent vehicles have been carried out in order to develop and implement ACC control logics for an experimental vehicle [4,5]. However, it seems that an integrated model of real-time multibody vehicle dynamics and control has not been developed yet.

The purpose of this paper is to propose an integrated vehicle dynamics and control model for a virtual reality vehicle simulator, especially in the application to ACC simulations. For the real-time simulations, the subsystem synthesis method [3] has been employed for the multibody vehicle dynamics model.

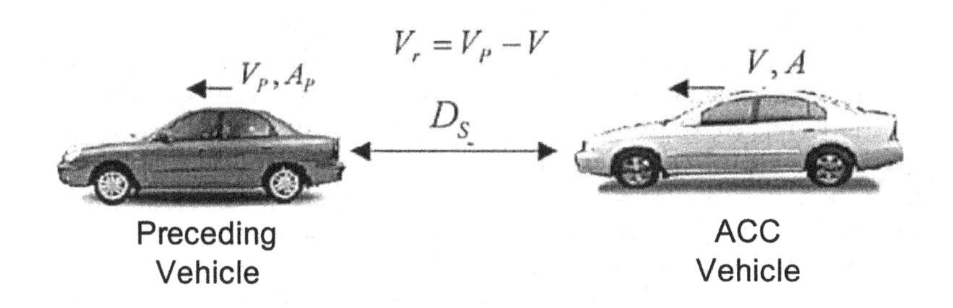

Figure 1. ACC Concept

W. Schiehlen and M. Valášek (eds.), Virtual Nonlinear Multibody Systems, 419–426.

2. Real-time Multibody Vehicle Dynamics and Control Model

The computational model for real-time vehicle dynamics and control consists of three modules as shown in Fig. 2, i.e., a preceding vehicle movement scenario module, a control module that includes the ACC logic and the throttle and brake switching logic, and a real-time multibody vehicle dynamics module. The preceding vehicle movement scenario module generates the preceding vehicle motion for the control module. With the preceding vehicle motion and the ACC vehicle states from the vehicle dynamics module, the control module computes required throttle angles and brake angles for the real-time vehicle dynamics module.

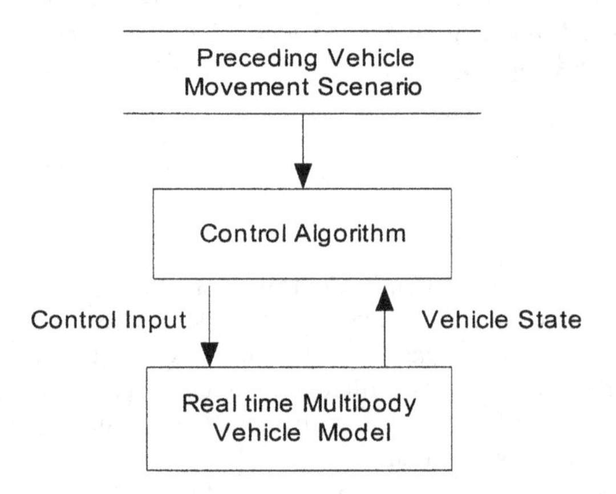

Figure 2. Computational Flow of the Real-time Dynamics and Control Model

2.1. REAL-TIME MULTIBODY VEHICLE DYNAMICS MODEL

The vehicle model used in this paper is HMMWV (High Mobility Multipurpose Wheeled Vehicle) [6]. HMMWV model consists of 4 independent SLA (Short-Long Arm) suspension subsystems, a power-train subsystem, STI tire subsystem [7,8], a rack and pinion steering subsystem, a brake subsystem, and a real-time multibody vehicle dynamics equation generator that is based on the subsystem synthesis method [3]. Due to the limitation of space, we only explain brief idea of the subsystem synthesis method and details are left to the reader by referring Ref.3.

In real-time simulations, equations of motion must be formed very effectively, and also solved efficiently. Thus, in the subsystem synthesis method, joint relative coordinates are used. The relative generalized coordinates of an independent suspension subsystem are not kinematically related to those of the other independent suspension subsystem. However, they are coupled with the generalized coordinates associated with the chassis, since each independent suspension subsystem shares a common base body (the chassis). This independent property is used to generate suspension subsystem equations of motion separately. When the suspension subsystem is synthesized to the chassis, dynamic effects of each subsystem are also added to the chassis' equations of motion.

In order to apply this subsystem synthesis method, a virtual reference body can be introduced. The virtual body does not have any dynamic property such as inertia and gravitational force, since it is only required for setting up the reference frame to describe the relative motion in the subsystem. If steering model is kinematically driven, then the HMMWV model can be decomposed into four subsystems with the virtual reference bodies and the chassis as shown in Fig. 3. It is noted that there is no relative motion between the chassis and the virtual reference body of the suspension subsystem.

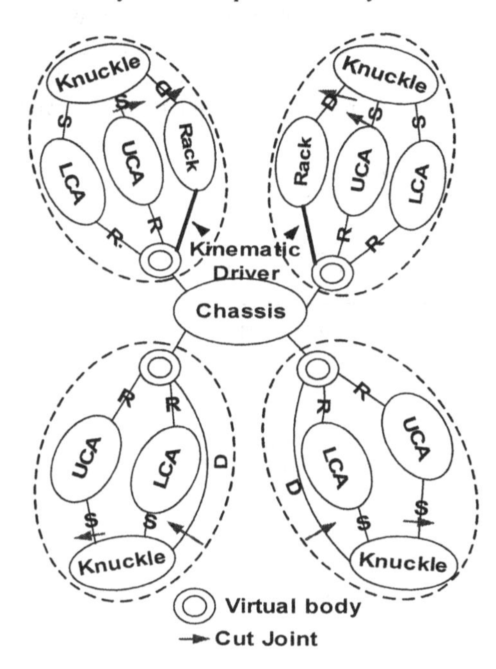

Figure 3. Topology of vehicle model with subsystems

Once the model as shown in Fig. 3 is established, the closed loop subsystem equations of motion can be obtained for each suspension subsystem by treating virtual reference body as a base body. Using the recursive formulation in multibody dynamics, the subsystem equations of motion can be expressed as [3]

$$\begin{bmatrix} \mathbf{M}_{qq} & \mathbf{\Phi}_{\bar{\mathbf{q}}}^{T} \\ \mathbf{\Phi}_{\bar{\mathbf{q}}} & \mathbf{0} \end{bmatrix} \begin{bmatrix} \ddot{\mathbf{q}} \\ \boldsymbol{\lambda} \end{bmatrix} = \begin{bmatrix} \mathbf{P}_{q} \\ \boldsymbol{\gamma} \end{bmatrix} - \begin{bmatrix} \mathbf{M}_{yq}^{T} \\ \mathbf{0} \end{bmatrix} \dot{\hat{\mathbf{Y}}}_{0} \tag{1}$$

For the chassis, 6x6 matrix form of equations of motion can be expressed as follows;

$$(\hat{\mathbf{M}}_{0}+\sum_{i=1}^{4}\breve{\mathbf{M}}_{i})\dot{\hat{\mathbf{Y}}}_{0} = (\hat{\mathbf{Q}}_{0}+\sum_{i=1}^{4}\breve{\mathbf{P}}_{i}) \tag{2}$$

where $\breve{\mathbf{M}}_i$ and $\breve{\mathbf{P}}_i$ (i=1~4) are the effective inertia matrix and the effective force vector of each subsystem, respectively [3]. In the subsystem synthesis method, 6x6 matrix form of chassis equations of motion, as shown in Eq. 2, is first solved. Then, for each suspension subsystem, equations shown in Eq. 1 are solved for suspension accelerations. Since several small size equations of motion are solved instead of solving large size equations, the subsystem synthesis method provides computational efficiency.

2.2. ACC CONTROL MODEL

Fig. 4 shows the computational flow in the control module. Sliding mode control is used to design a spacing controller for ACC [4]. According to relative distance between the preceding vehicle and the ACC vehicle, the required engine or brake torque is calculated. Using throttle and brake switching logic, either throttle or brake control is determined [5]. In the throttle control, inverse torque map is used to compute the desired throttle angle. In the brake control, necessary torque and the corresponding desired brake pedal angle are computed. These angles are the inputs of real-time multibody vehicle dynamics model.

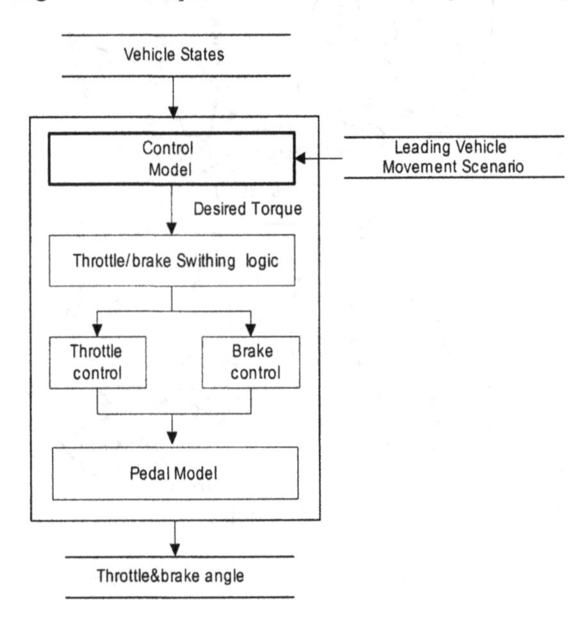

Figure 4. Control module flow chart

In this control model, the headway time control strategy is utilized to design the desired relative distance. The head way time is defined as the time that the ACC vehicle collides with the preceding vehicle, when the preceding vehicle stops immediately. According to the headway time strategy, the desired vehicle spacing, x_{r_des} is obtained from

$$x_{r\_des} = v_f \cdot t_h \tag{3}$$

where v_f is the following ACC vehicle speed, and t_h is headway time. The desired acceleration is also obtained as [4];

$$a_{f\_des} = \frac{v_r - \lambda\, Sat(\frac{s}{\Phi})}{t_h} \tag{4}$$

where v_r is the relative velocity, λ is a positive gain, Sat(.) is the saturation function, s is the sliding surface defined as $s = x_{r\_des} - x_r$, and Φ is the boundary layer thickness. The saturation function is used to have better riding quality by preventing wild throttle movement. In order to prohibit jerking motion of the ACC vehicle, the desired acceleration a_{f_des} is also bounded between 1m/s^2 and –2.5m/s^2.

The desired engine torque corresponding to the desired acceleration in Eq. 4 can be obtained using simplified vehicle equations of motion and the simple engine model [4].

$$T_{eng\_des} = [J_{eff} \cdot (\frac{v_{f\_des}}{H_r \cdot r_{drive} \cdot r_{gear}}) + H_r \cdot r_{drive} \cdot r_{gear} (F_a + F_r)] / r_{tq} \quad (5)$$

where H_r is the tire radius, r_{drive} is final drive gear ratio, r_{gear} is current gear ratio, r_{tq} is the torque ratio in the torque converter, J_{eff} is the effective total inertia from the engine, F_r is rolling resistance and F_a is aerodynamic drag.

After obtaining the desired engine torque, throttle and brake switching logic is applied to determine whether engine braking or active braking is necessary. If the amount of the desired torque is smaller than that of the torque obtained from engine brake with zero throttle angle, then active braking is applied. Once engine braking or active braking is determined, the desired throttle angle, α_{des}, is obtained from the inverse engine performance map.

$$\alpha_{des} = T^{-1}{}_{eng}(\omega, T_{eng\_des}) \quad (6)$$

For active braking, the necessary brake torque is computed in Eq. 7 and the brake pedal angle can be obtained from the linear relationship between the braking torque and pedal angles.

$$T_{b\_des} = \frac{T_{net} - T_{net\_des}}{r_{drive} \cdot r_{gear}} \quad (7)$$

where T_{b_des} is the necessary braking torque, T_{net_des} is the desired torque to maintain desired relative distance and T_{net} is the engine brake torque that is generated with zero throttle angle in the current engine speed.

3. A Low Cost Virtual Reality Vehicle Simulator

A PC based low cost virtual reality vehicle simulator is shown in Fig. 5. The simulator consists of two PCs (one for real-time vehicle dynamics and control, and the other for real-time graphic rendering), a visualization system such as a beam projector, and a vehicle cockpit. Two PCs are networked using UDP (User Defined Protocol). Every 40mili-seconds, vehicle states are transferred to the graphic PC to generate 25 frame/s display rate.

In contrast to a high-end driving simulator like NADS [10], the low cost virtual reality vehicle simulator, used in this research, does not include a motion base and a sophisticated visualization system. Thus, the application of this simulator is limited to moderate vehicle motion simulations and associated human factor analyses. In this paper, ACC simulations are confined to longitudinal motion control of the ACC vehicle with flat road situation. For this kind of simulations, the proposed low cost virtual reality simulator is effective enough to investigate performances of the ACC vehicle model and to provide realistic view of a driver.

To generate virtual reality environment, a graphic modeler 3D Studio Max, a real-time rendering scene generator OpenGVS have been employed. Fig. 6 shows the virtual environment for ACC simulations.

Figure 5. A low cost vehicle simulator

Figure 6. Virtual reality graphic scene

4. ACC Simulations

In order to verify the developed multibody vehicle dynamics and control model, typical simulations for the ACC vehicle have been carried out; i.e., stop-and-go, cut-in, and cut-out simulations. Fig. 7 shows the results of stop-and-go simulations. In this simulation, the preceding vehicle makes a complete stop and then restarts. The first figure of Fig. 7 shows that the ACC vehicle properly follows the preceding vehicle. The second figure of Fig. 6 shows the acceleration and brake pedal angles. It shows that the switching logic in ACC is properly working.

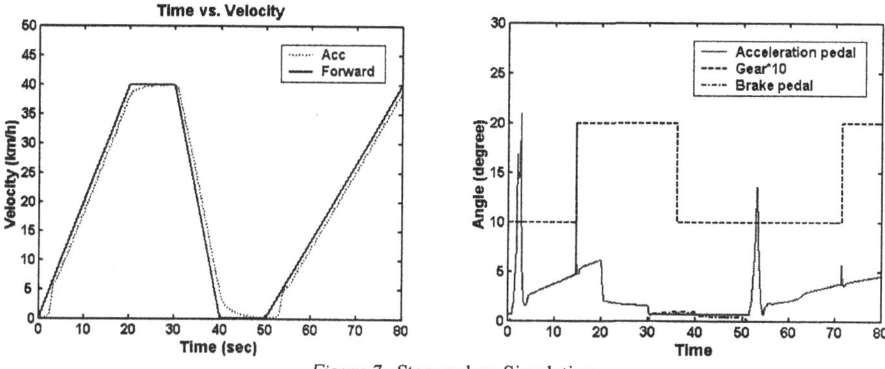

Figure 7. Stop-and-go Simulation

Cut-out simulations have been also carried out as shown in Fig. 8. In the cut-out situation, two preceding vehicles run ahead of the ACC vehicle. The second vehicle (V2), which is just in front of the ACC vehicle, changes lanes. Thus, the ACC vehicle suddenly has to follow the first vehicle (V1). In this case, the ACC vehicle experiences sudden changes in the relative position and the relative velocity between the ACC and the preceding vehicles. The first figure of Fig. 8 shows the velocities of the preceding vehicles and the ACC vehicle. After the lane change occurs, the ACC vehicle accelerates itself and then keeps the constant velocity that is larger than that of the preceding vehicle (V1) in order to maintain the desired distance. The second figure of Fig. 8 shows the relative distance between the preceding and the ACC vehicles. The ACC vehicle gradually recovers the desired distances.

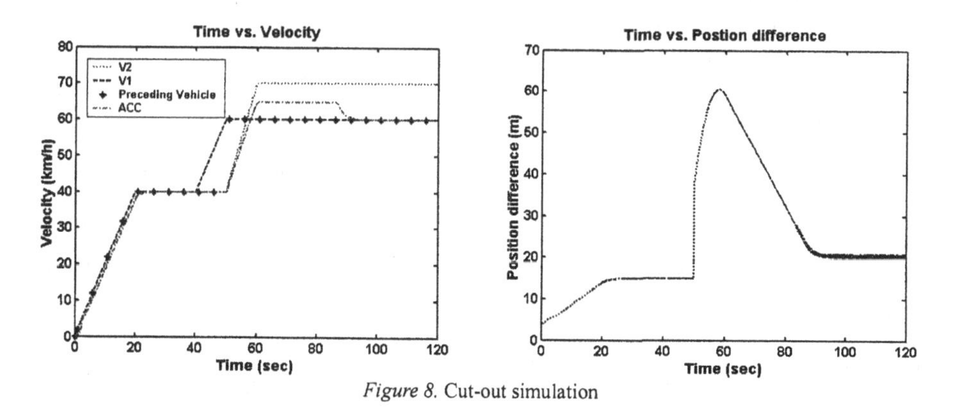

Figure 8. Cut-out simulation

Cut-in simulations have been also carried out. In this situation, during the ACC vehicle follows the preceding vehicle with constant speed, the other vehicle with the same speed in the next lane abruptly cuts in between the preceding and the ACC vehicles. The first figure of Fig. 9 shows the velocities of the ACC vehicle. After cut-in occurs, the ACC vehicle slows down a little bit. The second figure of Fig. 9 shows the relative distance between the preceding and the ACC vehicles. Due to cut-in at 50 sec., the relative distance is suddenly reduced. According to the headway time strategy, relative distance is increased back to the original distance within 10 seconds.

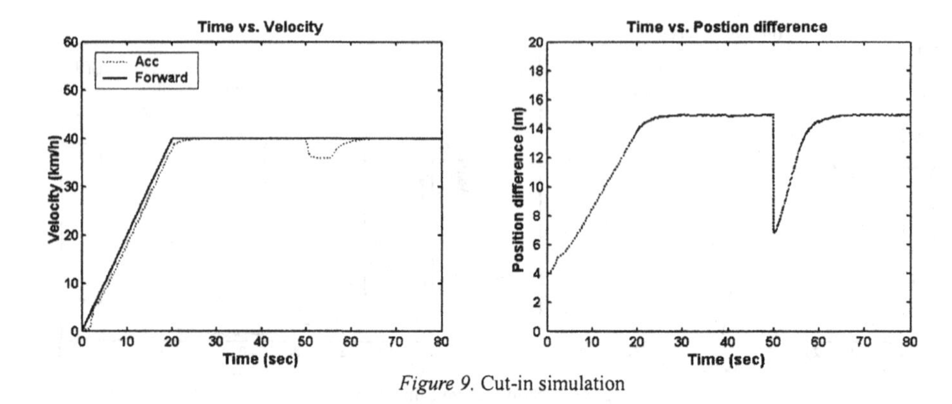

Figure 9. Cut-in simulation

To see the real-time capability of the developed model, CPU time has been measured in Pentium IV 1.6 Ghz CPU with RAM size 256 Mb. For 120 second overall simulation time, 5.23 seconds computational CPU time has been measured. For numerical integration, Adams Bashforth 3rd order formula is used with 8 ms integration step size [9]. This proves the developed model indeed has a real-time simulation capability.

5. Conclusions

A real-time multibody vehicle dynamics and control model has been developed for a low cost virtual reality ACC vehicle simulator. The subsystem synthesis method is employed to generate efficient multibody vehicle equations of motion. The sliding mode control and throttle/brake switching logic are used for the ACC. Several simulations such as stop-and-go, cut-in, and cut-out, show that the developed model is suitable for ACC simulations. The CPU time measure proves the real-time capability of the proposed model.

6. References

1. Freeman, J.S., Watson, G., et al., (1995) The Iowa Driving Simulator: An implementation and Application Overview, *SAE Paper No. 950174.*
2. Choi, G.J., Yoo, Y.M., Lee, L.P., and Yoon, Y.S., (2000) A Real-time Multibody Dynamic Analysis Method using Suspension Composite Joints, *Int. Journal of Vehicle Design*, **24(2/3)**, 259-273.
3. Kim, Sung-Soo, (2002) A subsystem Synthesis Method for Efficient Vehicle Multibody Dynamics, *Multibody System Dynamics*, **7(2)**, 189-207.
4. Won M., Kim S-S., Kang B.B., and Jung H.J. (2001) Test Bed for Vehicle Longitudinal Control Using Chassis Dynamometer and Virtual Reality: An Application to Adaptive Cruise Control, *KSME International Journal*, **15(9)**, 1248-1256.
5. Lee S., Hong J., and Yi K. (2001) A Modeling and Control of Intelligent Cruise Control Systems, *Transactions of the Korean Society of Mechanical Engineers A*, **25(2)**, 286-288.
6. Aardema J. (1989) Failure analysis of the Lower Rear Ball Jiont on the High-Mobility Multipurpose Wheeled Vehicle (HMMWV), *Technical report 13337*, RDE center, U.S. Army.
7. Allen R. Wade, Raymond E. Magdaleno, Theodore J. Rosenthal, (1995) Tire Modeling for Vehicle Dynamic Simulation, *SAE Paper No. 950312.*
8. Bernard J.E. and Clover C.L. (1995) Tire Modeling for Low-Speed and High-Speed Calculation, *SAE Paper No. 950311.*
9. Atkinson, K. (1988) *Numerical Analysis*, John Wiley & Sons.
10. Haug, E.J. et. al, (2002) Virtual Proving Ground Simulation for Highway Safety Research and Vehicle Design, Schiehlen, W., and Valasek, M. (eds.) *Virtual Nonlinear Multibody Systems NATO Advanced Study Institute*, Vol. 2, 87-95.

VIRTUAL RECONSTRUCTION OF IMPAIRED HUMAN HEARING

A. EIBER, H.-G. FREITAG and C. BREUNINGER
Institute B of Mechanics, University of Stuttgart,
Pfaffenwaldring 9, D-70550 Stuttgart, Germany
E-mail: ae@mechb.uni-stuttgart.de

1. Introduction

The hearing organ, a marvelous and highly developed part in the human body provides a very important sense in the daily life. In the middle ear there are pure mechanical mechanisms of sound transfer whereas in the inner ear a complex transduction from mechanical entities via chemical processes to electrical stimulation of hearing nerves take place. In case of hearing loss due to diseases or accidents a mechanical reconstruction has to be performed by inserting passive or active implants into the middle ear. Due to the complex mechanical structure of the middle ear with its spatially oriented ligaments, active muscles and the nonlinear cochlea amplifier of the inner ear, classical one–dimensional models are not sufficient to describe the dynamical behavior of the hearing organ. This interdisciplinary project together with physicians is focused on appropriate models to reduce the trial and test series on living patients by simulation.

2. Aims and Motivation

Virtual scenarios of the complex hearing process for pathological situations are becoming an increased meaning in the clinical practice to diagnose and to reconstruct an impaired hearing successfully:

- The *audiologist* needs a detailed insight into the dynamical behavior of the mechanical middle ear structures to give a correct diagnosis about a specific disease.
- For a successful reconstruction the *surgeon* must have a precise knowledge about the frequency dependent spatial motion of the ossicular chain during sound transfer through the middle ear to assess the effects of surgical manipulations. For reconstructions he has to choose an appropriate implant and to insert it in a suitable manner.

W. Schiehlen and M. Valášek (eds.), Virtual Nonlinear Multibody Systems, 427–435.

- The *developer* of passive and actively driven implants has to regard the dynamics of the implant itself together with the dynamical behavior of the remaining middle ear structure in the design and the optimization process.

By virtue of simulation of the hearing process and animation of the spatial motion of the middle ear structures, 'virtual hearing' can be carried out and demonstrated at the computer. This serves as a common base for the work of audiologists, surgeons and developers of implants. The procedure needs appropriate models with different complexity and confident parameters to describe linear and nonlinear effects of sound transfer through the outer and middle ear.

3. Modeling of the Hearing Mechanism

To describe the dynamical behavior of the hearing organ mechanical and electrical models are used. With models based on electrical circuits only the global sound transfer is described between the scalar input pressure p and the scalar output of 'hearing', e.g. Zwislocki [1] and Shaw and Stinson [2]. These models were derived from electro–acoustical and electro–mechanical analogies based on very strong simplifications. Due to the simplifications and the non–physical description they are less meaningful and less flexible than mechanical models.

3.1. NATURAL EAR

Finite element models of tympanic membrane and ossicles were published by Wada et al. [3], Beer et al. [4] and Prendergast et al. [5]. This approach leads to large systems with a huge number of parameters.

With three–dimensional multibody system models of the human ossicular chain presented in [6], the spatial motion of ossicles can be described. The methods of continuous systems, finite element systems and in particular multibody systems are used here to model the air in the outer ear canal, the tympanic membrane, the ossicles of the middle ear with its visco–elastic ligaments and the fluid of the inner ear. For the outer ear canal an air column in a rigid tube is considered as a continuous system. Derived form this, the air is represented by finite volume elements. They consist of mass points interconnected with springs and dampers. These elements allow the description of damping within the air and with the boundary layer as well as the dynamical interaction with the ear drum. For this a finer mesh is used in the vicinity of the spatially oriented ear drum. Modeling the air column as a lumped mass system allow the description of the natural frequencies in the frequency range of the human hearing

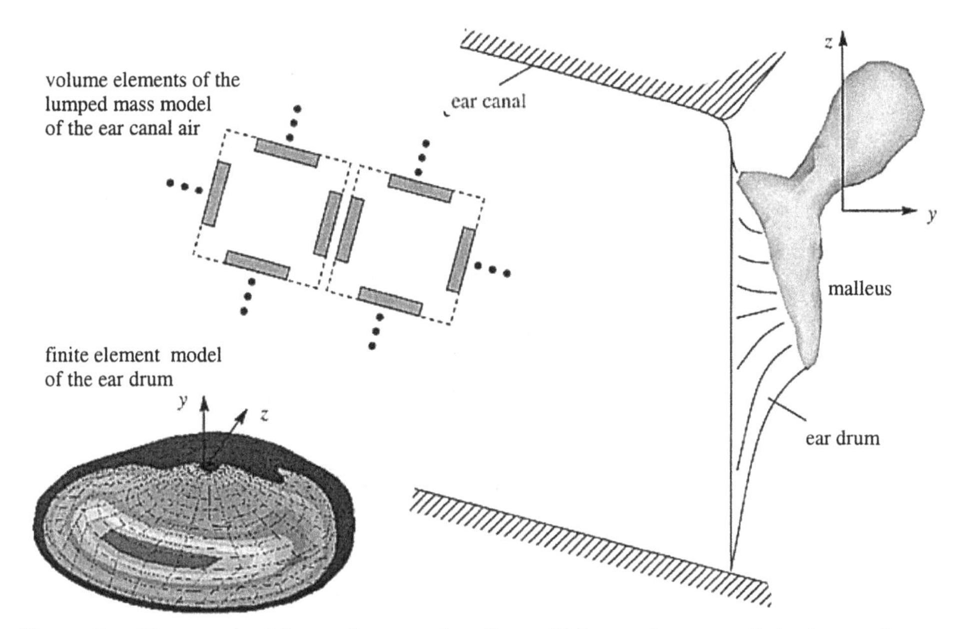

Figure 1. Ear canal with ear drum and malleus. Volume elements of the lumped mass model of the ear canal air. Finite element representation of the ear drum, first mode.

with a low number of the degrees of freedom. The finite element approach is used to describe the spatially formed membrane with its nonuniform thickness and bending stiffness, see figure 1, in order to derive a lumped mass model which is coupled with the air column and the ossicular chain. The three ossicles, malleus, incus and stapes, are modeled as rigid bodies spatially suspended by the ligaments in the air–filled middle ear cavity, see figure 2. The ossicles itself are coupled with visco–elastic elements allowing three translations and three rotations. These couplings are called 'incudo–mallear joint' and 'incudo–stapedial joint'. The incudo–mallear joint serves as a safety clutch in the case of higher sound pressure levels. The ligaments are considered as massless spring/damper elements [7]. The load of the inner ear is very roughly represented by single mass points. In figure 2 the hearing mechanism is shown with the ear canal, the ear drum, the rigid bodies of the ossicles and the inner ear fluid. Using a multibody systems formalism [8], the equation of motion can be generated in a symbolical form

$$\boldsymbol{M}\ddot{\boldsymbol{y}} + \boldsymbol{k} = \boldsymbol{q}\,, \tag{1}$$

with the generalized coordinates $\boldsymbol{y} \in \mathbb{R}^f$, the mass matrix $\boldsymbol{M}$, the vector of generalized centrifugal, Coriolis and gyroscopic forces $\boldsymbol{k}$ and the vector of generalized applied forces $\boldsymbol{q}$. The entire model of a normal ear has $f = 77$ degrees of freedom comprising the air in the ear canal, the tympanic membrane, the ossicles and parts of the inner ear.

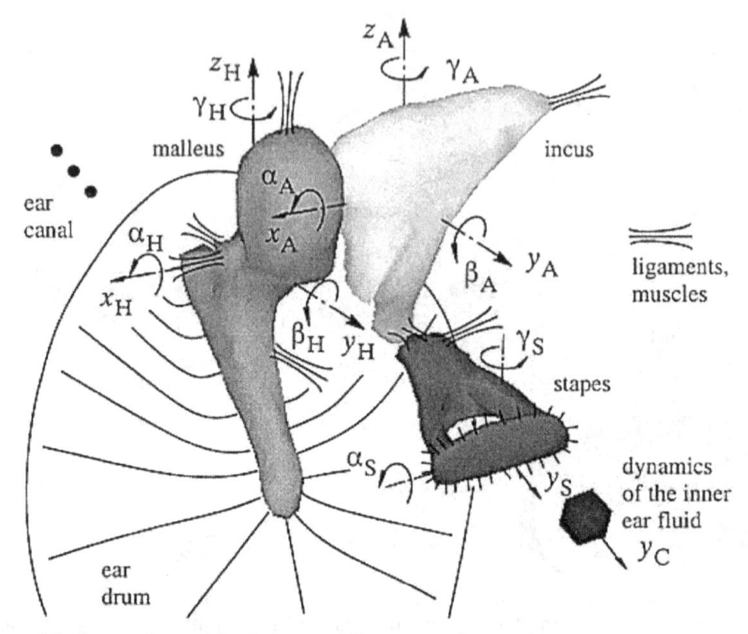

Figure 2. Mechanical model of the middle ear and its adjacent structures air of outer ear canal, ear drum, and fluid of inner ear. The ligaments and muscles are indicated as black lines. In particular the translations x, y and z and rotations α, β and γ of the malleus (index H) and incus (index A), translation y_S and rotations α_S and γ_S of the stapes are specified.

For moderate sound pressure level, linear kinematics and linear constitutive equations can be assumed and the joint between malleus and incus is considered as fixed [3]. Then the equation (1) can be linearized to perform eigenvalue analysis. For high sound pressure levels and in particular for impulsive noise, a nonlinear behavior of the middle ear occur. This is due to a nonlinear kinematic as well as to nonlinear constitutive equations of the ligaments and the ear drum. Their stiffness and damping coefficients are a function of the deformation s and deformation velocity $\dot{s}$ and can be represented with power series,

$$c = c(s, \dot{s}, t)\,, \tag{2}$$

$$d = d(s, \dot{s}, t)\,. \tag{3}$$

The assumed characteristic of the stiffening of the ligaments is sketched in figure 3 left. Since the deflection of the elastic elements in the middle ear particularly the ear drum and the annular ring are restricted, the elements become very stiff close to the border s_{lim}. Moreover, for high force intensity the joint between malleus and incus is articulated for decoupling both ossicles like a mechanical safety clutch to prevent the hearing from

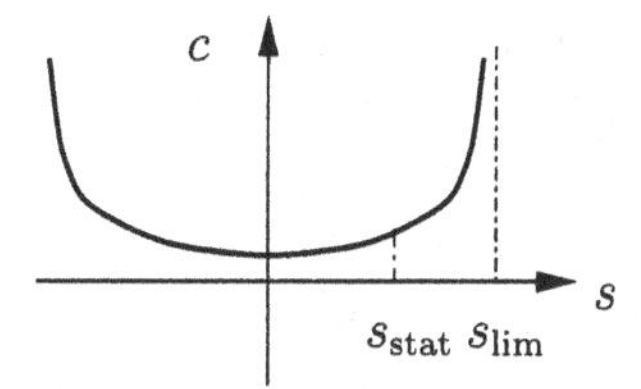

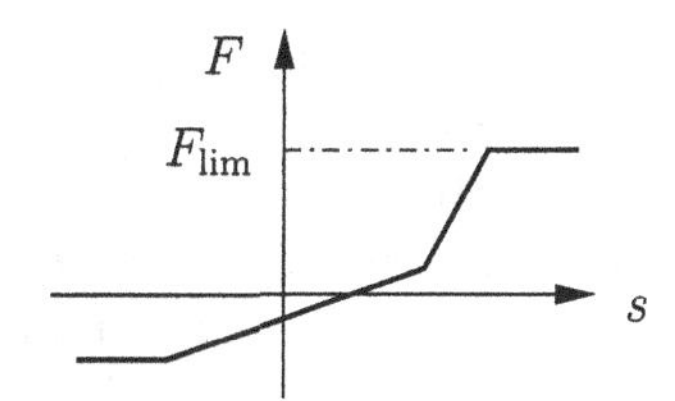

Figure 3. Stiffening of ligaments with static deflection s_{stat} and boundary value s_{lim} (left). Asymmetric stiffness and limited transfer force of coupling between implant and ossicle (right).

overload. Another source of nonlinear behavior are the active processes in the inner ear.

Distinct pathological situations like otosclerosis or partial malleus head fixation can be described with the established models by changing of specific parameters or the structure of the models.

3.2. RECONSTRUCTED EAR

Reconstructions of the middle ear structure are carried out to repair missing or disfunctional parts of the ossicular chain. *Passive* implants are used to replace ossicles in order to bridge a gap in the ossicular chain. Here the position of the attachment points of implant and ossicles determine the kinematics of the reconstructed ear. The coupling conditions between ossicles and implant influence the dynamic of the spatial motion of the reconstructed ear. *Active* implants additionally can drive the middle ear in order to compensate hearing losses due to imperfect sound transfer of the middle ear or damaged inner ear [9]. They can act as force or displacement transducers imposing a translational or a rotational motion. Beside the attachment points and coupling conditions between implant and ossicle, the spatial orientation of the excitation is essential for the dynamical behavior. A restricted coupling force and a distortion due to nonlinear coupling mechanisms may lead to unacceptable hearing results. On the base of the models of the natural ear, models of ears reconstructed with passive or active implants have also been established applying the multibody system approach.

Generally, the mechanism of coupling is highly nonlinear due to gaps or asymmetrical stiffness and damping behavior with respect to inward and outward movement. In particular the commonly used intermediate medium between prosthesis and ossicles like fascia show approximately a behavior with partially linear force/displacement relation and limited coupling forces as sketched in figure 3 right. In some cases the coupling between implant and ossicle is performed by pressing both parts against it. Then a so–called unilateral constraint is given which shows a highly nonlinear behavior. Such an asymmetric stiffness of the springs may lead to harmonic distortion in the sound transfer through the middle ear even in the case of moderate excitation. Due to the restricted space for inserting active elements in the human body the used actuators may have limited output forces or restricted output displacements leading to a distorted excitation of the hearing organ. But for powerful active implants, the applied excitation may cause high coupling forces which exceed a physical limit of the coupling region leading also to distorted sound transfer. By an intensive stimulation of the ossicular chain feedback effects may occur due to sound radiation from the ear drum to the microphone. This inhibits the full compensation of a particular disease.

4. Measurements and Parameter Estimation

Measurements in the clinical practice and in the lab have been conducted to determine the dynamical behavior of the hearing organ and to derive the belonging parameters of the models. Due to the very small displacements and velocities and the low masses, non–contact measurement principles like Laser–Doppler–Vibrometry (LDV) have to be applied.

Especially nonlinear effects could be detected by using LDV and extended procedures of the multifrequency–tympanometry (MFT). In the classical tympanometry a varying static pressure in the ear canal but only a fixed excitation frequency is used to determine the acoustical properties of the middle ear. Applying modern measurement and data processing techniques of the MFT, the nonlinear mechanical properties of the middle ear structures become apparent. The static pressure p_{stat} leads to a static deflection s_{stat} of ligaments and ear drum which causes increased stiffness and damping as illustrated in figure 3 left. Taking the simulated velocity of the tip of malleus handle (umbo) and calculating the spectrum contour plots result as shown in figure 4 right part. The corresponding plot resulting from measurement with LDV is shown on the left part of figure 4. These diagrams are called local multifrequency–tympanograms due to their similarity to those obtained from MFT [10]. The shift in the natural frequencies can be seen due to the nonlinear behavior of the ear drum and the middle ear ligaments. Comparing measurements of the MFT with simulations, the nonlinear behavior of the models could be approximated by power series. Based on virtual MFT diagrams the influence of particular pathologies like malleus head fixation or otosclerosis can be studied. The investigations for describing nonlinearities in the sound transfer and for parameter estimations are still ongoing.

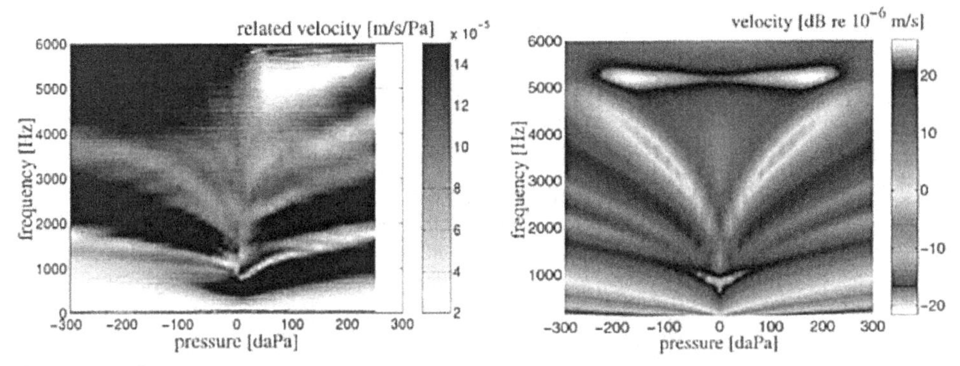

Figure 4. Spectrum of velocity of the umbo depending on static pressure p in ear canal for measurement (left) and simulation (right).

5. Virtual Prototyping

Based on the results of the simulations and animations the surgeon is able to predict the effect of particular surgical incisions. For developers of implants virtual 'test series' can be carried out by including the models of existing implants or new prototypes in the entire model. This allows the improvement of passive and active implants with respect to specific conditions in the middle ear. This is very efficient since particular parameters can be varied while remaining conditions are kept constant. Such virtual tests are helpful to shorten clinical trial series with living animals or human patients.

An optimization of hearing result is not only focused on the design of the implant itself but includes the dynamic behavior of the entire system with the particular way of inserting of the implant by the surgeon. One part of the parameters of the model are given by the specific anatomy. Another part can be chosen by the designer of implants and by the surgeon, they are denoted as design variables $\boldsymbol{p}$. Some parameters $\boldsymbol{q}$ may vary due to those surgical and operational conditions which cannot be influenced. Adjusting the design variables the sound transfer through the ear can be improved applying an optimization procedure. The goals of such an optimization are e.g. the best compensation of a particular disease, a good sound transfer in a particular frequency range and a low sensitivity against variations in the parameters $\boldsymbol{q}$ to get a sufficient robustness. For this, appropriate criteria of the kind

$$\psi_i = \psi_i(t, \boldsymbol{y}, \dot{\boldsymbol{y}}, \ddot{\boldsymbol{y}}, \boldsymbol{p}, \boldsymbol{q}), \quad i = 1(1)n \tag{4}$$

have to be formulated assessing the dynamical behavior of the system. The criteria can be formulated in the time and the frequency domain. In the time domain the response $\boldsymbol{y}(t)$ of the system for a particular excitation will be assessed regarding the nonlinear equations of motion (1). For criteria

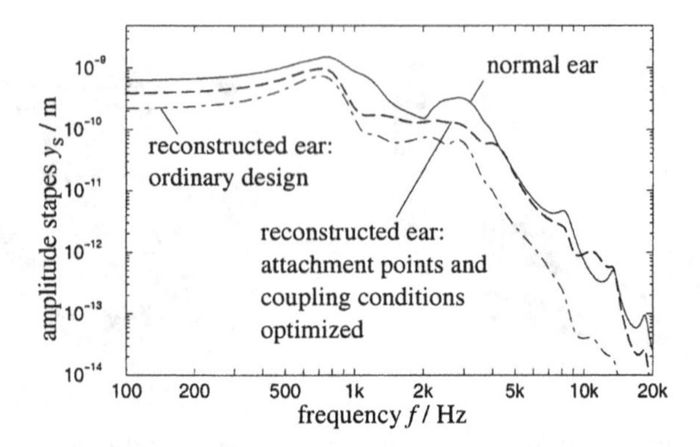

Figure 5. Comparison in the frequency response of normal ear and reconstructed ears.

formulated in the frequency domain the response $\boldsymbol{y}(f)$ for a harmonical excitation of the linearized model in the audible frequency range will be assessed. These criteria have to be minimized by means of optimization tools which are based on different optimization strategies and algorithms [11, 12].

To show the optimization of a passive implant, in figure 5 the frequency dependent response of the inward/outward motion of stapes for a normal ear and reconstructed ears are plotted. The sound transfer is strongly governed by the position of the points where the implant is attached to the ossicles and the coupling conditions which are set as design variables. The surgeon is able to choose the position of attachment as well as the coupling conditions by adjusting the statical preload between ossicle and implant. He may also use different materials of different thickness for the intermediate layer between implant and ossicle. For the optimization two criteria have been formulated: To assess the *loudness*, the amplitude y_S of inward/outward motion of the stapes should be large in the whole frequency range. To assess the *sound quality*, the transfer function should be similar to the normal ear. Taking both criteria into account and applying a multi–criteria optimization a better performance could be achieved as indicated in figure 5.

For an optimal design and use of passive and active implants reliable models of the entire hearing organ are necessary. They have to describe the three–dimensional motion and the nonlinear behavior of the mechanical structure.

Acknowledgments

This work was partially granted by the German Research Council under Ze 149/7–1 and Schi 119/27–1.

References

1. Zwislocki, J. (1962) Analysis of the middle ear function. Part I Input Impedance, *J Acoust Soc Am* **34**, 1514–1523.
2. Shaw, E. A. G.; Stinson, M. R. (1983) *The Human External and Middle Ear: Models and Concepts*, In: De Boer, E.; Viergever, M. A. (eds.), Mechanics of Hearing, University press, 3–10, Delft.
3. Wada, H.; Metoki, T.; Kobayashi, T. (1992) Analysis of dynamic behaviour of human middle ear using a finite element model, *J Acoust Soc Am* **24**, 319–327.
4. Beer, H.-J.; Bornitz, M.; Hardtke, H.-J.; Schmid, R.; Hofmann, G.; Vogel, U.; Zahnert, T.; Hüttenbrink, K.-B. (1999) Modelling of Components of the Human Middle Ear and Simulation of Their Dynamic Behaviour, *Audiol Neurootol* **4**, 156–162.
5. Prendergast, P. J.; Ferris, P.; Rice, H. J.; Blayney, A. W. (1999) Vibro–Acoustic Modelling of the Outer and Middle Ear Using the Finite–Element Method. *Audiol Neurootol* **4**, 185–191.
6. Eiber, A.; Schirm, W. (1991) *Mechanics of Hearing -Dynamics of the Middle Ear-*, Proc. 2nd Polish–German Workshop March 1991 Paderborn. Bogacz, R.; Lückel, J.; Popp, K. (eds.) IPPT, Polish Academy of Sciences, Warsaw.
7. Eiber, A.; Freitag, H.-G.; Burkhard, C.; Hemmert, W.; Maassen, M.; Zenner, H.-P. (1999) Dynamics of Middle Ear Prostheses—Simulations and Measurements, *Audiol Neurootol* **4**, 178–184.
8. Schiehlen, W. (ed.) (1993) *Advanced Multibody Systems Dynamics*, Kluwer, Dordrecht.
9. Suzuki, J.-I. (ed.) (1988) *Middle Ear Implant: Implantable Hearing Aids*, Karger, Basel.
10. Eiber, A.; Freitag, H.-G.; Hocke, T. (2000) *On the Relationship between Multifrequency Tympanometry Patterns and the Dynamic Behavior of the Middle Ear*, In: Rosowski, J.; Merchant, S. (eds.), The function and Mechanics of Normal, Diseased and Reconstructed Ears. Kugler, The Hague.
11. Bestle, D. (1994) *Analyse und Optimierung von Mehrkörpersystemen.* Springer, Berlin.
12. Bestle, D.; Eberhard P. (1992) Analysing and optimizing multibody systems, *Mechanics of Structures and Machines* **20**, 67–92.

INDEX